The Incredible Machine

The Incredible Machine

National Geographic Society

The Incredible Machine

Published by
The National Geographic Society

Prepared by
National Geographic Book Service

Charles O. Hyman
Director

Ross S. Bennett
Associate Director

Margaret Sedeen
Managing Editor

David M. Seager
Art Director

Susan Eckert Sidman
Director of Research

Richard S. Wain
Production Manager

Staff for this book
Robert M. Poole
Editor

Leah Bendavid-Val
Illustrations Editor

Linda S. Glisson
Karen Jensen
Edward Lanouette
Carol Bittig Lutyk
Esther Mackintosh
Elizabeth L. Newhouse
Barbara Seeber
Leah Bendavid-Val
Lynn Addison Yorke
Editors and Writers

Jennifer Gorham Ackerman
Lydia Howarth
Catherine Herbert Howell
Mary Luders
Melanie Patt-Corner
Maura J. Pollin
Judy A. Reardon
Penelope A. Timbers
Editorial Researchers

Jennifer Gorham Ackerman
Art Coordinator

Jean Shapiro Cantu
R. Gary Colbert
Gwenda L. Hyman
David Ross
Illustrations Researchers

Diana E. McFadden
Illustrations Assistant

Lydia Howarth
Style

Charlotte Golin
Design Assistant

Karen F. Edwards
Traffic Manager

Leslie Adams
Andrea Crosman
Production Assistants

Georgina L. McCormack
Teresita Cóquia Sison
Editorial Assistants

John T. Dunn
David V. Evans
Ronald E. Williamson
Engraving and printing

George I. Burneston III
Indexer

Lennart Nilsson
Photographs inside the human body

Jan Lindberg
Consultant for Nilsson photography

Nathan Benn
Dan McCoy
Photographers

Scott T. Barrows
Robert J. Demarest
Jean A. Miller
Kirk Moldoff
Carl-W. Röhrig
Medical Illustrators

William A. Knaus, M.D.
Director of ICU Research
The George Washington University Medical Center
Medical Consultant

Frederick C. Robbins, M.D.
University Professor
Case Western Reserve University School of Medicine
Medical Consultant

Greta Arnold
Louisa Clayton
Norman Cousins
Gillis Hägg
John A. Kasbarian
Susanna Nicholson
Cynthia Riggs
Jean Kaplan Teichroew
Contributors

Page 3: Special optics transform the human body into pixel-like facets. Photograph by Michel Tcherevkoff. Special effects, Oscar Guzman for Optifex.

 Library of Congress CIP data page 384.

First edition: 300,000
384 pages, 443 illustrations, including 399 photographs and 44 paintings and diagrams.

Contents

Foreword

By Lewis Thomas, M.D.

There was a time, not very long ago, when a book like this one, designed to be examined, read, and pored over by an enchanted public-at-large, would have been inconceivable. When I was a medical student, back in the 1930s, I believed that I had entered an arcane world, filled with secrets known only to my masters in the profession. The human body and its strange contents were special matters reserved for the authorities, the keepers of the temple. What was more, these mysteries were not thought to be a proper interest of the public. And, as I now recall, they were not. It used to puzzle me that so many people of my acquaintance outside the medical profession, especially my learned academic friends in other parts of the university, knew so little about the plainest facts of their own anatomy and physiology, had so much of that very little all wrong, and wanted to know no more. It was as though they were afraid of catching something.

In those years, a book like this would have been impossible for another, better reason. Although the professionals of 50 years ago felt secure in their knowledge, and the mass of detailed information about human form and function seemed enormous, times have changed. Most of the things contained in the pages ahead are items of information that are entirely new, many of them coming from research done in just the past decade. Moreover, some of the astonishing pictures that illustrate the book are the result of advances in the technology of photography made quite recently, beyond imagining just a few years back.

Times have changed indeed. We are in the early stages of a genuine revolution in biology, which had its beginnings some 40 years ago with the opening up of the field of molecular genetics. The discovery that genes are made of deoxyribonucleic acid (DNA), followed within a few years by the elucidation of the Watson-Crick model of that molecule, was one of those remarkable events in science that changed everything—not by turning things upside down or overheaving any earlier doctrine, but simply by bringing in a totally new way of looking at living processes. Deep matters in the interior of cells became approachable by experimentation.

Meanwhile, medicine continues to move along, bobbing in the wake of biological science. Or perhaps this is an unfair way to view the relationship, since many of the new discoveries in biology have come about as the direct outcome of experiments that began as proper medical problems. In fact, the discovery that genes are made of DNA was made in the course of research aimed at finding new ways of treating pneumonia, back in the 1940s. Many of the latest revelations in molecular biology and immunology, now flooding the scientific journals each month, are the direct result of experiments intended to achieve a clearer understanding of cancer, heart disease, the autoimmune disorders, AIDS, multiple sclerosis, and any number of other human diseases. Biology and medicine are now all of a piece.

Much of the new science described in this book can be attributed to a single institution, perhaps the greatest social invention of the 20th century, the National Institutes of Health (NIH). Beginning in the years just after World War II, NIH began to expand across its campus in Bethesda, Maryland, constructing new buildings, recruiting distinguished scientists from this country and abroad, and taking on as its mission the whole field of human disease.

The basic research adventure sponsored by NIH, accompanied by an increase in the support of biomedical research by many of the country's major philanthropic foundations, has reflected the growing interest of the public in all matters pertaining to public health. But more than this, it reflects a general interest in science itself, for its own sake. The appearance on newsstands of one new science magazine after another, the commitment of whole sections of national newspapers to the latest advances in research, and the new attention paid to science in television and radio news programs, are in themselves evidence that the public wants to know more about what is going on in science—not just as it relates to the treatment and prevention of human disease, but because of a fascination with nature itself and how it works. Most of all, how the human species relates to the rest of nature, and how we fit in.

Biology has begun to provide clear answers to problems that once seemed beyond the reach of science, and each of the new answers has brought along new and more puzzling questions, especially about the nature of man and his place in the living world. The universe has become a far more strange place for the physicists, and now the life within that universe is turning into an even stranger phenomenon for the biologists. At the center of all puzzles is the connectedness of the Earth's life. Ever since Darwin, we have known that the wide varieties of species on the planet are in some sense related to each other, but now we must face the fact that they comprise, all together, one form of life, a coherent *system* of life, a living mass in which we humans have the look of working parts.

Our lineage is the greatest puzzle of all, even something of an embarrassment. Somehow or other, everything around us today—all animals, ourselves, all plants, everything alive—can trace its ancestry back to the first manifestation of life, approximately 3.5 billion years ago. That first form of life was, if we read the paleontological record right, a single bacterial cell, our Urancestor, whose progeny gave rise to what we now call the natural world. The genetic code of that first cell was replicated in all the cells that occupied the Earth for the next 2 billion years, and then the code was passed along to nucleated cells when they evolved, then to the earliest multicellular forms, then to the vertebrates some 600 million years ago, and then to our own human forebears. The events that gave rise to that first primordial cell are totally unknown, matters for guesswork and a standing challenge to scientific imagination. But the events that must have followed, taking life all the way from a solitary microbial cell to the convolutions of the human brain and the self-consciousness of the human mind, should be sweeping us off our feet in amazement. No matter how it happened, whether by the plodding gradualism of random mutations and natural selection, or by sudden leaps from one form to the next, the plain fact remains: We are a part of the Earth's life and we have cousins all over the place.

From our point of view, the human being is the highest achievement of the natural world, the best thing on the face of the Earth. Or anyway this is the way we have always tended to view ourselves and our place, masters of all we survey. There are risks for us in this point of view, however. We need reminding that we are a very young species, only recently down from the trees, still preoccupied by the new gift of language, and still trying to figure out what we mean and what the world means. We are juvenile, as species go. We seem to be a stunning success in biological terms, already covering more of the Earth than any other single form of life since the famous trilobites, whose fossils abound everywhere, but we should be going warily into our future. We may be error-prone at this stage of our development, apt to fumble and drop things, too young to have our affairs in order. If we get things wrong, we could be leaving a very thin layer of fossils ourselves, and radioactive at that.

But so far, so good. We are indeed a splendid invention, capable of learning, insatiably curious about ourselves and the rest of nature, anxious to be loved, hoping to be useful. As social animals, we have a long way to go, a vast amount to learn, and an eagerness to run risks. Perhaps we are right to view ourselves as nature's finest accomplishment thus far, with appropriate reservations.

One promising thing about us is the way we make language and then use it for thinking about ourselves. The title of this book is *The Incredible Machine*. It is nice to reflect that that word "machine," which seems to put the mind off at first thought, comes straight down to us from the ancient Indo-European word *magh* used thousands of years ago, meaning "power," from which we made the cognate, closely related word "magic." Incredible magic, that's what we are, and that's what this book is about.

The Cosmic Creature

A lone white blood cell prowls through tissue, stalking bacteria. It devours its prey. The cell is a living organism, a wondrous, throbbing entity, yet alone it would not exist. The cell's significance lies in its role as part of a larger, far more wondrous organism—the living body. And the genius of the living body—of human life—is born of the greater genius of life itself.

Why is life so precious? We are special because we are integral parts of infinity. Within our bodies course the same elements that flame in the stars. Whether the story of life is told by a theologian who believes that creation was an act of God, or by a scientist who theorizes that it was a consequence of chemistry and physics, the result is the same: The stuff of stars has come alive. Inanimate chemicals have turned to living things that swallow, breathe, bud, blossom, think, dream. The living beings of Earth are cosmic creatures, products of celestial events—atomic collisions, stellar explosions, molecular unions—that were cataclysmic yet fortuitous. We are children of the universe. And we are children of chance.

Whatever the steps may have been in life's history, it seems evident that human life only narrowly missed nonexistence. Probably, the sequence that produced life began with nothing more than hydrogen and helium eddying through space, collecting into vast clouds. The forces of gravity within the clouds sent atom crashing into atom to trigger nuclear reactions, set the clouds afire, light the stars. One by one, the large stars exploded, sowing their elements throughout the universe.

In a spinning cloud of gas and dust, our star, the sun, formed. Then grains and clumps of matter crashed together to form a ball which produced great heat at its center. Thus, over millions of years, was born the smoldering ember we call Earth. Within the seething infant planet, rock melted, bursting up through the wrinkling crust, belching out stellar vapors that had been trapped inside. The vapors created a primitive atmosphere. Rain fell. Earth's first oceans grew. Ocean water gathered molecules washed from Earth's crust and from the atmosphere. Bolts of lightning or ultraviolet radiation fused some of the molecules into chains: one of amino acids, the material of proteins; the other of nucleotides, the material of genes. Somehow, somewhere, these chains entwined, and life on Earth began. This singular phenomenon could survive the present by absorbing molecules to nourish itself, and it could persist into the future by arranging its own molecules into offspring.

What was it, in that primordial sea, that turned chemical to creature? Modern researchers have tried to unravel the mystery. They assembled the supposed ingredients of the early atmosphere—hydrogen, methane, ammonia, and water vapor—in their laboratory flasks. They electrified this brew with flashes of man-made lightning. They watched the liquid in one flask turn pink, then muddy red, as amino acids took form. But there the transformation ended. These scientists and others since have produced the components of life but not life itself. Some think that the missing ingredient is time. Others believe that our grandchildren will live to see life made in the laboratory.

Millennia of chemical interactions appear to us in fossils. Molecular life, pummeled by lightning and solar radiation, underwent metabolic modifications, mutating into light-sensitive cells. They took energy from the sun and began to produce oxygen as a by-product of photosynthesis. Oxygen, a gas new to the planet, revolutionized life processes. Oxygen-breathing cells developed with nuclei that held genetic material—leading to sexual reproduction and the ancestors of plants and animals. The simple cells proliferated, joined, then developed into complex multicelled organisms. Those that evolved ways to use the powerfully corrosive oxygen now found it safe to move from sea to land. Those that multiplied not just by division but also by sexual combination of genetic material were no longer dependent on haphazard strikes of lightning for cellular refinements; now life could diversify with every birth.

Earth's green blanket of flowerless conifers and ferns exploded into a many-colored brocade of blossoms. Creatures developed in synchrony, shaping themselves to each other and to the world around them. Sweet nectars evolved in concert with the hummingbird; magnolias and buttercups appeared with the beetles that thrived on their pollen. Birds, bats, butterflies, and pterosaurs developed wings fashioned to similar aerodynamic principles, though they shared no common flying ancestor. They had not inherited wings; they had inherited the wind. Squid, octopuses, vertebrates, and some spiders independently developed eyes with the same basic structure. Whales shaped their flesh to the sea, molding themselves into the form of fishes even though, as mammals, their forebears probably resembled sheep or cattle.

Five hundred million different species have appeared on Earth. That genes shuffled from a single cell could produce such diversity and, in time, a human body, may seem as impossible as heaving bricks to build a palace or casting metal filings to the wind to make a Rolls-Royce. But life has had time on its side.

Between the appearance of life on Earth and the debut of humanity, more than 3.5 billion years elapsed. This span of time—some 35 million centuries—is almost meaningless without comparisons. About one century ago Wilhelm Röntgen discovered X rays. Christianity has inspired its followers for about 20 centuries. The wheel has been used for 55 centuries. Modern humans—physically similar to us—appeared roughly 300 centuries ago.

Yet time does not explain the complexity of even a single organism. Where, in the vast beat of millennia and mutation, is the direction, the orientation, that produced the intricate and interacting systems of our bodies?

"Everything existing in the Universe," wrote Democritus, "is the fruit of chance and of necessity." Jacques Monod, the Nobel prizewinning biochemist, later used this philosophy as the theme of his theory of evolution. Mutation supplies the endless chance, Monod wrote; natural selection supplies the necessity. Creatures adapt to nature's demands or they perish. Natural selection thus lends life its course: It accepts the ingenious and executes the inadequate. Of the 500 million species that have appeared on Earth, nearly 99 percent have faltered and vanished, failing nature's incessant testing.

Only those that can elaborate survive. Evolution accumulates the successful. It also invents incredible machines, building living organisms into wonderful contraptions, adapting existing structures to new conditions. Selection works like an eccentric tinkerer, says François Jacob, the geneticist. The tinkerer "uses whatever he finds around him whether it be pieces of string, fragments of wood, or old cardboards," Jacob has written. "What these objects have in common is 'it might well be of some use.' For what? That depends on the opportunities. . . . From an old bicycle wheel, [the tinkerer] makes a roulette; from a broken chair the cabinet of a radio. Similarly evolution makes a wing from a leg or a part of an ear from a piece of jaw."

Evolution is addition or subtraction—reproduction with a garnish. Among the most precise of selection's refinements is "the wisdom of the body," in the words of physiologist Walter Cannon. This wisdom is apparent to some degree in every living organism. Billions of years, changes, deaths, and lives have created a remarkable self-adjusting balance called homeostasis. By this process, biological mechanisms work in unison to protect the body's internal stability from the threats of nature. These forces automatically activate, some within a fraction of a second, some more slowly: If a man hemorrhages, his body pulls water from tissues into the circulatory system; this keeps blood pressure from dropping below critical levels. When a woman is suffering frostbite, her body has automatically adjusted by slowing the blood flow to her toes and fingers, ears and nose, reserving heat and oxygen for the all-important brain and the organs in her chest and abdomen. This internal equilibrium can adapt itself to a temperature change of even a fraction of a degree, bathing the body in cooling perspiration when it becomes too warm, and, when cold, prompting shivers to convert energy to heat.

Once people said that natural selection favored the most aggressive animals, those red in tooth and claw that survived because they tore apart their competitors. But ages of mutations have produced many species that flourish because of alliance rather than aggression. The human mother who perishes to protect her offspring, the female monkey who chooses to mate not with the dominant male but with a gentler partner, these live on in the genes of the surviving family.

One of the most intricate of life's creations is the human brain. Its growth rate began to increase about a million years ago. Just five hundred thousand years ago, the brain's growth rate peaked at ten cubic inches—ten heaping tablespoons—of sparking electrical tissue every hundred thousand years. The human race has reduced the size of jaws and teeth, extended infancy and childhood, postponed maturity and sex life. All this has reserved time and fuel for development of the brain.

With that exquisite advantage—by the strength of thought and hope rather than muscle—humans lifted the lineage of life to the capacities to cherish, to ponder, to succor, to guarantee the future of the species not just by proliferation but by nurture. Fossils, including footprints, show us that the first humanlike beings stood erect some four million years ago. The fossils suggest that these hominids probably stood up not to free their hands for weapons but to carry home food to share. Hundreds of thousands of years after that act of generosity and imagination, Neandertals took time from the demands of their brutal world to support the crippled and maimed of their groups, to anguish over the sepulchers they dug for their dead, to leave blossoms in the graves. By showing reverence for life, they extended life.

In the last hundred thousand years, while the evolution of our brain size has slowed down, our gains in knowledge have not. Generations of learning people add ideas to their inherited body of wisdom, boosting themselves up, in the words of Bernard of Chartres, "like dwarfs seated on the shoulders of giants," where they can "see more things than the Ancients and things more distant." The most breathtaking views we have had up there, on the shoulders of our sages and scientists, are of the distant past. Discoveries about genetics, about cosmic molecules, enrich existence for all of us.

Preserved in the fossil record and in our bodies is evidence that all life is related to all other life, that we take as our common ancestor the first single cell. The same four nucleotide bases compose the genetic code of all living beings—tree and germ, man and amoeba, whale and ant. The same twenty amino acids build our proteins; the same combustion systems change food to energy. The same chemical compounds that modulate the human brain also ripple through protozoa.

The human being, in fact, is a community of other beings. Each of our cells has from one to thousands of tiny power plants called mitochondria, which turn food into cellular energy. Each mitochondrion has its own genes. Can the mitochondria reproduce on their own? Scientists think so; they believe the mitochondria are primordial bacteria, swallowed whole by early cells that did not digest them, but instead became hosts and beneficiaries. In addition, many of the cells in or on our bodies are not our own; some are microbes living on our skin, in our intestines, and elsewhere, supporting our lives as we support theirs.

Living things are confluent rivers of matter, their molecules continually flowing from one life form to another. Carbon, hydrogen, oxygen, nitrogen, phosphorus, sulfur, sodium, and all the other materials on which life depends, move from Earth to body and back. Some complete a cycle in months; others take millions of years. Molecules pass from ground to grass and gooseberry, to scampering squirrel, to strolling human, not interrupted even by death. Molecules act in such harmony, their cycles maintaining Earth's temperature and atmosphere, the oceans' alkalinity and salinity, that the entire planet functions as a single organism.

These molecules bind us to the scale of time as well as matter. The nucleic acids of the first single cell reside, at this moment, in one of us, or in the gooseberry, or the squirrel. The salt in our bloodstreams and the calcium in our bones flowed in the first sea. We will in time pass these components on again, not just to sons and daughters but to unimaginable life forms in an unimaginable future.

Infinity and eternity stand like sentries at both ends of existence; just as the minerals in our bodies originated in a star, they will waft back into space someday when our sun dies. The elements of Earth, of our bodies, will then flame up into some other star.

These evolutionary origins do not diminish us; they exalt us. Human life is indeed an accident, but it is a celestial accident, an accident so intricate it will probably never be repeated, even though the precursor molecules of life are seeded throughout the universe.

The Copernican revolution, it has been said, humbled mankind by displacing Earth from the center of the universe. With our human bodies—biological galaxies of stellar dust—we are not only the *center* of the universe, we are more. We *are* the universe. We are its nuclei, its electrons, its atoms, combining and recombining; we are its past and we are its future.

Susan Schiefelbein

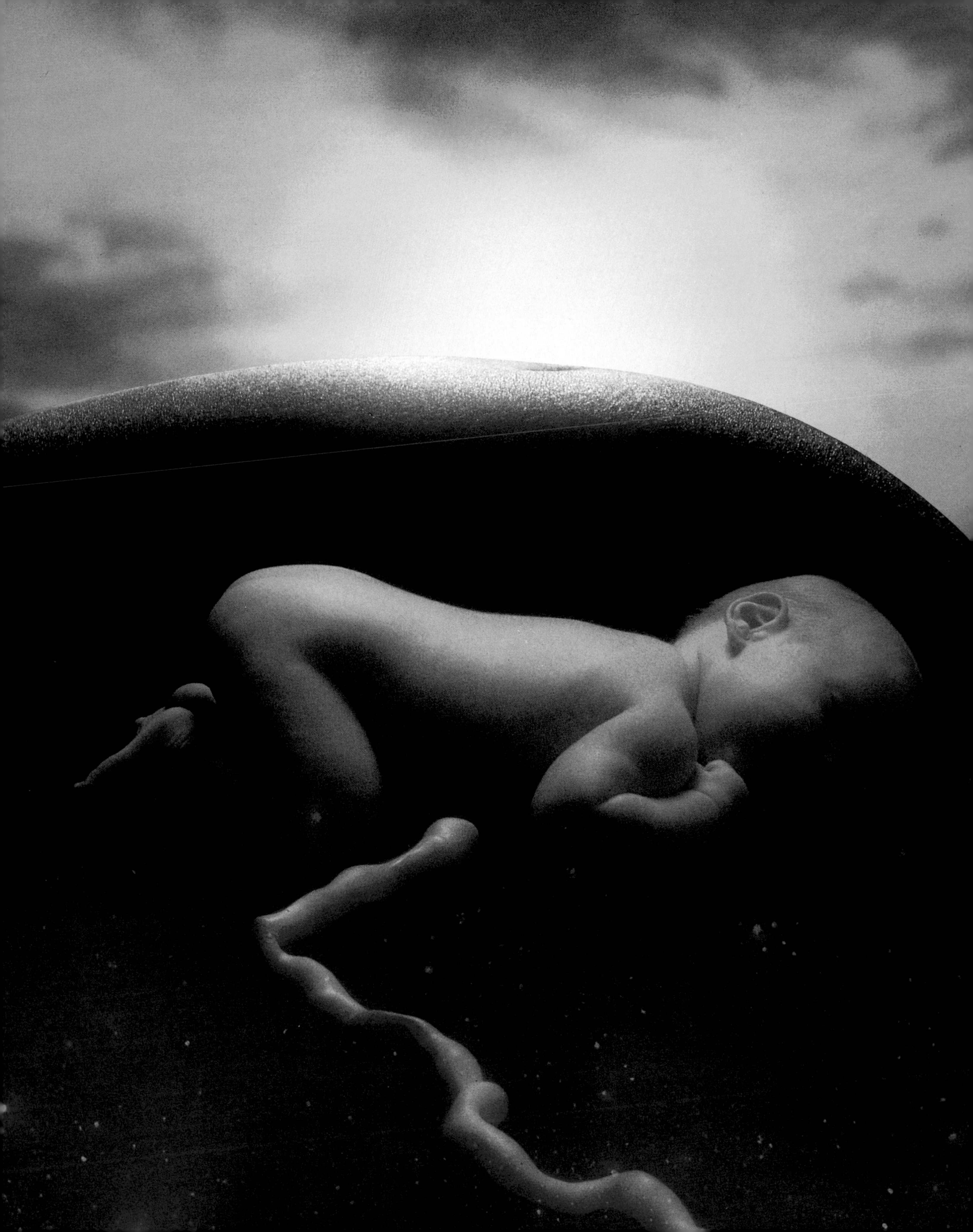

Beginning the Journey

The newborn baby embodies innocence, yet conceals the most taunting of all riddles: the generation of human life. The story begins with sperm and egg as they combine to form a single cell. Sheltered in the mother's womb, the cell multiplies. Soon there are hundreds of different cells able to make some 50,000 different proteins to control the work of all our cells—collagen to build skin, insulin to control energy use, hemoglobin to supply oxygen.

Before long, the groups of cells are gathering into layers, then into sheets and tubes, sliding into the proper places at the proper times, forming an eye exactly where an eye should be, the pancreas where the pancreas belongs. The order of appearance is precise, with structures like veins and nerves appearing just in time to support the organs that will soon require them. In four weeks the progeny of the first cell have shaped a tiny beating heart; in only three months they are summoning reflex responses from a developing brain.

Nothing more than specks of chemicals animate these nascent cells as they divide. Yet in just nine months, some twenty-five trillion living cells will emerge together from the womb; together they will jump and run and dance; sing, weep, imagine, and dream.

A single cell engenders a multitude of others, but the multitude acts as an entity, as a community. The appearance of life in the womb, like the appearance of life on Earth, cannot be completely explained, even by our burgeoning scientific knowledge. Biologists have identified the sequence of reproductive events; yet they know merely *what* happens. They do not yet know exactly *how* or *why*.

For centuries sages explained reproduction with various proposals. Some thought that miniature human beings were tucked within the seed of Adam or Eve—all the generations to come. Others speculated that only men played an active role in fertilization, while women contributed nothing but nourishment for the developing child. Not until the improvement of the microscope could scientists see the wondrous realities. Then they understood that the body is made of cells—and that the fertilized egg is just a particularly important cell.

By 1881 researchers, studying starfish and sea urchin eggs and sperm, had found in them a threadlike hereditary material called nuclein. When sperm and egg joined, their nuclein merged. Though these researchers did not know it, they had witnessed the manner of human reproduction, for what is true of starfish and sea urchins is true of humans: Mother and father each contribute a cell filled with genetic patterns. When the patterns in the two cells fuse, no fully formed little ladies or gentlemen curl within the fertilized egg, just a fragile thread spun of chemical memory. The instructions for life are written

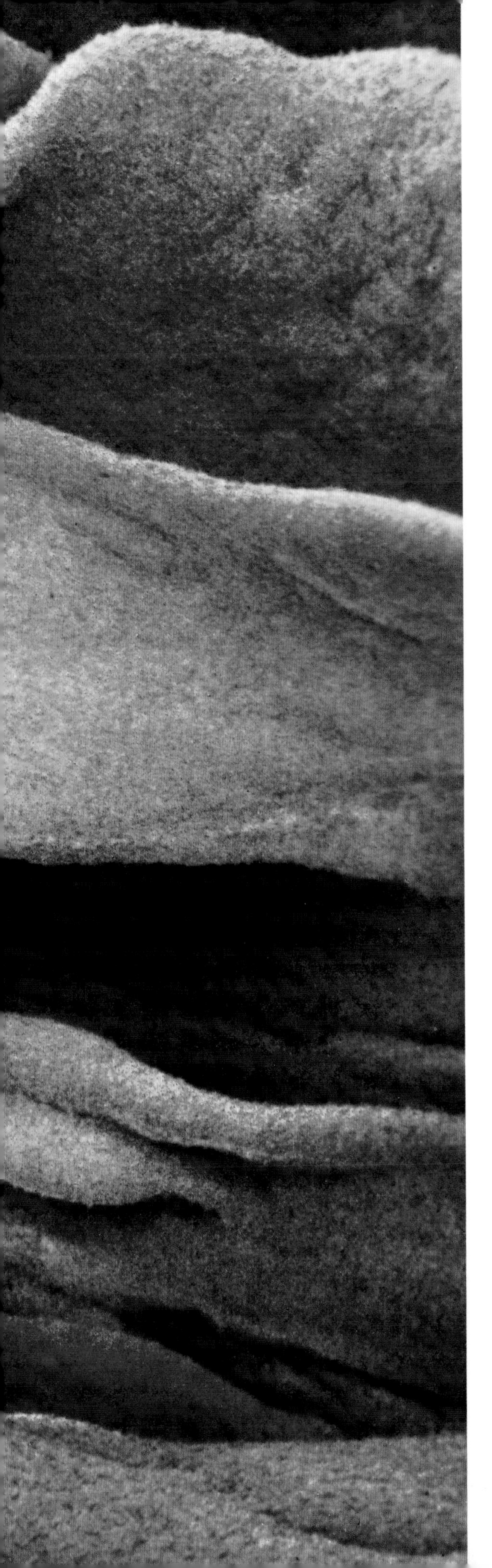

The journey begins for a new human life. An egg cell, swept from the mother's ovary, lies in the folds of a fallopian tube. In this four-and-a-half-inch long tunnel to the uterus, a sperm cell from the father will fertilize the egg to create a new human being, perhaps tall like the mother, blue-eyed like the father, red-haired like a grandparent, but unique and unmatchable. Not just one in a million, each new life carries genetic material from all its forebears mixed in an individual combination.

A coating of bumpy nurse cells supports and surrounds the egg, or ovum. The nurse cells, which will soon be shed, may also help to nourish the egg. As the fertilized ovum travels, it divides, then divides again, once or twice a day, until it is a cluster of sixteen cells. After about four days in the fallopian tube, the cell cluster reaches the uterus and settles in. In just nine months a baby is born.

Magnification: 450 times

on those fibers to be passed silently from generation to generation, like a flame exchanged from burning wick to wick, the spark of life itself—first animating the living, then guarded by them, as they too reproduce so as not to let the sacred fires flicker out.

Messengers of Heredity

Since 1881 we have learned much. We know the code by which these gossamer filaments send their messages. Heredity is written on a chemical ribbon that twists like a spiral staircase, the steps built of four chemical bases attached to chains of sugars and phosphates—DNA, or deoxyribonucleic acid. Thousands of these steps make up a single gene. Tens of thousands of genes, arranged along structures called chromosomes, transmit the instructions for existence, dictating eye color, hair texture, vulnerability to disease, perhaps even stuttering and altruism.

Some six billion steps of DNA in a single cell record one life's blueprint. This DNA plan for a single human life can be stretched six feet, yet it is coiled in a repository just 1/2500 of an inch in diameter—the cell's nucleus.

Genes not only tell a creature how to reproduce; they compel it to meet its match. The story of reproduction is not just one of two people drawn to each other; it is the story of a miniature odyssey: Two sets of DNA, one cradled by an egg, the other ferried by a sperm, embark on an urgent mission to sustain the species.

So forceful is the command to reproduce that, when the female embryo is only six weeks old, she makes preparations for her motherhood by developing egg cells for future offspring. (When the baby girl is born, each of her ovaries carries about a million egg cells, all she will ever have. She will lose most, before puberty and after, through degeneration.) In prenatal life each primitive egg cell begins a process known as meiosis. The DNA is randomly shuffled in preparation for the day when a sperm will add a complementary set of hereditary material. The wait can be a long one, as many as fifty years for some eggs. Meiosis guarantees that each egg and sperm is genetically unique. So vast are the possible combinations of DNA that the odds are infinitesimal—about seventy trillion to one—that the same two parents will, in separate fertilizations, produce an identical offspring.

In the female fetus, meiosis proceeds only partway before birth. Chromosome division hangs suspended, the immature eggs in limbo, until they are needed, one each month for ovulation. At puberty menstruation begins, but there is yet no release of eggs ready for fertilization. Several years may pass before a regular ovulation cycle establishes itself.

Before the egg leaves the ovary at ovulation, it

Magnification: 6 times

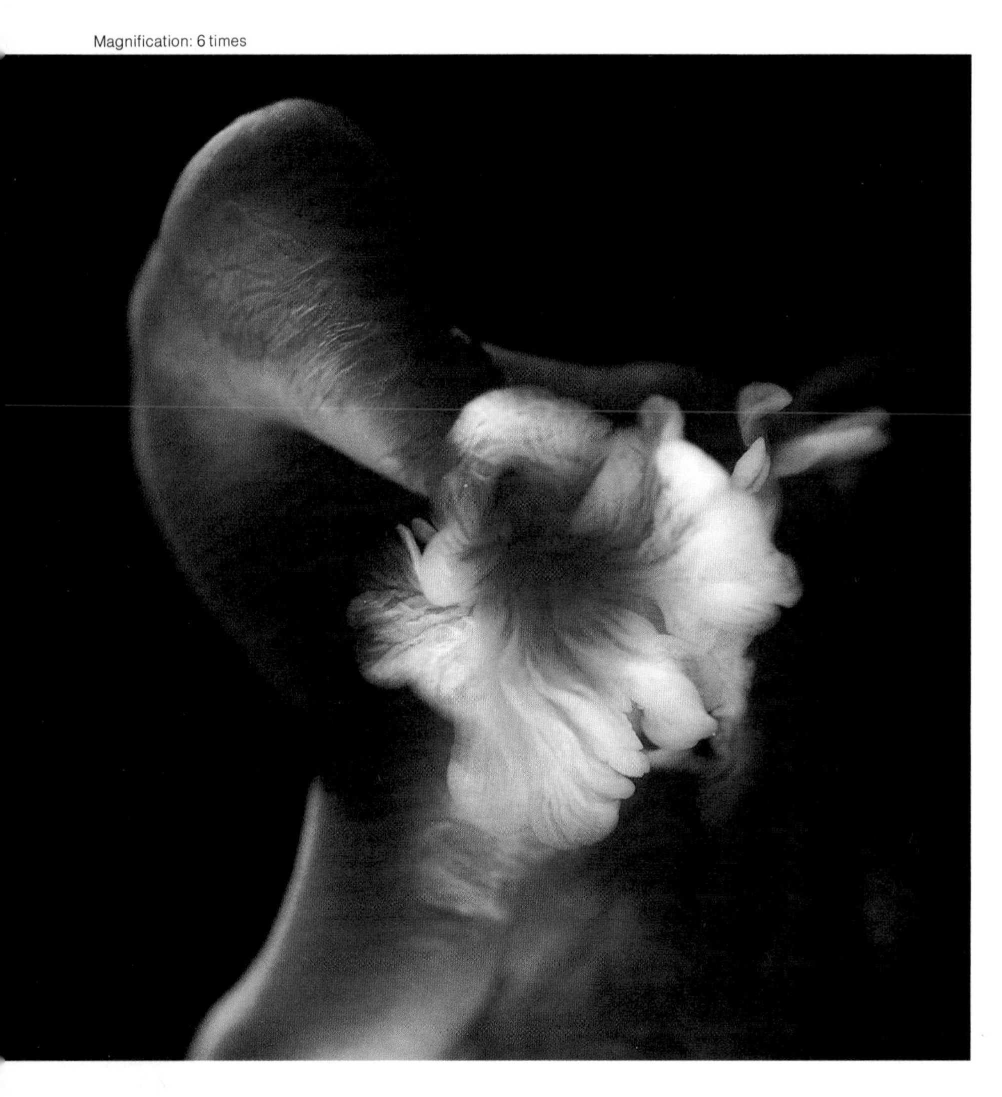

undergoes a process called maturation. It divides once, reducing the number of chromosomes by half, and casts off a cell with little more than a nucleus, called a polar body. But it still has twice the genetic material it needs. It divides again and casts off a second polar body when the sperm fertilizes the egg.

Each month, about two weeks after the menstrual period begins, hormones within the woman's body stimulate ovulation. Of the thousands of stored eggs, several may mature, but usually a single one bursts from a follicle in the ovary and enters into the abdominal cavity. Supporting nurse cells surround it.

Fringelike petals edge the fallopian tubes, tunnels to the uterus. At ovulation these fimbriae reach out to sweep the egg from the cavity into a tube, where it will await a sperm cell.

If two mature eggs find their way into the fallopian tube, and are fertilized by different sperm, the mother will carry fraternal twins—babies that resemble each other in a general family way, just as any siblings resemble each other. About one pregnancy in ninety produces twins. Two-thirds of these will be fraternal. In the other third, a single fertilized egg splits into two genetically identical parts that produce identical twins. They will be of the same sex and will look almost exactly alike. Most fertilizations—about 98 percent—result in a single embryo.

The male embryo, too, prepares for future reproduction of the species. When only six weeks old, no more than half an inch long, the embryo develops primitive sperm cells. At puberty, in convoluted tunnels within the testes known as the seminiferous tubules, the cells begin to divide, to supply sperm throughout the male's reproductive life. As they divide, their offspring—immature spermatozoa—undergo meiosis, after which their mature form emerges: shaped bullets of DNA in the sperm's head, midsections for energy, and tails for propulsion.

Along the next tract, a torturous, twenty-foot duct called the epididymis, sperm develop the ability to move. Finally they reach the vas deferens, a tube that empties into the urethra. The relentless process churns a thousand sperm each second into the holding areas. There many die, but the survivors wait, quiescent, for the sudden nervous system signal to begin the most treacherous and consequential phase of their mission —to deliver their DNA to the egg.

The Difficult Journey of a Sperm

When a man and a woman copulate, the man's pelvic muscles shudder into a rhythmic series of contractions. The clenching muscles propel a grayish white, viscous fluid known as semen. This fluid carries as many as 200 to 500 million sperm into the woman's vagina. The fluid of the semen helps the sperm accomplish their task: It contains citric acid to dissolve the woman's cervical mucus, alkalies to buffer the vagina's acidity, sugars to give the sperm energy, and hormones to stimulate vaginal and uterine contractions that will fling some sperm far up into the woman's reproductive tract.

But the vagina's buffered acids still are hostile to sperm. A hundred million die almost instantly or leak from the vagina. Survivors quickly leave the now coagulated semen and forge into the channels of cervical mucus. Normally this mucus is thick and abounds with white blood cells, which devour any foreign substance that invades the womb. However, during ovulation, the mucus changes. The number of white blood cells

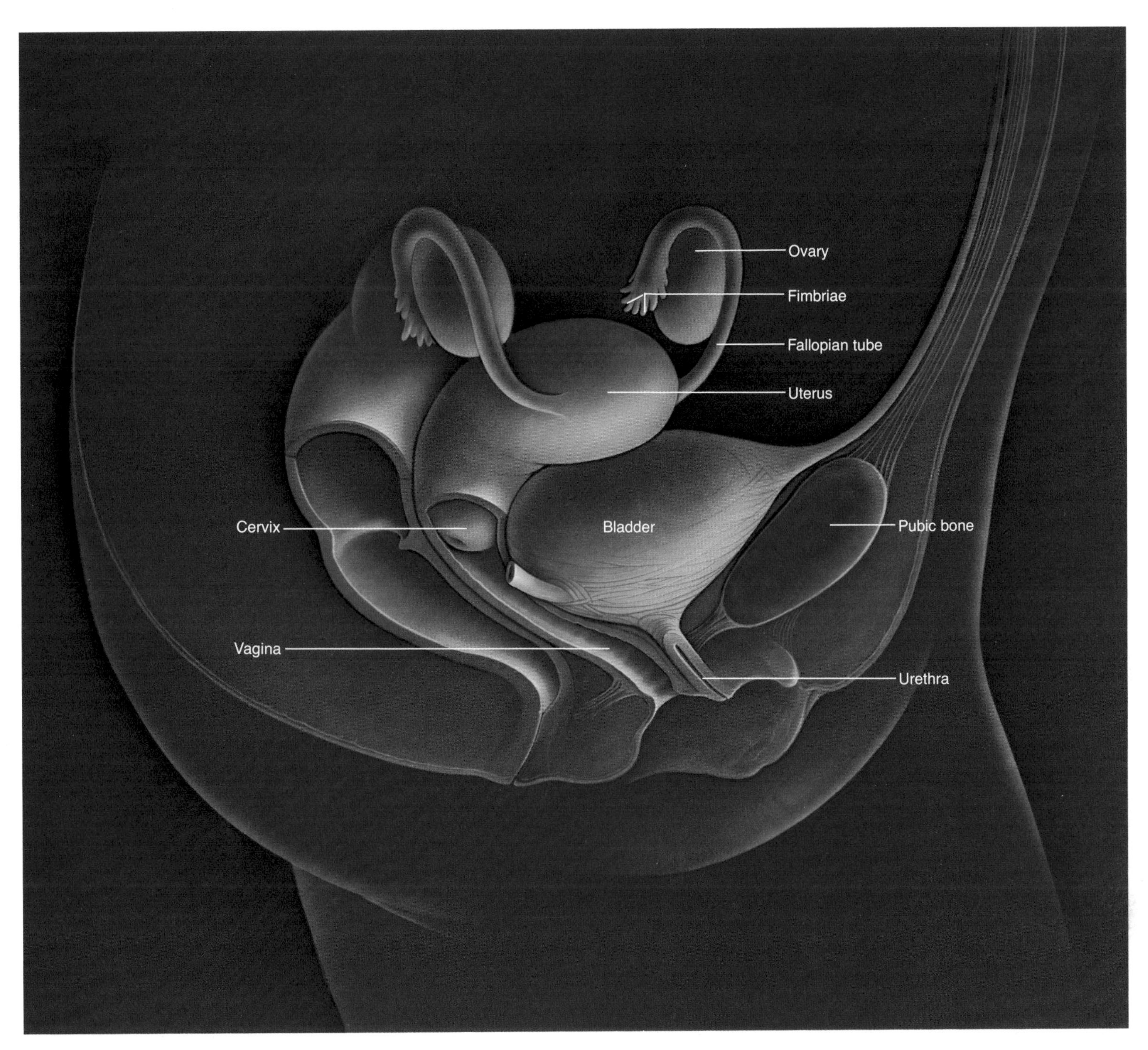

drops. The water content increases and the structure of the mucus changes—from an almost impenetrable web into rows that open just enough for sperm to move through. They wriggle along at a rate equivalent to a swimmer's covering twelve yards a second.

Perils loom at every turn. White blood cells attack and destroy many sperm. Only a few thousand sperm enter the uterus. Fewer than five hundred—sometimes as few as ten—reach the part of the fallopian tube where the egg waits.

When sperm enter the woman's reproductive tract, they encounter secretions there, which help dissolve the protein sheathing that coats each sperm's bulletlike head. Once the sperm sheds this covering, its ability to penetrate the

The female reproductive organs lie deep within the abdomen, protected by the bony pelvis (above).

These organs produce the egg—their contribution to conception—and make a place for the product of conception, from fertilized egg to full-term fetus. Ovaries also produce the female hormones, estrogen and progesterone.

Each ovary, about one and a quarter inches long, usually releases one developing egg every twenty-eight days. Fimbriae help sweep the egg into the funnel-like opening of a fallopian tube (opposite). The tube transports the egg to the uterus, a hollow, muscular organ three inches long.

Each month during a woman's fertile years, her uterus prepares a blood-rich lining to support a pregnancy; if she does not become pregnant, the woman sheds this lining through the cervix and vagina—the same elastic path a fetus takes at birth.

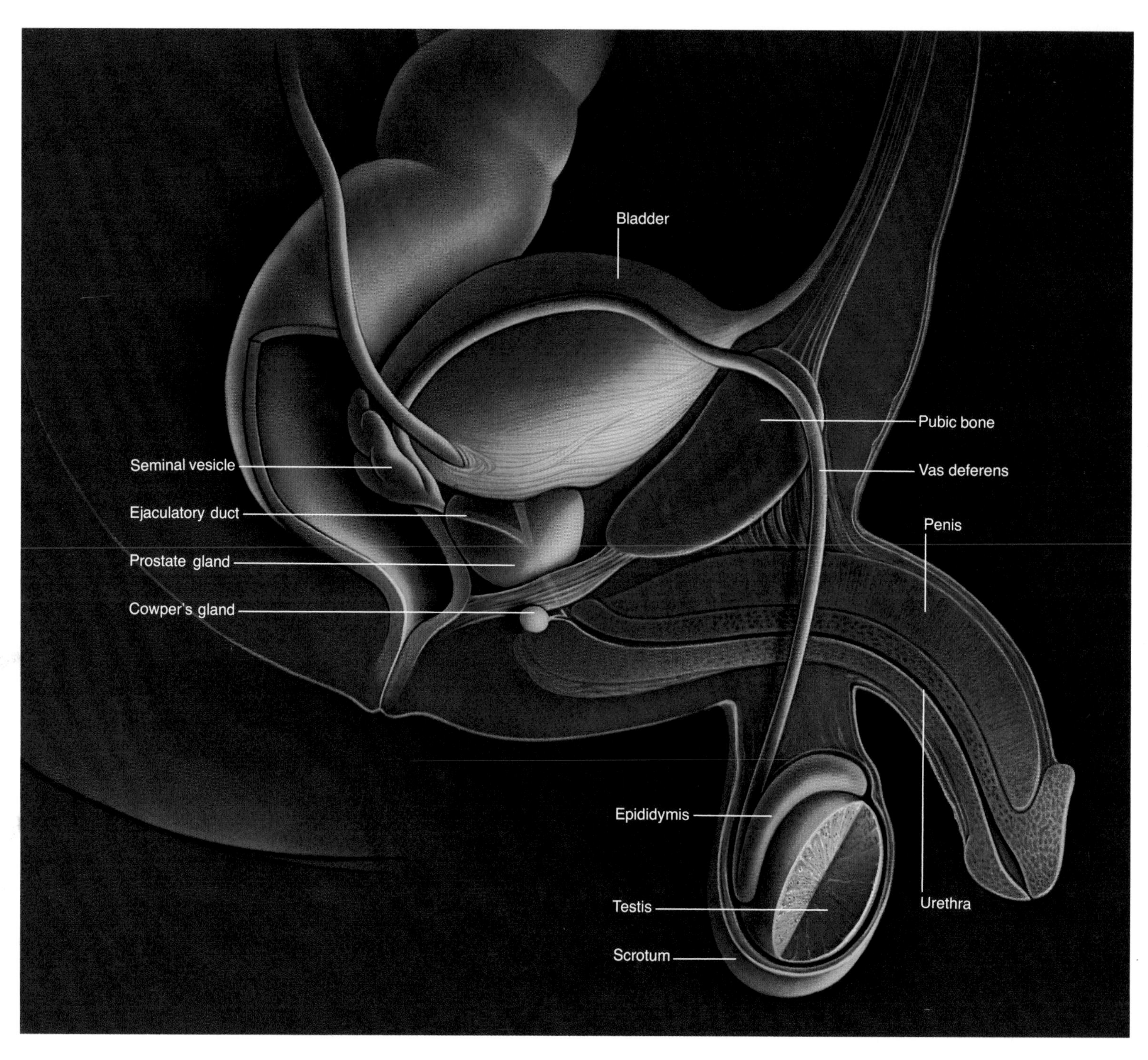

The male reproductive organs (above) have two roles—to produce sperm cells and to send them toward the egg. The two testes hang away from the body in the scrotum. In this pouch of skin, the temperature is about 4°F cooler than normal body temperature. Ducts carry the sperm, manufactured in each testis, to the epididymis, a holding area where the sperm mature. From there sperm travel to the vas deferens, one of two "pipelines" that lead to the ejaculatory ducts; these join the urethra, which also carries urine.

During sexual intercourse, the sperm are joined by secretions from the seminal vesicles, Cowper's glands, and the prostate to form a mixture known as semen. Semen pours into the urethra. Muscles in the base of the penis contract and propel the liquid out of the body with great force.

egg, and thus fertilize it, is enhanced.

The sperm approaches like a tiny space voyager meeting a huge planet. The egg is the largest of human cells, 85,000 times bigger than the sperm that rockets closer. It meets a layer of nurse cells surrounding the egg, and then the egg's jelly coat—the zona pellucida. What happens next, we are only beginning to understand. As the sperm enters the coating of nurse cells, a reaction may occur inside the sperm's head: Enzymes leak from the tip of the sperm. The enzymes blaze a trail, digesting a path through the outer layers of the egg for the sperm to follow. When the sperm reaches the deepest layer of coating, the egg's vitelline layer, the sperm pierces through to reach the plasma membrane

Magnification: 2 times

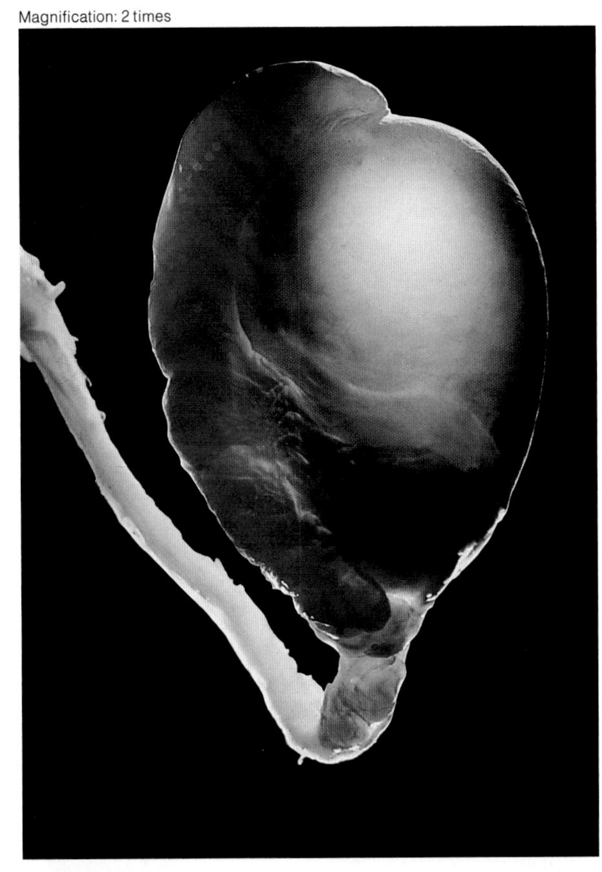

Magnification: 10 times

of the egg cell, and fuses with it.

The egg shivers into action. An electrical charge shoots across the membrane. Microvilli on the membrane sprout around the sperm and pull it inward. Generally the electrical charge keeps out other sperm. It lasts about one minute, time for the egg to devise another blockade. Now the egg raises its vitelline layer above the membrane and infuses the layer with enzymes so that no more sperm enter. Then the egg secretes yet another chemical that hardens this barrier, making it into an armor coat. Usually these elaborate defenses protect the egg from the additional sets of chromosomes that would come with more than one sperm. The additional set or sets of chromosomes brought in by extra sperm would lead to abnormalities in the developing embryo that are nearly always incompatible with life. When they do occur, these "confused" conceptions usually expire early in the pregnancy.

Inside the armor coat, the sperm loses its tail, and the egg completes the final reduction of its chromosomes. Now the egg nucleus—the mother's DNA—moves toward the sperm nucleus—the father's DNA. Seven to ten hours after fertilization began, the nuclei fuse. The two are one. Life and life alone gives birth to life.

Toward the Shelter of the Uterus

After the fertilized egg—the zygote—begins the division of mitosis some fifteen hours later, cilia carpeting the lining of the fallopian tube start to beat and move it toward the uterus. Muscle contractions of the tube wall probably help it along. About three days later (four days after ovulation), the egg has divided into a solid spherical mass of about sixteen cells—the morula, from the Latin word for mulberry, which it resembles. The morula enters the uterus.

There the sphere floats free, cells multiplying, changing in form to a fluid-filled globe called the blastocyst. The outer cells of the blastocyst will form the placenta and fetal membranes. A tiny cluster of cells huddled against one inside pole of the globe will become a baby.

Until now, the sphere has drawn nutrients from stores it carried with it from the ovarian follicle and from secretions in the fallopian tube and uterus. But by six days after fertilization, the blastocyst must adhere to the mother's tissues for sustenance and support. Following ovulation, the uterine walls have swelled with water and an extra blood supply, creating a folded lining with many pockets that will provide the proper environment so that the blastocyst will anchor there rather than pass through in a spontaneous abortion—a miscarriage. If the sphere brushes against the walls, its outer cells tap like tiny roots into uterine tissues and excrete substances that open maternal blood vessels. Soon the blastocyst embeds itself in the nourishing uterine lining. The woman is now pregnant. A protective sac—the amnion—forms around the embryo. By the eighth day the embryo and the uterus begin to secrete into the sac the amniotic fluid that will cushion it against jolts for the next nine months.

With the world's burgeoning population, it is easy to lose sight of just how chancy an event conception is. On the average, only 84 of every 100 eggs that meet sperm are fertilized. Only 69 of these ever make it to the uterine wall. After one week, only 42 are alive. Often these incipient lives expire in spontaneous abortions without the woman's *(Continued on page 29)*

A powerhouse in profile, one testis (upper) can produce millions of sperm cells a day.

Its secret lies inside the scrotum, beneath a covering of connective tissue. Here, beginning in puberty, the seminiferous tubules (lower) produce mature sperm cells. The fine tubules coil for several hundred yards in each testis—a structure only about one and a half inches long. Cells between the tubules yield androgens, the male hormones.

A New Life Begins

Sperm cells thrash around an island of calm—the human egg (right). When an egg bursts from a ripe follicle on the ovary (below), ovulation occurs (1).

Sperm meets egg in the female body if sexual intercourse occurs within a day or two of ovulation. Drawn into the fallopian tube, the egg is fertilized (2). It begins to divide, and travels down the tube (3) to the uterus. It enters the uterus as a compact sphere of some sixteen cells, the morula (4).

Still dividing, the cluster floats free and becomes a nearly hollow globe, the blastocyst (5).

Inside, a mass of cells develops on one side of the blastocyst: This will be an embryo. When the cells of the blastocyst number about a hundred, its outer layer burrows into the uterine lining (6).

The follicle that held the egg becomes a yellowish corpus luteum, which manufactures hormones; they help maintain a pregnancy in its early stages.

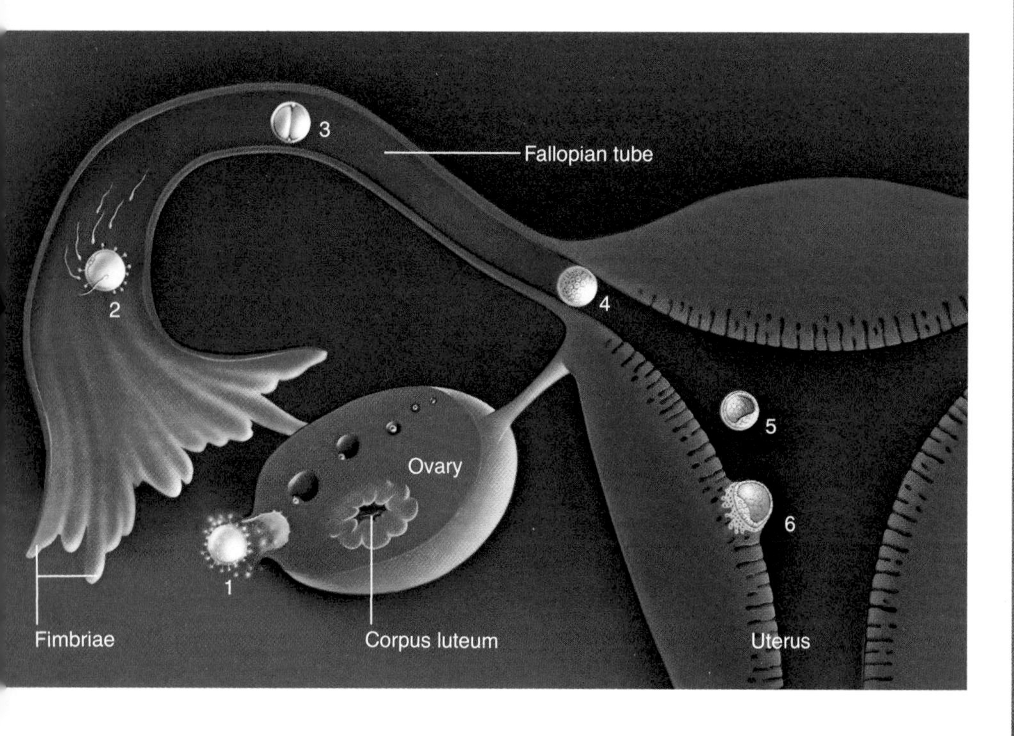

Magnification: 2,200 times

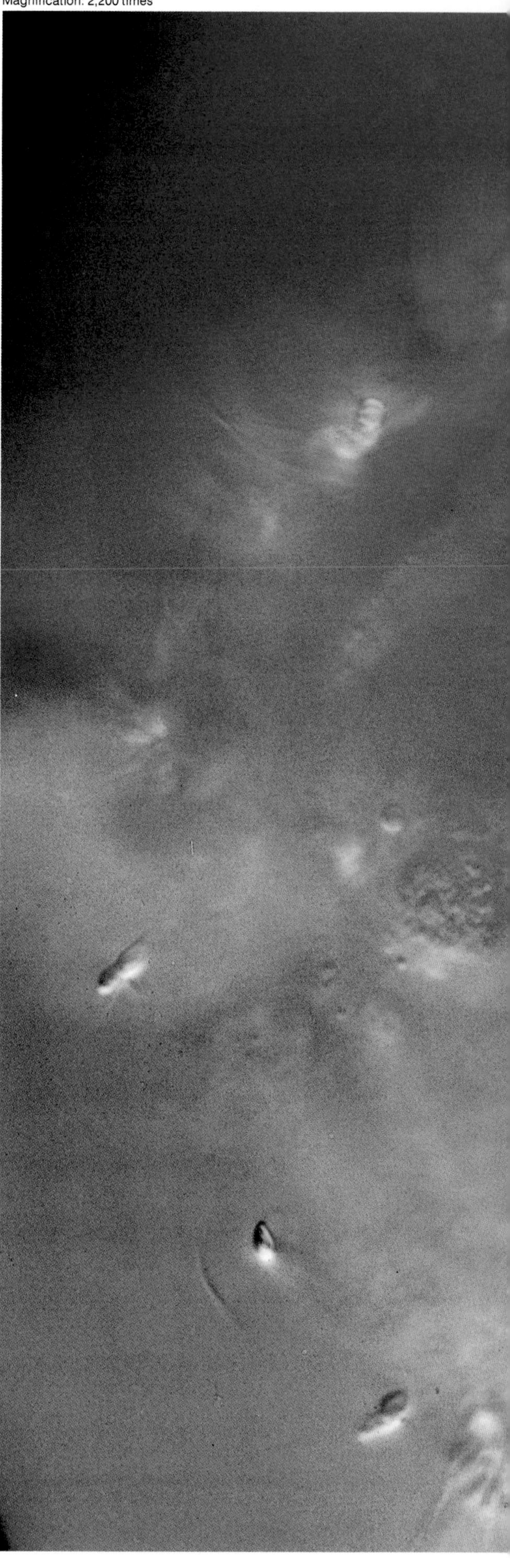

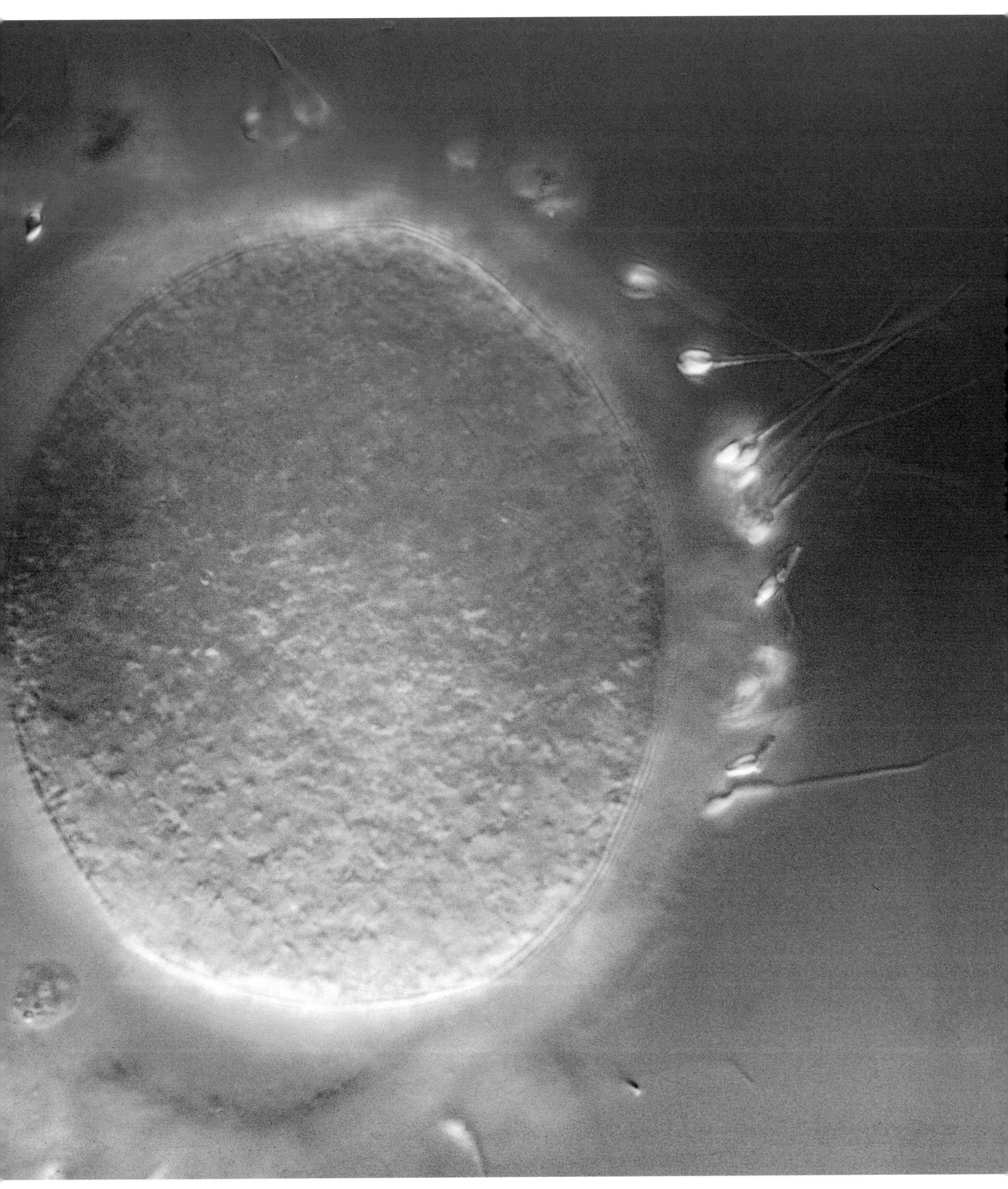

Magnification: 12,000 times

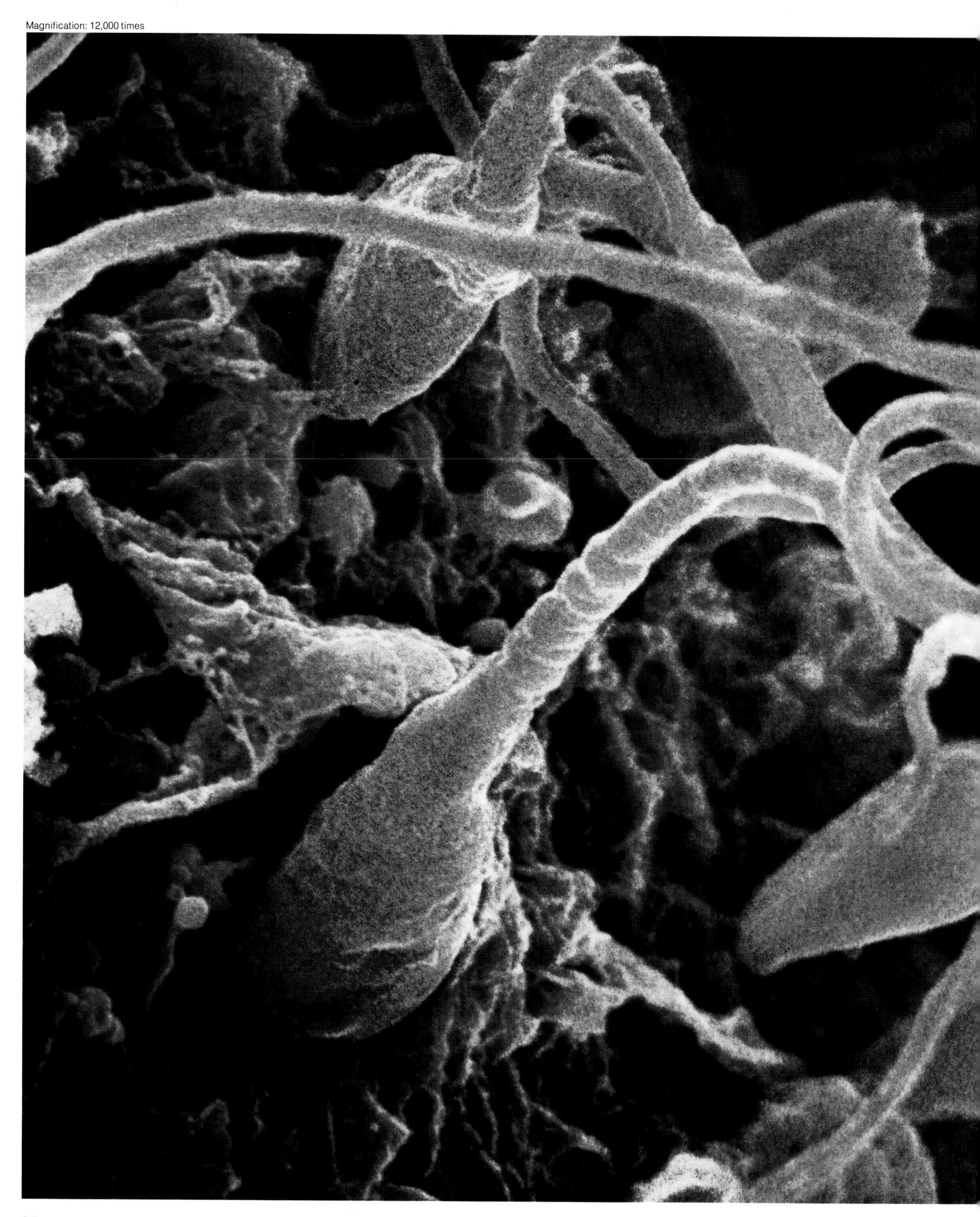

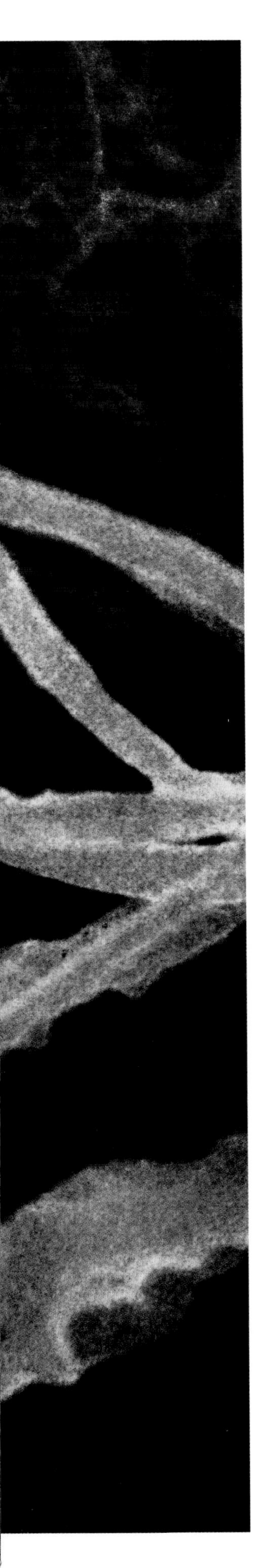

With their tails lashing, sperm cells (left) pummel the layered cells that surround an egg.

Although the sperm from one ejaculation may outnumber the egg by some 500 million to 1, fewer than 500 sperm reach the upper part of the fallopian tube where fertilization occurs.

Of these, usually only one can fertilize the egg, plunging in headfirst as it penetrates the egg's plasma membrane (right, upper). Tiny projections on the egg's surface pull the sperm inside the egg. The sperm's tail disintegrates, leaving only its head and mid-piece within the ovum (right, lower). Now a packet of genetic information, the sperm will merge with the genetic information of the egg, marking the moment of conception.

The sex of a baby is determined by the father. Each sperm carries either an X or a Y chromosome; an X chromosome contains instructions for female characteristics, the Y for male. The female egg carries only an X chromosome. At conception, if a father's Y sperm reaches the egg first—then the baby is a boy, an XY combination. If an X sperm fertilizes the egg, it joins with the X from the mother, making an XX pair, and the baby is a girl.

Magnification: 12,000 times

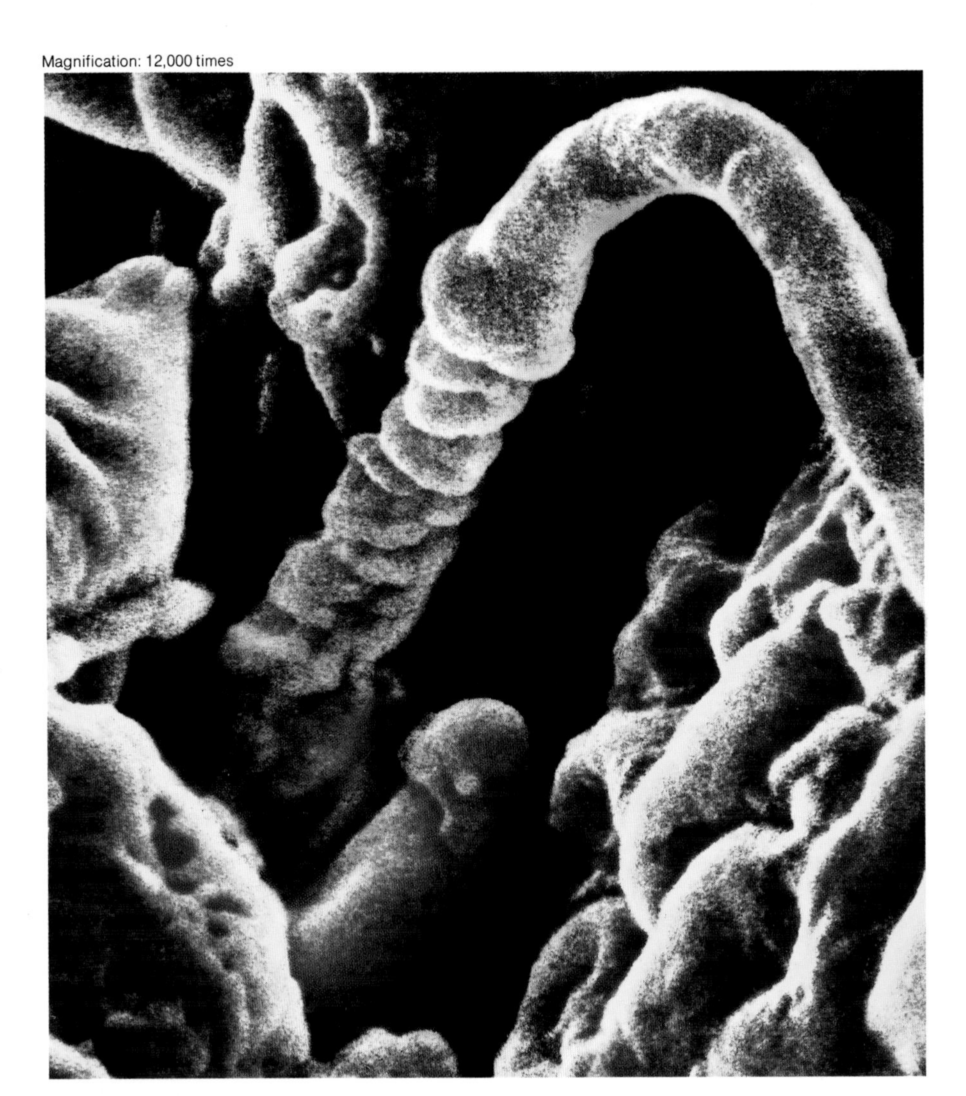

Magnification: 12,000 times

Cell by cell, new life grows. For the first few days after fertilization, each division creates similar, smaller cells (below) that become part of a sphere, the morula. Weeks pass, the cells begin to specialize, and a form emerges: A groove (right, upper) signaling the development of the nervous system is visible in an embryo just three weeks old—and no larger, at this point, than a sesame seed.

Beginning in the third week, cells (right, lower) form folds that come together across this neural groove, creating a tube. The folds first enclose the embryo's midsection (opposite), then "zip up" the remainder. The top of the tube swells to become a brain; the spinal cord emerges from the lower portion. At week's end the foundation for a human nervous system is in place.

Magnification: 750 times (below); 200 times (right)

Magnification: 300 times

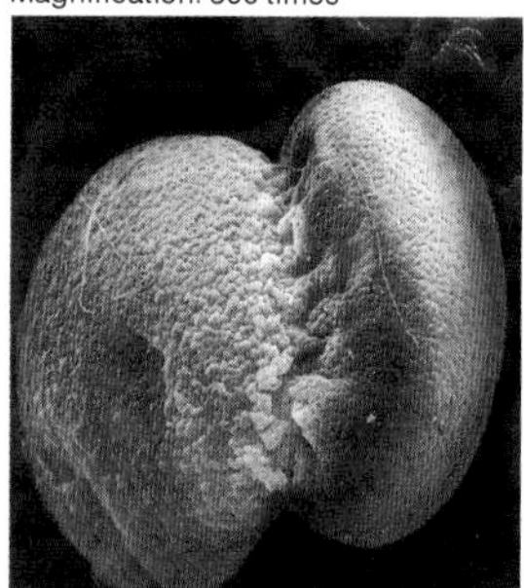

Magnification: 2,500 times

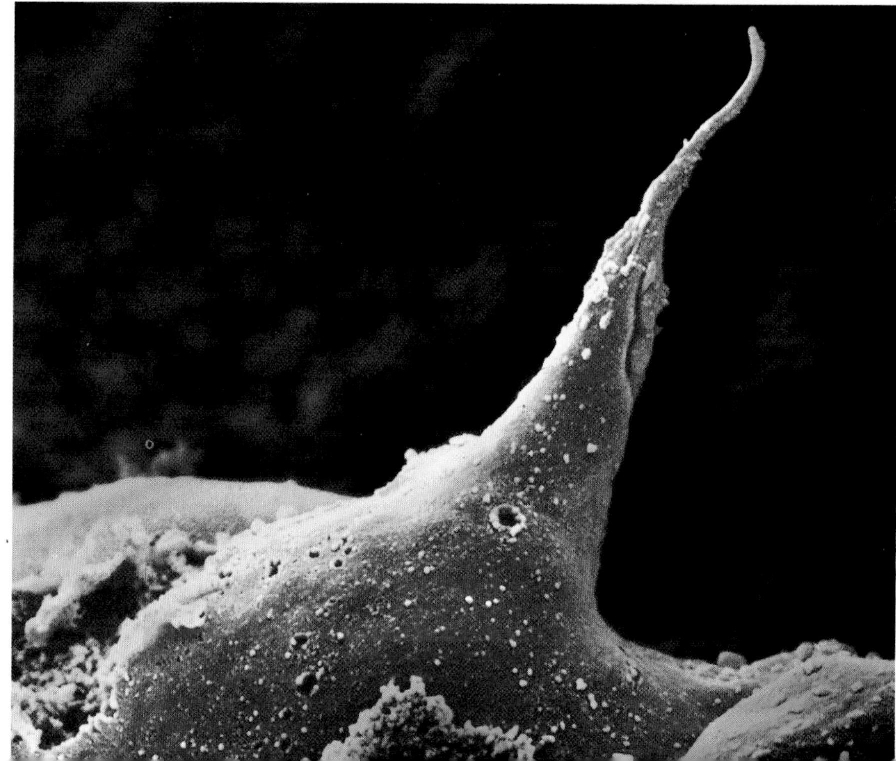

Magnification: 20 times

A bulging head and vestigial tail mark the embryo at four weeks (opposite). It curls around an already beating heart.

Arches appear—like gills—just beneath the embryo's head. Although these seem similar to the gills in fish, the human embryo never uses them for breathing. In the next weeks they will change shape, forming parts of the head and neck.

At six weeks (right, upper) a translucent form reveals the dark mass of heart and liver at its center. Now a half inch long, the embryo floats in the cushioning waters of a sac known as the amnion (right, lower). Still tethered like a balloon to the embryo, the yolk sac is a measure of the embryo's progress: The sac manufactured blood for the embryo until its own liver assumed the task during the fifth week; the bone marrow will ultimately take over this function. The yolk sac, now useless, will soon detach and begin to shrink and harden.

Magnification: 10 times

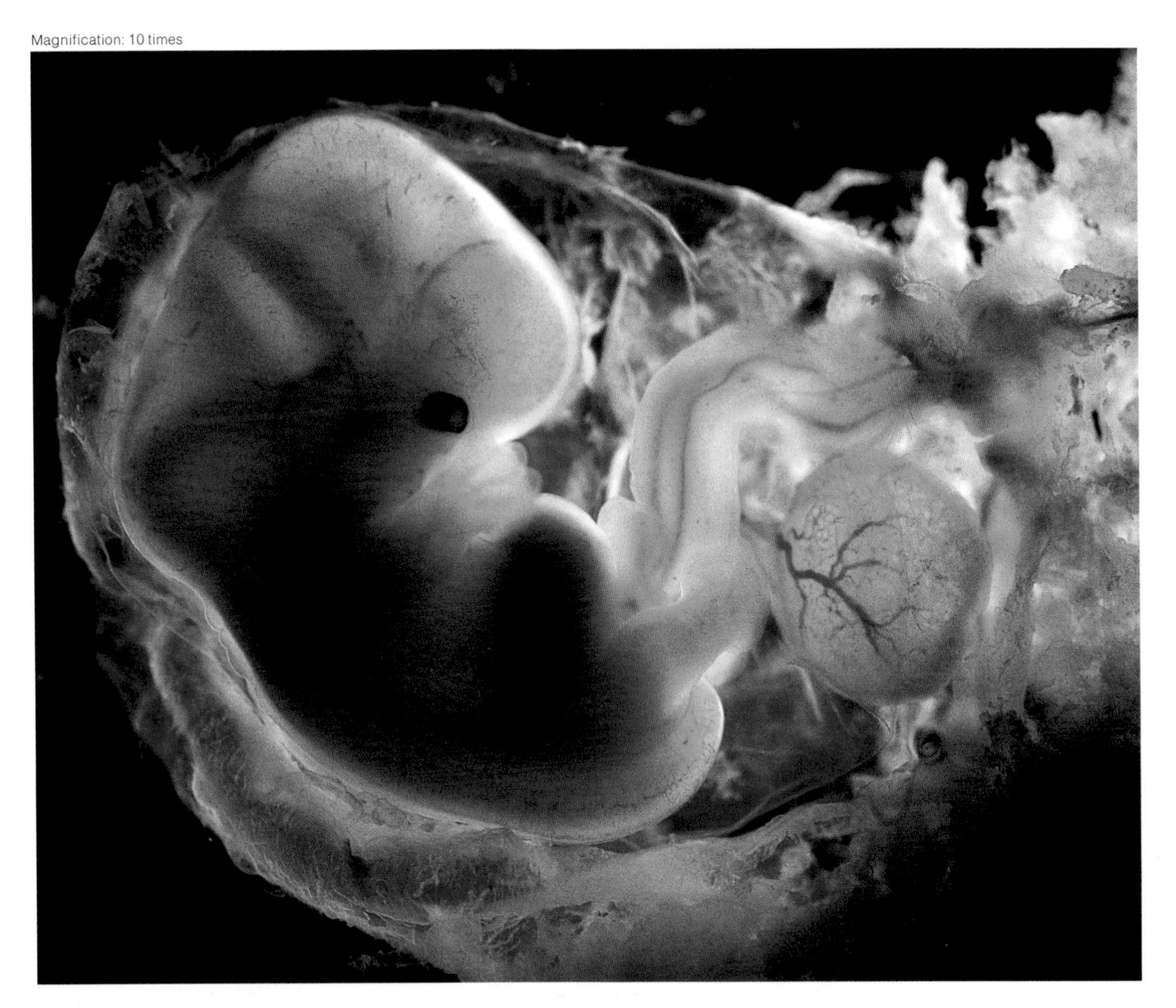

Magnification: 5 times

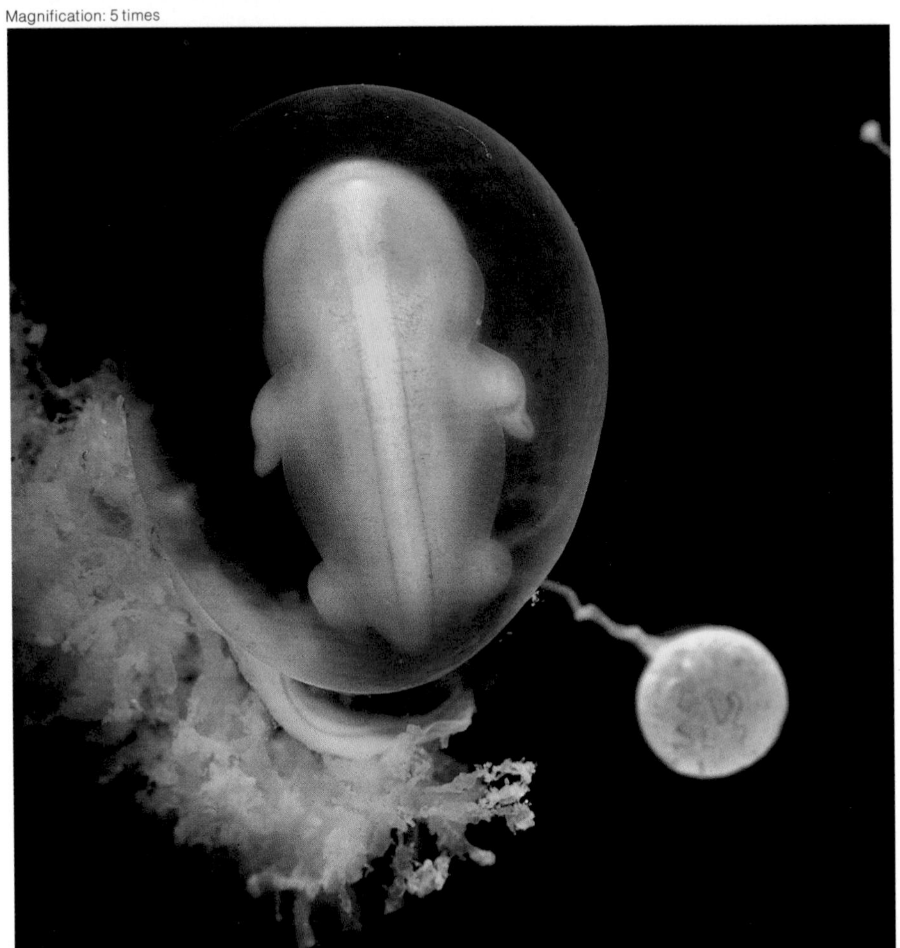

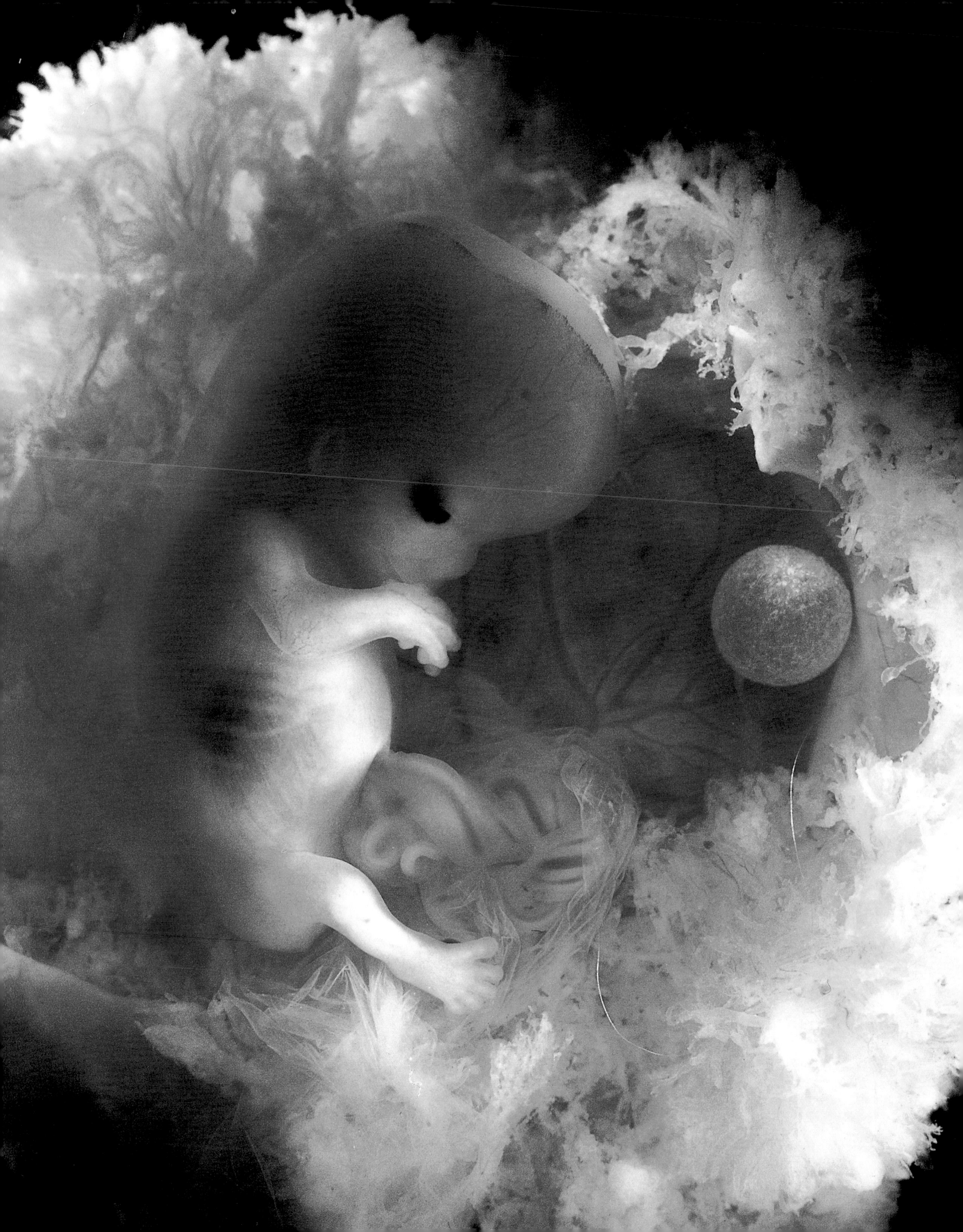

Cradled in a nest of membranes, the embryo celebrates a milestone at eight weeks. Though only the size of a walnut, it has all of its basic organs and systems. Now the embryo begins to test its muscles, which are growing daily. Its heart has been beating for one month, and the embryo takes on a more familiar, babylike appearance. Its head, now almost half the embryo's total size, signifies the rapid brain development that is taking place.

It will be known, from the ninth week until birth, as a fetus. Anchored to the life-supporting placenta by an umbilical cord, the fetus now enters a phase of dramatic growth: During the remaining seven months of prenatal life, its weight multiplies 3,600 times, and its structures will be refined.

Magnification: 7 times

knowledge, before the next menstruation begins undelayed. About half the lost conceptions are genetically abnormal. Nature erases many of its mistakes before they can be discovered.

When Cells Begin to Specialize

The sphere that nestles in the womb faces a far greater challenge than merely avoiding death; it must also sustain life. Its 100 nonspecialized cells must somehow multiply into the trillions of cells that make up the most complex vertebrate on Earth—the human being. The phenomenon of differentiation—how cells assume their different forms—remains one of the most baffling questions of science.

How are DNA's orders issued to the cells? Is the story of life written once, in the fertilized egg? Are its orders parceled out to offspring cells as it divides? To find out, scientists about a hundred years ago took a sea urchin's egg shortly after it had divided into two cells. They wondered if each cell, if separated from the other, would produce one-half of a sea urchin. Yet each cell developed into a complete creature. Even when the fertilized egg had divided 16 times, the separated cells turned into 16 sea urchins.

Now there are new mysteries. Today we know that all cells are derived from generalized cells and each one contains the genetic code for the whole body. But how does a cell, supplied with the voluminous script for all of life, come to play its exact role in the development of a human? How does a cell that could turn into anything turn into something? How is a rod cell in the retina, for example, designated to obey only orders to manufacture a light-absorbing protein? How does a pancreas cell know to produce insulin?

These questions speak to the essence of differentiation. We cannot fully answer them, but there is a trail of fascinating clues. One revealed that the ability of each cell or group of cells to produce a whole animal seems to end eight days after the blastocyst has embedded itself in the uterine wall.

This stage is gastrulation. The embryo, no longer made of indistinguishable cells, flattens into a disk which arranges itself into two and then three layers of distinct kinds of cells. Each of the layers generally will produce its own group of body parts, with layers sometimes cooperating to form parts. The top layer becomes the nervous system, sense organs, and the outer layer of skin; the middle layer forms the heart and other muscles, the dermis of the skin, blood, lungs, sex organs, and skeleton; the bottom layer will become tissues for breathing and digesting.

The layering shows that each generation of cells gets more sophisticated. The layers sculpt themselves in various ways, thickening and detaching to form an eyelid, folding into a tube for the spinal cord, creating a pocket as the cavity of an organ, all the while casting off dead cells as the details form.

With a few slips and slides of their facing sheets, the layers bring separated cells into contact. Other cells respond to hormonal directions given at a distance. This communication is a startling secret of differentiation. Cells converse. Whispering continually in a yet undeciphered chemical or electrical language, cells collaborate, telling each other what to be and when to be it. In the laboratory, if a barrier as frail as a sheet of cellophane is placed between interacting layers of an embryonic disk, the whispers go unheeded and normal development ceases.

The phenomenon is aptly named induction because one group of cells induces a neighboring group to take the shape of a supporting part. An eye, for example, begins to form when tissues of the forebrain push toward the surface of the head. The advancing part of the forebrain begins to fold in on itself, forming a cup that will turn into an eye. A chemical message in the cup induces the skin on the surface of the head to specialize into a lens.

In the early stages of embryonic development, all cells adapt readily to new surroundings; cells from a skin-forming area, for example, can turn into brain cells, and vice versa. At some time, not yet pinpointed by scientists, the cells reach, it seems, a point of no return. The final command is given and accepted. After this point, if a cell is transplanted to another part of the body, the transplant will stubbornly follow the orders for its original position.

What timer controls the induction schedule? What decides how the embryo will organize its cells into recognizable organs and tissue? The exact answers remain unknown. But it is possible that as differentiation proceeds, a cell's genes probably switch on and off, here telling the cell to build a heart, there to make a finger, there to become a nerve cell. Scientists are beginning to find what they believe are "master switches," genes that contain a certain sequence of DNA that helps to regulate the activity of other genes. While a cell's environment often tells it which orders to obey, its DNA may tell it when to perform each metamorphosis and when to stop.

Watching cells transform under a microscope, of course, is not nearly so moving as touching an expectant mother's swelling form, feeling the shape of a tiny hand or the thrust of a foot. The real wonder of differentiation is that it builds actual human parts.

At eight weeks all the organ systems are established, and the embryo is now called a fetus.

From the outer cells of the blastocyst grows an organ which is essential in the womb and useless outside it. At birth it dies. This is the placenta. The placenta's size and health directly affect the baby's well-being. It is *(Continued on page 36)*

Magnification: 7 times

A Human Being Takes Shape

The budding hand charts the steady growth of a human form.

From a scalloped paddle at five weeks (upper), fingers emerge by week eleven (right) on a hand no bigger than a teardrop. At this stage the body is no heavier than an envelope and a sheet of paper. Nail beds are in place, and muscles move the fingers and hand. Soon those muscles will carry the thumb to the mouth. The skeleton still consists largely of cartilage, but the steady encroachment of bone that begins in the ninth week is readily visible at sixteen weeks (opposite). Bone continues to grow until we are about age twenty-five.

Magnification: 14 times

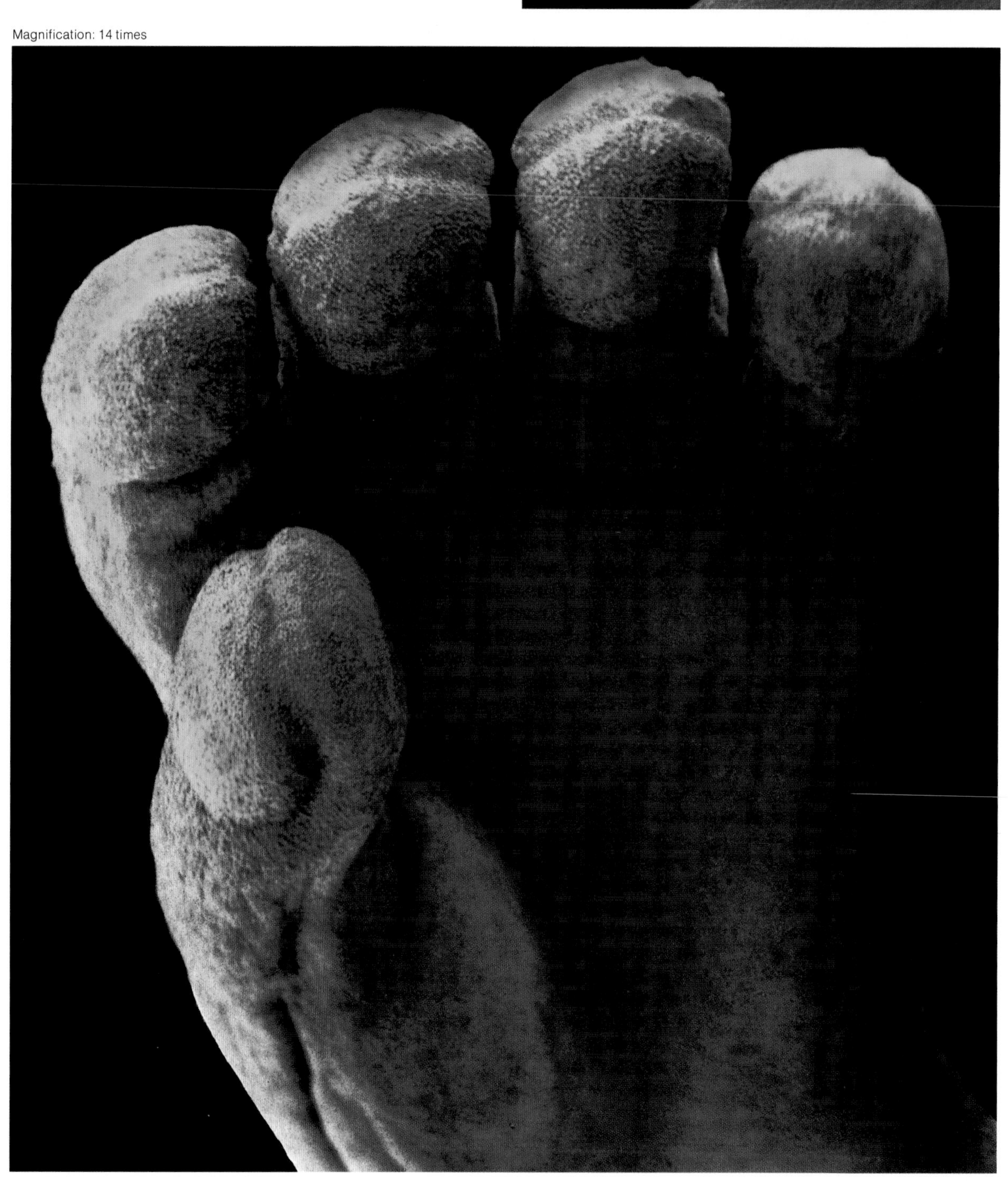

Magnification: 10 times

Magnification: 30 times

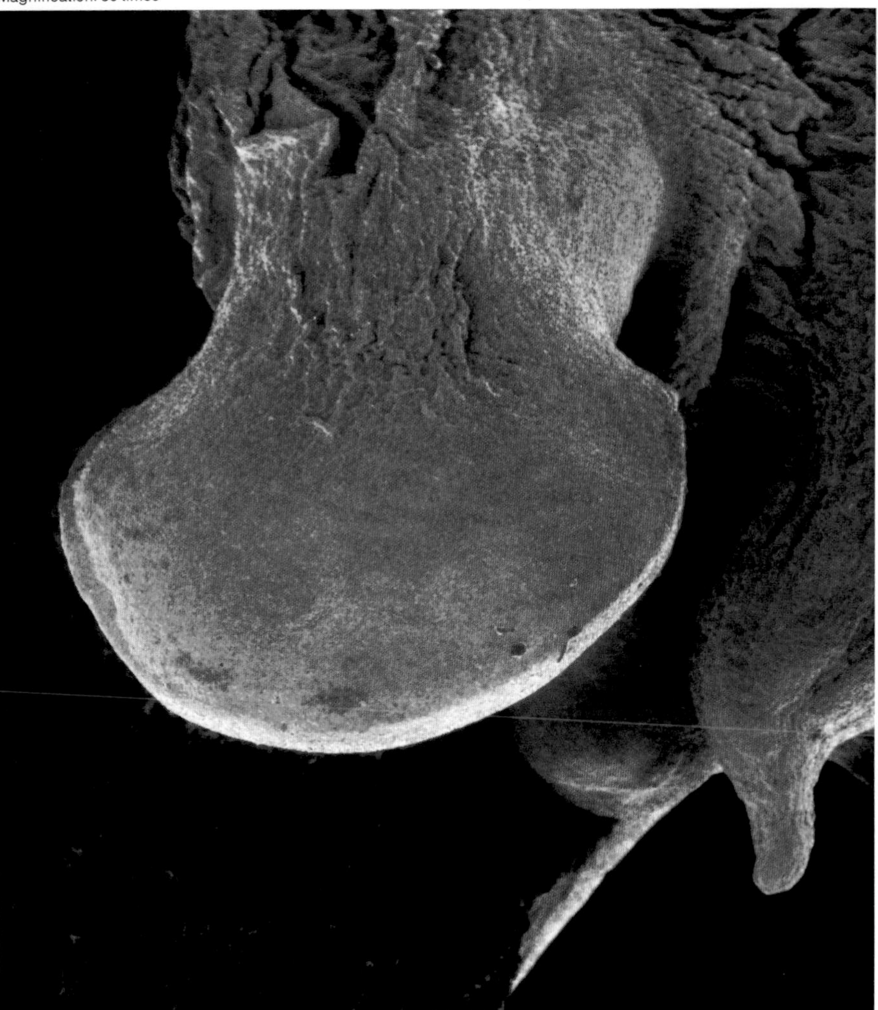

Magnification: 30 times

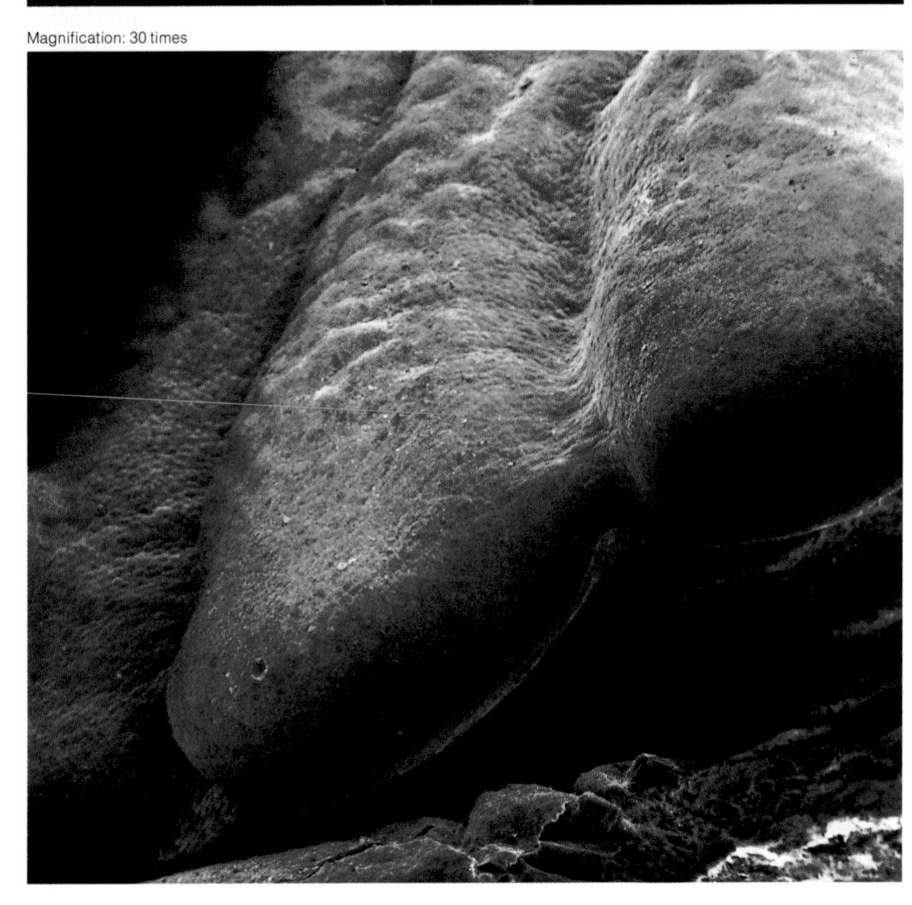

Following the lead of the hand, a spatula-shaped flipper at five weeks (left, upper) becomes a foot in a mere six weeks more.

Undulating ridges foreshadow toes to come at seven weeks (left, lower), a design apparent only four weeks later (right) in a well-formed foot.

The development of the leg and foot lags behind that of the hand and arm by several days as the embryo develops from top to bottom. The outer layer of skin continually sheds and renews its bumpy cells.

Magnification: 40 times

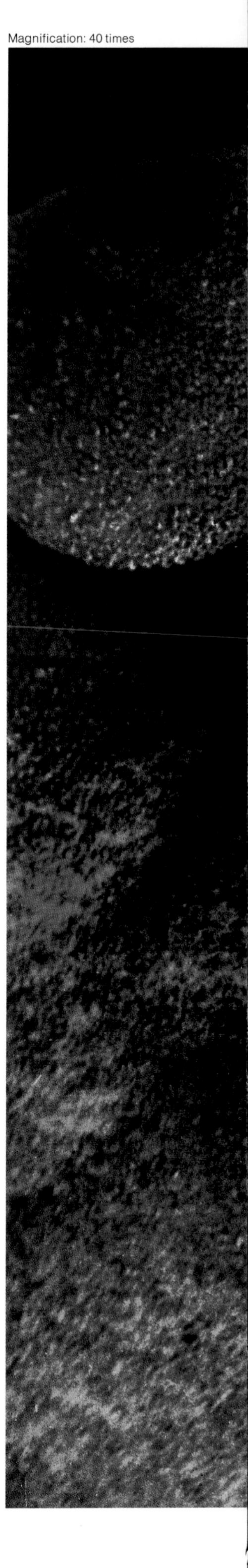

Magnification: 30 times

Magnification: 50 times

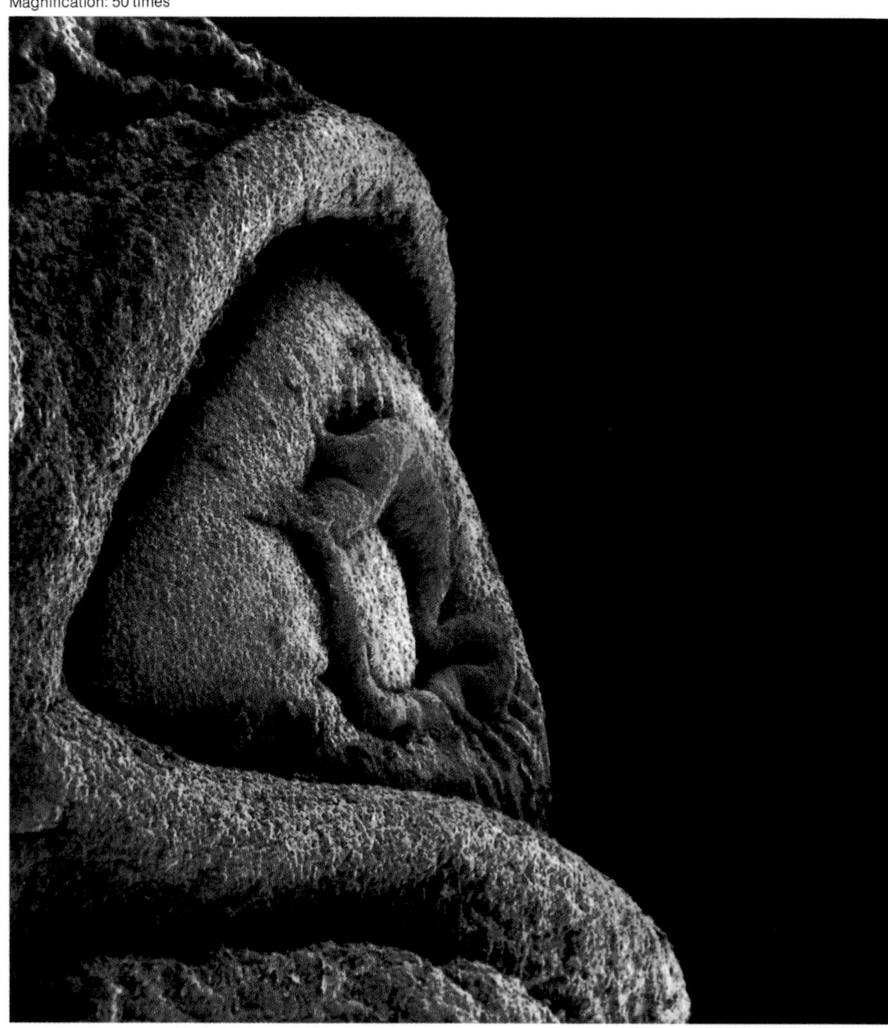

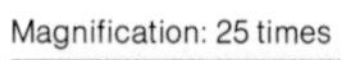

Magnification: 25 times

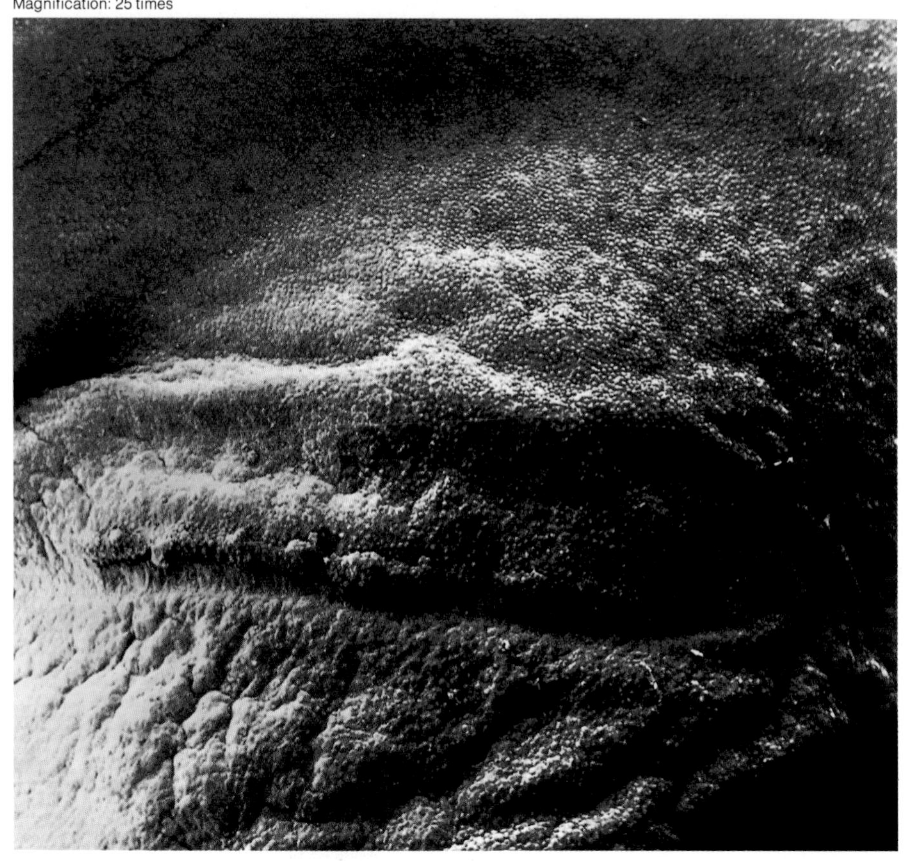

A portrait of the embryo at five weeks (opposite) shows the face of a stranger: Mouth and nose form a single cavity. The small holes on the sides of the head are the eyes, which develop from the front part of the brain. Their groundwork is well laid by the sixth week (left, upper), one week before they migrate to the front of the face, changing shape and traveling somewhat like an amoeba.

By the twelfth week (left, lower) lids have closed over the eyes, which remain shut for about three months more. Around the twenty-fifth week, they open.

a soft disk about 8 inches across and a pound in weight at birth. It grows from the cell layers that first tap into the mother's uterus. These "taproots" and their successors develop a surface area more than 400 feet square, convoluted like the florets of a head of broccoli.

By means of this great area, the placenta carries out the functions of lungs, intestines, kidneys, liver, and endocrine system for the fetus. It absorbs oxygen and nutrients from maternal blood, filters them across a thin, cellular barrier into the fetus's separate bloodstream, then returns the fetal waste products to the mother for excretion. This exchange uses the spiraling vessels of the umbilical cord, a conduit about two feet long, containing two arteries and one vein.

The placenta requires enormous amounts of oxygen and nutrients to do its work, and itself consumes as much as a third of the oxygen and glucose that come to the embryo. No other organ except the brain seems to operate at such a high metabolic rate. Developing rapidly in the first week after implantation of the blastocyst, the primitive placenta is ready to go to work when blood vessels connect and the rudimentary heart begins to beat.

Life Inside the Womb

The fetal circulatory path and blood change at birth. With the placenta, blood need not be filtered through lungs and liver. The fetus's heart and blood vessels develop temporary shunts that guide much of the blood past these organs which are not yet fully at work.

Floating in its amniotic waters, "breathing" only through its placental "scuba gear," the fetus lives in an environment so low in oxygen that an adult would be asphyxiated. But fetal red blood cells have a molecular structure that allows them to bind oxygen atoms more tightly than adult blood can. In an emergency—should the mother faint, for example—the fetus draws oxygen at a higher rate from the dwindling supply. If necessary the fetus can increase its production of red blood cells and can reduce blood flow to the extremities and other less essential organs, thus increasing the amount of blood that goes to placenta, heart, and brain.

Nerves appear just three weeks after the egg is fertilized. By the eighth week, the brain has taken shape, and brain waves register on an electroencephalograph. Nerve fibers have penetrated limbs, ears, eyes, and trunk. By the twelfth week, the fetus moves spontaneously, and may even be turning somersaults.

About this time, nerves, lungs, and diaphragm begin to synchronize. The fetus, a tiny, floating form, breathes in and out, moving tracheal fluid instead of air, exercising its lungs for the first breath after birth. This intermittent mechanical

From the realm of the expanding womb, the fetus prepares to enter the outside world.

At twenty-two weeks a girl, about ten inches long, already looks and may act much like a newborn: Pale, wispy eyebrows are showing. She may squirm, suck her thumb, and even turns somersaults in the one-quart pool of amniotic fluid. In the fifth month the mother usually detects her baby's movements and she may even grow aware of its cycles of sleeping and waking.

Yet several major organs are still too underdeveloped for a fetus of this age to survive outside the womb. The remaining four months are a time for the finishing touches.

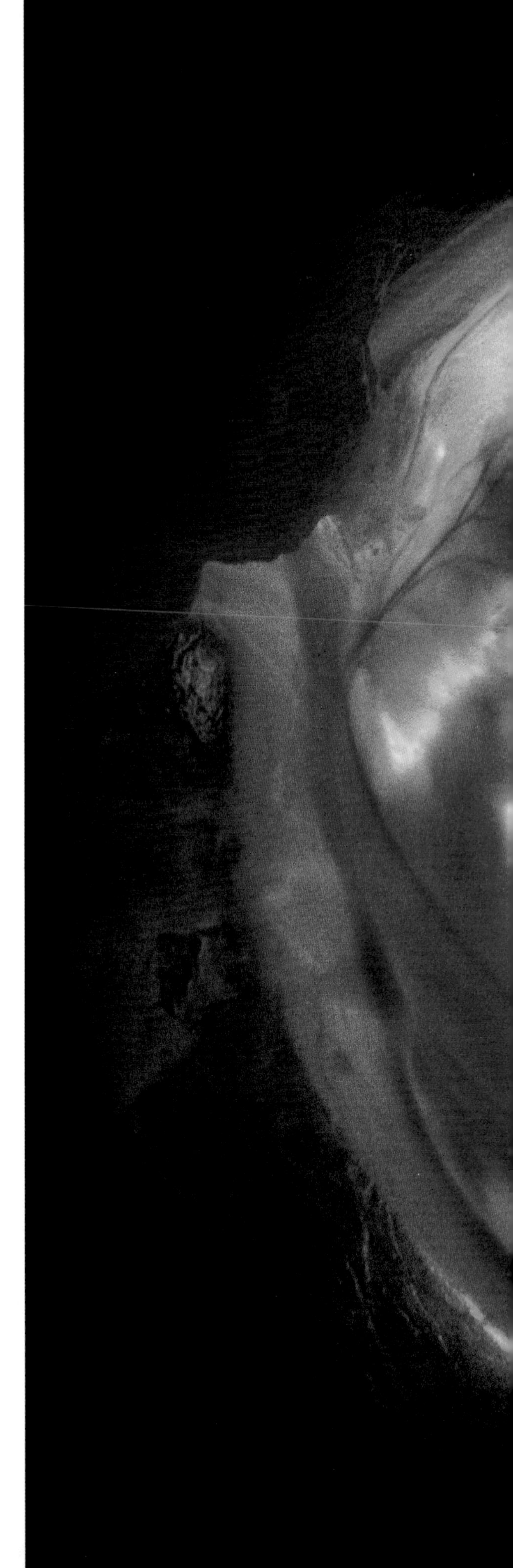

Magnification: 2 times

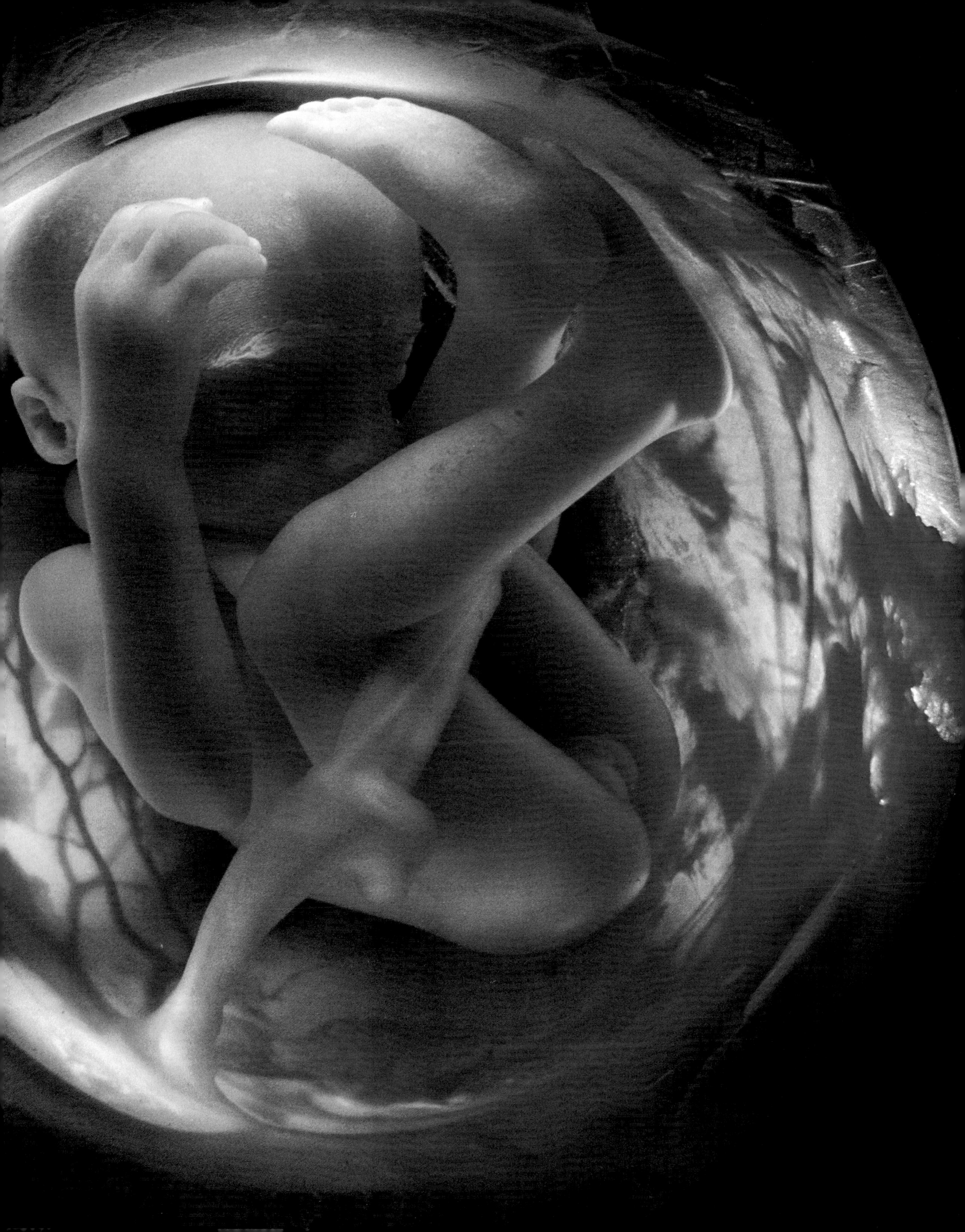

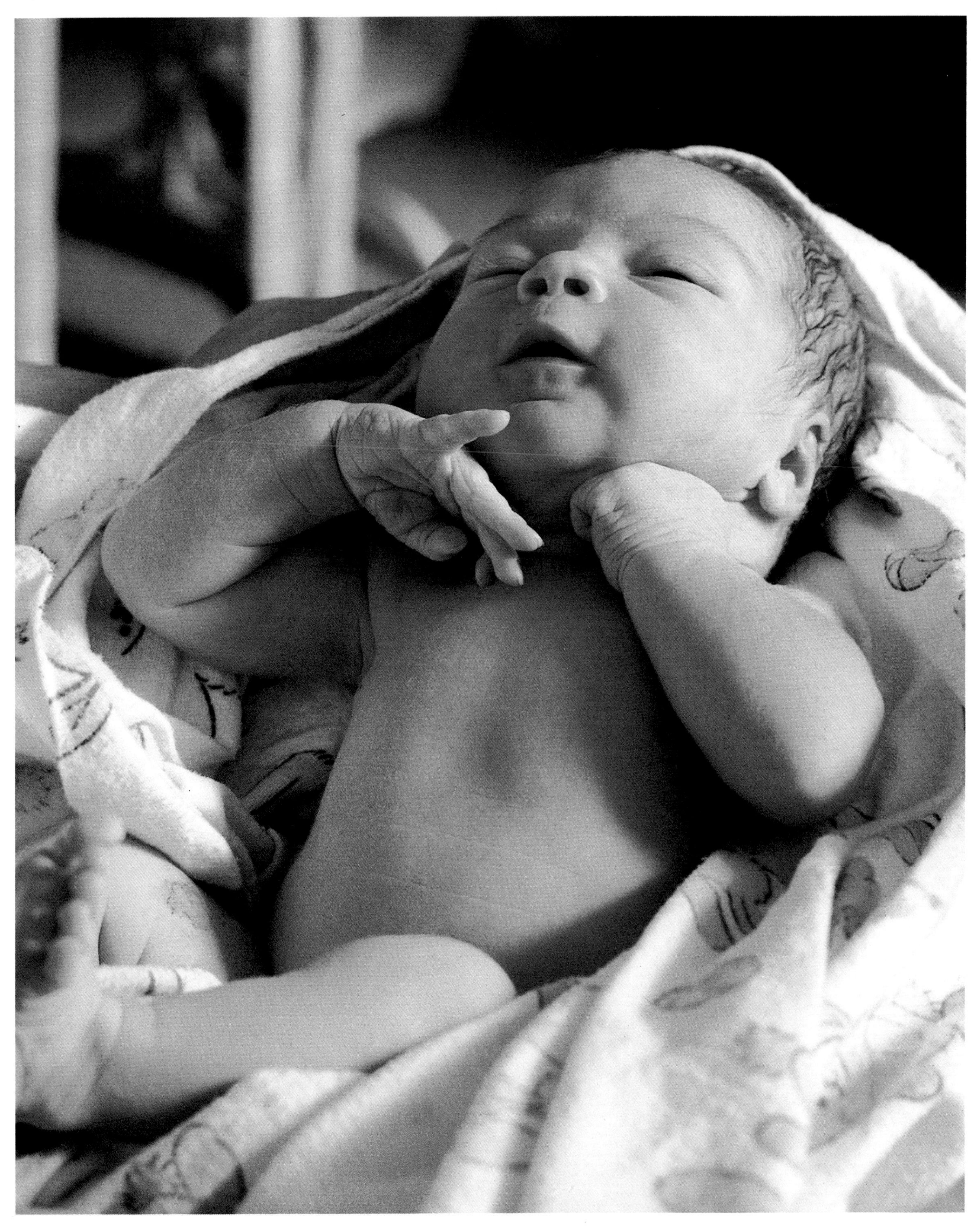

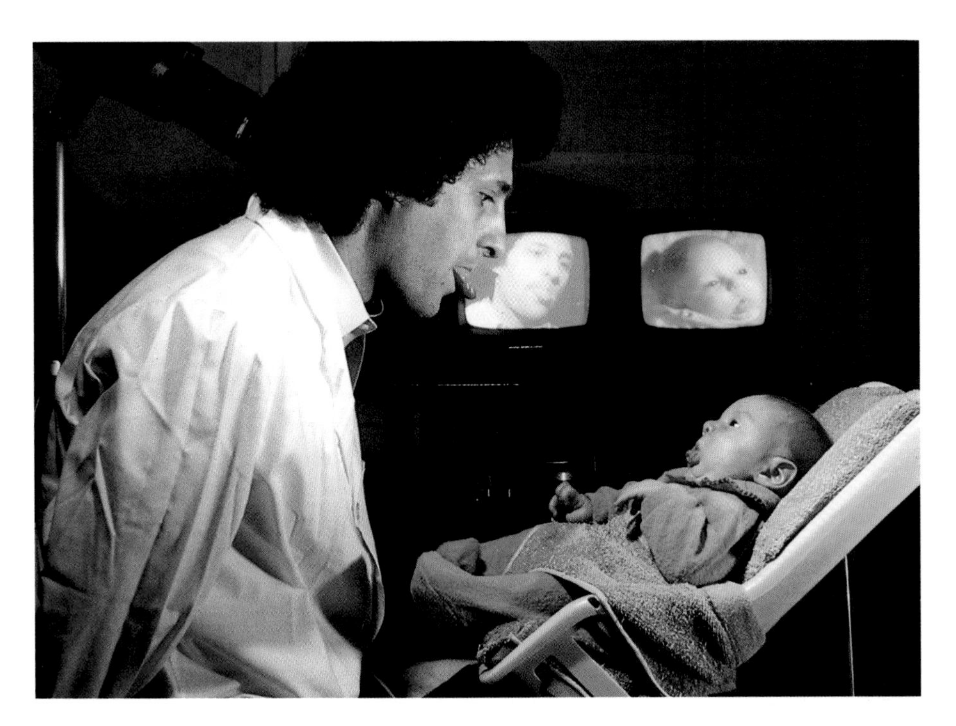

Fresh to the world, a baby only minutes old (opposite) already possesses an active intellect.

Newborns show evidence of memories from the womb by what they prefer to hear—maternal heartbeats, their mothers' voices, even stories read to them before birth.

They are also capable of complex thought processes, as Dr. Andrew Meltzoff learned. When Meltzoff shows his tongue to an 18-day-old boy, he responds in kind (above). With M. Keith Moore, his associate at the University of Washington, Meltzoff found that even babies less than an hour old can imitate the facial expressions of adults. This skill requires the young mimics to translate visual perception into muscular action, even though they have never seen the movements of their own faces.

action supplies no oxygen, but healthy lung growth depends on fetal breathing movements. And brain connections for post-birth life and emergencies have already been established.

By the fifth month, about the time the mother appears pregnant, other responses show up. The fetus reacts to temperature change. Now the fetus begins to swallow amniotic fluid, taking in as much as twenty-five ounces a day in late gestation. If a sweetener is added to the fluid, the fetus swallows twice as fast.

By twenty-four weeks the fetus demonstrates a most interesting neural development. During much of its sleeping time, it displays rapid eye movements, an indication of dreaming in children and adults. Just six months after a single cell cleaves in two, its offspring may be practicing to dream in a world where it has no companions, and has yet seen no visions.

But they do hear sounds, transmitted through the mother's abdominal wall. Fetuses of this age seem to respond to music—they blink their eyes and move, as though dancing to a beat.

By the seventh month, fetuses show movement that may be something other than reflex action. Earlier, a fetus moves reflexively, without prior intention. The act begins first, then the heart accelerates. At seven months the heart rate quickens before the fetus begins some movements. Does the anticipatory acceleration mean that the fetus has contemplated its action? We do not yet know.

Eluding the Body's Defenders

The mother's recognition of the fetus's stirring in the womb may forge a bond between her and her child. Yet the fetus is not completely dependent, passively accepting the mother's contributions of nutrients and shelter; it seizes them. To do this, it must devise a ruse to conceal its identity and sidestep the mother's immune system. Some of her cells (as in all human bodies) can read and recognize chemical identity markers on other substances. By reading these markers, her body can distinguish between its own parts and foreign intruders like viruses, bacteria, certain other body cells, and inorganic matter. White blood cells attack and destroy these invaders. Tissue and organ transplants often fall prey to these defenses. Yet the fetus somehow manages to elude attack. Why?

Is it spared because the genes it inherited from its mother mean that its tissues are read as part of the mother's body? If this were the case, mothers would be able to tolerate skin grafts and other tissue transplants from their own children, but they cannot.

Does a mother's immune system retire during pregnancy? Again, no. The fertilized egg displays foreign markers almost immediately, and the mother's body counterattacks. Lymph nodes near the uterus swell with increased production of white blood cells. Antibodies and white blood cells targeted at the invading zygote pour into her bloodstream.

Yet the new life survives, even thrives, in the face of this onslaught. Perhaps placental hormones help protect it. During pregnancy the fetus produces progesterone, estrogen, and other chemicals that may inhibit the killing capacity of the mother's white blood cells. But the fetus's main line of defense probably comes from the trophoblast cells, those cells that first invade the lining of the mother's uterus. The trophoblast seems to have several strategies at its command, including the ability to trigger the production of special blocking antibodies that help blunt the response of the mother's immune system. This allows the temporary truce called tolerance.

Still, the fetus faces many demands. Not only must it grapple with the mother's immune system; for healthy growth it must increase its weight 2.4 billion times during its gestation. The building material has to come from somewhere. A mother can affect her fetus by what she does—or does not—swallow or inhale. Poisons in drugs, alcohol, and cigarettes can cross the placenta and damage the baby.

There has been some evidence that chronically malnourished mothers produce babies who are significantly smaller than average. But short-term malnutrition does not seem to affect birth weight seriously. During a World War II famine, undernourished mothers in the Netherlands had babies who weighed only about 11 ounces less than normal. One reason may have been that a pregnant woman's body, while needing more nutrients, has an enhanced ability to use them. And the fetus also has *(Continued on page 48)*

The Cell, Human Dynamo

A human cell reveals our inner architecture. Five cells, opened up for an inside view, cluster together (right). Nuclei appear as large yellow spheres; mitochondria are colored red. Irregularly distributed ribosomes show up as pale dots. Throughout each cell runs the ropelike maze called the endoplasmic reticulum.

Not all cells look alike. An adult human, in fact, is an assemblage of some 100 trillion cells, with groups of cells specialized in structure to perform a particular job.

A chromosome (below) conveys a detailed set of genetic information from parent to child. It also gives cells the directions they need for reproducing themselves. Forty-six chromosomes appear in each body cell; only twenty-three each exist in sperm and ovum.

Magnification: 20,000 times

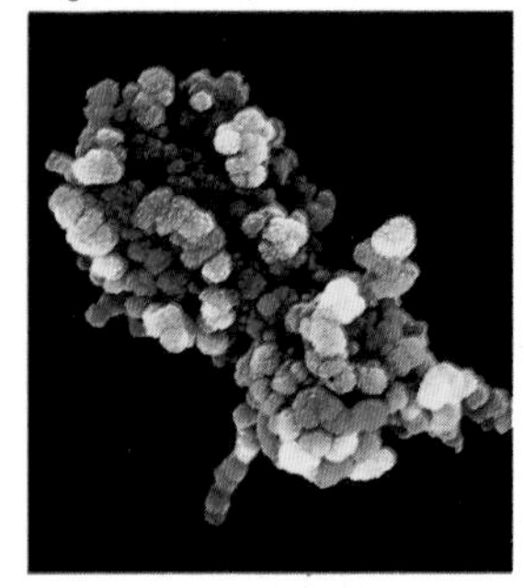

Magnification: 10,000 times

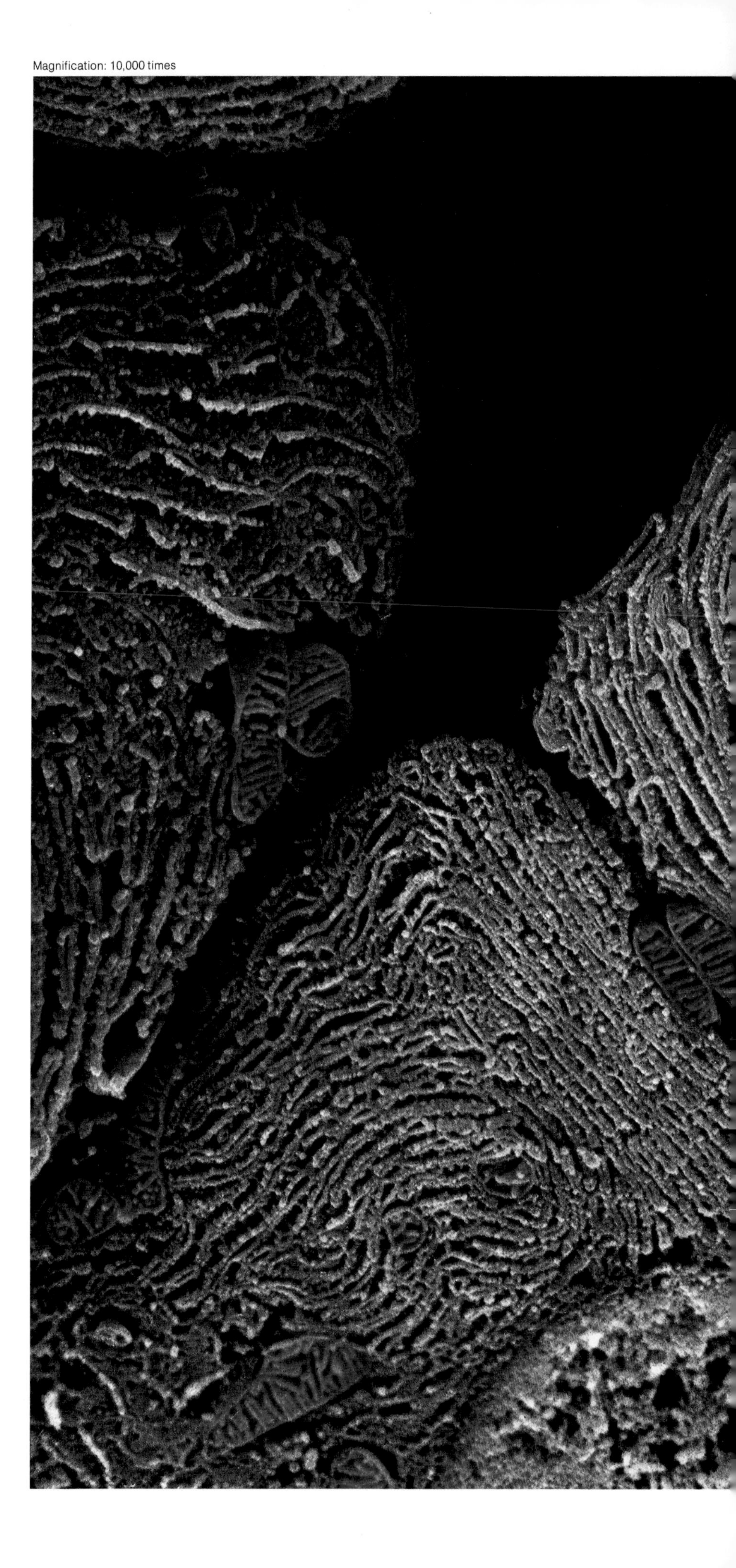

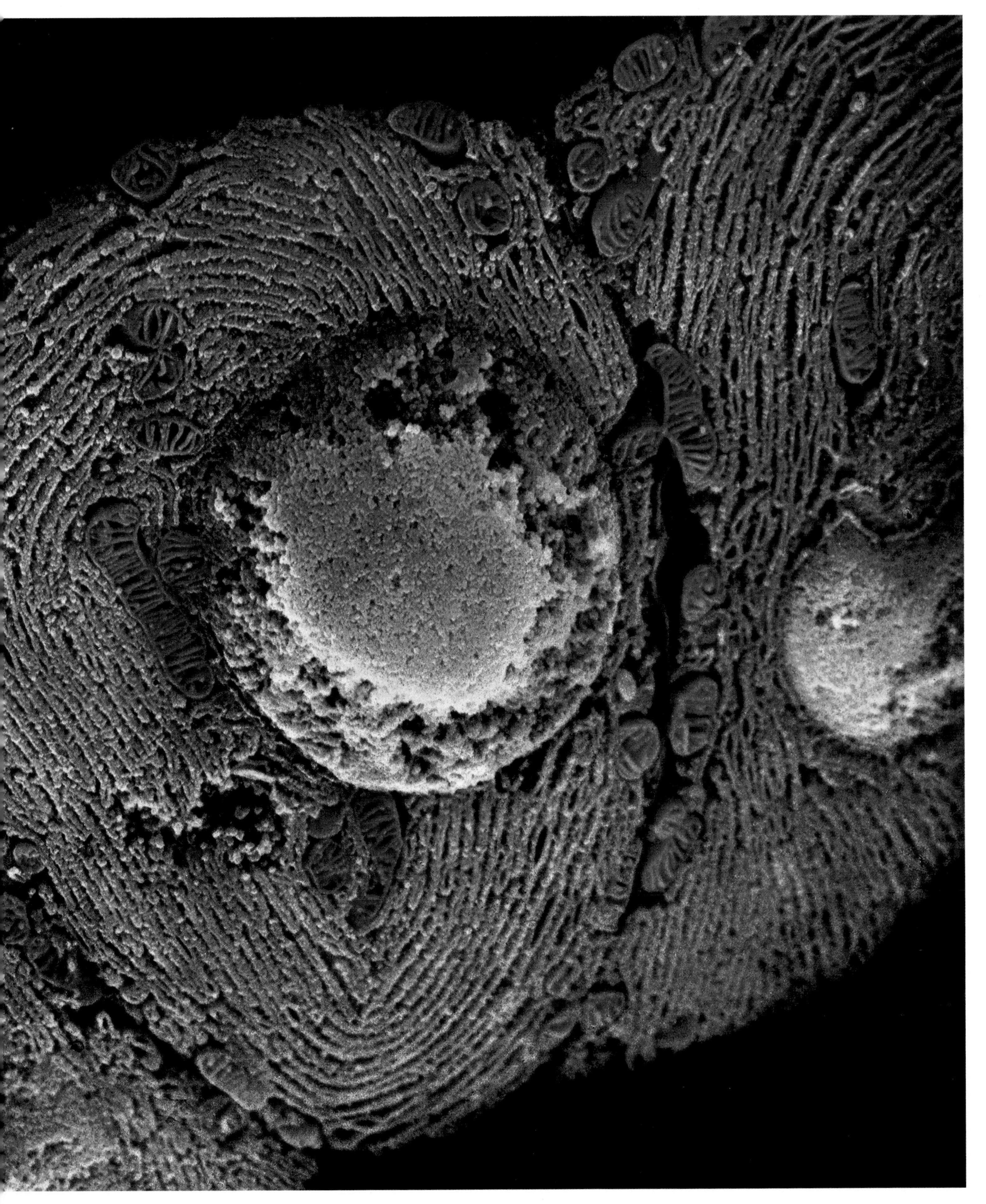

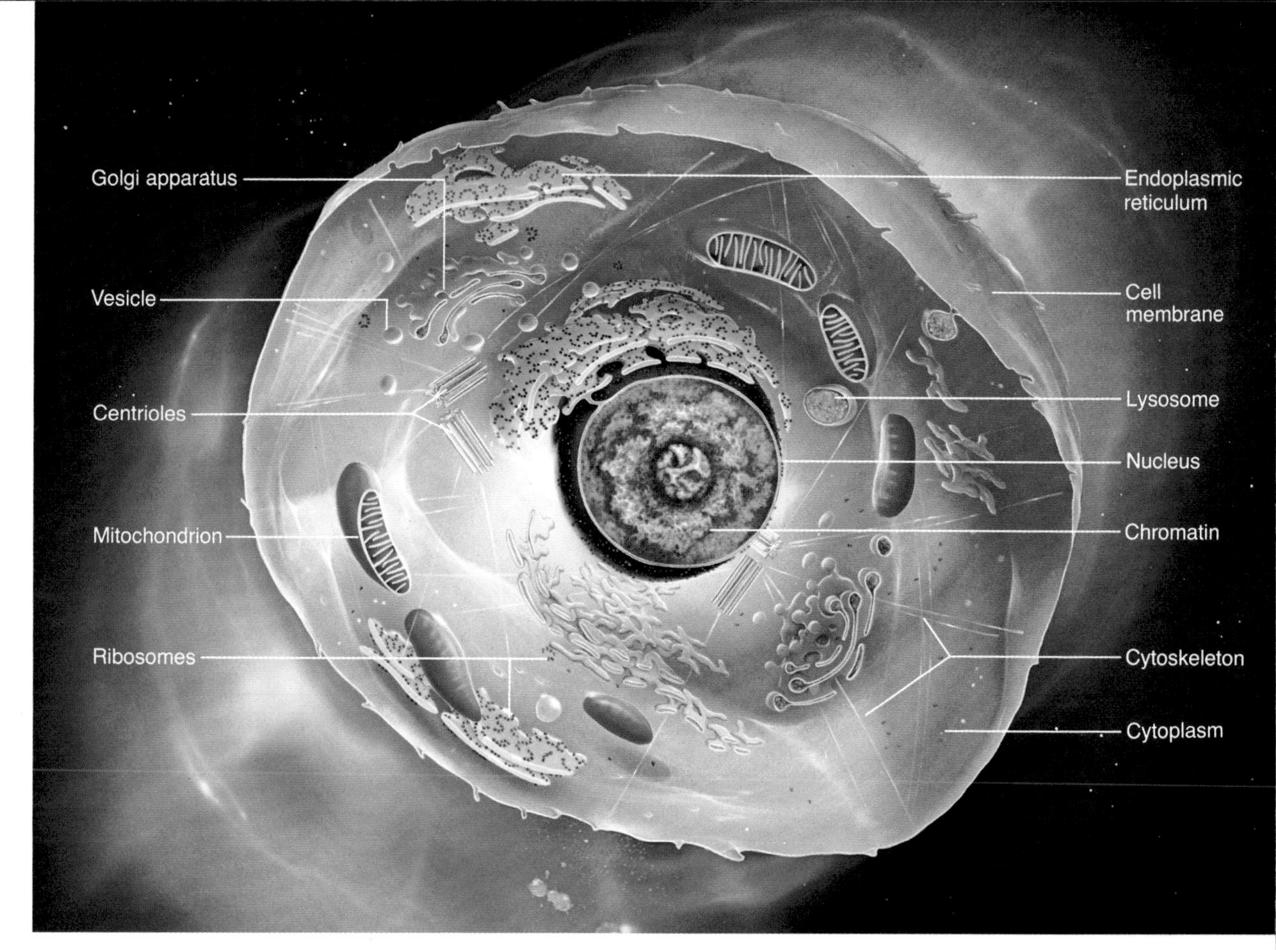
Golgi apparatus
Vesicle
Centrioles
Mitochondrion
Ribosomes
Endoplasmic reticulum
Cell membrane
Lysosome
Nucleus
Chromatin
Cytoskeleton
Cytoplasm

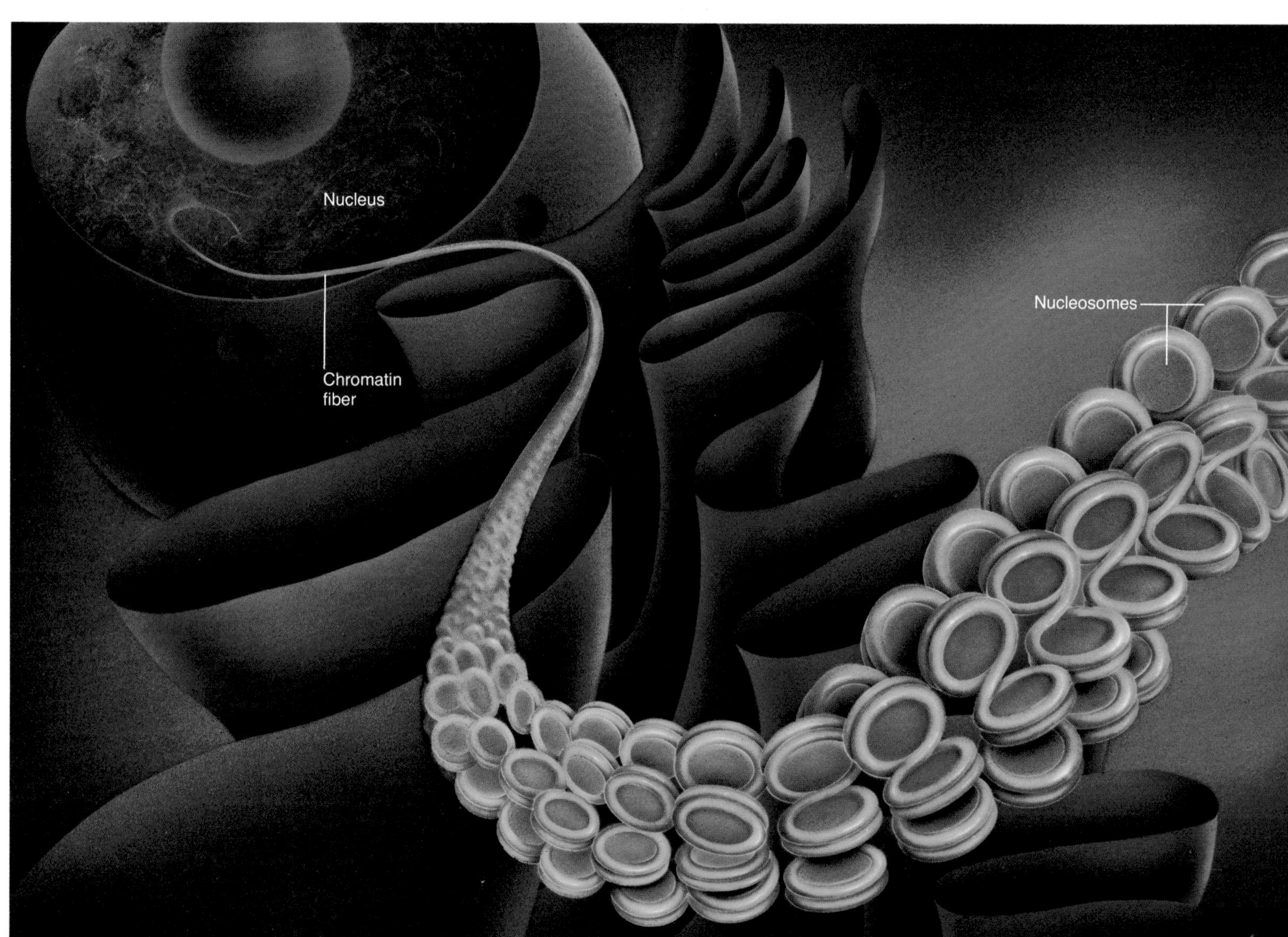
Nucleus
Chromatin fiber
Nucleosomes

The human cell bustles with activity (left). In this microscopic world, a highly organized system of specialized cell parts, known as organelles, conducts the many tasks that keep the body working.

An ultrathin membrane surrounds the cell's gelatinous contents. The filaments of the flexible cytoskeleton buttress the cell from within; they give it the ability to move and change shape. Mitochondria act as power plants, supplying the cell with energy. Thus fueled, ribosomes manufacture some of the cell's main products—proteins.

One group of ribosomes produces proteins for the cell itself; another group makes proteins for export. As each new protein destined for secretion comes off the production line, it travels through the endoplasmic reticulum for processing. Packaged in vesicles by the Golgi apparatus, the protein moves to the cell's surface. The vesicles fuse with the cell membrane and discharge their contents.

The Golgi apparatus may also create lysosomes, bags of enzymes that digest bacteria and other material entering the cell.

A nucleus (left and below) controls and coordinates these and many other activities. The plan for this work lies stored within the chromatin fibers that pack the nucleus. Before a cell divides to form two new cells, forty-six chromosomes take shape from these fibers.

Each chromatin fiber consists of a thread of deoxyribonucleic acid, or DNA. Through segments called genes, DNA determines the makeup of every cell and the hereditary traits of each one of us.

End to end, a cell's full complement of DNA would measure six feet. An efficient packing mechanism (below) allows it to fit and function in a space only 1/2500 of an inch in diameter: Each strand of DNA curls around cores made of eight protein molecules, forming a tightly coiled string of repeating units called nucleosomes.

The DNA molecule is a miracle of organization, structured like a twisted ladder. The sides of the ladder—alternating sugars and phosphates—form the molecule's backbone. The "rungs" are interlocking pairs of four chemicals called bases: adenine, thymine, cytosine and guanine. Only two different pairings are possible: Adenine (A) always bonds with thymine (T), and cytosine (C) always bonds with guanine (G). Using this four-letter alphabet, DNA dictates the protein mix that fulfills our genetic inheritance.

Elegant in structure, DNA (pages 44-45) is also vibrant. Any still portrait of this molecule conveys only part of its nature, for motion characterizes the rest. DNA bends and twists a billion times a second while its ladder sides "breathe" in and out. This dance likely arises as DNA engages in its two key roles: to direct the creation of protein and to duplicate itself.

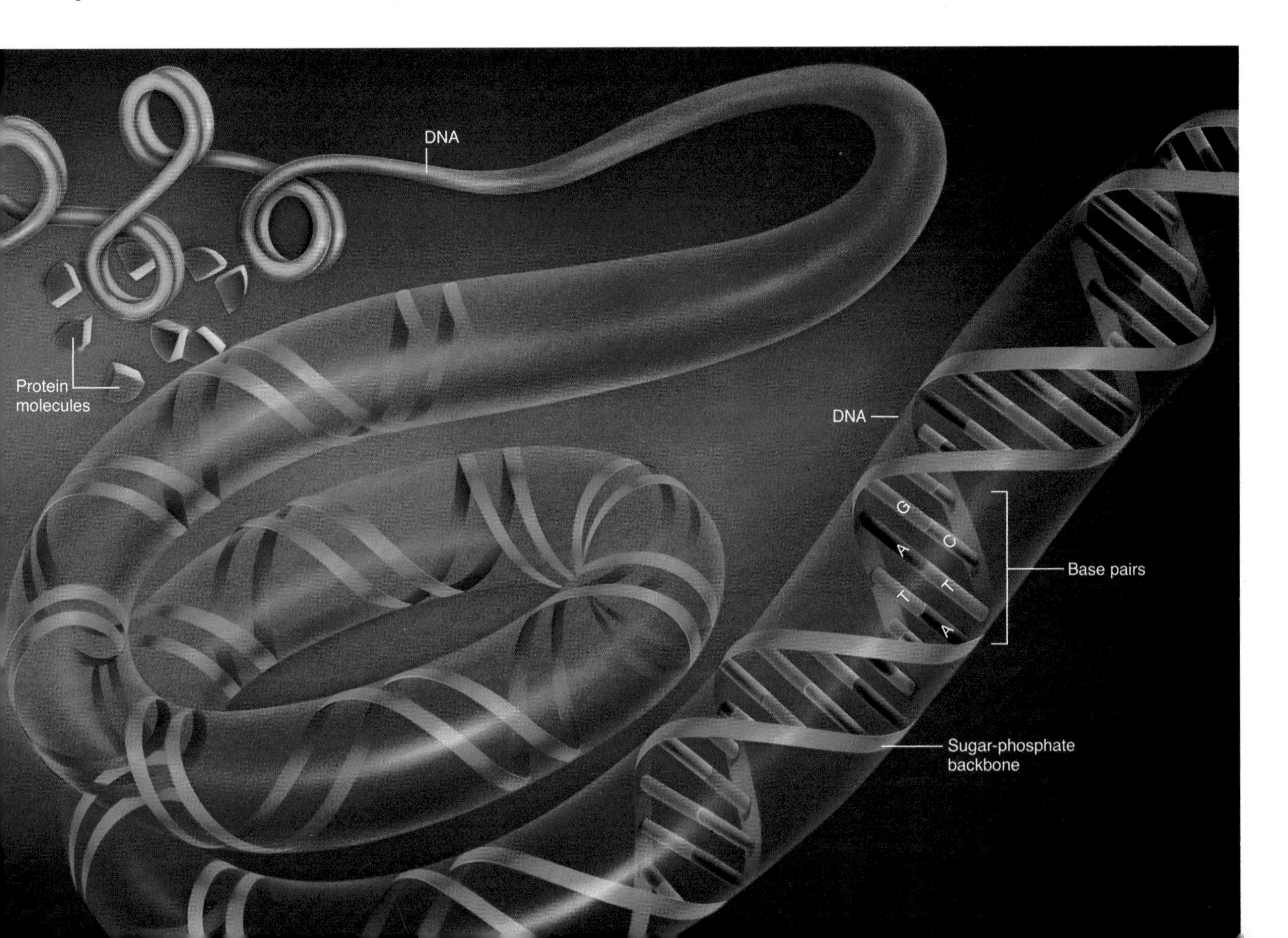

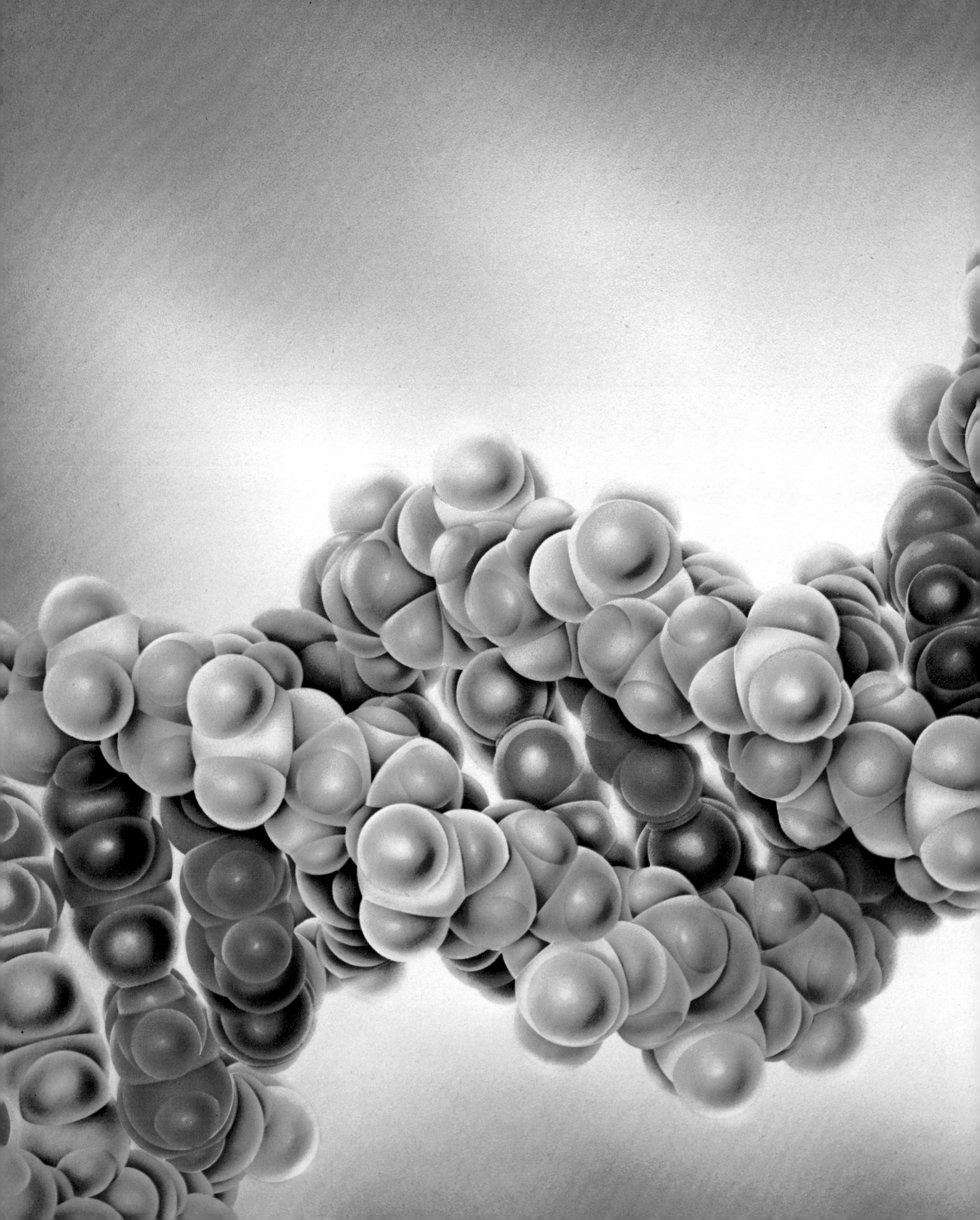

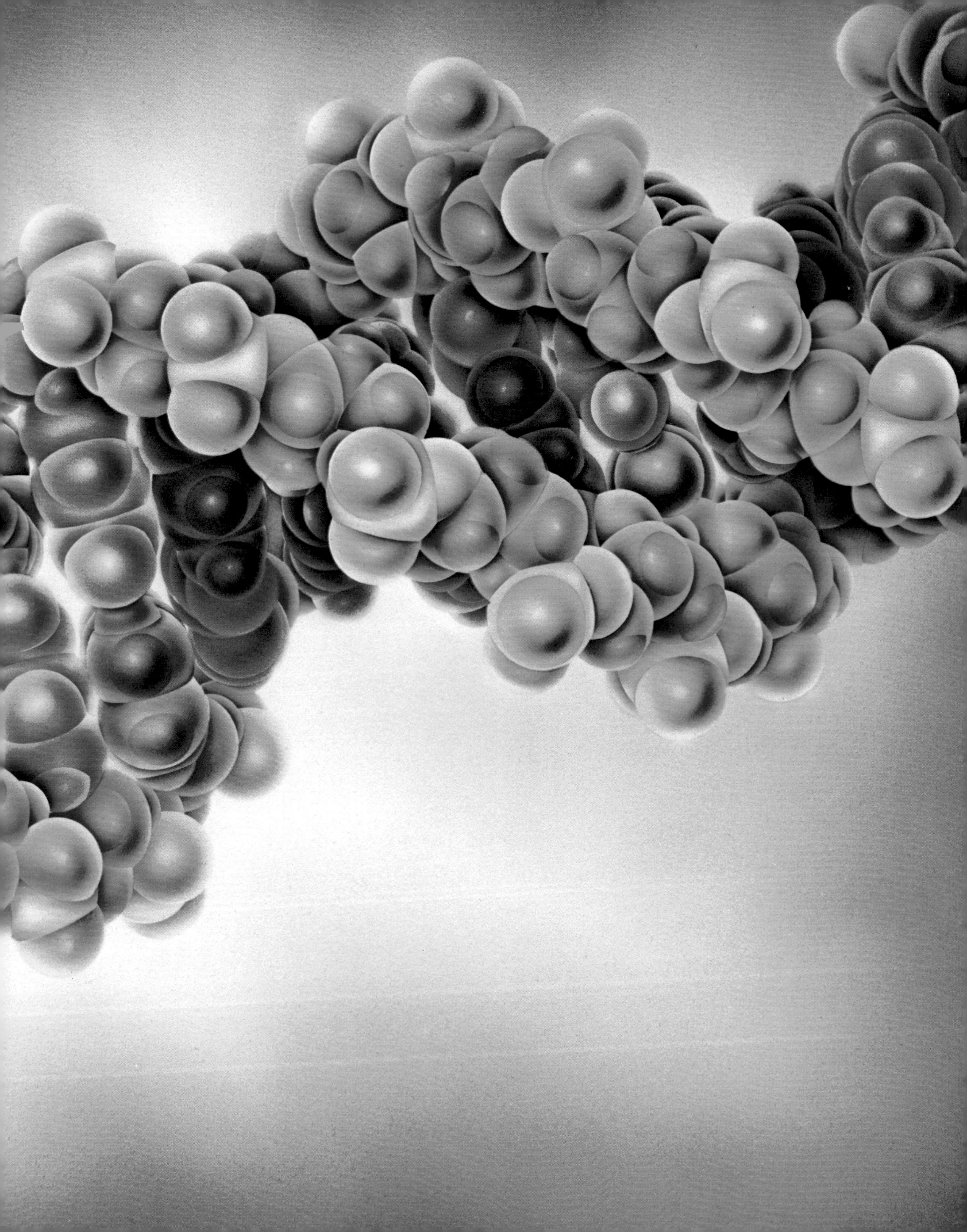

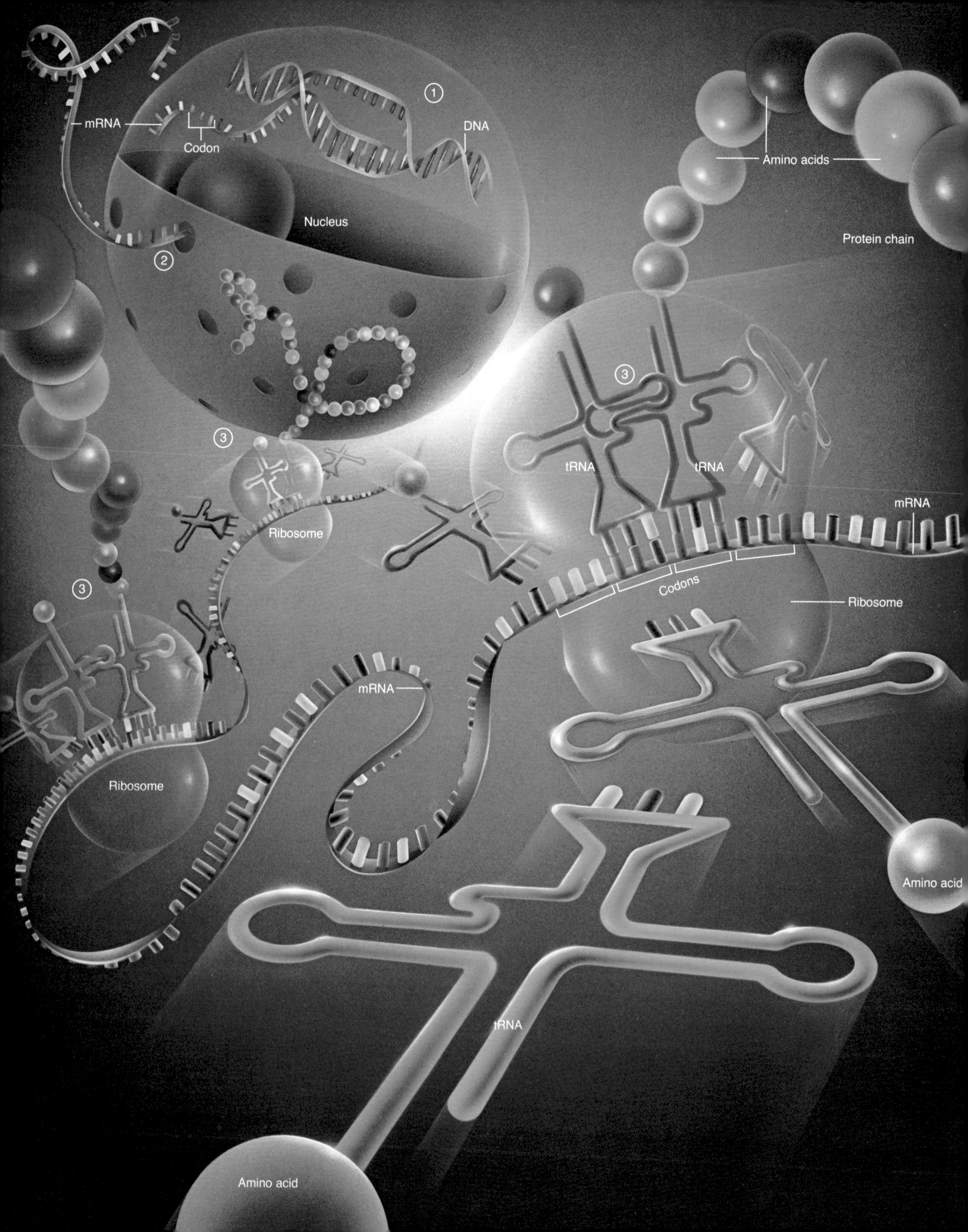
mRNA
Codon
1
DNA
Amino acids
Nucleus
2
Protein chain
3
3
tRNA
tRNA
mRNA
Ribosome
3
Codons
Ribosome
mRNA
Ribosome
Amino acid
tRNA
Amino acid

With accuracy and control, a cell manufactures protein under the direction of DNA (left).

The process begins in the nucleus as DNA unwinds in the region of a gene (1), a segment of DNA that holds instructions for building a particular protein. The exposed segment of the DNA acts as a pattern for the formation of a chemical relative of DNA, messenger RNA (mRNA).

The protein message is encoded in the sequence of bases that form the mRNA. The bases fall into three-part units called codons. Each codon represents one of twenty different amino acids, the basic material of protein.

The mRNA, bearing its coded message, travels out of the nucleus and into the cell's gelatinous cytoplasm (2). Here it encounters assembly units called ribosomes, and the protein synthesis begins (3).

Small molecules called transfer RNA (tRNA) decode the message. These tRNA decoders move toward the assembly site. One end of each tRNA matches a codon on the mRNA; the other end tows an amino acid corresponding to that codon.

A ribosome moves along the mRNA and translates the message one codon at a time, calling upon the appropriate tRNAs to add their amino acids to the growing protein chain.

When the ribosome reaches a "stop" signal on the mRNA, the protein chain is released.

Ribosomes operate with remarkable efficiency. In just one second, the human body completes about 500 trillion faultless copies of hemoglobin, a protein containing more than 570 amino acids.

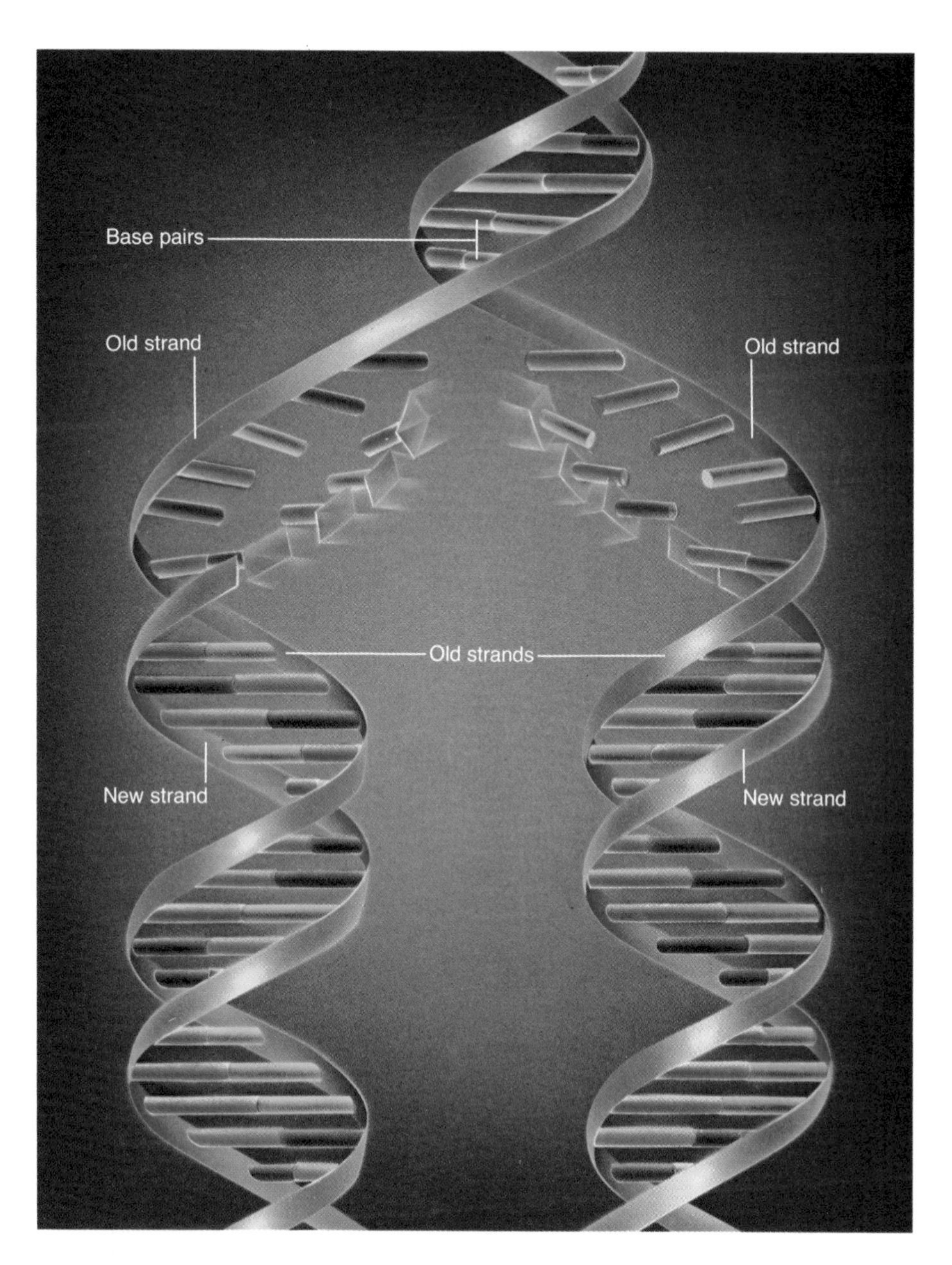

One strand of DNA unfurls to become two (above.)

This process, replication, is the first step in ensuring that a cell will pass along identical copies of its hereditary information to the new cells created when it divides.

Replication starts with the unwinding of the DNA molecule and the separating of its two strands between the paired bases.

Each separate strand serves as a template, or positive mold, for a full, new strand. Because of the bases' specific affinity for one another, an AGCC sequence on the original strand will attract only complementary bases, creating a TCGG sequence in the new strand and, eventually, an exact duplicate of the missing strand.

An ensemble of enzymes directs this activity, initiating its action and checking for mistakes. Although fifty base pairs unite almost every second, few errors occur—about one for every one billion base-pair replications. Thus does the body preserve life's integrity from one cell to the next—and from one family generation to the next.

Arnold Newman

a way of guarding its larder. It can raid the mother's tissues. When the mother's diet is insufficient for both, the fetus robs its mother, extracting protein stores from her muscle and minerals from her bone.

The fetus spends its final two months or so preparing to leave the maternal shelter. So shocking will be the move from warm womb to cold air that the newborn's survival depends on its ability to maintain body temperature. To prepare for the transition during these last months, the fetus will lay down a large store of a special kind of fat—called brown fat—that easily burns to generate heat. Oxygen stokes these fires. The battle of birth requires a great deal of oxygen—at first, the newborn consumes it at nearly twice the adult rate. Birth also brings an instant change of oxygen sources, from umbilical-borne blood to lung-filling air.

Inflating the tiny lungs makes an energy demand from 10 to 20 times that made by quiet adult breathing. Newborn lungs can meet this challenge only if they have received a final touch, a chemical coat known as pulmonary surfactant. The surfactant, triggered by one of the same kind of hormones as involved in labor, enables the lungs to retain precious air after each breath. Its function is to prevent them from collapsing. Without it, the infant would have to repeat the effort of the first breath with every intake of air and would soon die of exhaustion.

No one knows exactly what signals labor to

A likeness of features plays across the faces of this Italian-American family gathered for a reunion in the Registry Room at Ellis Island, New York (far left).

The unity and diversity common to all families stem from the mixing of genetic information. It occurs as egg and sperm, each carrying a separate chemical message, mature and unite.

Identical twins (left) emphasize the unity. These twins develop from a single fertilized egg, and thus share the same genetic endowment. Sometimes during the egg's division into two cells, the cells become separated and two individuals form, always of the same sex. Identical twins can also develop during later cell divisions. The chance that the same two parents could produce identical children in separate conceptions is virtually nonexistent.

begin. Throughout pregnancy the muscles of the uterus have been rehearsing for this moment by slight, intermittent contractions. Now, fetal hormones induce the production of maternal hormones that thrust the uterus into regular, strong contractions. Aided by the muscles of the abdomen, the uterus exerts as much as 50 pounds of pressure per square inch as it propels the fetus on its often bruising passage through the pelvis and the birth canal.

Nature prepares in other ways for the ordeal. As delivery approaches, the fetus usually assumes a final, head-down exit position. Because several bones on top of the fetus's head have not yet fused, they can be compressed as the child squeezes through. Also, hormones soften the ligaments of the pelvis and allow the pubic bones to separate a bit, slightly enlarging the narrow passage. During pregnancy the role of the cervix has been to keep the fetus within the uterus but during labor, hormones induce the cervix to soften and stretch and let the baby go.

As labor begins, the muscles of the uterine wall contract mildly once every 15 to 20 minutes. They squeeze the fetus, in its amniotic sac, downward toward the dilating cervix. First-time mothers may experience these contractions—labor pains—for as long as 18 hours. For women who have had children before, the period of labor can be shorter than an hour.

The contractions, now longer and stronger, come closer and closer together. By the time they are a minute or two apart, the cervix has stretched to an opening of four inches and the sac has broken. Soon the baby's head appears. As the head pushes through, it rotates and usually the baby's body does the same, to slip more easily along its narrow course. The child is born.

Within seconds the baby draws in air. The shunts in its now obsolete fetal circulatory system begin to seal off and collapse. Blood surges on a new, lifelong path through lungs and liver. The infant's skin brightens with the pink of oxygen intake. For the first time eyes see unshaded light, ears hear unmuffled sounds, and skin knows direct touch. A human being is now liberated from isolation to a strange and busy world.

The baby has left its watery, nine-month home but is still, briefly, attached to it by the umbilical

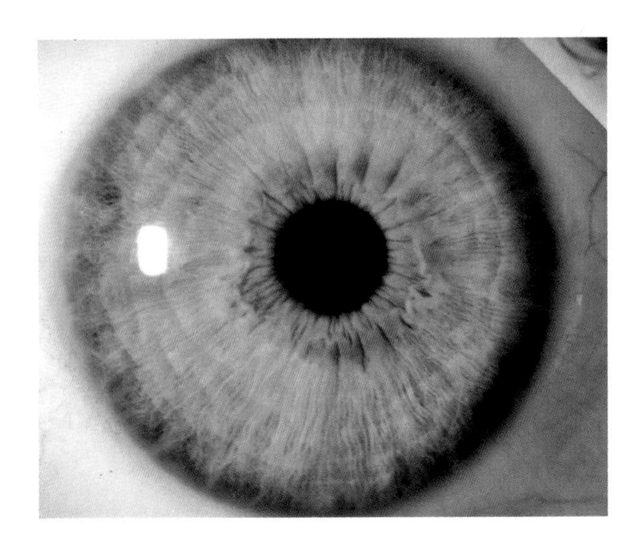

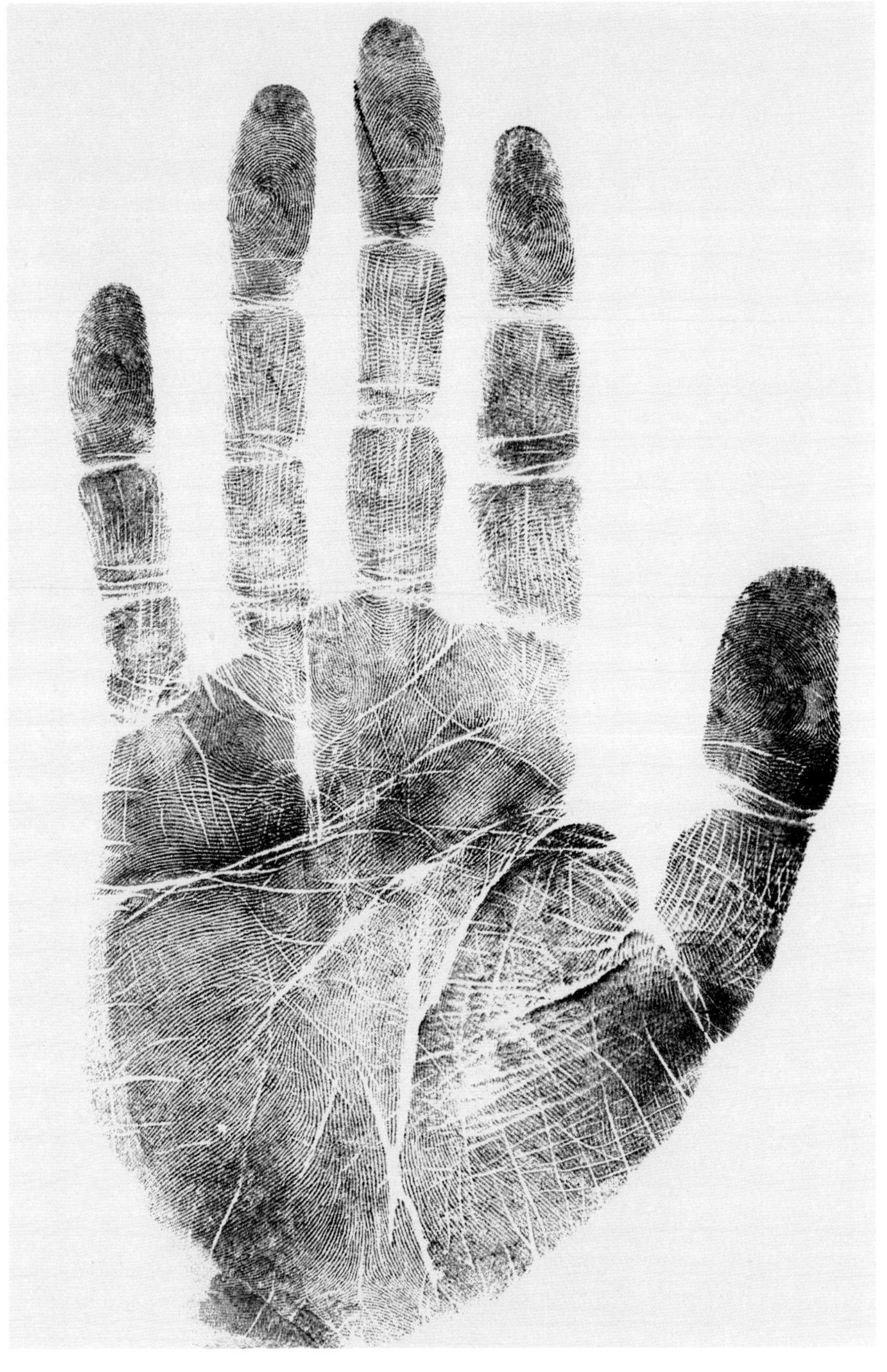

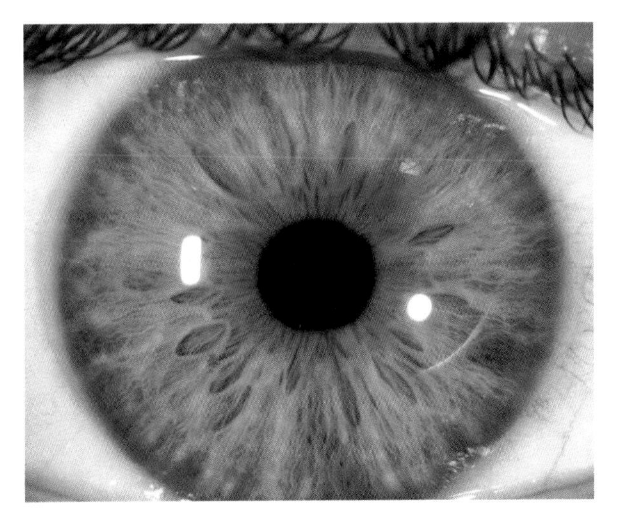

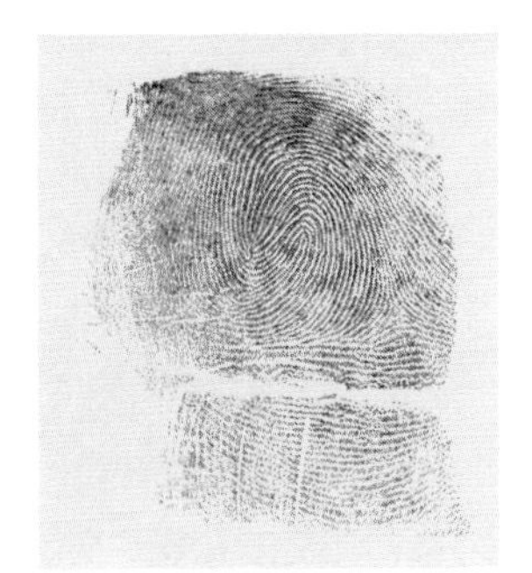

Human diversity and uniqueness show up in prints found on fingers and palm, and perhaps in the pattern of the iris (above). Even identical twins differ in these areas.

Variety begins with meiosis, the cell division that takes place in the ovaries and the testes.

Sperm and egg carry genetic information, stored in the 23 chromosomes of each. At conception the information merges, creating a new cell, which has the normal complement of 46 chromosomes.

This illustration (opposite) shows how chromosomes undergo meiosis. For clarity, only 2 pairs appear instead of the usual 23. Each pair contains one chromosome from each parent: blue for one parent, yellow for the other.

The original cell duplicates its chromosomes (1, 2). The duplicate structures link at a bridge called the centromere (2). Then the doubled chromosomes from each parent exchange genetic information at points known as chiasmata (2A, 2B). Tiny fibers seem to pull one doubled chromosome to the opposite poles of the cell (3). Other filaments tighten the cell's girth. The cell divides (4).

The cells, however, still have twice the genetic material they need, so the chromosomes are tugged apart at the centromeres (5). The 4 new cells each have half the number of chromosomes of the original cell, and a unique genetic endowment (6).

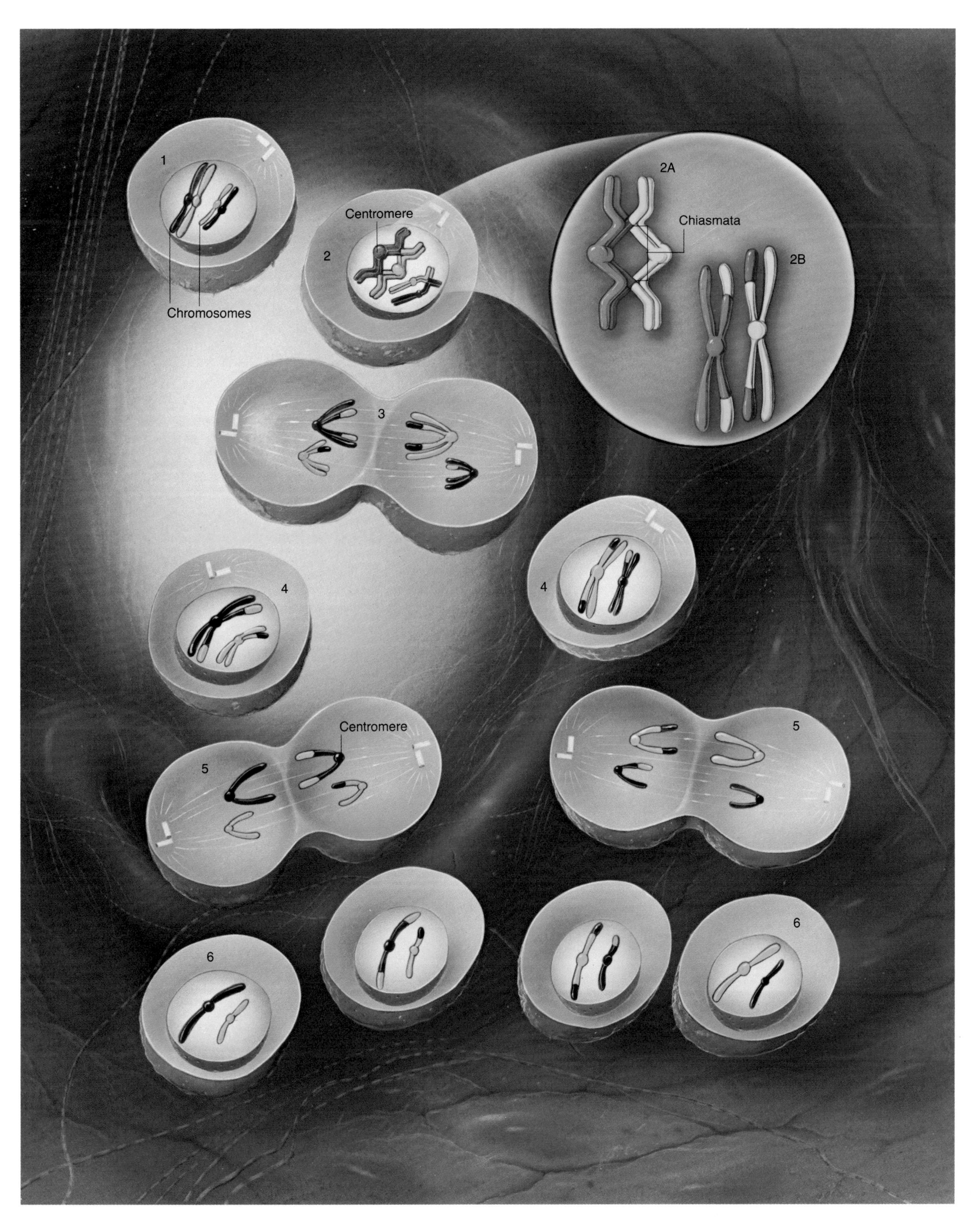
1
Chromosomes
Centromere
2
2A
Chiasmata
2B
3
4
4
Centromere
5
5
6
6

Sweet Remembrance

Journeying into the past, a woman and child search genealogical records for clues to the identity of their ancestors (above). The records, kept in Salt Lake City, Utah, belong to the Church of Jesus Christ of Latter-day Saints and form the world's largest collection of genealogical information.

Vast stores of microfilm hold billions of names, from records gathered around the world. One book from 1895 chronicles the history of the Leach family (left) who donated it; such contributions offer a palpable sense of the past—and of the ties that link all humans.

cord, which must be cut. Milder contractions help expel the placenta.

A newborn baby displays some amazing behavior. It turns toward its mother's voice. It seems to harbor a memory of intrauterine sounds, such as the mother's heartbeat. Tapes of a heartbeat appear to make it cry less and gain more weight than usual. A baby seems to recognize music that its mother played frequently during pregnancy. Many of us seem to think that these complicated creatures burst into existence at birth, though all their abilities attest to an intricate evolution from the moment of conception.

This developmental continuum reaches into childhood and beyond. The baby's body has many immature parts. Its immune system will develop. The intestines will continue to grow, as will the bones of the skeleton. The newborn's brain will triple in weight during the first year. Billions of brain cells called neurons will continue to make the multiple connections that persist and develop for life.

In some ways the human baby is still a fetus. While some other animals can walk at birth, a newborn human cannot even crawl until eight months or so. If we were sheltered in the womb for all the stages of development that other primates are, human pregnancy might last a year and a half. Our brain, so much larger than any other animal's in relation to our size, would make delivery impossible at eighteen months.

For human beings to finish gestation outside the womb, carried in their parents' arms, may be regarded as an advantage. Even as newborn fetuses continue their physical development, they learn from the world of sights, sounds, smells, language, and emotion—for the making of a human being is more than just assembling differentiated cells. Our genes issue orders for our bodies and our brains, but our humanity is the embodiment of concern provided or begrudged, education offered or withheld, love denied or bestowed.

Susan Schiefelbein

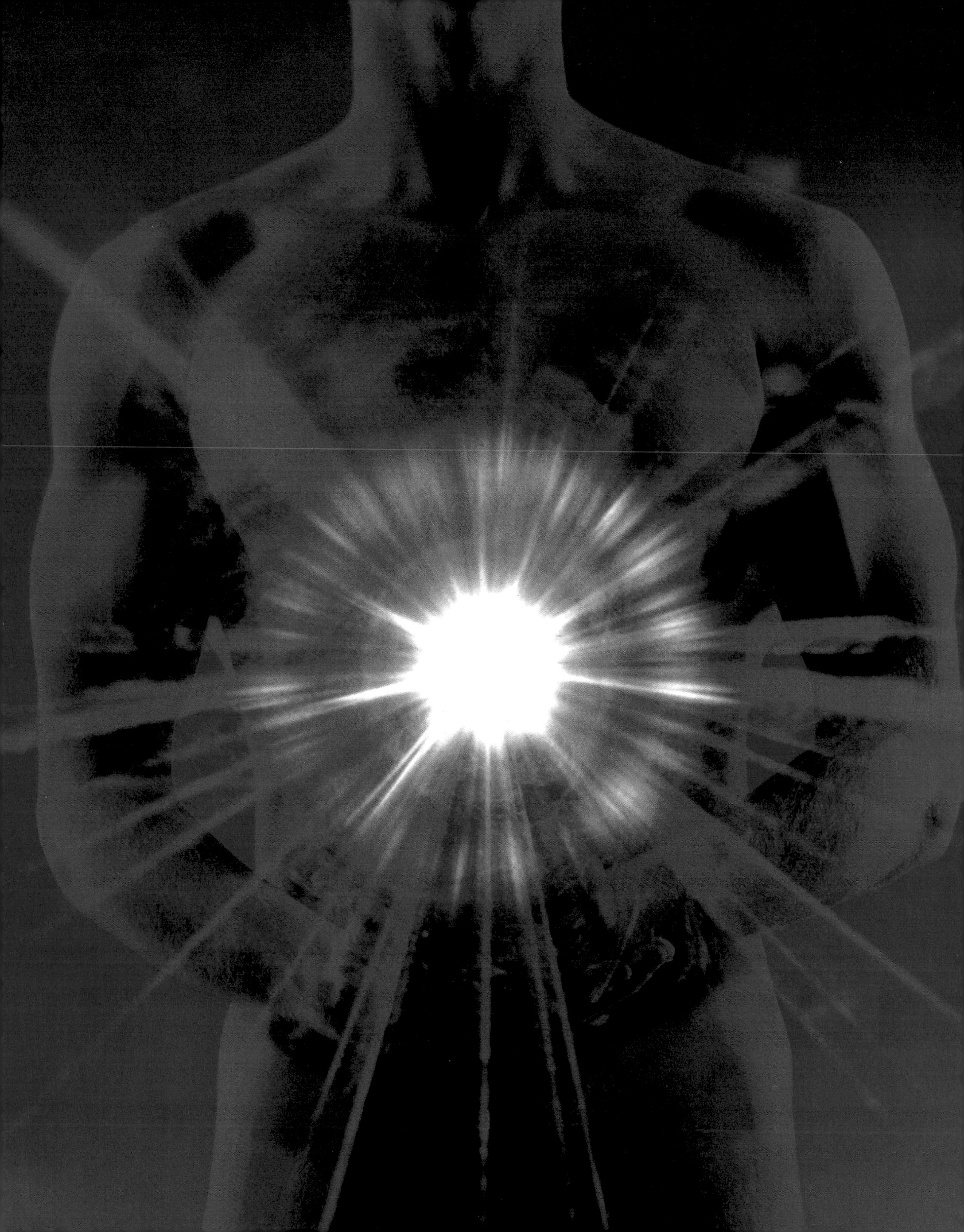

The Cycle of Energy

The alchemist of old achieved renown for claiming to turn base metals into gold. After each meal, we exercise a power much less celebrated but far more astounding: We convert plants and animals into the very blood and bone and flesh of our living bodies. We eat a peach; the peach is us. The summer squash that once lay on our plate becomes part of the substance of our eyes. The pat of butter we smoothed onto our bread turns into the silken membranes of our cells. Our bodies process food not just into tissue but also into energy: The grain of wheat once waving in a field exists for a fleeting moment as energy expended by the wave of our hand.

Through the process of nutrition, the body breaks down food into simple building blocks, then recombines them into living human tissues. The cells within our bodies draw much of the energy for this transformation from the nutrients that we call essential. These nutrients are crucial to our well-being and we must get them from outside sources—our foods. Fortunately, these substances are widely distributed in a variety of foods—the toast we eat at breakfast, the lunchtime soup and sandwich, the evening meat and potatoes. From these nutrients, the body creates every eyelash, hormone, blood cell, and toenail in the human anatomy.

The nutrients divide neatly into five categories. The principal ones—carbohydrates, fats, and proteins—make up the bulk of what we eat. Vitamins and minerals, while also necessary for good health, are required in smaller amounts. Most foods offer some combination of these. The lunchtime bowl of cream of mushroom soup, for example, contains all the major nutrients, in addition to some vitamins and minerals. A diet that consists of several different kinds of foods, and few processed foods, normally guarantees adequate amounts of all the nutrients.

Carbohydrates, a major source of energy, occur in fruits and vegetables—and also in the multitude of snack foods that add less to the body's well-being than to its girth. Plants harness the sun's energy to manufacture carbohydrates from air and water. They first arrange carbon, hydrogen, and oxygen atoms into the simplest carbohydrate, a single molecule of sugar known as a monosaccharide. Disaccharides are made by linking two sugar molecules together; a common one is sucrose, or table sugar, made by the sugar-beet and sugarcane plants. Plants can link hundreds or even thousands of separate sugar molecules to create a wide array of carbohydrate chains. Starches, for instance, found in grains, legumes, and in root vegetables such as the potato, link many molecules of the monosaccharide glucose. Finally, plants build their main structural scaffolding from cellulose, the most abundant such complex carbohydrate, often called fiber.

For most people in the world, carbohydrates make up the largest part of the diet. Starches and milk offer other nutrients—vitamins and minerals, and some protein. But high-carbohydrate foods such as candies and soft drinks provide only energy and few valuable nutrients. In the United States, each of us consumes almost a hundred pounds of refined sugar every year. Fats, or lipids, are made of the same chemical elements found in carbohydrates, but their arrangement differs. The major building blocks of lipids are called fatty acids, and there are several types, depending on their molecular arrangement. In one configuration they form saturated fats, which come from animal sources, including whole milk and most other dairy products, and from chocolate. Another takes the form of unsaturated fats present in seeds, nuts, and most vegetable oils. Like some carbohydrates, fats have fallen into disrepute in affluent societies where food is plentiful and obesity is a problem. But in fact fats are essential for several reasons: They can provide energy, they help our bodies absorb certain vitamins, and they cushion the internal organs against the jostles of everyday life.

Building the Body's Proteins

Mention protein, and a hungry diner may visualize a hamburger, a glass of milk, or a fish fillet. But to the biochemist a protein is nature's finest necklace—a long, intricate strand made up of smaller units called amino acids. The amino acids—the diverse beads on this necklace—are in turn formed of combinations of carbon, hydrogen, oxygen, nitrogen, and sometimes sulfur. Our bodies can assemble the atoms of half the 20 or so amino acids. We must acquire the rest—the essential amino acids—from the foods we eat. Those foods, moreover, must provide us with all the essential amino acids in relative proportion to one another. Meat and animal products such as eggs, milk, and cheese are known as complete proteins because they contain all the indispensable amino acids in the amounts needed for growth and maintenance. Vegetables contain partial arrays and must be eaten in combination with other protein food to provide all the essential amino acids. Staple dishes of many cultures, such as the beans and rice of Mexico, accommodate this need, combining plant foods into complete proteins.

Vitamins were discovered only this century, and there still are things we do not know about them. Although they provide no energy themselves and do not serve as building blocks, they are crucial in helping the body use those nutrients that do.

Most knowledge about vitamins grew from experiments in which symptoms of deficiency diseases were induced in test animals. The effects of vitamins were then identified by their ability to alleviate the deficiency. Early scientists could

The portrait of a nutritious diet reflects our changing knowledge of the links between nutrition and health.

Fruits, vegetables, and whole grains should make up the bulk of our diet, according to guidelines published by the American Institute for Cancer Research. Lean meats, fish, poultry, and low-fat dairy products may be eaten in moderation, with an occasional helping of eggs, nuts, fatty meats, and cheese. Rich desserts, fats, oils, and alcohol should be indulged in very sparingly, say Institute researchers.

Long-term studies indicate that what a person eats can affect certain cancers and heart disease. Today's ideal diet emphasizes fewer calories; less fat and cholesterol; less sugar and salt; more fiber and starches. Older—and still valid—dietary standards focused on the need for the essential nutrients, and counseled a balanced diet of the "basic four" food groups: dairy products, meats, breads and cereals, and fruits and vegetables.

only guess that most foods contained some unknown substance that sustained health. Later, researchers called them vitamins. Today, even children know that eating carrots is good for the eyes, a simple application of the link between night blindness and the need for vitamin A. The body manufactures the vitamin from carotene found in yellow-orange vegetables and fruits and dark green vegetables.

Most vitamins help the body break down and absorb the major nutrients or convert their components into substances needed by the cells. Through a complex series of chemical reactions, for example, one of the B vitamins, thiamine (B_1), helps the body make use of carbohydrates; another, pyridoxine (B_6), helps break down certain amino acids.

Of the two categories of vitamins, the fat-soluble ones—A, D, E, and K—are stored in the body. The water-soluble vitamins—C and the B complex—are excreted by the kidneys and therefore may be safely consumed more frequently. Though most people obtain adequate amounts of all the vitamins if they eat a balanced diet, the market for vitamin supplements thrives.

Minerals, like vitamins, yield no energy but work in conjunction with other nutrients. The body often incorporates minerals into structures that need strength. Calcium from milk and cheese hardens the teeth and also strengthens the skeleton. Iron, plentiful in meats, liver, eggs, leafy green vegetables, and dried fruits, helps carry oxygen to the cells. The body requires an adequate supply of six primary minerals, along with minute or trace amounts of about a dozen others. Like vitamins, minerals are sometimes lost when foods are processed.

The Transformation Begins

Before these nutrients can enter our tissues, they must be freed, radically changed from their original form as chicken, or noodles, or turnips. The digestive tract, the coiling 30-foot tube that winds through the center of the body, brings about this molecular liberation. Its passageways serve as a kind of anteroom in which unusable matter is separated from food and rejected, with only the useful substances granted entry to the innermost reaches of the body.

Imagine a dinner of chicken, potatoes, a lettuce-and-tomato salad with oil-and-vinegar dressing, carrots, and milk. All the basic nutrients are there: proteins in the chicken and milk; carbohydrates in all the vegetables and the milk; fats in the meat, oil, and milk; and vitamins and minerals in all the foods. Before it even reaches our mouths, the sight or aroma or just the thought of this meal sets off a series of processes that lasts several hours. In the end, the food molecules will have been *(Continued on page 66)*

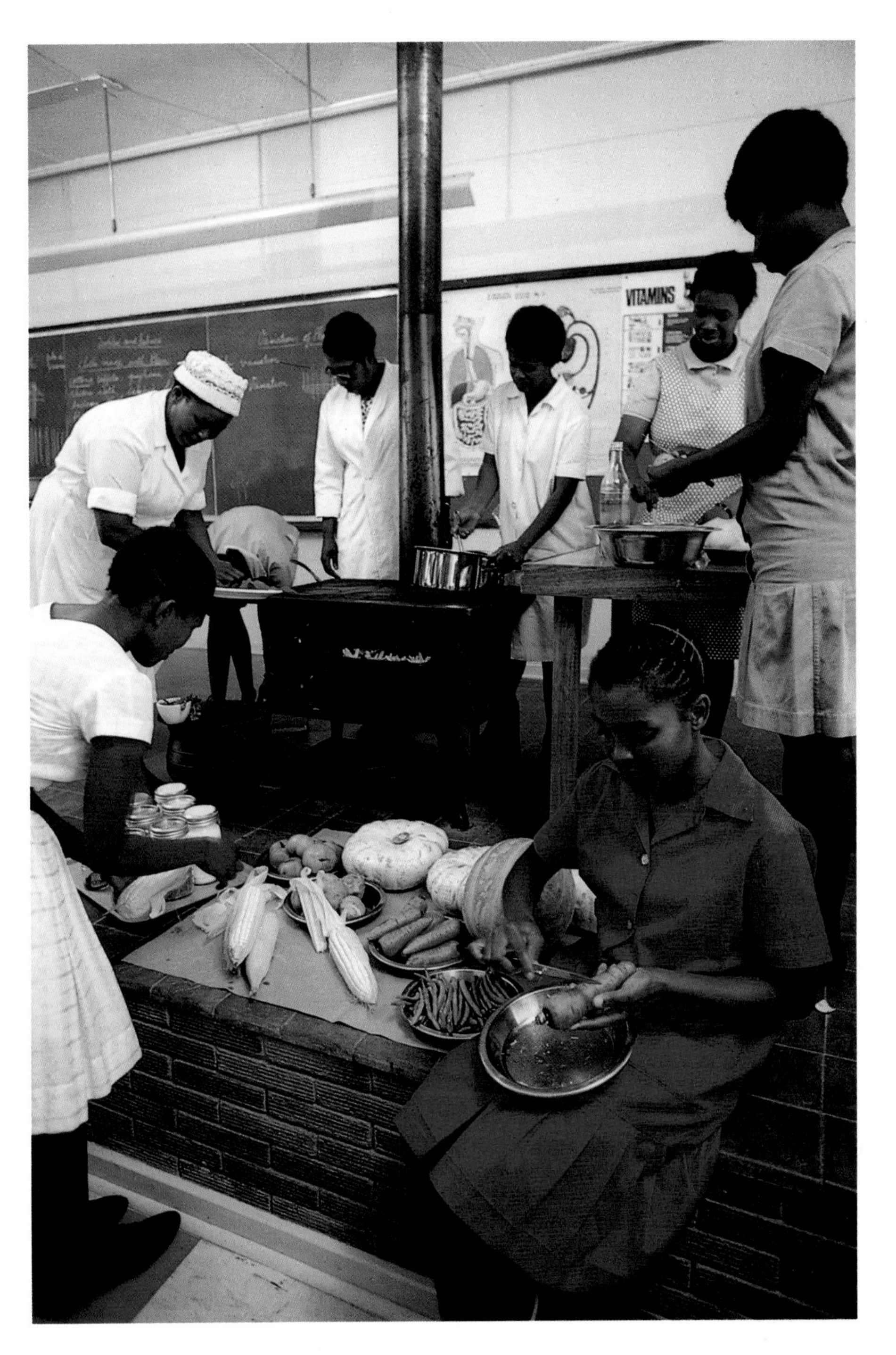

Breast-feeding provides a tranquil moment for mother and child in a Zimbabwe village (opposite). Mother's milk is a complete source of nutrition for the first six months of life. It also helps protect children from illness until their immune systems mature enough to take over. Human milk, unlike commercial formula or cow's milk, contains antibodies to fight gastrointestinal infections—the leading cause of infant death in developing countries.

Geography and economics influence what we are taught about proper diet. In affluent societies, people must learn to choose wisely from a vast selection of foods, and to avoid overeating. In developing countries such as Swaziland (above), students in a cooking class learn how to get the maximum benefit from limited food resources.

The Varied World of the Foods We Eat

Carbohydrates, which provide the primary source of energy for most of the world's people, take the form of sugars (simple carbohydrates) and starches (complex carbohydrates). Sources of starch include cereal grains, beans, pasta, potatoes, and bread.

Traditionally the "staff of life," bread has long been a staple around the world. It comes in a variety of shapes and sizes, including the *baguettes,* or wands, carried like cordwood by a young Parisienne (below).

Chinese diners obtain similar food value from their noodles, made in long strands to symbolize long life. By deftly swinging, stretching, and doubling a fist-size lump of flour-and-water dough (right), a master noodle maker can create 2,048 strands of *long xu mian*—dragon's beard noodles—each as fine as a human hair. The delicacy, deep fried and wrapped into thin pancakes, gets its elasticity from long-chain protein molecules.

神光普照
關聖帝君
BOITES
COGNAC
COURVOISIER
THE BRANDY OF NAPOLEON
12 B
12 B ETUIS
COGNAC

High-quality protein—in this case an armload of carp pulled from China's Yangtze River—brings a smile of triumph to a fisherman. The chief component of many of our body's tissues and chemicals, proteins come from an array of foods that include meat and fish, dairy products, and legumes.

A butcher shop in Hungary (above) displays cuts of pork that, along with beef, make up a large part of Western diets. The abundance of meat in many Western nations may be too much of a good thing. Americans eat about twice the amount of protein they need, much of it in the form of meat and whole milk products. Such a diet brings with it large quantities of the saturated fats and cholesterol now implicated as a cause of some heart diseases.

Fats come from a variety of sources including chocolate (above), whole milk (opposite, upper), and a wheel of cheese (opposite, lower). They provide the body with the most concentrated source of energy—more than twice the energy of carbohydrates and proteins. Fat also helps keep the skin smooth, maintains the cell membranes, and cushions the body's internal organs.

But a little fat goes a long way. Fats account for about 40 percent of the daily energy intake of Americans, about 10 percent more than needed.

Americans get most of their fat calories from the high level of saturated fats

contained in whole milk, butter, cheese, meats, and other foods of animal origin. Polyunsaturated fats, found in nuts, seeds, and vegetable oils made from sunflower, safflower, corn, and cottonseed, help to lower cholesterol levels in the blood whenever they are used as substitutes during cooking.

The body's daily requirement of vitamins and minerals is minuscule—less than a thimbleful. But without them all the carbohydrates, proteins, and fats we eat would be useless. The ability of vitamins to work wonders in the body has led to misconceptions about their therapeutic value. For example, the claim remains unproven that large doses of vitamin C, plentiful in citrus fruits like limes (right, lower), will prevent colds.

Overloading on vitamin supplements such as those being processed in New Jersey (opposite) may be useless at best, toxic at worst. Vitamin A, believed to protect against some kinds of cancer, may also, if taken in large doses, bring on bouts of nausea, insomnia, headaches—and result in hair loss.

Minerals, too, require a fine balance in the body. Sodium, a component of salt (right, upper), may contribute to high blood pressure if taken to excess. And so may too little calcium and potassium, according to recent studies.

converted to a form that the body's cells can use.

The first step is salivation. Salivary glands pour out saliva—a mixture of water, mucins, and enzymes—in the mouth. Enzymes are the body's catalysts; they direct and accelerate chemical reactions. The enzyme in saliva, amylase, starts to break down carbohydrates by shattering the bonds that link starch molecules. As the teeth break down the food mechanically, amylase begins to disintegrate the potato and other vegetables. The saliva also moistens food, and the tongue manipulates each chewed-up bit into a soft, lubricated ball that will easily slip down the pharynx, or throat.

As food passes from the mouth, a "trapdoor," the epiglottis, shuts, closing off the windpipe. The only route normally left open is the esophagus, a ten-inch tunnel behind the windpipe.

Two layers of muscles line the esophagus, one arranged in circles, the second running up and down its length. Squeezing and relaxing in concert, these muscles mix the food and propel it

along, much as your hands knead and squeeze a tube of caulking compound. These peristaltic waves are so powerful that they will move swallowed food even if you stand on your head. Below the esophagus, the digestive tract expands to form the stomach, a J-shaped sac. Hanging in folds when empty, the stomach can stretch to accommodate more than a quart of food. Sphincters around the entry and exit of the stomach act much like purse strings. One muscle prevents backflow into the esophagus and the other keeps food from proceeding into the small intestine before there is room.

The stomach picks up the demolition job begun by saliva, further breaking down the food by kneading it and dousing it with gastric juice. Millions of gastric pits, each containing several glands that spew a potent mixture of juice, dot the stomach's inner surface. Pepsin, the stomach's main digestive enzyme, begins the process of pulling apart proteins, fracturing the chains of amino acids in the chicken and milk. Hydrochloric acid, among the most corrosive substances known, provides the medium in which pepsin works best. The acid also kills most bacteria and other living cells that entered with the food.

From Stomach to Intestine

Eventually the food, mixed and emulsified, begins to leave the stomach through the pylorus, an exit valve surrounded by a ring of muscle. This valve enables the semiliquid mass to pass through a little at a time, at a rate dependent on the small intestine's capacity to handle it. Carbohydrates are the first to be released. Fats and proteins stay somewhat longer, but the stomach usually empties completely about four hours after eating a meal.

Beyond the pyloric gateway lies the convoluted tube of the small intestine. Its 15 to 20 feet of twisting passageway, coiled within the abdomen, serves as the body's major digestive organ. Here, in a process that will last several more hours, usable food is prepared for its ultimate journey into the cells of the body.

Though food arrives in the small intestine as a semiliquid mixture unrecognizable as the chicken, milk, and vegetables that entered the mouth, the nutrients are still more or less intact. Some links in the amino acid chains that make up proteins have been broken, and the starches and sugars have been split apart into simpler compounds, though there has been no chemical modification of fats. The small intestine completes the process of chemical digestion, reducing the large, complex molecules to small, simple ones that can pass through the wall of the small intestine and into the bloodstream. All carbohydrates, whether from vegetables or milk, must be converted *(Continued on page 81)*

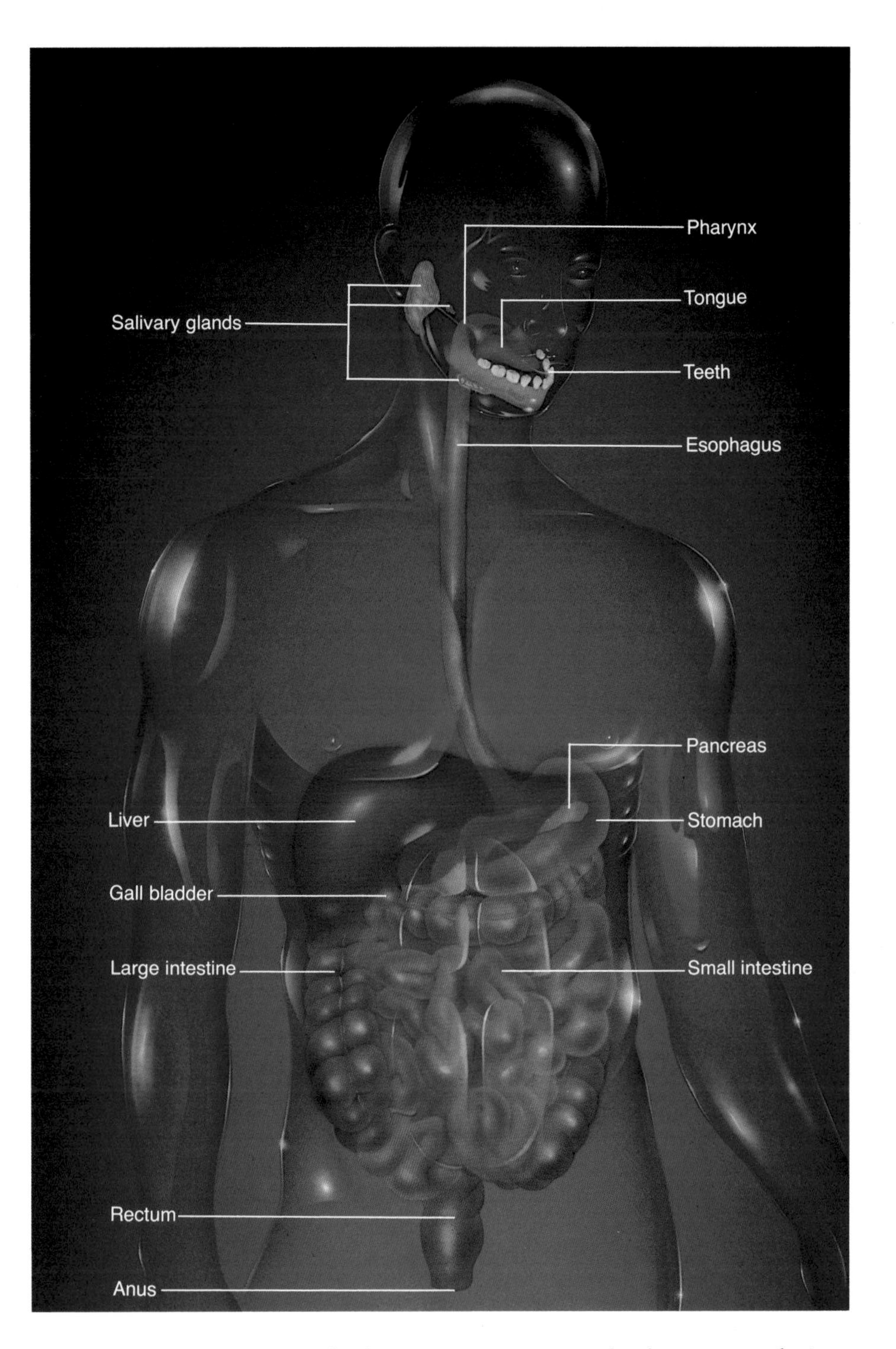

The digestive system (above) physically and chemically converts the food we eat—such as the huckleberries (opposite)—into substances our bodies need. The system includes the alimentary canal, a continuous tube made up of mouth, pharynx, esophagus, stomach, small intestine, and large intestine. As food enters the tract, the teeth cut and grind it into morsels that can be swallowed.

Once the food has been swallowed, muscles lining the alimentary canal mix and propel it by mechanical action. Enzymes in the stomach and small intestine chemically break down the complex food molecules. Other organs—the salivary glands, pancreas, liver, and gallbladder—also contribute enzymes and other digestive substances. Reduced to simple forms and processed in the liver, food enters the bloodstream and is carried to the cells.

Digestion: Stoking the Body's Flameless Furnace

The breakdown of food begins in the mouth, where teeth cut and grind it into smaller pieces.

To help protect teeth from the wearing action, enamel, the hardest substance manufactured by the body, sheathes the visible surfaces of each tooth. But enamel cannot fend off every assault. Plaque, a colorless, sticky film alive with bacterial colonies, continually forms on the teeth. Feeding on the sugary residues left on a tooth after a meal, the bacteria

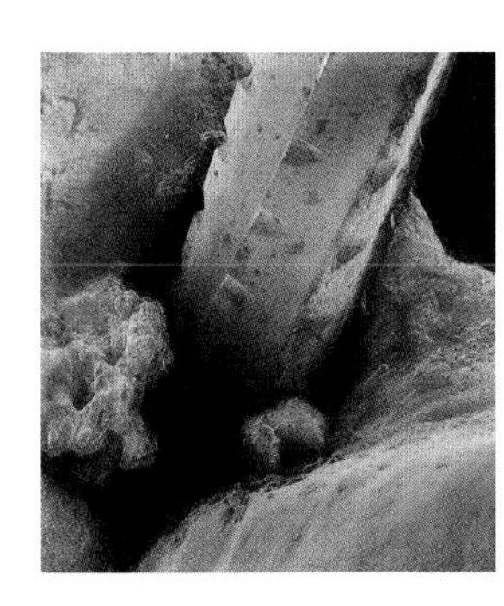

secrete acids that can destroy enamel. The result: a cavity (dark spot in center of the tooth below). When a cavity forms, the decay must be drilled out (left) and replaced with a filling to prevent further damage.

Bacteria living in the plaque become their most destructive when organized into colonies—a process that takes about 24 hours. The bristles of a toothbrush (right), scour the scum and constitute the best line of defense against plaque buildup. Dental floss helps remove bacteria from between the teeth and under the gum line where the brush cannot reach.

Magnification: 10 times (above and below)

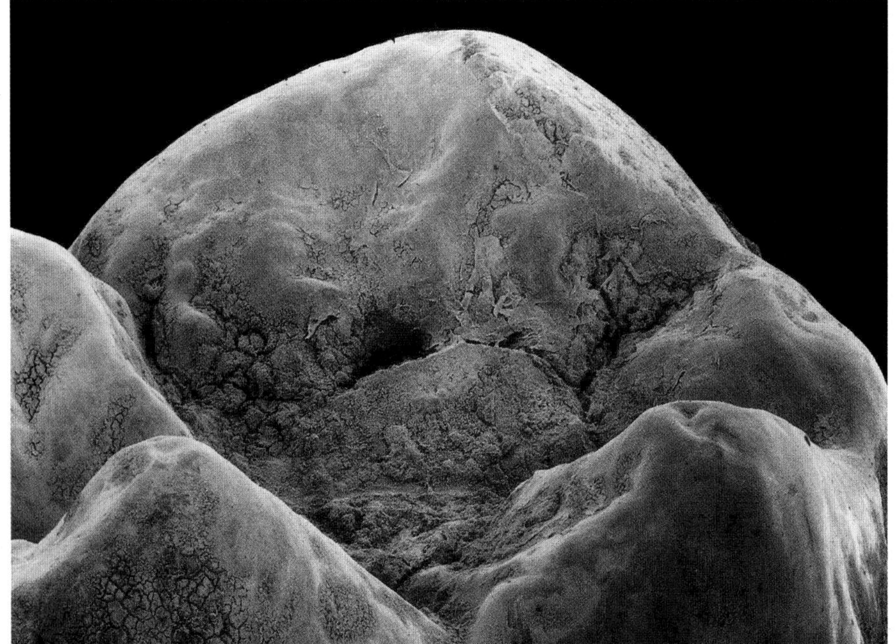

Magnification: 75 times

Magnification: 6 times

Useful for tasting, essential for speaking, the tongue also serves as the chief manipulator of food inside the mouth.

Interwoven bundles of striated muscle, anchored to the jawbones and running in three planes, enable the tongue to grow shorter or longer, thinner or thicker, making it one of the body's most versatile sensory organs.

The tongue's upper surface is carpeted with small projections called papillae (below). They provide a rough surface for handling and processing food.

As we chew, the muscular tongue guides food toward the throat and shapes it into a ball, or bolus. Swallowing begins when the tongue pushes the bolus back into the throat, or pharynx.

To ensure that food gets to the right place, the soft palate—which extends from the roof of the mouth—rises reflexively to block the nasal passages. At the same time, the larynx, or upper part of the windpipe, rises, automatically closing off the windpipe.

The epiglottis, a flap of tissue open during breathing (right), folds over the larynx. With nowhere else to go, food enters the esophagus and begins its journey to the stomach. Olive-like nodes lining the root end of the tongue are lymphatic tissues that help guard against infection.

Magnification: 15 times

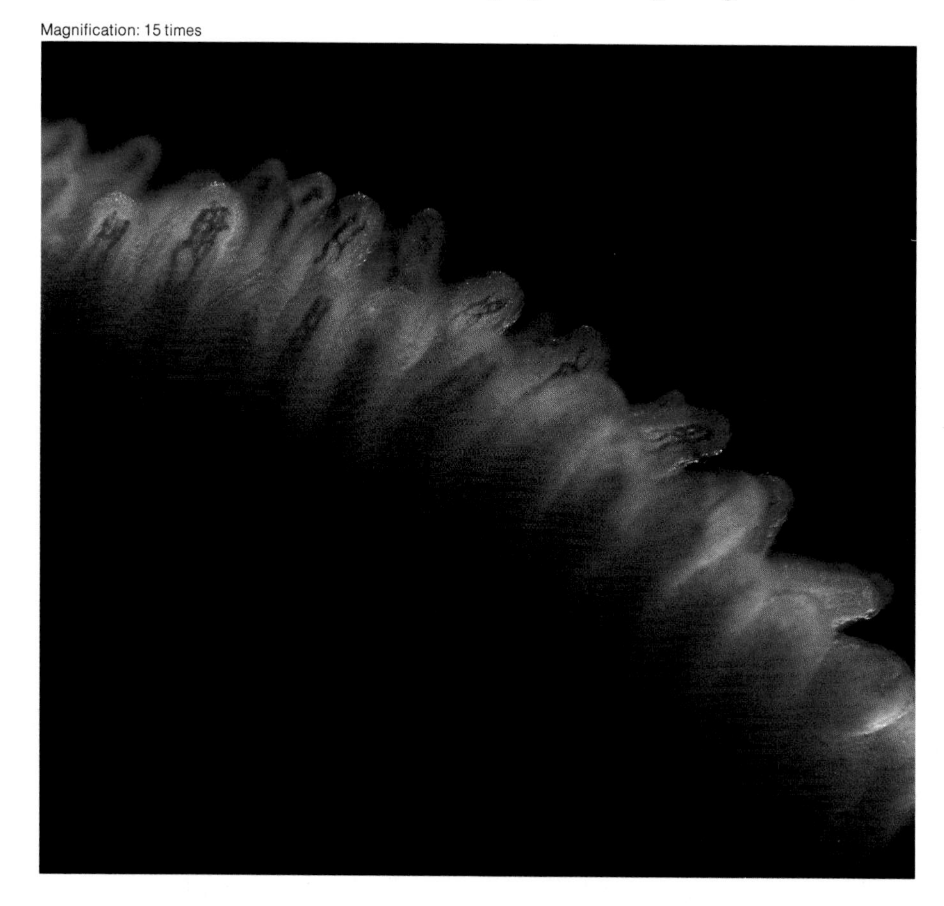

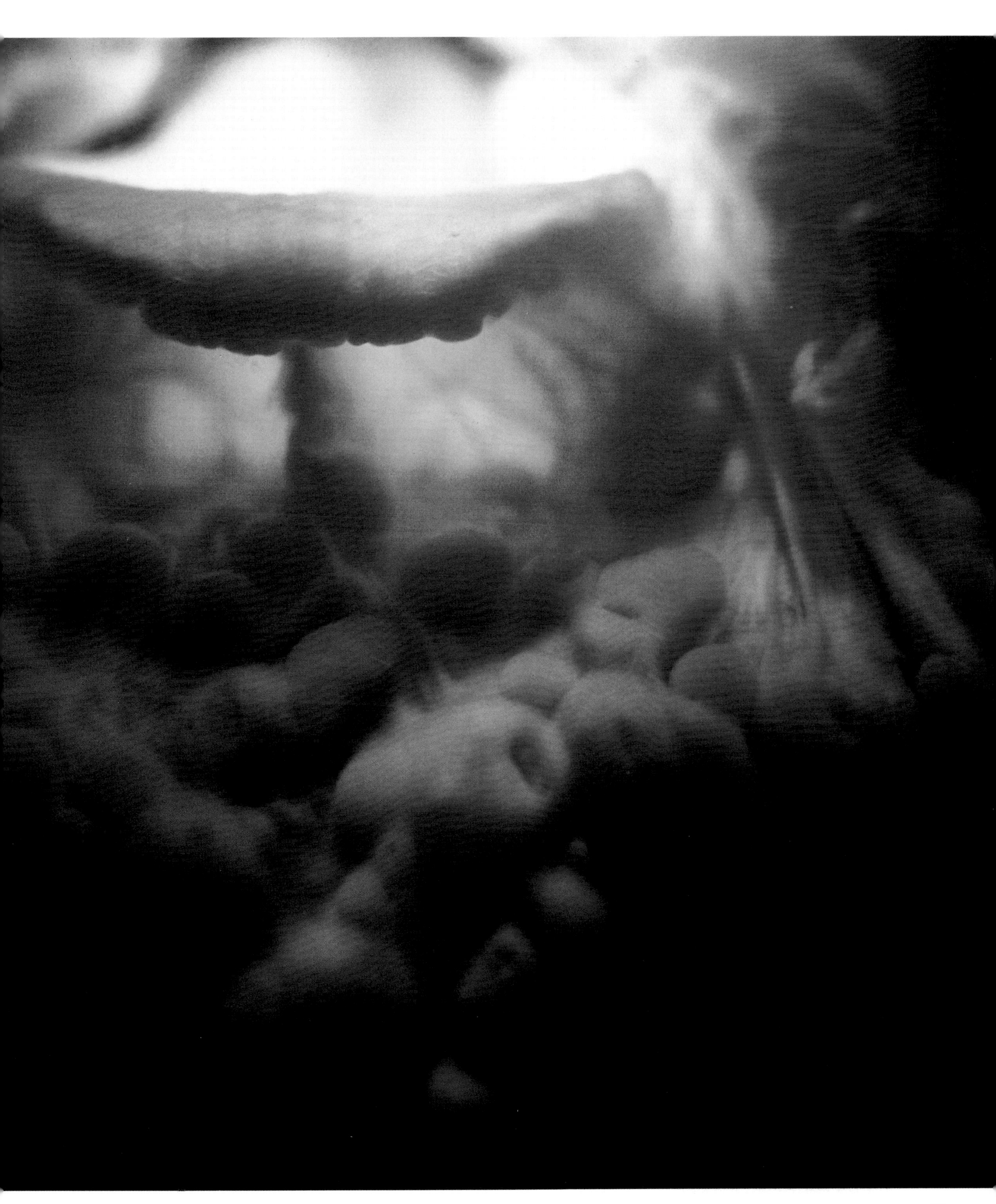

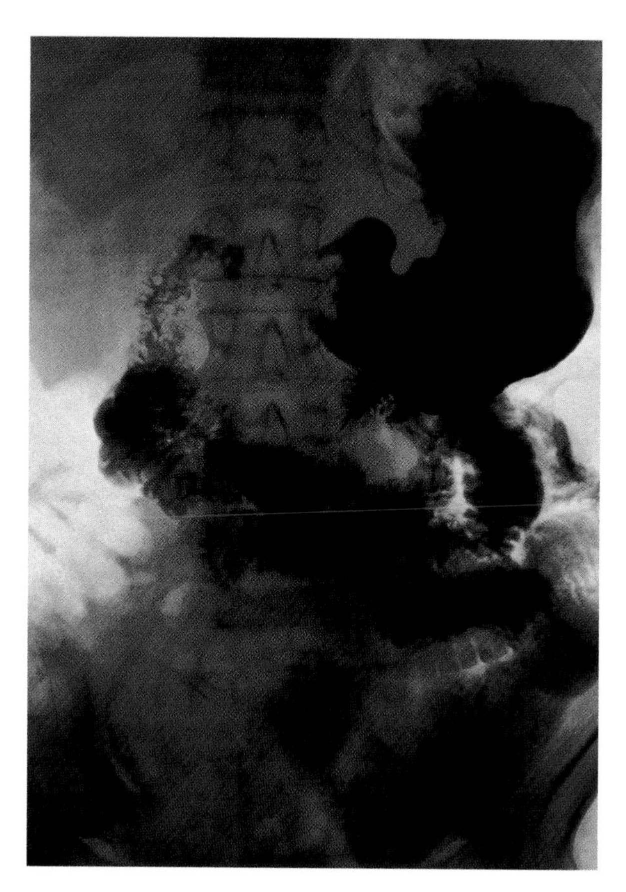

Silhouetted by X-ray photography, the stomach appears as a dark blob (above) as it rides atop the convoluted twists and curves of the small intestine.

The hills and valleys of the stomach's inner lining (right) will flatten out as the organ fills with food. As the stomach wall expands, it triggers the release of the hormone, gastrin. The gastrin then stimulates production of gastric juice.

Such hormones produce their effects by binding tightly to receptors on the filaments of a target cell (opposite, lower) in the stomach. In response, glands in the stomach lining spew forth a mixture of other digestive substances such as hydrochloric acid, and the enzyme, pepsin.

To shield the stomach from the effects of its own corrosive juices, the cells and glands located in the stomach wall secrete a layer of protective mucus. The mucus (tinted yellow; opposite, upper) erupts from a gastric pit; within the pit floats a single red blood cell, perhaps a sign that the stomach has been irritated by aspirin, alcohol, or another substance.

Magnification: 15 times

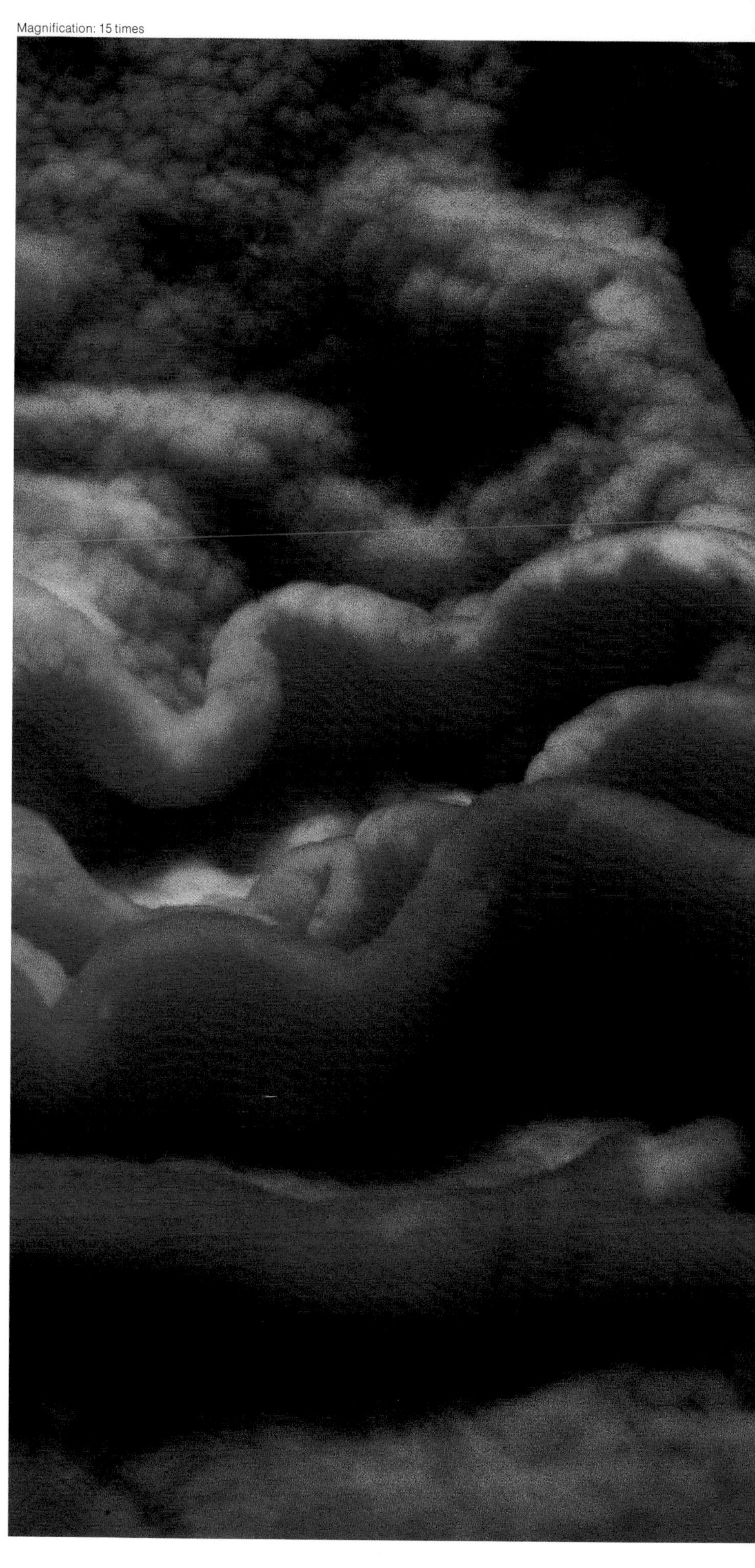

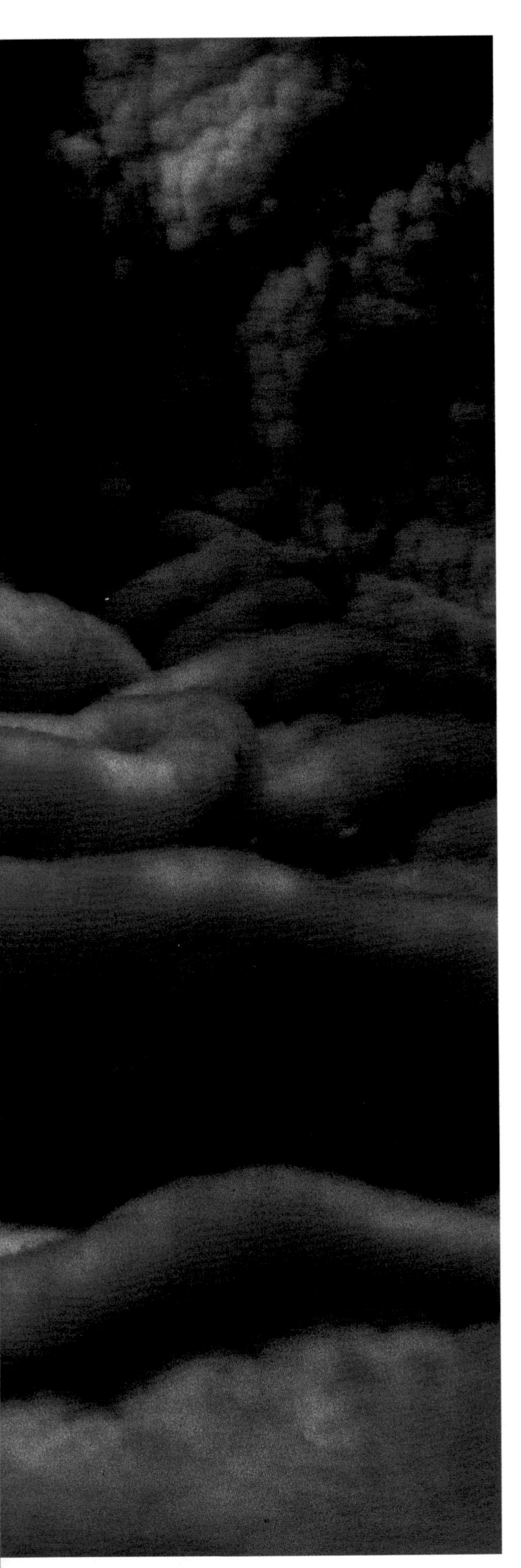

Magnification: 3,000 times

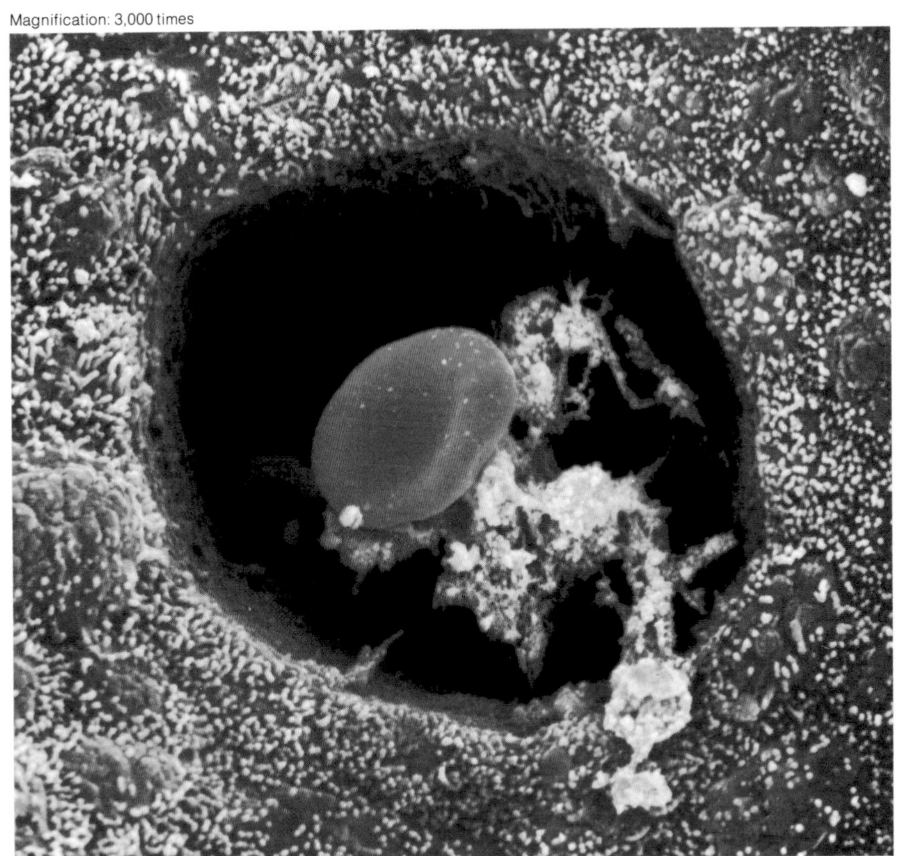

Magnification: 10,000 times; Lennart Nilsson, © Boehringer Ingelheim International GmbH. (also left and above)

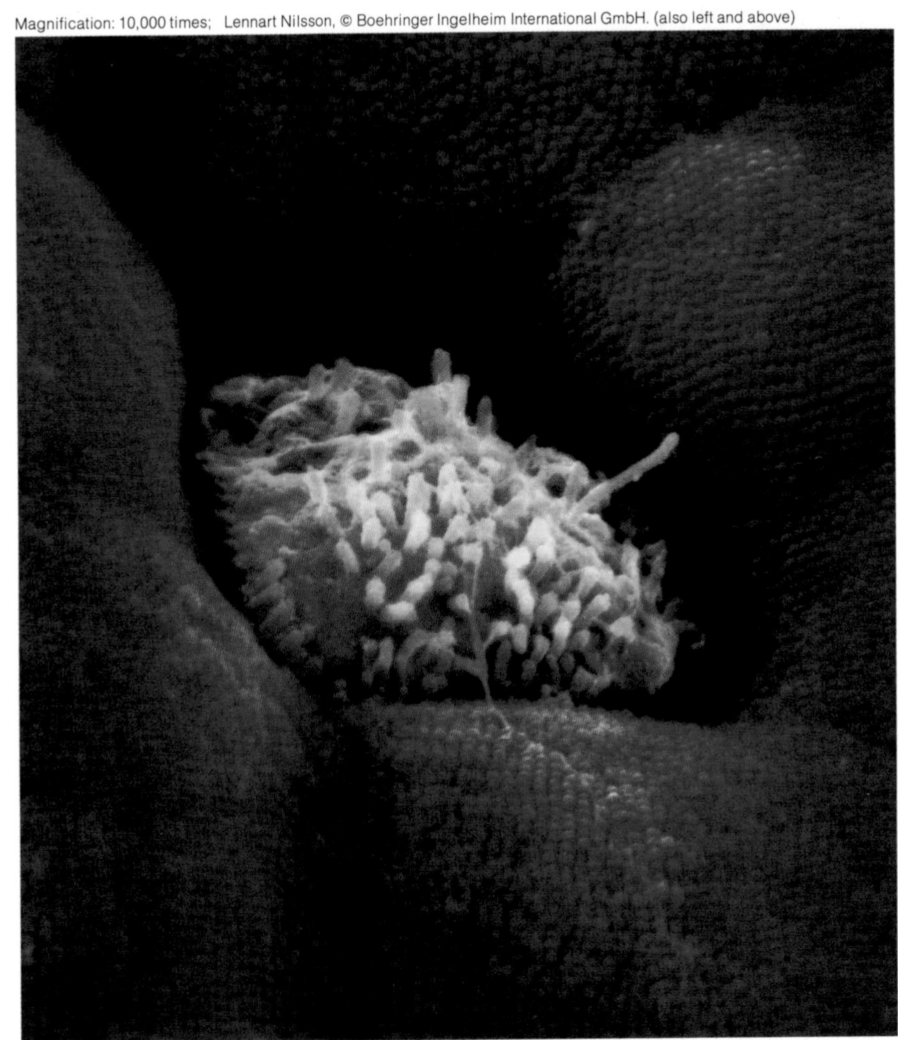

Magnification: 2 times

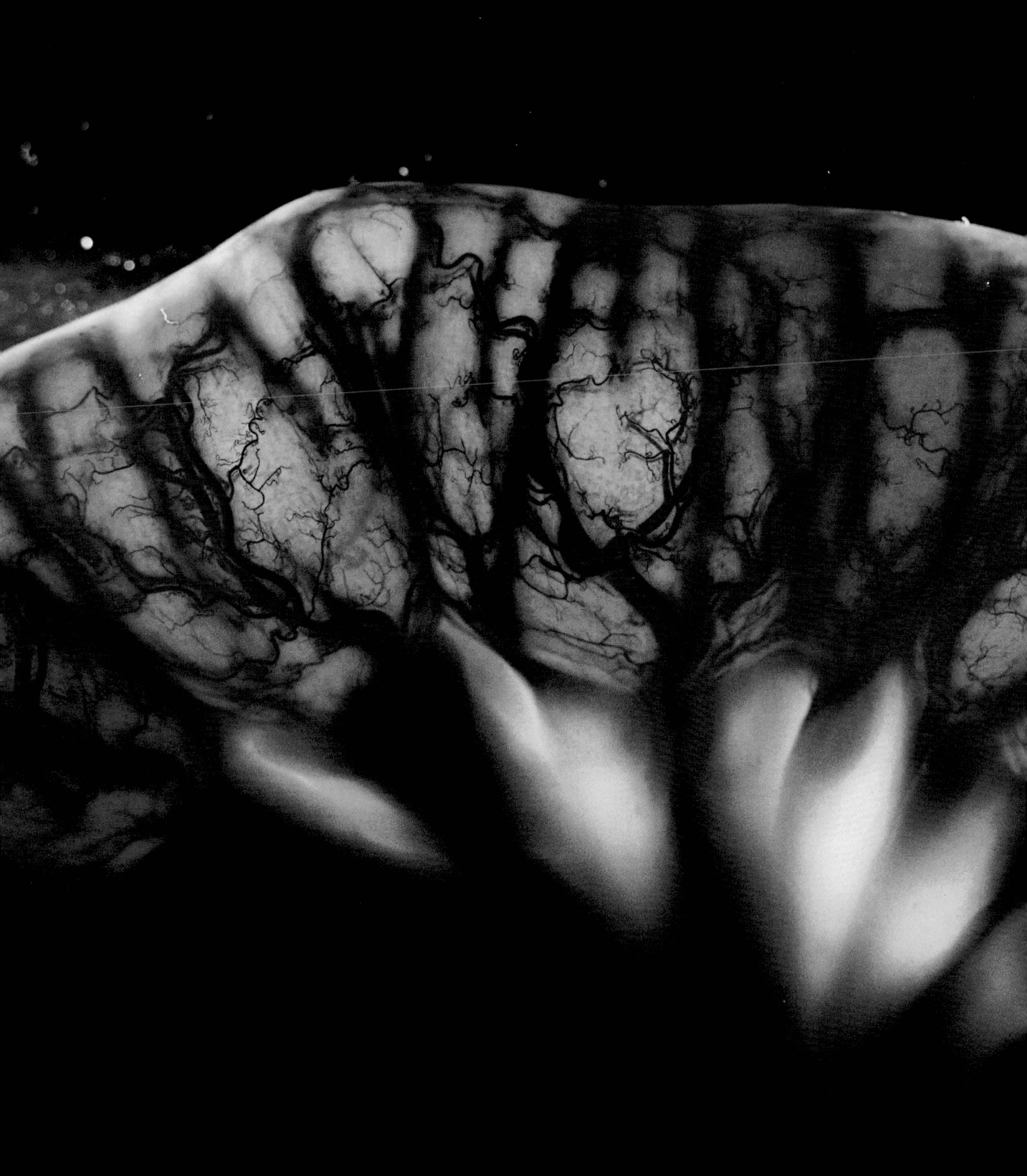

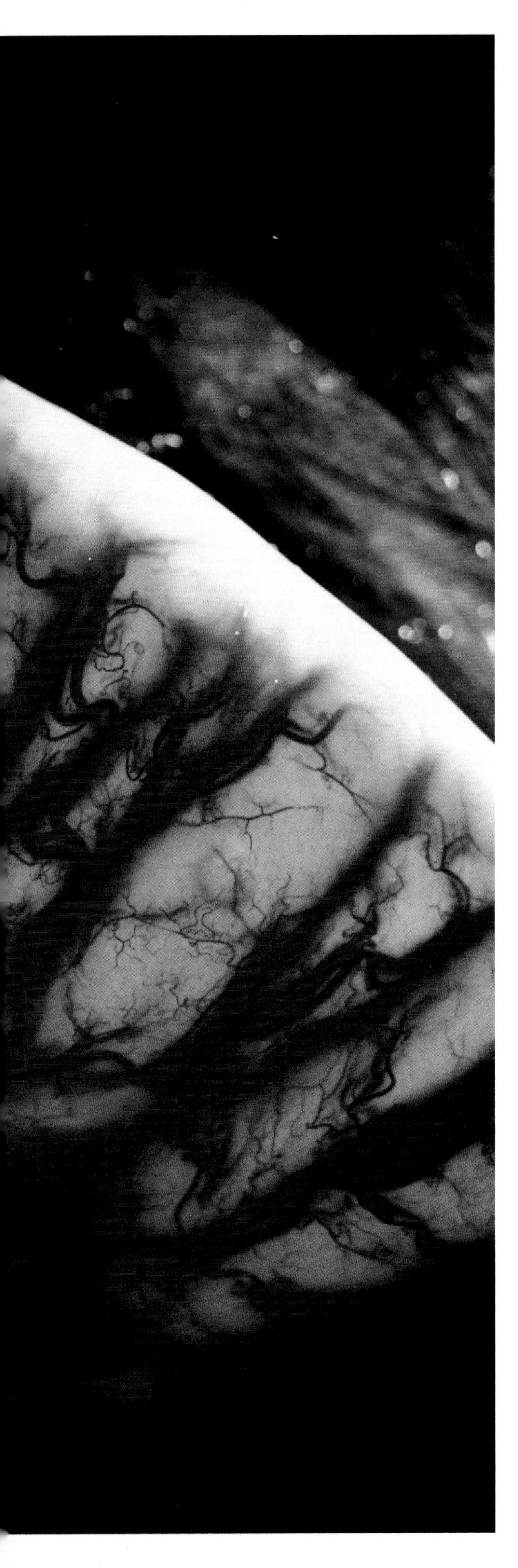

Magnification: 300 times; Lennart Nilsson, © Boehringer Ingelheim International GmbH.

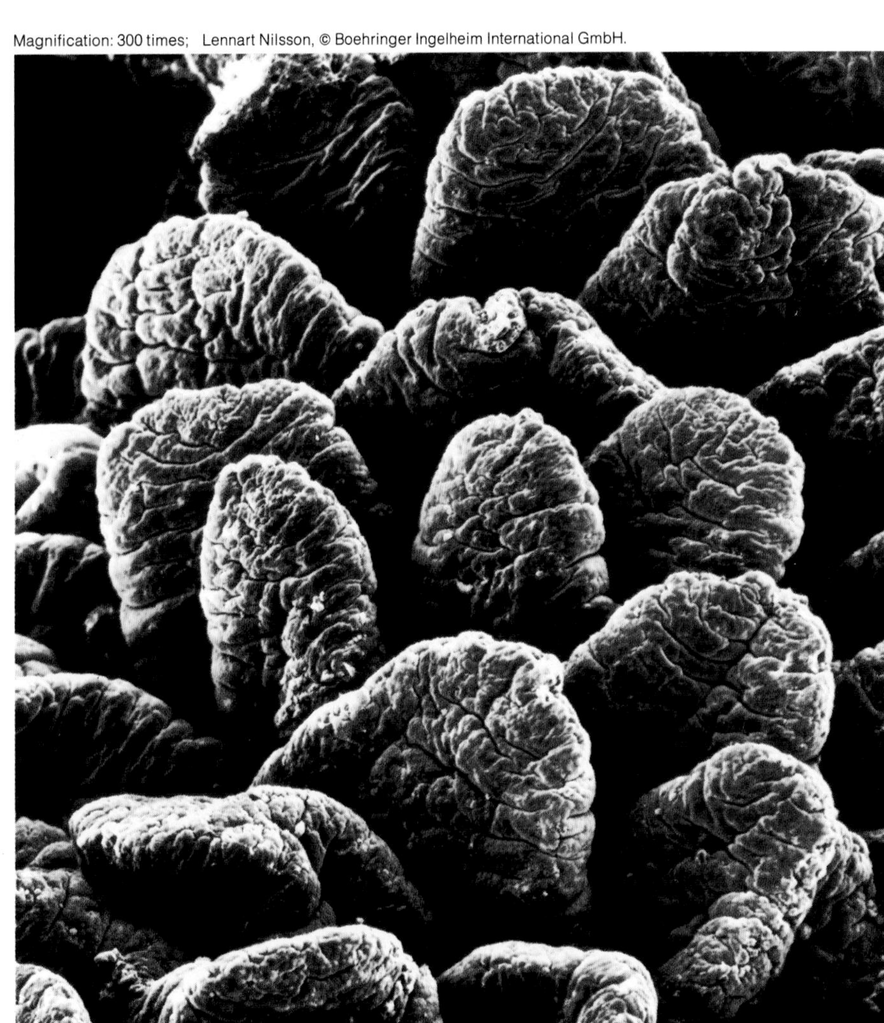

A blanket of absorptive structures called villi (above) lines the small intestine—thousands of them per square inch. Rippling through each villus, a dense capillary network radiates into the larger blood vessels of the small intestine (left).

Partly digested food from the stomach enters the small intestine, where nutrients can be broken down and absorbed. Powerful muscular contractions mix the food with digestive juices, push the food along, and bring it into contact with the intestinal wall. The small intestine's 15-to-20-foot length, combined with the circular folds of its lining and the millions of villi, provide an extensive surface area for absorption.

Magnification: 150 times

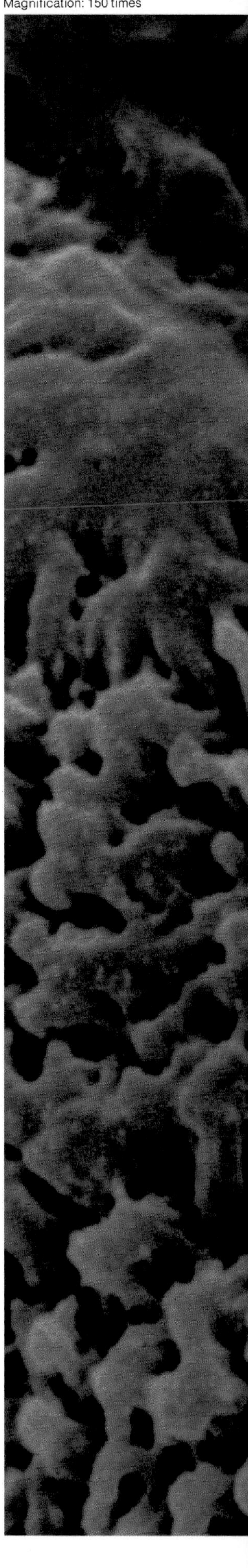

Magnification: 10,000 times; Lennart Nilsson, © Boehringer Ingelheim International GmbH. (also right)

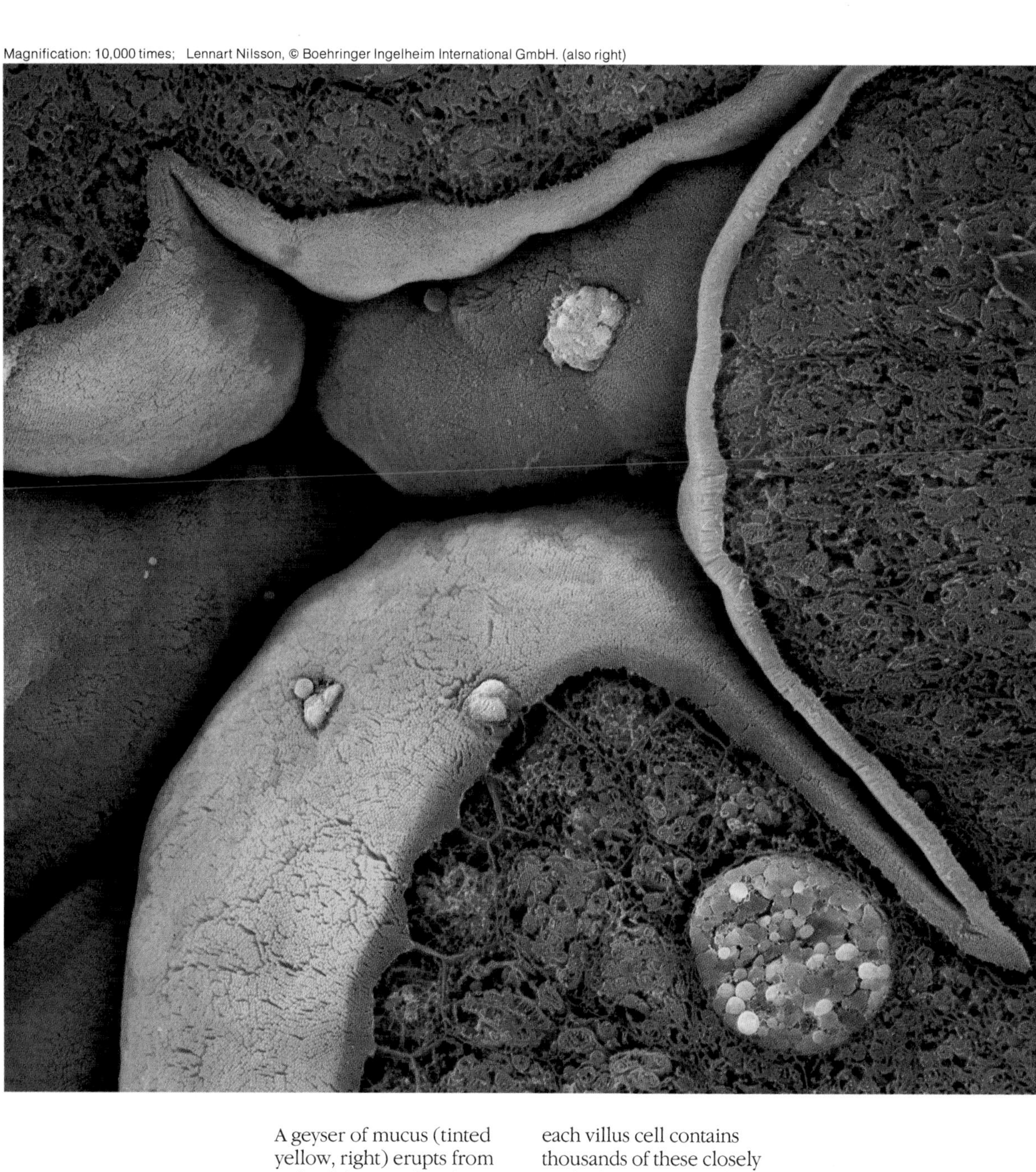

A geyser of mucus (tinted yellow, right) erupts from a gland deep within the small intestine. The mucus, originating in goblet cells, helps lubricate the food and provides a protective coat for the intestinal lining.

As food moves through the small intestine, nutrients filter into the absorptive cells of the villi (above). These cells, here photographed in cross section, are fringed with hairlike microvilli (tinted yellow). The surface of each villus cell contains thousands of these closely packed projections, which greatly increase the absorptive area of the small intestine. A mucus-like droplet appears at bottom right in the photograph.

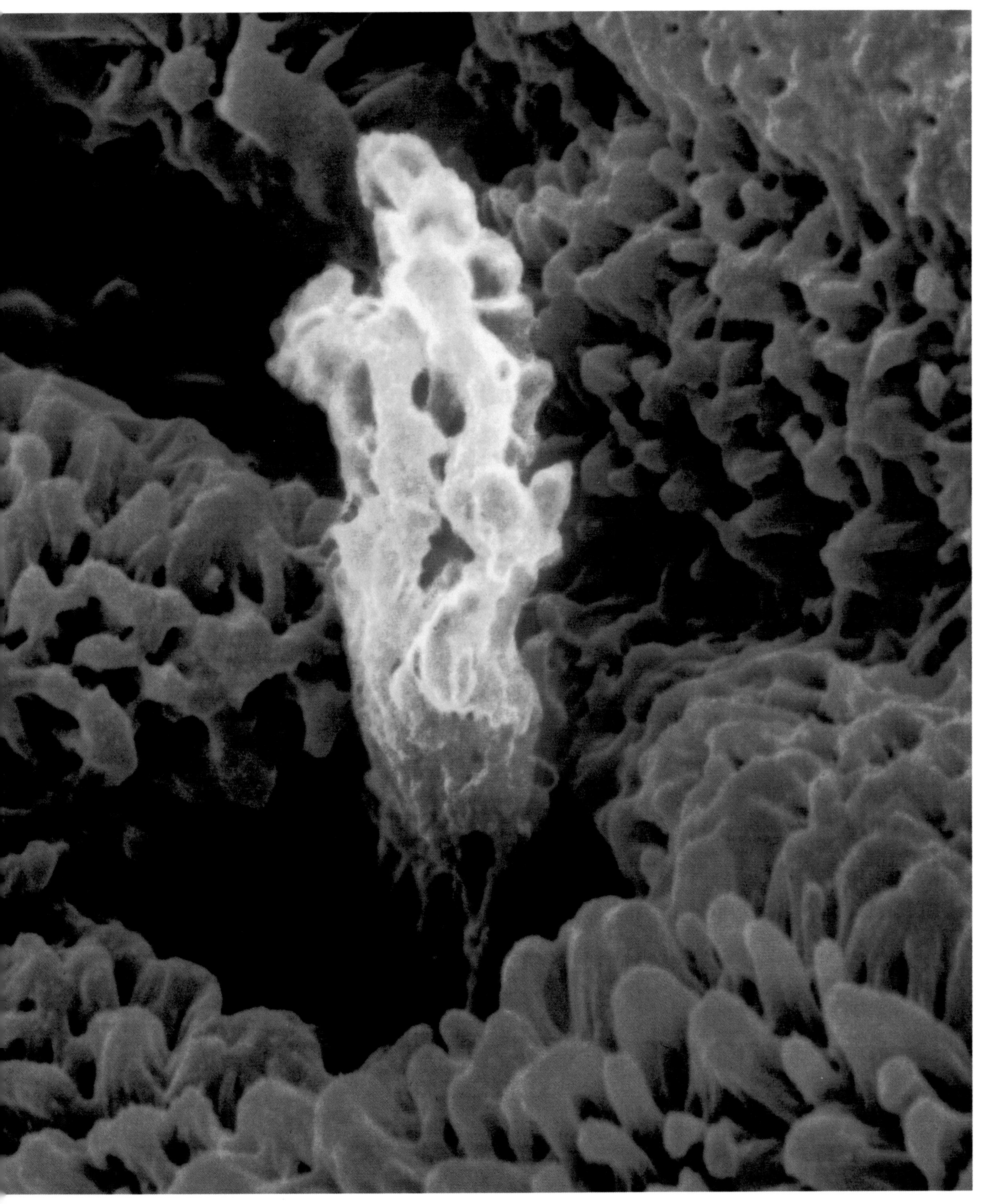

Magnification: 400 times; Lennart Nilsson, © Boehringer Ingelheim International GmbH. (also left, lower)

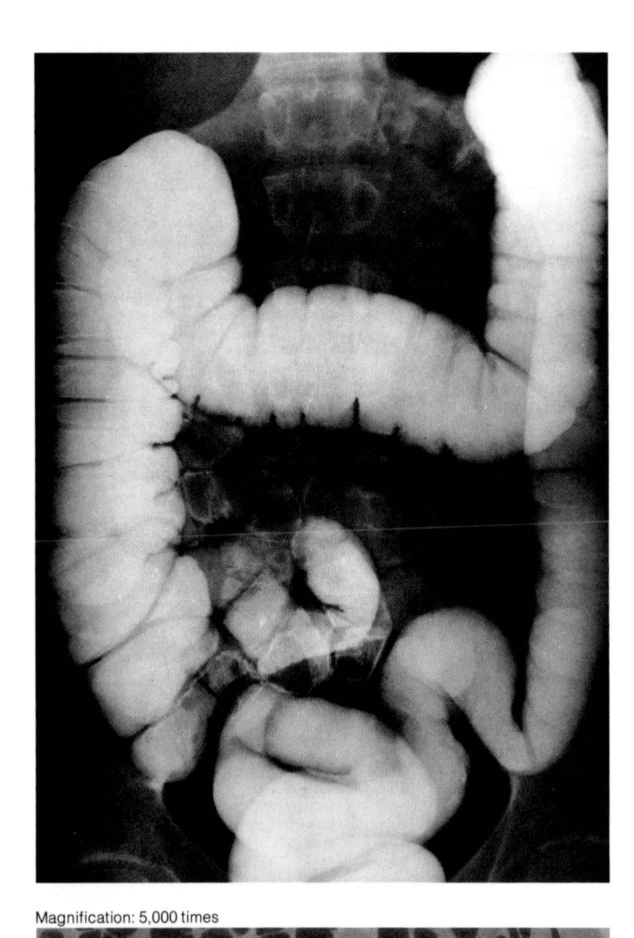

Magnification: 5,000 times

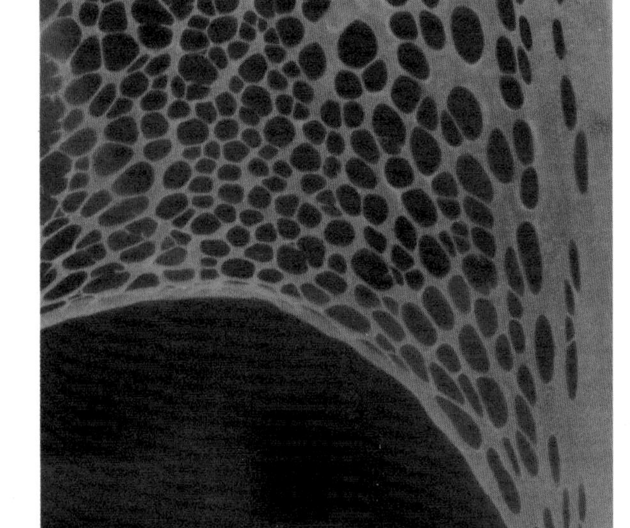

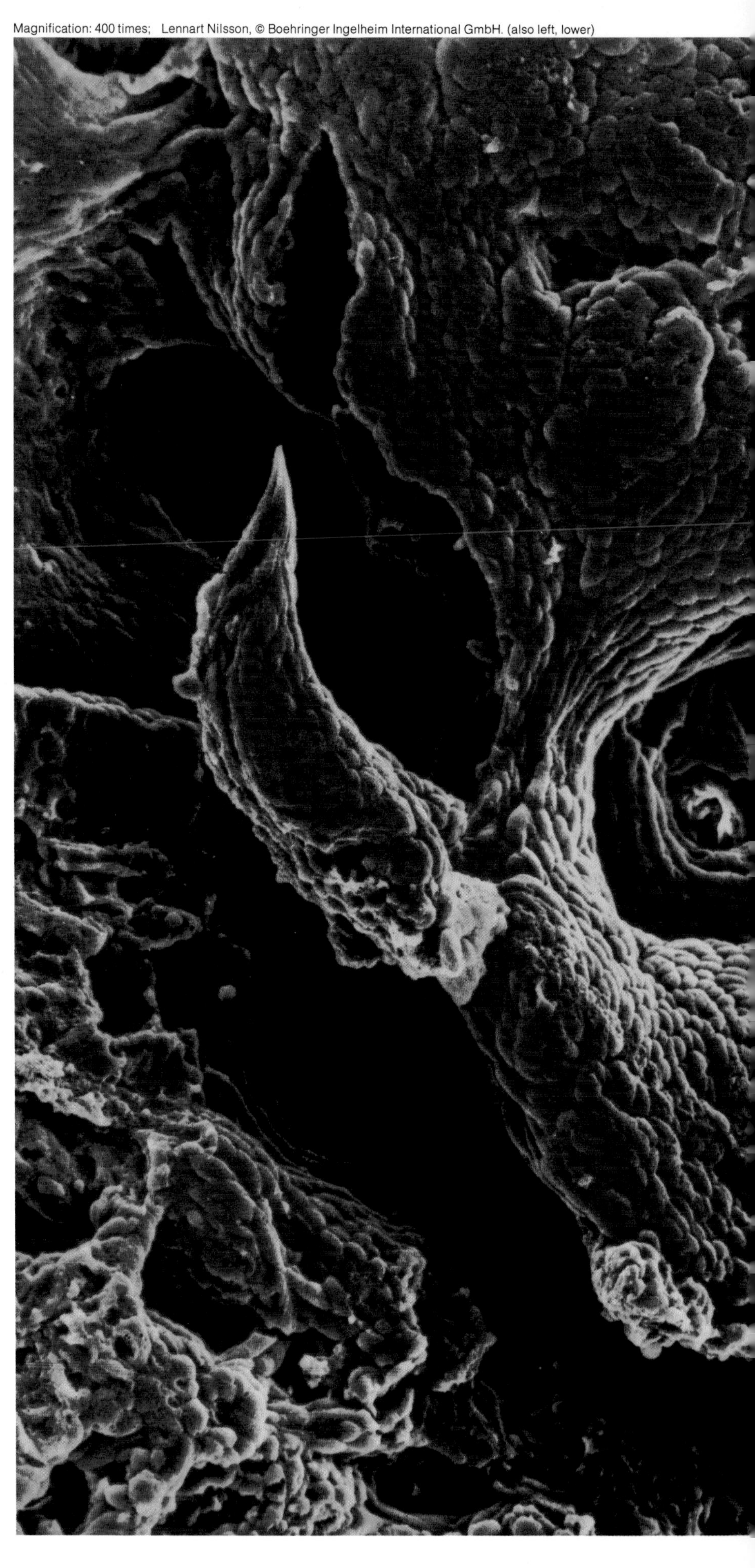

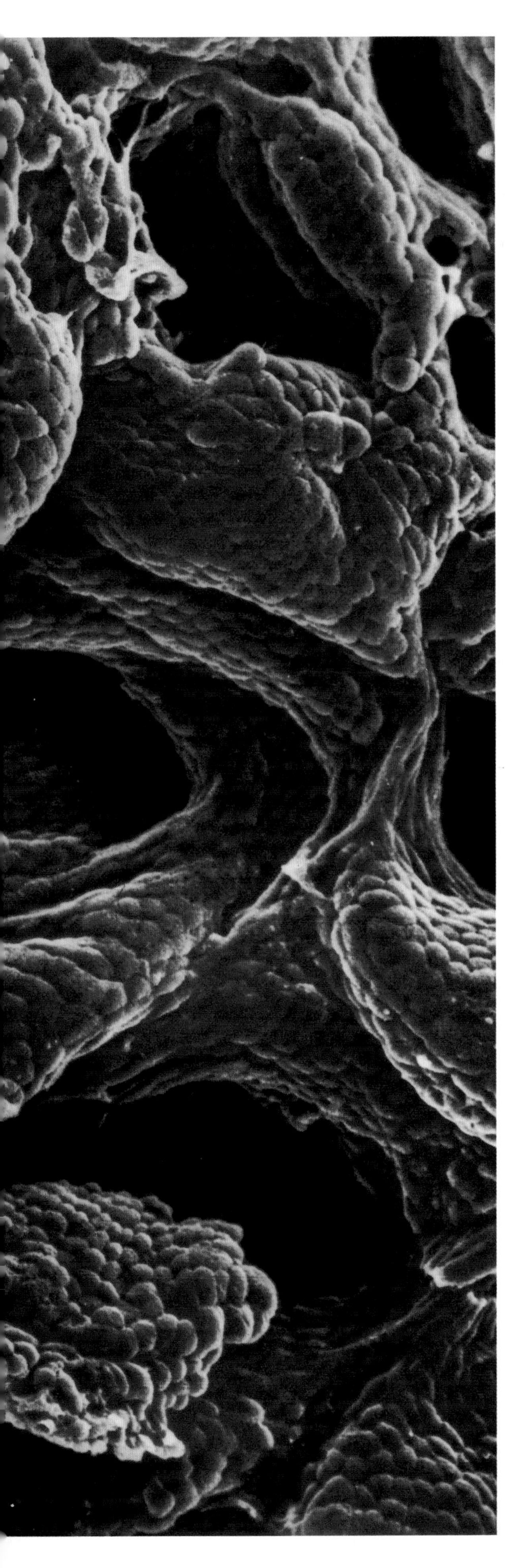

The large intestine—named for its diameter, not its length—frames the abdominal cavity (opposite, upper). By the time food reaches the large intestine, the body has absorbed most of the usable nutrients. The meal now consists mostly of water and undigested residue such as cellulose. The colon, the longest part of the large intestine, absorbs most of the remaining water, leaving firm fecal matter.

The lining of the large intestine (left) reveals a phantasmagoric landscape of cobbled tunnels and crevices. Here, deep pits reach down into glands that contain an abundance of mucus-making cells. Secretions from these cells appear as a lacy froth (opposite, lower). The mucus helps to bind waste materials together and lubricates the feces for the final passage into the rectum and out of the body.

to monosaccharides, or simple sugars; fats from the milk, oil, and chicken must yield fatty acids and other simple molecules. Most proteins must break into individual amino acids.

The job of digesting and absorbing all the nutrients needed by the body requires great surface area, far more than the small intestine, with a diameter no more than an inch and a half, would appear to offer. But the lining of this tube doubles into numerous folds, and millions of densely packed, fingerlike projections called villi poke up about four-hundredths of an inch into the passageway. Each villus, in turn, bristles with hairlike microvilli, all increasing the intestine's inner surface some six hundred times. Spread flat, the small intestine's lining would wrap around the body three times.

In the small intestine, the food mass is sprayed again, this time with powerful, corrosive mixtures from two accessory digestive organs. The pancreas, a soft, pink gland behind the stomach, pours alkaline juice into the intestine to neutralize any remaining hydrochloric acid in the food. The quart or so of this pancreatic juice that flows daily into the small intestine also contains enzymes that break down food. The other gland, the liver, secretes bile that is first stored in the gallbladder and then released to the small intestine. This provides the detergent-like action that breaks up fats so that enzymes can act on them.

Together, pancreatic juice and bile, plus enzymes secreted by the small intestine's own cells, attack the nutrients. They snap off simple sugars from disaccharides or more complex carbohydrates, reducing the milk, the potato, and other vegetables into smaller and smaller molecular chains. Specific enzymes attack assigned links of protein until most of the amino acids in the chicken and milk are liberated. Pancreatic juice and bile join to reduce the fats in the chicken, milk, and salad oil into fatty acids and other fat components, the forms in which fat passes through the intestinal wall.

Wringing Out the Nutrients

While the enzymes are chemically pulling apart the nutrients, several kinds of mechanical action in the small intestine perform a complementary task. Here, as in the stomach, peristaltic waves push the food along, and the many folds of the intestine prevent it from moving so quickly that nutrients will be lost. Meanwhile, circular muscles around the intestine squeeze rhythmically, mixing the mass and pressing it against the surface of the villi, which provide entry to the body's interior. The villi wave and contract, each one wagging independently like an index finger, constantly bringing new molecules into contact with the intestinal wall.

Running up the center of each villus is a

Magnification: 15 times; Lennart Nilsson, © Boehringer Ingelheim International GmbH. (also below)

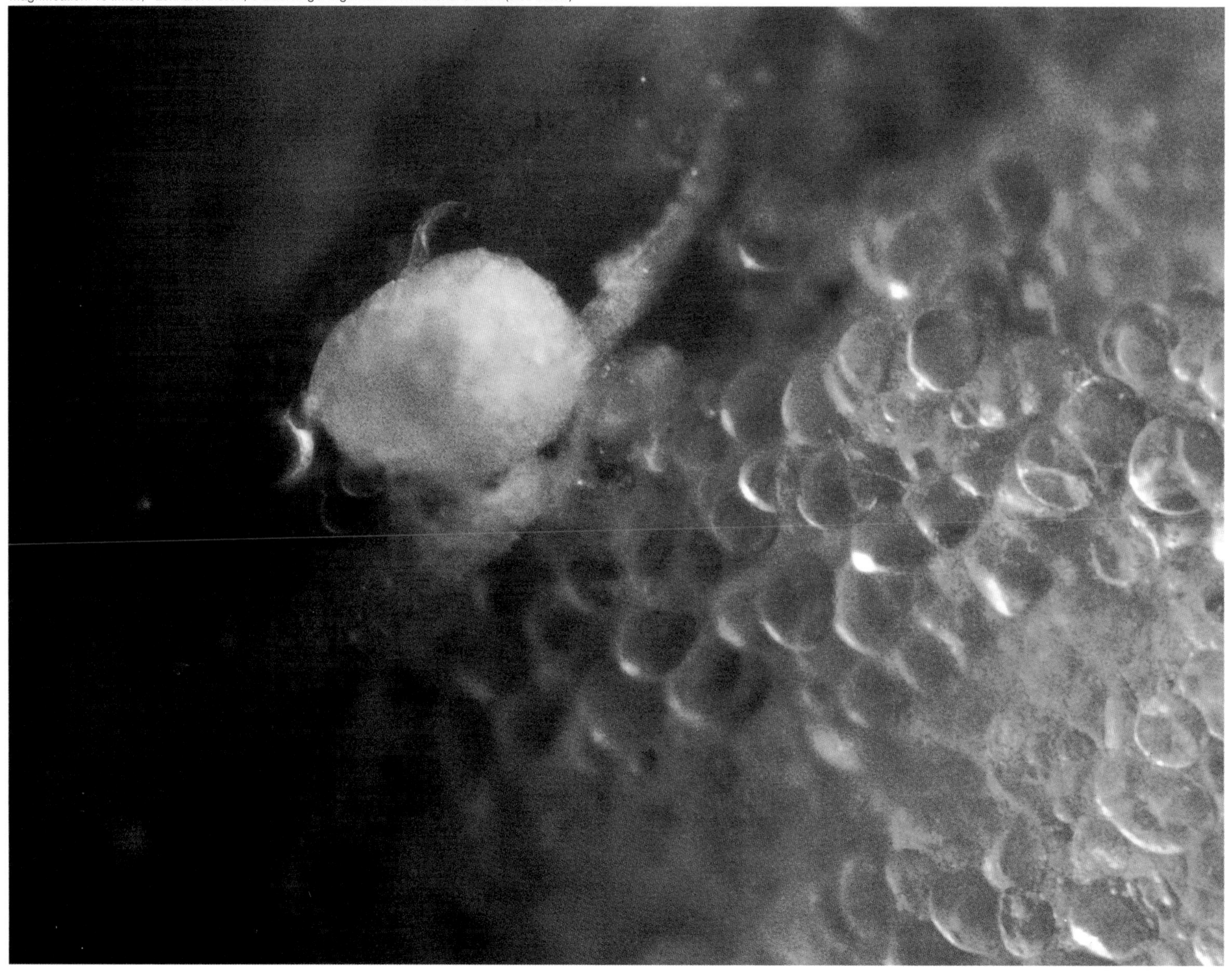

Magnification: 100 times

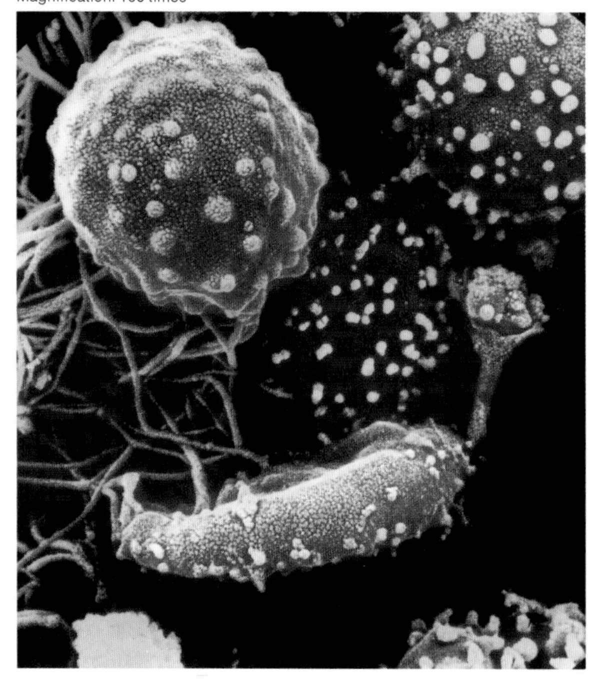

Grapelike clusters in the pancreas, acinar cells (tinted yellow, above) produce enzymes that travel through a duct to the small intestine. There the enzymes prepare carbohydrates, fats, and proteins for absorption into the body by shattering the chemical bonds that link the molecules together.

Scattered among the acinar cells, other cell clusters called islets of Langerhans (tinted white, above) secrete their products directly into the bloodstream. Within one of these islets (left) appears a weblike backdrop of connective tissue and spherical cells dotted with secretions—probably insulin droplets.

In addition to insulin, the islet cells produce glucagon as a major product. These hormones work in concert to regulate the level of glucose, the body's chief fuel. Insulin reduces the glucose, or blood sugar, level. Too little insulin or faulty insulin action causes the blood sugar levels to rise. The result: a case of diabetes.

network of tiny blood vessels. When the carbohydrates, proteins, and some fats are reduced to small enough molecules, they pass into this network and enter the bloodstream. Vitamins and minerals move through intact.

Most of the usable nutrients have joined the exodus into the villi by the time our dinner has advanced just a third of the way along the small intestine. The wastes, consisting largely of dead digestive tract cells and undigestible matter such as cellulose, move into the large intestine.

Though only about five feet long, the large intestine has a much greater diameter than the small intestine. It takes about a day to complete the process of extracting nutrients and preparing wastes for elimination. During this time bacteria act on the wastes, digesting some food residues and manufacturing certain vitamins. The colon walls absorb water, solidifying and compacting the residue into feces. The process is almost complete when muscular action moves the feces to the rectum to be expelled.

The nutrients that passed through the walls of the small intestine, meanwhile, have at last entered the body. The components of our dinner have been absorbed into the blood. Nutrients from the chicken, salad, potato, carrots, and milk, now in strikingly different form, ride the bloodstream directly to the liver.

The Liver's Vital Functions

Resting just behind the right lower ribs, the liver stretches across almost the width of the body, occupying a space about the size of a football. Ancient Babylonians, on seeing the liver's rich supply of blood in sacrificial rites, proclaimed the organ the seat of the human soul. In a sense, it is. The liver is one of the body's most vital organs, processing nearly every nutrient that comes from the intestines.

Up to this point, the whole thrust of the digestive process has been to break down the meal into simple components. In the liver the earliest stages of the building process begin. Here the basic substances of the food are prepared to meet the body's needs—building and repairing tissues and moving muscles.

Each minute more than a quart of blood, mainly from the digestive tract, flows to the liver. Nutrients easily slip from the liver's porous capillaries into individual liver cells, called hepatocytes, which carry out more than 500 different functions. The liver cells package fatty acids into forms that can be stored and used for fuel. They convert amino acids to blood proteins and other products. They rearrange the structures of the sugar molecules, breaking complex sugars into glucose, or blood sugar. They consume worn-out red blood cells and recycle the iron. And they detoxify poisons such as alcohol and drugs.

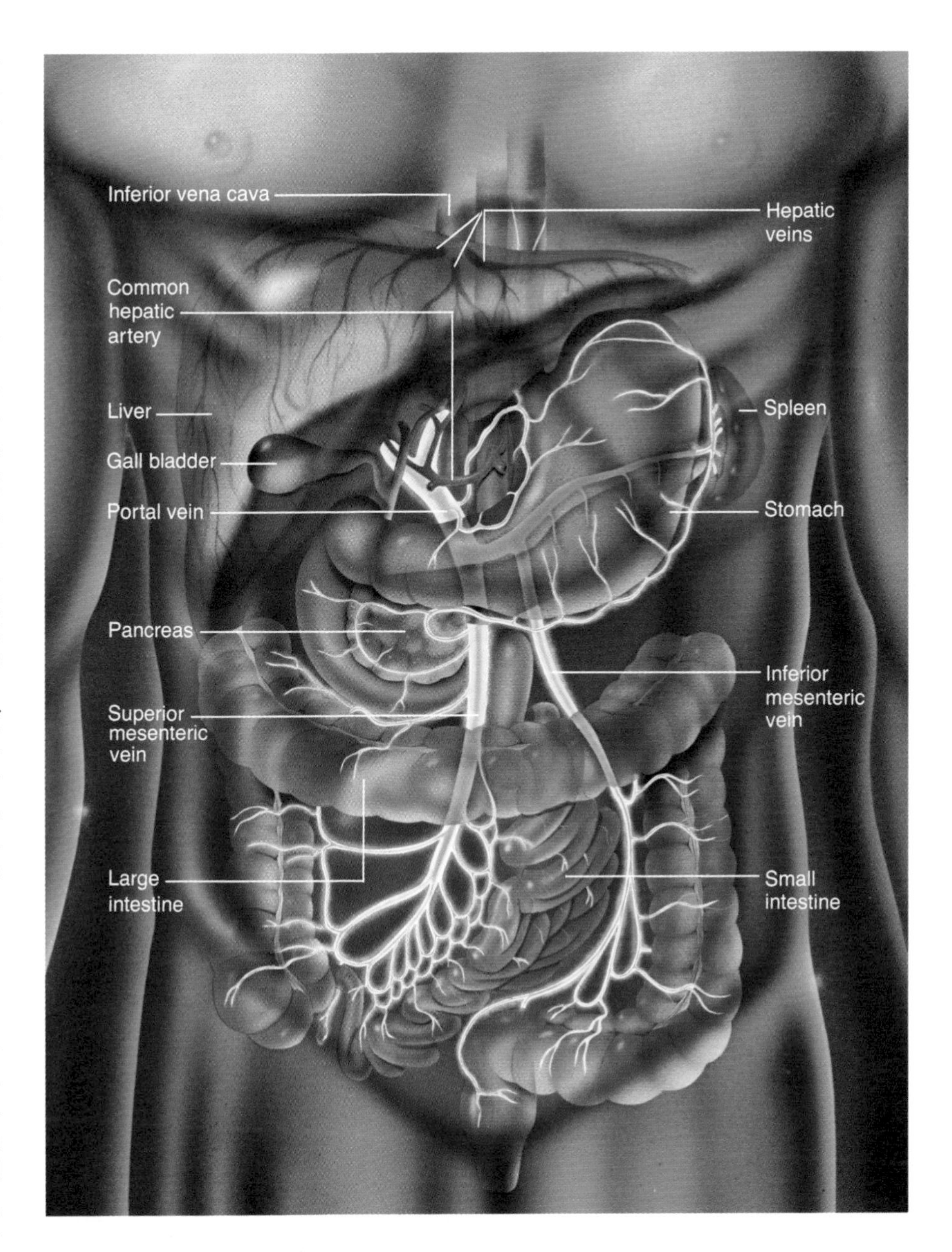

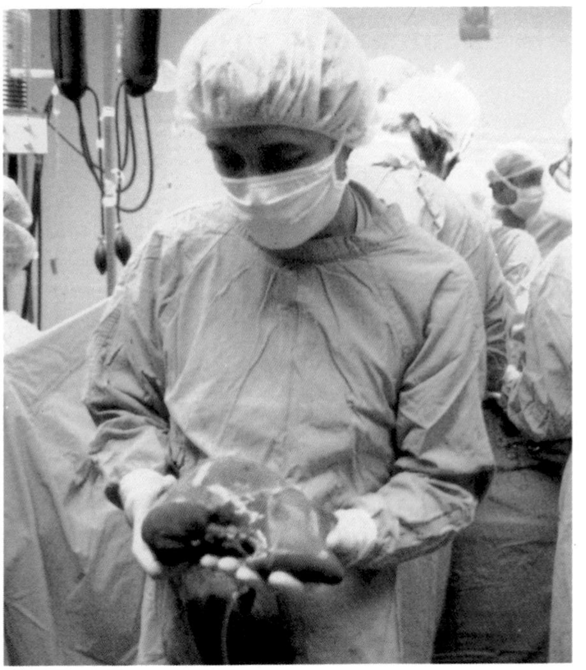

The liver, here seen on top of the digestive system (above), receives blood from two sources: Blood charged with oxygen from the lungs arrives via the hepatic artery. Blood carrying nutrients from the digestive tract enters the liver from the portal vein. The liver, the body's main processor of absorbed nutrients, converts the nutrients into glucose and other substances needed by cells and releases them into the bloodstream as they are required.

The liver's large size (it weighs more than three pounds) and thick network of blood vessels make it one of the most difficult organs to transplant (left).

Liver lobules
Central vein
Central vein
Hepatocytes
White blood cell
Kupffer cell
Sinusoid
Fat-storing cell
Bile canaliculi
Fenestrations

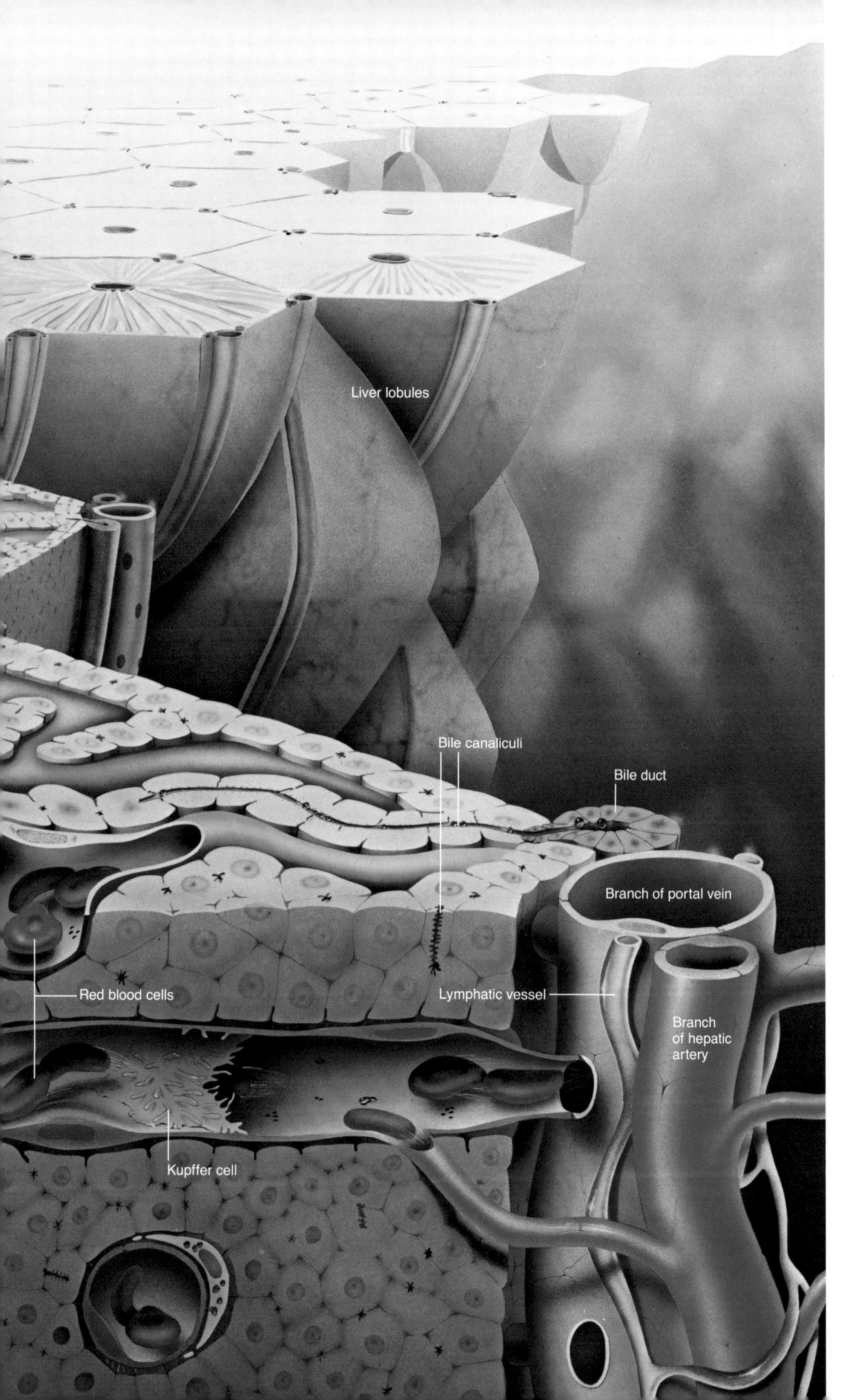

A landscape of the liver, here cut away by the artist and magnified more than a thousand times, reveals the hexagonal lobules that make this organ one of the body's most vital and versatile. Blood-borne nutrients enter the lobules from branches of the portal vein. Branches of the hepatic artery carry oxygen-rich blood.

Inside the liver, the blood flows through a spongelike warren of specialized capillaries called sinusoids. Kupffer cells embedded in the sinusoid walls and roving white cells clean the blood of impurities. Tiny holes, or fenestrations, in the sinusoids allow nourishment to reach the liver cells (hepatocytes). Here the nutrients, converted to glucose and other liver products, reenter the bloodstream via the porous capillaries and the central veins.

Bile secreted by the hepatocytes collects in tiny canals, the bile canaliculi, that feed into the network of bile ducts.

Largest of the body's internal organs, the liver is made up of some 50,000 lobules and processes about a quart of blood a minute. Its multitude of functions include producing blood sugar, storing excess nutrients, and building proteins. The liver also secretes bile to aid digestion and helps rid the body of poisons.

The liver also functions as a storage and distribution center for the body. For example, the body requires a constant supply of glucose for energy. Immediately after a meal has been digested, the level of glucose in the blood rises slightly, exceeding the body's needs. But the level soon drops back to normal as the liver converts excess glucose into a storable form of energy called glycogen. As blood-sugar levels drop during the hours between meals, the liver continuously reconverts the stored glycogen to glucose, assuring a relatively constant and stable level of glucose in the blood. The liver also stores iron and some vitamins, gradually releasing them as the body needs them.

When the liver has done its work, the repackaged nutrients ride the blood flow from the liver to the location where the next series of conversions will occur: the body cells. Here the goal of the food intake is at last achieved. Some nutrients will make new substances for maintenance or growth of body tissues, but most will meet the body's energy needs.

The individual cell represents a world in itself, manufacturing goods and burning fuel in its own infinitesimal industries. The cell's membrane, the filmy coat that encompasses its world, serves as a living barrier, not just keeping out unwanted molecules but also selectively allowing necessary nutrients in. This membrane is constructed so that certain molecules—fats and water, for example—enter as through a sieve. Others are

Food rich in fats—especially saturated fats (right, upper)—or the fatlike lipid known as cholesterol, raises the blood's cholesterol level. Continuously high levels raise the risk of coronary disease. But that risk may be reduced in most people by cutting back on dishes rich in saturated fat and cholesterol.

Diet may play a role in decreasing the risk of certain cancers. Vegetables such as cabbage (left), brussels sprouts, and broccoli provide abundant supplies of vitamins A and C, possible anticancer agents. Such foods of the cruciferous family may stimulate the protective action of enzymes, or they may assure normal cell development.

Controversy surrounds claims that special diets can cure cancer. Dr. Anthony J. Sattilaro (right, lower) believes that a diet heavy on brown rice and other whole-grain foods helped him recover from prostate cancer. Few specialists support Sattilaro's claim, but most agree that sensible nutrition is good preventive medicine.

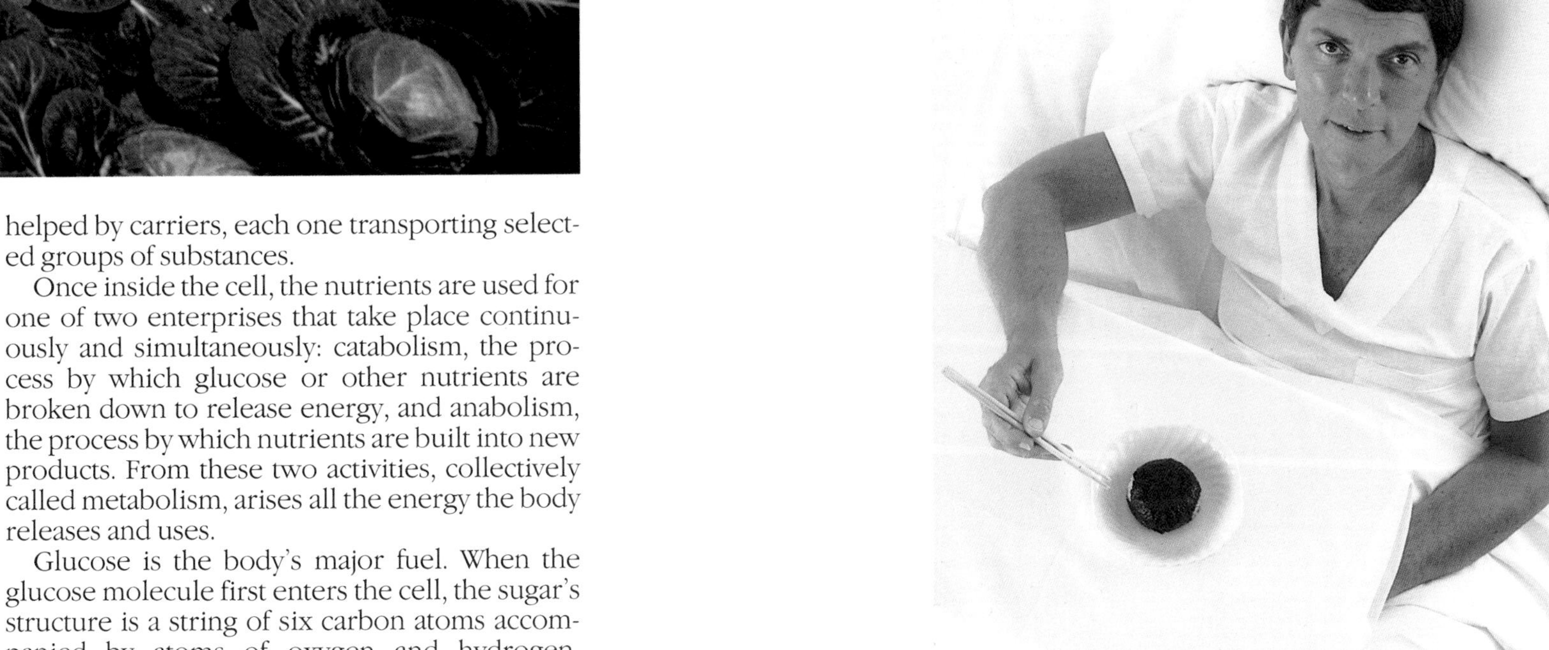

helped by carriers, each one transporting selected groups of substances.

Once inside the cell, the nutrients are used for one of two enterprises that take place continuously and simultaneously: catabolism, the process by which glucose or other nutrients are broken down to release energy, and anabolism, the process by which nutrients are built into new products. From these two activities, collectively called metabolism, arises all the energy the body releases and uses.

Glucose is the body's major fuel. When the glucose molecule first enters the cell, the sugar's structure is a string of six carbon atoms accompanied by atoms of oxygen and hydrogen. Chemical energy is locked within the bonds of

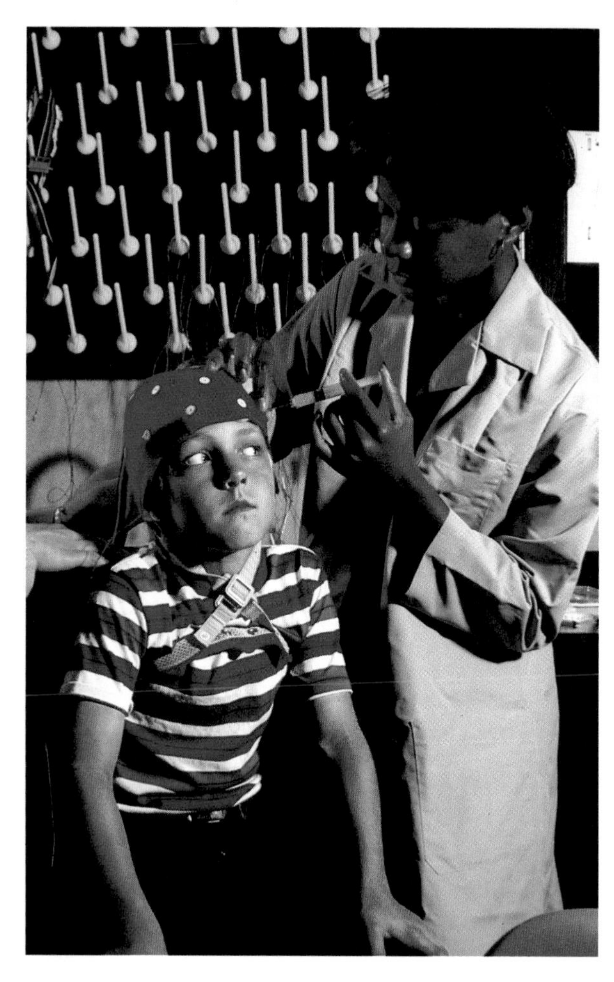

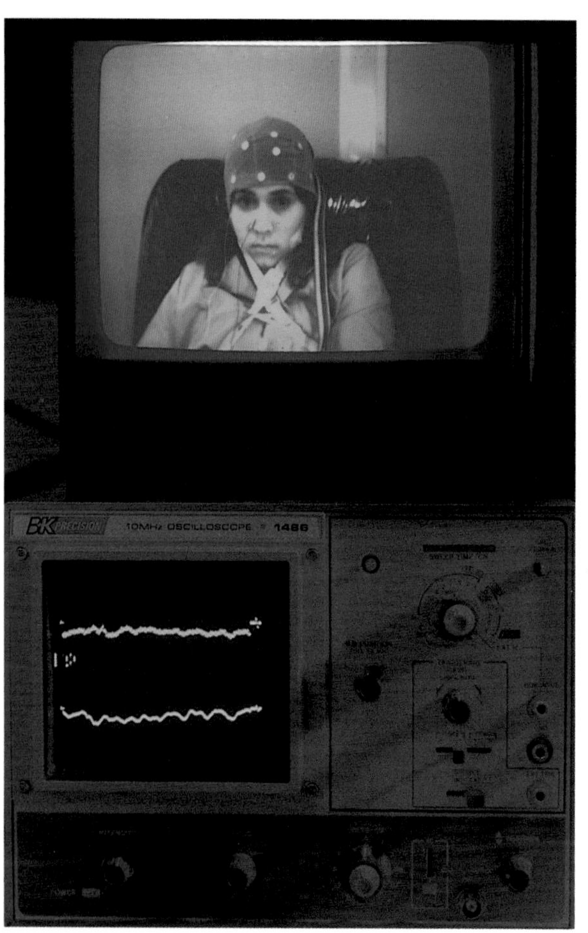
10MHz OSCILLOSCOPE
1486

How does food affect the brain? To find out, a technician at the University of Maryland prepares to wire a young volunteer for an electroencephalogram (opposite, upper). As images flash on a screen (left), the EEG measures the boy's brain wave response. An oscilloscope (opposite, lower) checks the reliability of the measurements.

Studies such as these indicate a link between diet and intellectual performance. Children whose diets contain relatively large amounts of sugary foods tended to score lower on intelligence tests and to fare more poorly in school than their more adequately nourished classmates. Scientists conclude that improper diet contributes to learning disorders.

By studying their subjects' eating habits and monitoring electrical activity in their brains, scientists hope to identify specific foods that inhibit or stimulate mental performance.

the sugar molecules, and this chemical energy is what the cell needs to do its work.

The first step in releasing this energy is to cleave the glucose molecule into two smaller molecules of pyruvic acid, which then move to compartments of the cell known as mitochondria. Occurring in a wide variety of shapes, each mitochondrion is only a thousandth the size of the cell, but each one engages in a remarkable amount of activity.

When a fuel is burned, it consumes oxygen and releases energy in the form of heat; when our cells burn fuel, they too use oxygen and release heat. We measure the energy value of food by the amount of heat it produces when burned—a calorie represents the heat needed to raise the temperature of a gram of water by 1° Celsius (or by 1.8° Fahrenheit).

The Body's Flameless Furnace

This flameless combustion takes place in the cell's mitochondria. Here the pyruvic acid molecules are dismantled in a series of carefully controlled steps that depend on a complex of enzymes. Each enzyme breaks a particular bond, reducing the molecule, carbon atom by carbon atom. Each of these splits releases energy. But the energy is preserved, stored in bonds of yet another chemical compound, adenosine triphosphate, also known as ATP. Every glucose molecule yields many molecules of ATP, which can easily be broken down to release their energy when a cell needs it.

The oxygen in the mitochondrion meanwhile rips off hydrogen atoms one by one, again releasing energy with each break. Finally, the free carbon, oxygen, and hydrogen atoms combine, forming carbon dioxide, which we breathe out through our lungs, and water, which we excrete in urine or secrete in perspiration.

The molecules of ATP produced during catabolism are the body's "currency," the means by which the food we eat can be traded in for action. When the muscles of the face receive a signal from the brain to contract, for example, the bonds of the ATP molecules in those muscles are broken, the energy locked in the bonds is released, and a smile is born.

But ATP is crucial in many less obvious ways as well. The cells spend much of their ATP supply not in walking, lifting, and smiling, but in maintenance activities. It keeps the heart beating, the lungs breathing, the kidneys filtering. Everyone needs a minimum of calories simply to sustain such life processes.

Because these maintenance activities continue throughout life, the conversion of nutrients to energy must also take place continually. It is so important that if the supply of glucose runs out, the relentless ATP factories within the cell carry

MIDWAY
Ex. 3333
Thank You
for not SMOKING!

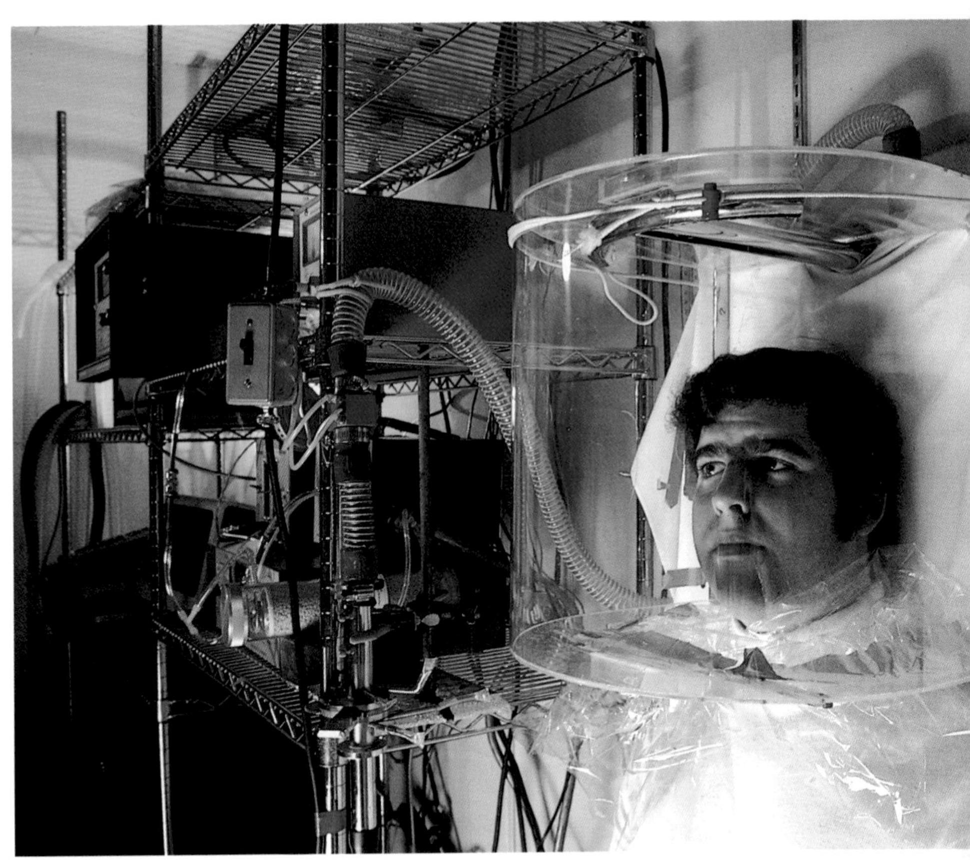

Playing a video game (left) burns up as much energy as would mild exercise for this volunteer in a metabolism study. By comparing oxygen intake and carbon dioxide output, researchers at Columbia University can measure a subject's energy expenditure.

Scientists at St. Luke's-Roosevelt Hospital Center in New York City use a similar hookup (above) to determine the basal metabolic rate of a volunteer at rest. Even when fast asleep, our bodies burn calories to maintain automatic life processes, such as breathing, digestion, and heartbeat.

on. Unlike atomic power plants or coal furnaces, the body can adapt to burn any energy source available. Fats and proteins contain many of the same elements as carbohydrates and therefore can be broken down to fill the energy gap.

The total store of muscle glycogen, plus the liver glycogen that can be quickly converted to glucose, provide only about 2,000 calories. About a day's worth of normal activity will deplete these glycogen reserves, but because glucose is needed to fuel the nervous system, the body will not exhaust its supply before turning to other alternatives. As the glycogen level falls, the body begins to mobilize fat stores from the adipose tissue under the skin. This is the main source of energy for as long as it lasts. If both glycogen and fat supplies are expended, the body draws on the ultimate fuel reserve: the proteins of its own tissues. Even in this dire situation, which occurs only in starvation or prolonged fasting, nature sets a priority system. Muscle tissue is the first to go because movement is not as important to survival as wound healing or maintenance of the immune system. But as long as life continues, so does ATP production, which governs the essential life processes.

Although energy production consumes most of the nutrients, the cell uses some for building structural components, for growth, and for replacing worn-out parts. The most important nutrient for this purpose is protein, which is synthesized on cell particles called ribosomes. So small that their detailed structure remains a mystery, ribosomes are barely visible even with an electron microscope. Yet their role in protein manufacture is essential.

A molecule of any protein in the body is built by linking together in a specific length and sequence some combination of the 20 or so different amino acids. To help carry out this process the patterns or instructions for all proteins are inscribed on deoxyribonucleic acid, or DNA, the material within the cell nucleus from which our genes are made. Like rare books vaulted in a library, these instructions never leave the safety of the nucleus.

The first step in manufacturing a protein is to copy out these instructions. A molecule called ribonucleic acid, or RNA, molds itself to a DNA segment for a particular protein. Once it has transcribed the pattern, this messenger RNA (mRNA) carries the instructions out of the nucleus to the ribosomes. The ribosomes translate the message carried by mRNA. In this process, other RNA molecules called transfer RNA (tRNA) carry the amino acids to the mRNA. The ribosomes move along the mRNA, reading its message. The tRNA adds amino acids to others, and a protein chain starts to grow. The number of amino acids and the amount of time needed to complete a molecule vary from one type of protein to another; a hemoglobin molecule, for example, consists of 574 amino acids and is completed in a minute and a half.

From Protein to Body Parts

During the first few years of our lives, our cells synthesize 500 to 1,000 pounds of protein—more than 5 tons in a lifetime. We need not consume all that protein, however, to supply our cellular factories. The parsimonious liver synthesizes the nonessential amino acids for much of the pound or so of protein our cells manufacture each day. We need to eat only a small amount—about 2 ounces, or as much as would be provided in a generous serving of fish and a glass of milk. With this and the amino acids the liver produces, thousands of varieties of proteins can be created—patterns that make up muscle tendrils, spin the fibers of hair and skin, make the hemoglobin that transports oxygen from the lungs to the tissues, and create the very enzymes that facilitate metabolism.

While only a small amount of dietary protein is needed per day, a prolonged deficiency can bring on serious health problems. An adult who does not eat enough protein may lose only muscle mass. But a child who suffers from kwashiorkor, a severe protein deficiency common in some developing countries, may stop growing.

Excess food, on the other hand, including excess protein, can be converted from any other form and stored as body fat. A fat content of

The bathroom scale may tell how much you weigh, but it does not measure how much fat your body contains. To find out, a researcher at Columbia University weighs a volunteer in a water-filled tank (opposite). Because lean tissue is denser than fat, the dunking enables researchers to calculate the body's total density, and from that, to estimate its percentage of fat.

The "whole body counter" (right, lower) gauges fat content by another method—measuring the quantity of potassium found in lean body tissue. The slight radioactivity emitted by potassium within the body enables the machine to assess lean tissue mass.

The TOBEC method (right, upper) measures total body electrical conductivity. Because lean tissue is more conductive than fat, the machine can estimate lean body mass. Researchers can then calculate the ratio of lean to fat tissue within the body.

Such instruments offer clues to an individual's nutritional status. By analyzing body composition, experts devise treatment for obesity and other weight-control disorders such as anorexia nervosa and bulimia.

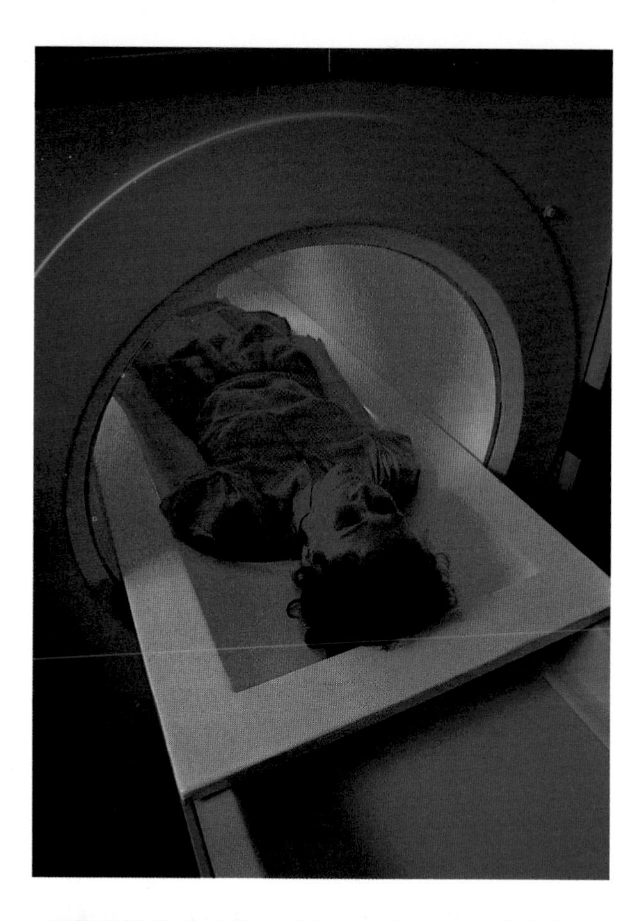

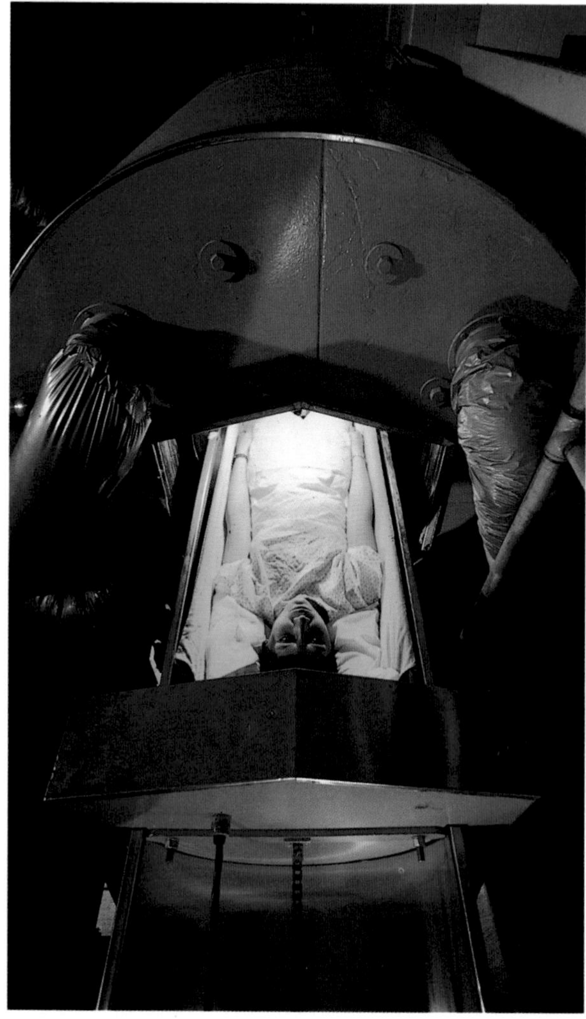

about 10 percent of body weight is normal for adults, but too great an accumulation produces the unhealthy condition known as obesity. This is a major and increasing health problem in many affluent countries. Evidence gathered over the years suggests that obesity, which affects 20 percent of the population in the United States, is linked to a variety of health problems, including heart disease.

By studying selected societies, scientists have discovered a link between various cancers and certain environmental factors, including diet. Some believe that the death rate from cancer in the United States could be lowered by 35 percent if Americans modified their diet. Part of the evidence comes from studies of communities

Why do some people snack on cake, pizza, and other foods rich in carbohydrates? For the answer, researchers at the Massachusetts Institute of Technology are studying the brain chemical, serotonin. Volunteers with a preference for carbohydrates live in a dormitory (left), where researchers monitor their food choices. Snacks from a computerized vending machine record each subject's eating habits (opposite, upper). Leftovers are weighed after every meal (opposite, lower) to assess how much was actually eaten.

Observers concluded that most subjects chose carbohydrate-rich foods for snacking. At mealtime, they selected varied dishes. Research suggests that eating a small amount of carbohydrate-rich food triggers the production of serotonin, which, in turn, shuts off the desire for more carbohydrates.

Those who snack excessively may do so because of abnormal serotonin production or because they follow a stringent, low-carbohydrate reducing diet. Such a diet depletes the brain of serotonin, inducing uncontrollable carbohydrate cravings. When test subjects take a drug that prompts serotonin production, their craving usually wanes. Presumably this confirms the chemical's ability to control carbohydrate appetite.

with distinctive dietary practices. One such group, the Seventh-day Adventists, have lower rates of breast, colon, and some other cancers than the general population. A possible reason: their vegetarian diet, which contains less fat and considerably more fiber.

Fiber is not the only dietary element associated with lower cancer rates. Certain vitamins in our food are now believed to play a protective role as well. Some evidence suggests that vitamins A, C, and E act as anticancer agents. Cabbage, broccoli, and brussels sprouts, rich sources of vitamins A and C, are associated with reduced cancer risks.

The nutrients we eat have these visible results, but they also have other, less tangible effects that we are only beginning to understand. Researchers are now discovering the ways in which what we eat might affect brain development and performance. A child who is severely undernourished early in life may fail to develop the normal neuron connections in the brain. The effects for such children can be pronounced, including poor school performance as a result of diminished mental capabilities.

Even factors in a normal diet can affect mental and emotional states. Recent research indicates that a number of chemicals that flow within the brain are influenced by diet; that is, the production of these chemicals, called neurotransmitters, depends on the availability of certain nutrients that act as precursors. Some neuro-

transmitters affect mood, memory, pain sensitivity, sleepiness, aggression, and appetite.

The complex relationship between nutrition and brain activity is illustrated by the neurotransmitter serotonin, manufactured in the brain from the amino acid tryptophan. When present in the brain at elevated levels, serotonin has been shown to diminish pain sensitivity, decrease the craving for carbohydrates, and induce sleep. Because tryptophan is an amino acid present in many foods, it would seem that eating protein would raise the level of serotonin. But this does not appear to be the case. Amino acids have only one way of getting from the blood to the brain—they must compete for space on carriers that transport them across the blood-brain barrier. When we eat a lot of protein, other amino acids compete with tryptophan, so less of the amino acid wins space on the carrier. Food rich in carbohydrates, on the other hand, lowers this competition, so more tryptophan can enter the brain. As we learn more about foods, we may be able to use them to influence our behavior and emotional states.

Food for Further Thought

Because many of these diet studies are so tentative in their conclusions, the question remains: Just what are we to eat to help maintain our mental and physical well-being? So far, it seems that, under normal conditions, we should limit our use of vitamin supplements because a balance of nutrients comes with an adequate diet. Perhaps animal studies can teach us something about the factors that affect our food choices.

When rats in experiments are offered a palatable diet of varied snack foods—including cookies, salami, and marshmallows—some prefer the variety to a bland, but more adequate, diet of animal chow. Another study reveals that rats so assiduously seek variety in food that they do not want to eat from a single food source, even if it is nutritionally adequate.

How does this relate to humans? Maybe not at all, but both studies reveal that animals seek variety in food. Why? Perhaps it is an adaptation, one that guarantees the needed nutrients. In other words, the more different kinds of foods you eat, the higher your chances of obtaining the nutrients your body requires.

Some humans are fortunate enough to be able to choose from a wide selection of foods. Even though we do not always do so, we can make reasoned choices in food selection. Life on Earth sprang from varied elements. If, through nutrition, we wish to enhance our health and well-being, we can best do so by drawing from the great variety of food our planet offers us.

SUSAN SCHIEFELBEIN

Cross-pollinating a soybean flower (opposite), a laboratory worker wears magnifying goggles to assist him in his task. Meticulous work like this has helped farmers reap bountiful crop harvests in recent years.

In the United States alone, the introduction of hybrid corn contributed to a fivefold increase in crop yields between 1930 and 1984. Other selective breeding programs since the mid-1920s produced hens that have more than doubled egg output. Egg sizes, too, can be influenced (above), depending on the breed, size, and age of the hen, and by altering feeds and hours of light.

But traditional breeding techniques alone may not meet the growing challenge of world hunger. Genetic engineering shows greater promise. By manipulating the genetic material of plants and animals, scientists hope someday to create more and better foods for a hungry world.

The Powerful River

Within the human body flows a river unlike any other earthly river—a crimson stream that courses through every organ, twists past every cell on a journey that stretches sixty thousand miles, enough to circle the planet two and a half times. Earthly rivers refresh the land with water; the body's stream nourishes and cleanses, delivering food and oxygen to every cell, removing wastes, regulating the human environment. Earth's rivers flow through inorganic rock and sand; the body's river travels through living tissue. The powerful heart that propels this stream and the vessels that guide it are all alive. The human river can regulate its own velocity, its banks widening or narrowing to control the shifting tides. And it can change its own course, instantly channeling its swift currents to meet new demands: Swimming or sleeping, contemplating, celebrating, running a race or rocking an infant—each alters the flow of this powerful river.

The body's river retains an age-old tie to Earth's waters, a link through time to primal seas. The first life-forms, single cells, fueled their existence by absorbing oxygen dissolved in seawater through their membranes and excreting carbon dioxide, waste product of metabolism, back into the ocean. As cells joined together to function as a single organism, many lost contact with this oceanwide circulatory system. With life's increasing complexity an inner stream evolved to nourish every cell. Yet within that stream we still carry a bequest from the beginnings of life: In our blood flows the same balance of minerals and salts that existed in ancient Cambrian seas, a heritage half a billion years old.

Like its prototype the ocean, blood is a crowded sea, teeming with a diverse society of cells that carry out specific tasks and coexist in strict proportions. So critical is this balance that a decline in the population of any one element can endanger life.

Blood is a liquid tissue, little more than half fluid. Poured into a test tube and treated with salt, it separates into three distinct layers. The thickest layer, plasma, which floats on top, is a clear, golden liquid; next comes a narrow, solid band of white blood cells; and, finally, settling to the bottom in a thick band of crimson are the red cells that give blood its color.

Erythrocytes, the body's red blood cells, constitute about 45 percent of blood's volume; yet they are the most abundant cells in the body. Some 25 trillion red cells course through our bloodstream, their numbers corresponding to the urgency of their task. These tiny cells—so small that a stack of 500 would only measure .04 inches high—are the body's cellular lungs, designed solely to ferry oxygen to every tissue and remove carbon dioxide. Shaped like plump, round cushions dimpled in the center, red blood cells consist primarily of water and a red protein called hemoglobin. It is the power of this protein that gives the red blood cell its vast oxygen-carrying capacity.

Constructed from more than 10,000 atoms, a hemoglobin molecule consists of 4 elaborately entwined strands of amino acids called globin. In the middle of each strand is the heme, a disk of carbon, hydrogen, and nitrogen atoms with a single iron atom wedged in the center. The iron atom in heme acts as a magnet, snapping up oxygen, then clinging tightly to it.

The magic of hemoglobin, however, lies not in its power to latch onto oxygen but in its ability to release it. Were iron atoms to float freely in the bloodstream, they would bind irrevocably to oxygen, keeping it from the body's other tissues. But embedded in the heme disk and surrounded by the tangled folds of globin, iron forms only a temporary bond with oxygen. The hemoglobin molecule can thus tighten or loosen iron's grip according to the pressures of surrounding gases.

Where oxygen is plentiful—as in the lungs—iron exercises its full powers, vacuuming up oxygen molecules. Conversely, where the oxygen supply is low—as in muscles after exercise—the hemoglobin molecule eases iron's hold, forcing it to surrender the gas to other tissues. Deprived of oxygen, hemoglobin instantly draws in carbon dioxide, turning a breathless blue as it carries the waste back to the heart and lungs.

The oxygen-transport capacity of our blood is enormous. Although one hemoglobin molecule can carry only four oxygen molecules—one clamped to each heme disk—every red blood cell contains about 270 million of these complex proteins. And, in turn, so many red cells crowd the bloodstream of a single human that stacked, these cells would reach 31,000 miles into the sky.

The Life Cycle of a Red Cell

The healthy body meticulously maintains its bounteous supply of red cells. Deep within the bone marrow, primitive cells called erythroblasts continually divide. The resulting pair of cells in turn divides until each has produced 16 red blood cells. As it matures, a red blood cell expels its nucleus, thereby giving up the ability to repair or reproduce itself. Launched into the bloodstream, each cell will live only 4 months, traveling between the lungs and other tissues 75,000 times before returning to the bone marrow to die. In the second it takes to turn a page of this book, we will each lose about 3 million red cells. Yet during that second the marrow will have produced the same number.

So attuned are our bodies to the urgent demand for these oxygen carriers that we automatically alter production to meet changing needs. If you donate a pint of blood, the marrow will speed its manufacture of red cells to replace the loss within a few weeks. And if you need more

The throbbing rhythms of an aerobics class strengthen the heart by training it to pump with more power and efficiency. Aerobic exercise, in which the muscles demand increased oxygen, includes swimming, brisk walking, jogging, and jumping rope. Most experts prescribe such vigorous activity for healthier hearts and circulatory systems. But they advise easing into strenuous workouts and keeping them regular. Studies show that besides promoting fitness, 15 to 30 minutes of aerobics at least 3 times a week can reduce the risk of heart disease by lowering blood pressure and cholesterol levels.

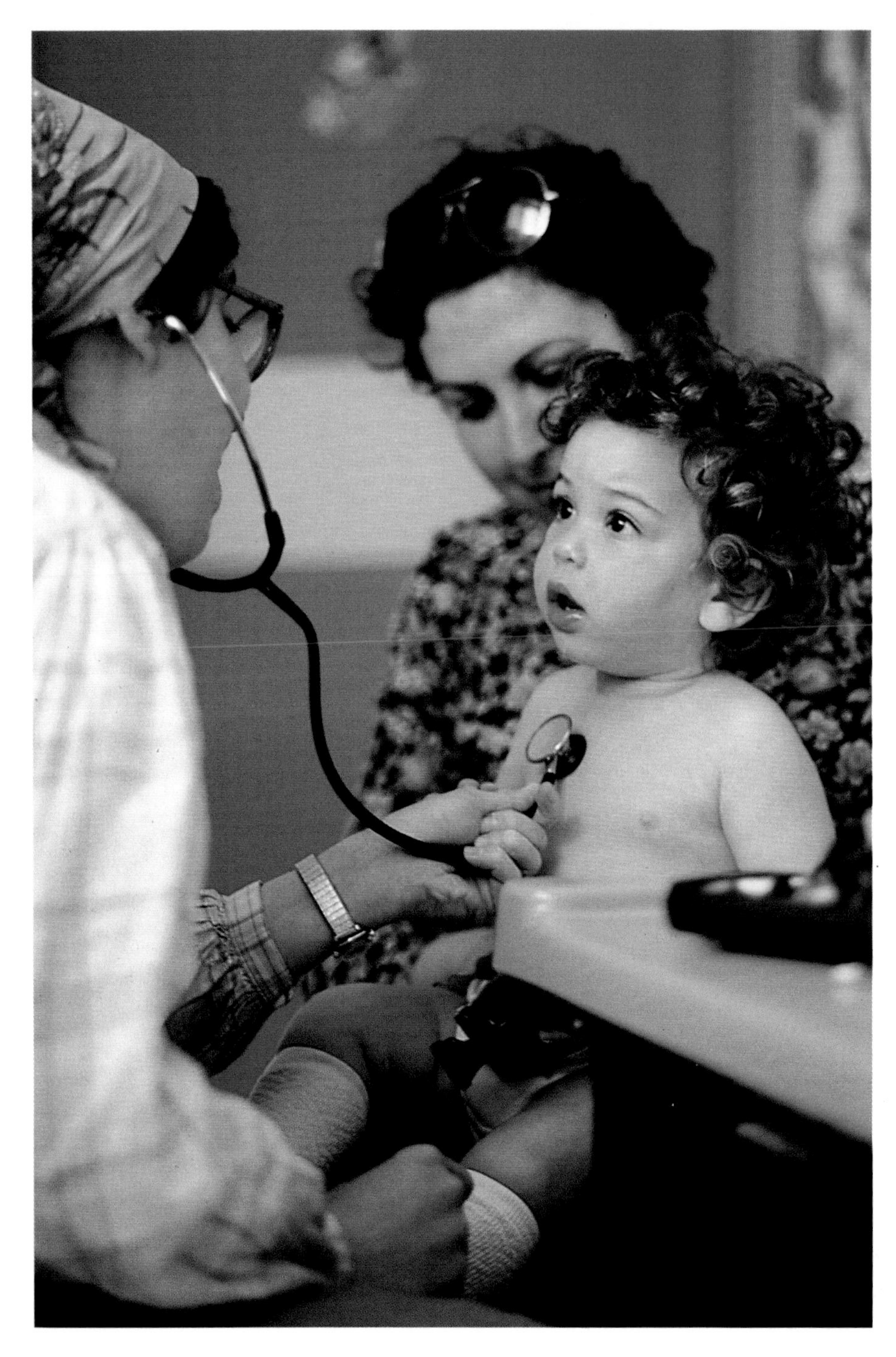

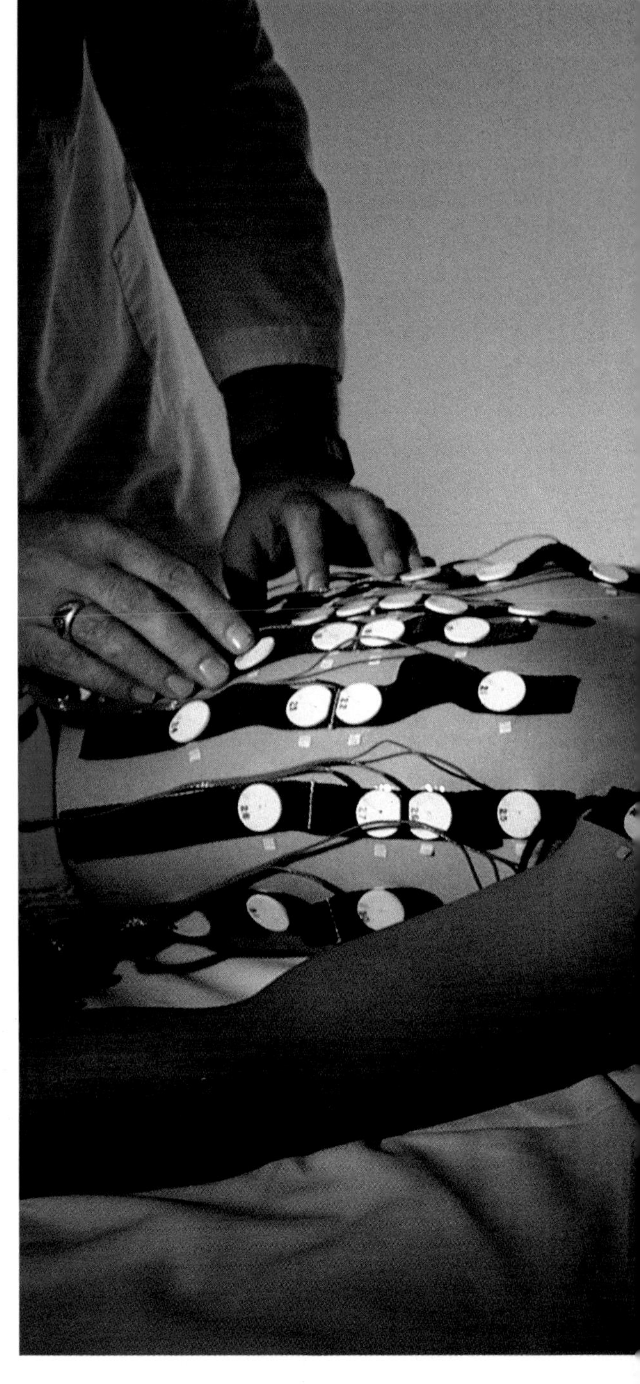

Through the stethoscope, a doctor can hear heart valves as they open and close (above). A hollow tube links the earpieces with a diaphragm which transmits vibrations from the patient's chest. A trained listener can recognize the sound of blood backing up instead of flowing forward—a sure sign of heart disease.

The same acoustic principle was at work in 1816, when a French physician put his ear to one end of a rolled-up sheaf of papers, placed the other end on a patient's chest, and drew inspiration for the device we call a stethoscope.

An ultramodern descendant of those rolled-up papers carries a formidable name—the electrocardiographic body surface potential map (far right). It picks up the heart's reverberating electrical current at 32 points on the body, instead of at the standard 12 of the electrocardiogram, and reveals the heart's size and shape and the condition of its tissues.

oxygen to sustain a rigorous exercise program, your heart will pump more oxygen-rich blood to meet the rising demand.

With its steady supply of oxygen-carrying red cells, our blood has evolved far beyond the primitive ocean from which it developed. If we still depended, as did primal single cells, on oxygen dissolved in water, our bodies would require some 90 gallons of the liquid, and we, less than agile at 800 or 900 pounds, would not have survived as a species. So valuable is the gift of hemoglobin that, throughout evolution, the greater the red cell's oxygen capacity, the more complex the species became. Mammals, possessing the greatest power and stamina of all creatures, also possess the red cells most able to hold oxygen.

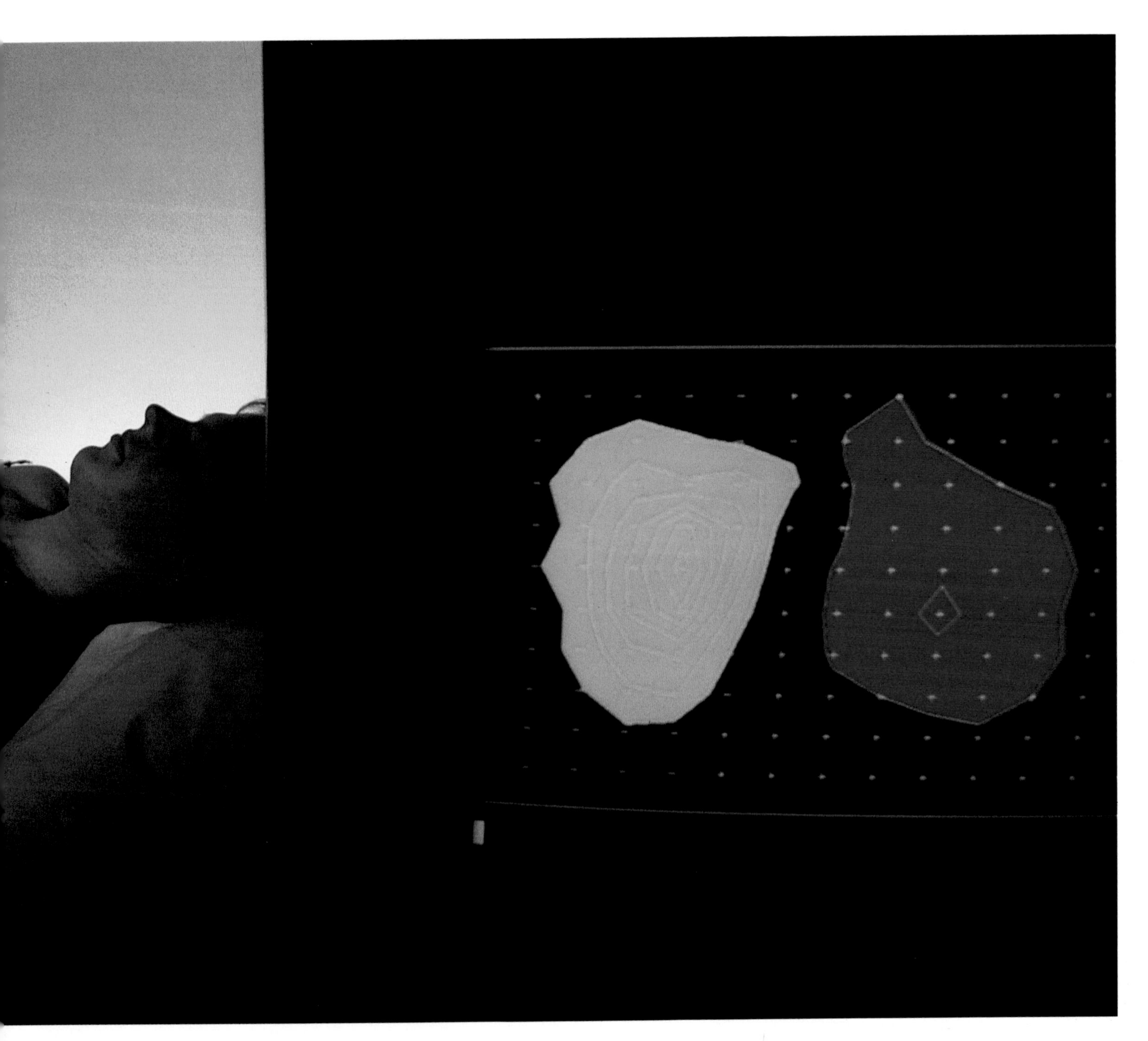

White cells and platelets make up the next and narrowest layer in our test tube of separated blood. Platelets, or thrombocytes, the smallest of blood's cells, are also the most versatile. If the red cells are the bloodstream's cargo vessels, designed expressly to shuttle oxygen, then platelets are its repair force, equipped with special tools to ply their trade. More than a trillion of these minute, colorless cells patrol the body's river, repairing rents and tears in its banks to ensure the smooth, uninterrupted flow.

Like red blood cells, platelets form in bone marrow, produced from huge cells called megakaryocytes. As these giant marrow cells mature, they shatter into thousands of platterlike platelets. Manufactured at a rate of almost 200 billion a day, these flat, grainy blood cells are so small that a single drop of blood oozing from a pinprick brims with tens of thousands of them. Not only small but also short-lived, platelets will die after only 10 days in the bloodstream.

When a blood vessel is cut or ruptured, nearby platelets in the bloodstream stick to collagen fibers in the torn vessel wall. The flat cells swell and shoot out tiny spikes to plug the break; at the same time they send a chemical cry for help. Adenosine diphosphate (ADP), released by the first platelets to arrive, attracts others to the injury. Platelet plugs formed by this process mend millions of tiny rents each day but can only temporarily staunch blood loss from larger wounds.

If the rip in the *(Continued on page 113)*

More Than Two Billion Beats in a Lifetime

Held in awe by the ancients, the human heart sustains life. From beneath the breastbone this fist-size masterpiece (below) sounds its beat more than 2.5 billion times in a 75-year lifetime. It drives 5 quarts of blood a minute to every cell in the body, constantly cleansing and nourishing with its ebb and flow.

To nourish the heart itself, two encircling coronary arteries draw blood from the nearby aorta. The arteries sink their tributaries deep into the walls of muscle to deliver oxygen to each heart cell. Glowing red under ultraviolet light, the network of the coronary arteries reveals itself to be curiously thin; if even one thread clogs, the dependent section of heart muscle will die.

From inside the heart, oxygen-rich blood rises up through the aortic arch (right) and enters three major arteries that carry the life-giving liquid to the upper body.

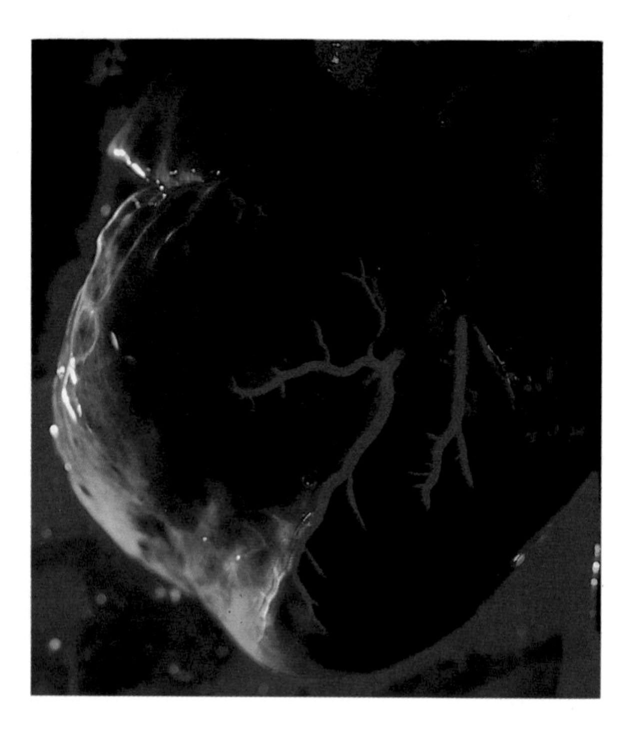

Magnification: 3 times

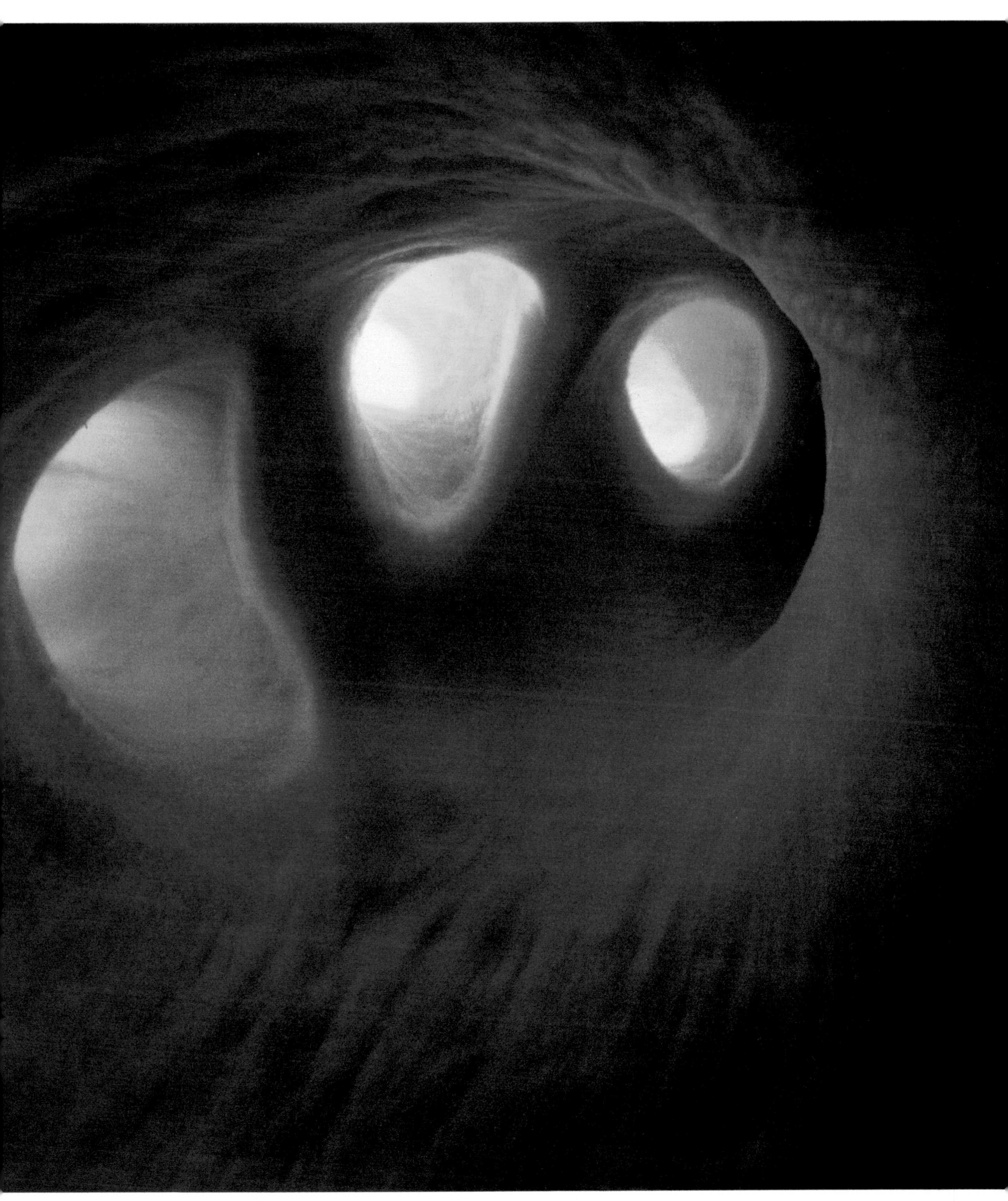

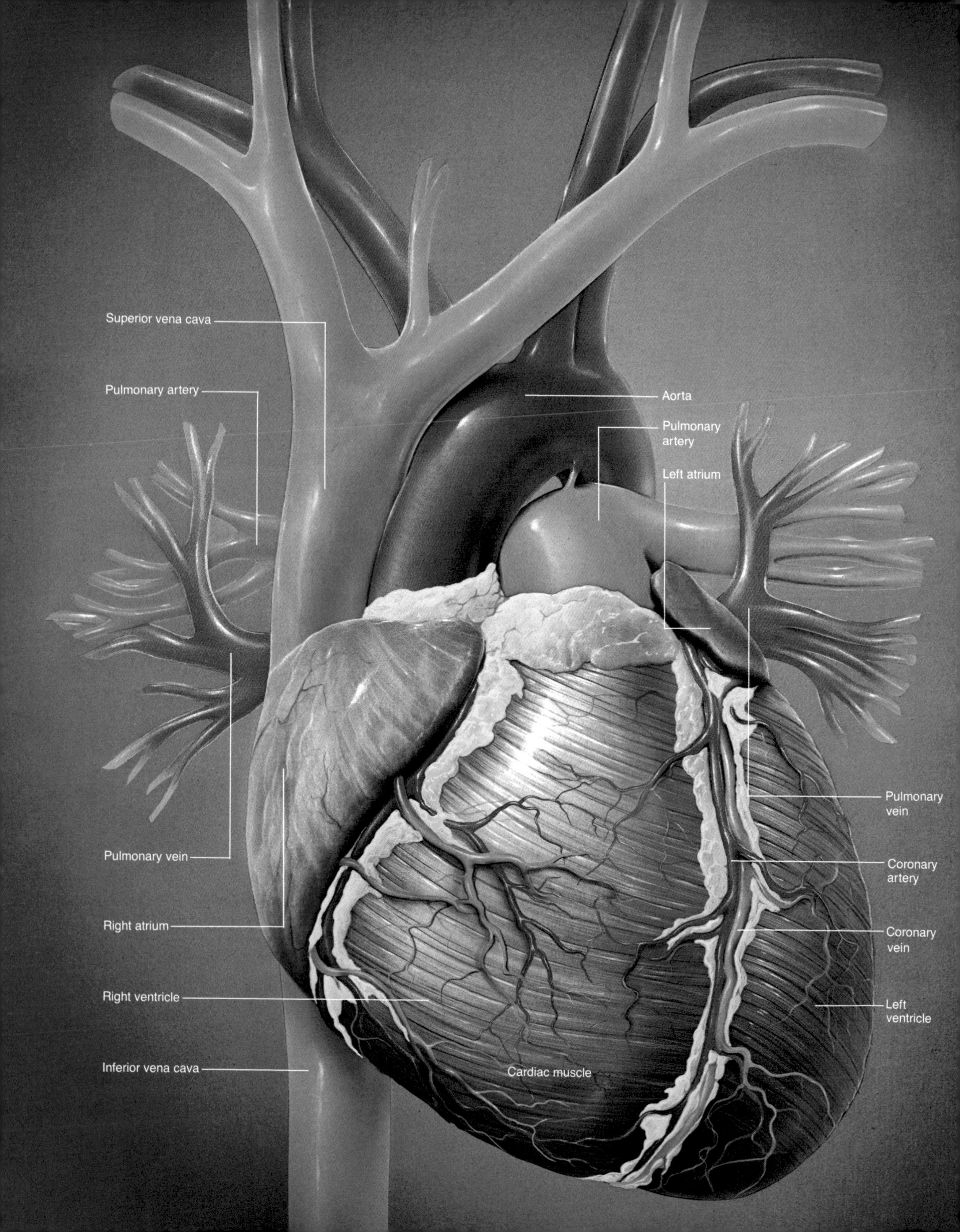

Superior vena cava
Pulmonary artery
Aorta
Pulmonary
artery
Left atrium
Pulmonary
vein
Pulmonary vein
Coronary
artery
Coronary
vein
Right atrium
Right ventricle
Left
ventricle
Inferior vena cava
Cardiac muscle

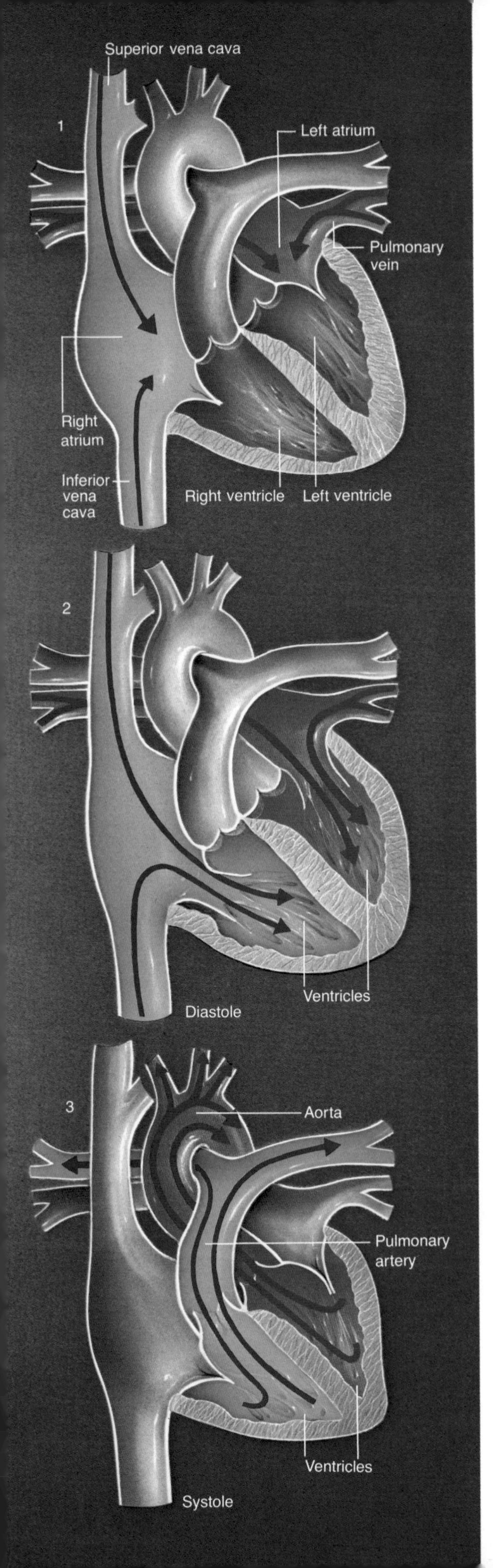

A map of the heart and a sequence of diagrams help locate the different parts and explain the pumping cycle. Four hollow chambers fortified with blood vessels make up this muscular organ (opposite). Two thin-walled anterooms, the left atrium and the right atrium, serve as holding tanks for blood entering the heart; thick-walled ventricles, one under each atrium, perform the serious labor of pumping blood. An impermeable wall divides the right and left sides of the heart.

The heart fills and empties in a second-long rhythmic cycle of contraction and relaxation, a heartbeat. Dark oxygen-spent blood (blue) from the body converges in two great veins, the superior and inferior venae cavae, en route to the right atrium, while bright red oxygenated blood moves toward the left atrium from the lungs via the pulmonary veins (1). As the heart relaxes, a phase called diastole (2), the atria fill with blood and force open two valves through which the blood drains into the left and right ventricles. In the pumping, or systole, phase (3), the ventricles contract, closing off the two valves. The right ventricle forces the oxygen-spent blood into the pulmonary artery; the left ventricle pumps oxygenated blood into the aorta and on into the body's tissues.

Inside each heart cell, tiny structures called mitochondria power a heart's contractions. When dye-stained and lit by fluorescence, the mitochondria's energy levels can be read. A bright flash (below, left) shows that mitochondria are spending their energy supply. Low fluorescence (below, right) indicates a resting state.

Magnification: 2,500 times (below, left and right)

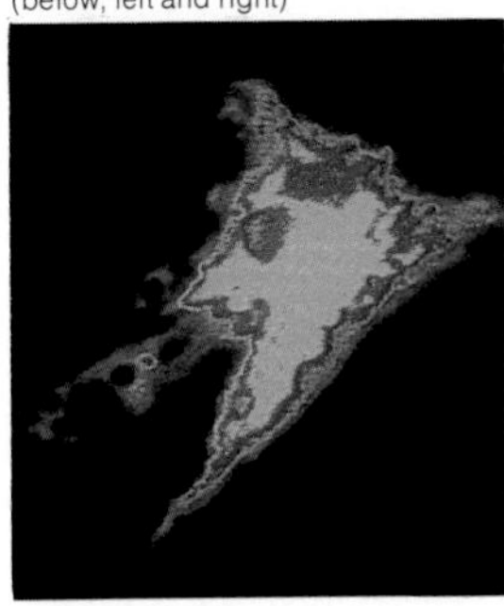

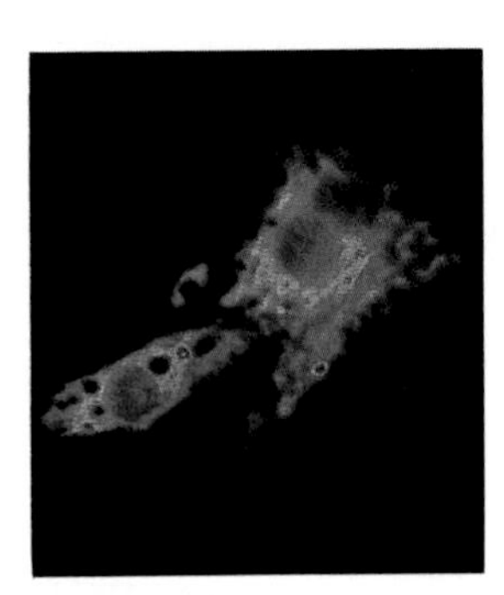

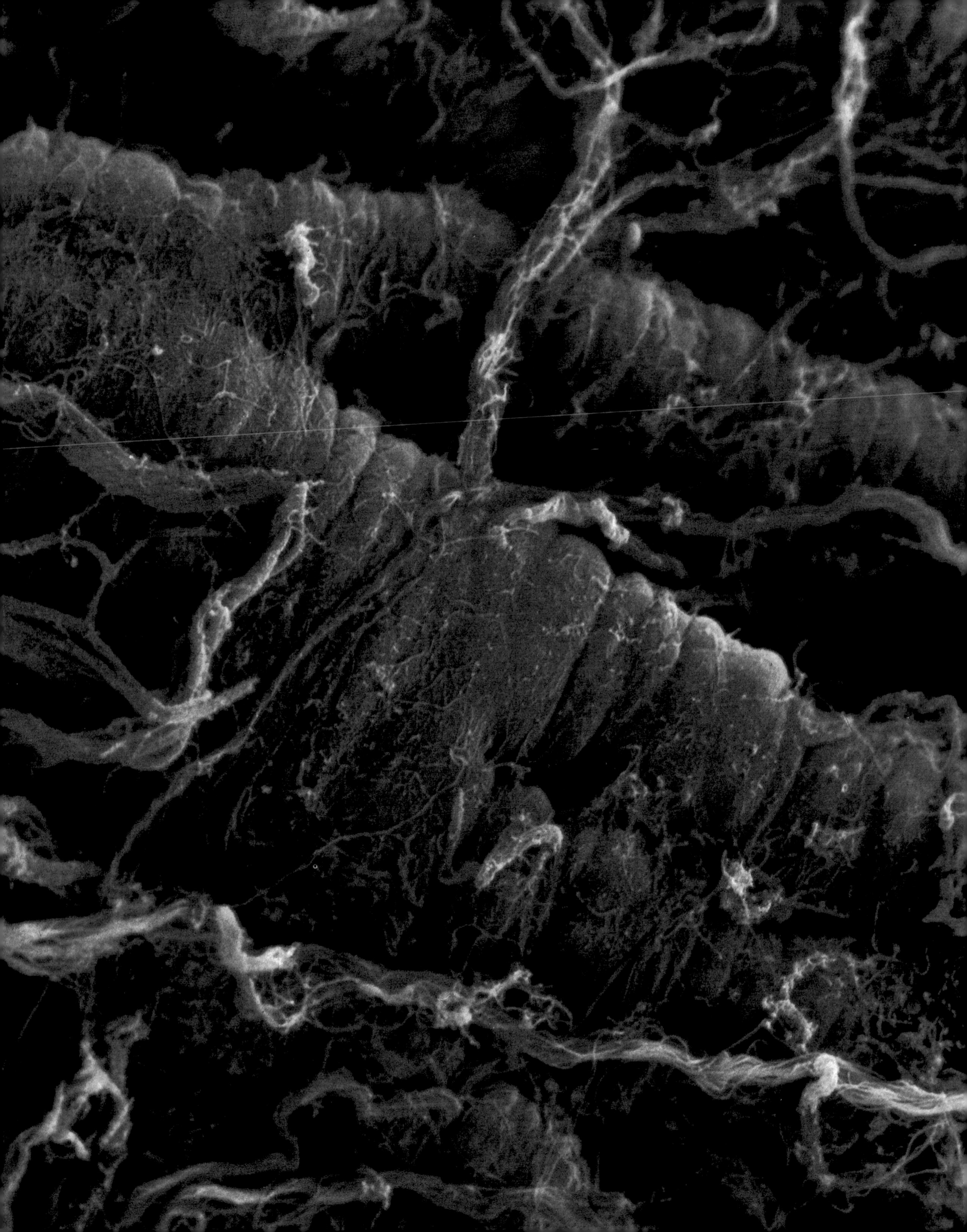

A natural pacemaker sends the spark that starts a heartbeat: High on the right atrial wall, a tiny bundle of nerve tissue called the sinus node (right, upper) ignites an impulse that races across the wall and down to the atrioventricular (AV) node, a cell cluster at the gateway to the ventricles. In its wake a contraction ripples the atria, sending blood to the heart's lower chambers.

The AV node, in turn, flashes the spark through conduction pathways into a nerve network that lines the ventricles (opposite). The spark leaps across the ventricle's muscle fibers at almost seven feet a second. The resulting contraction sends blood flowing from the heart.

A set of backup devices sustains the heart's electrical system in times of need. If the sinus node fails, the AV node begins the heartbeat; even special muscle cells can deliver an impulse, and will do so if the AV node does not.

But when the system short-circuits, the results can be dire: Electrical disruptions cause a large percentage of heart disease deaths. A mechanical pacemaker can sometimes take over the work of a weak or faulty natural one. The battery-and-timer model (right, lower) regulates the heartbeat by sending electrical impulses directly to the right ventricle.

opposite: Lennart Nilsson, © Boehringer Ingelheim International GmbH.

Magnification: 30,000 times (opposite)

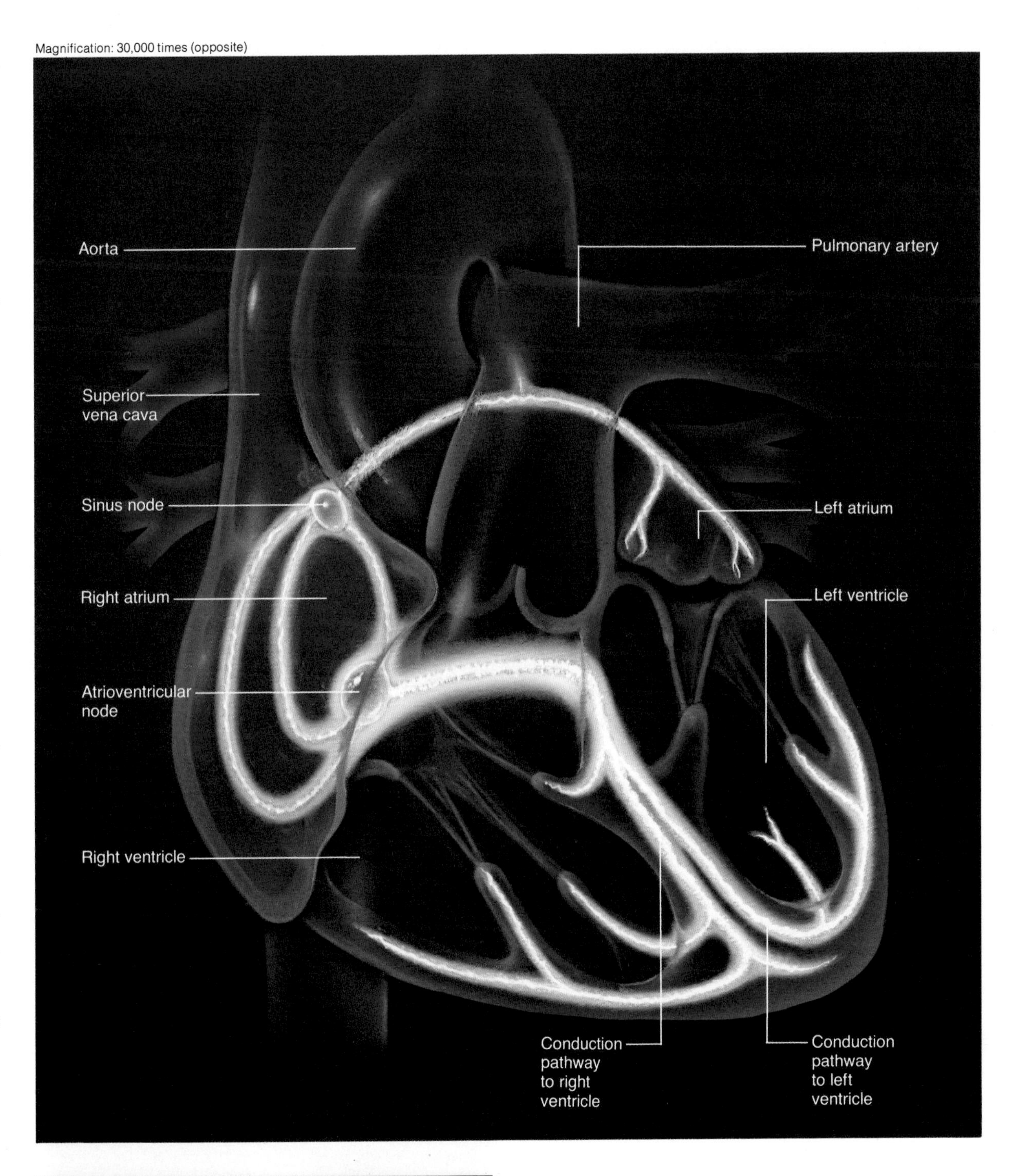

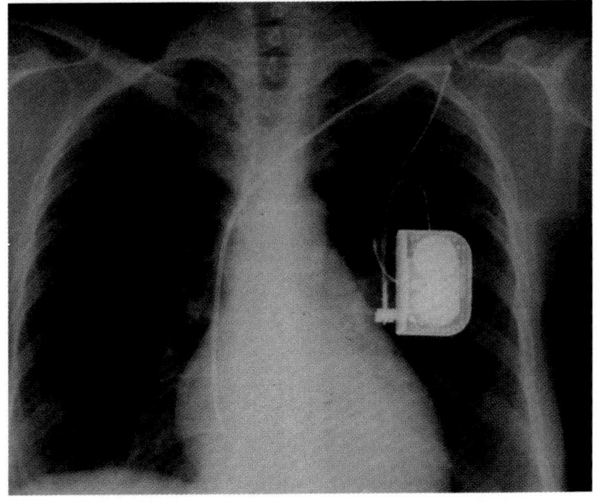

Magnification: 2 times (below); 4.5 times (right)

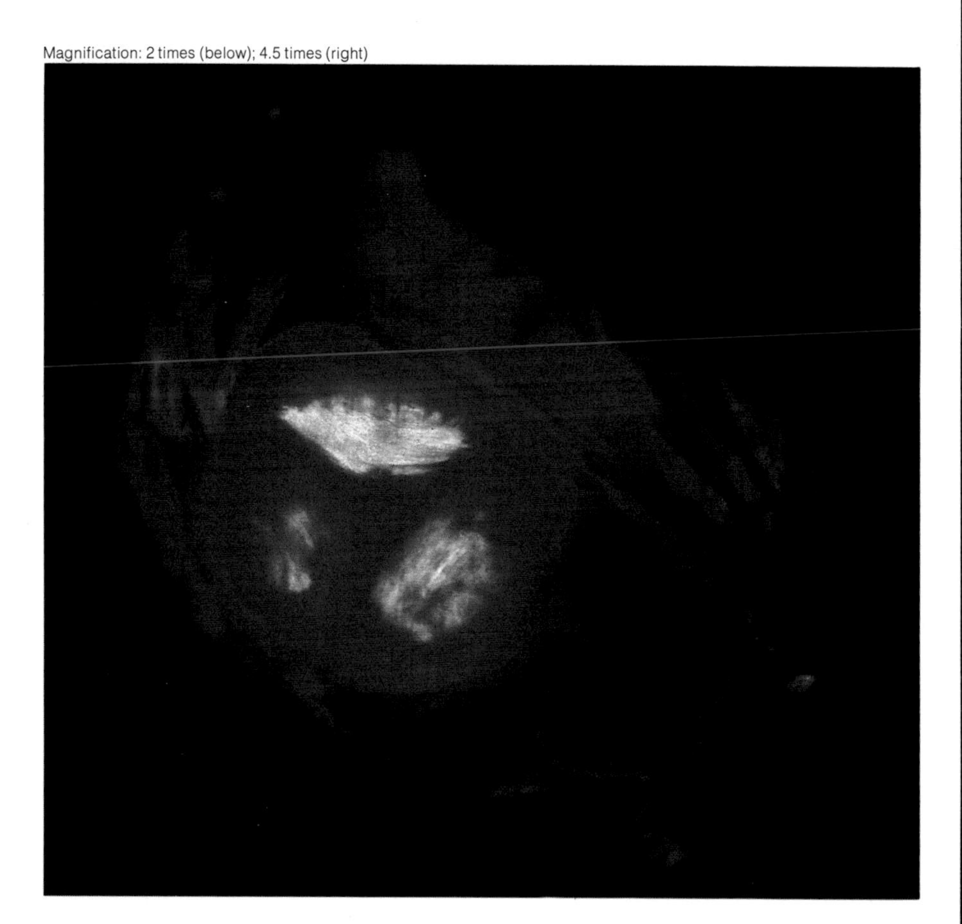

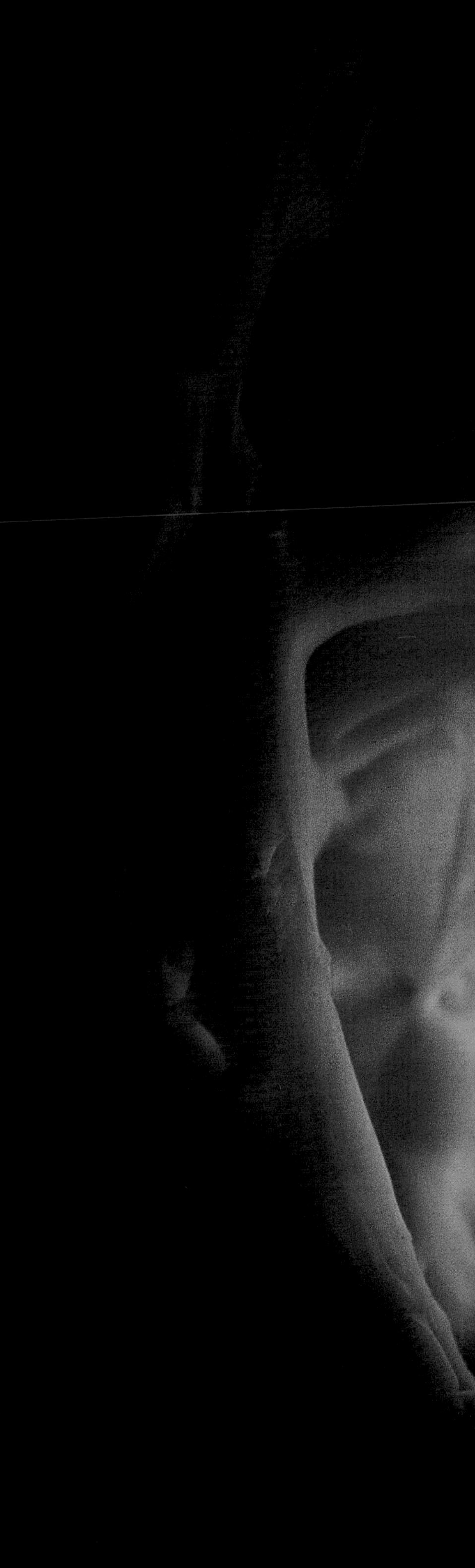

Valves as thin as tissue paper and sturdier than iron hinges open and close with each heartbeat, controlling the blood's passage through the heart. Anchored to the ceiling of the right ventricle, a tricuspid valve (right) opens wide during an atrial contraction. The valve allows blood to gush into the lower chamber. Tendons of the supporting muscle stalk reach upward to the valve flaps to prevent them from collapsing into the atria when they snap shut. During a contraction of the left ventricle, the blood presses the leaflets of the aortic valve (above) against the entrance walls of the aorta, then rushes forward into the artery.

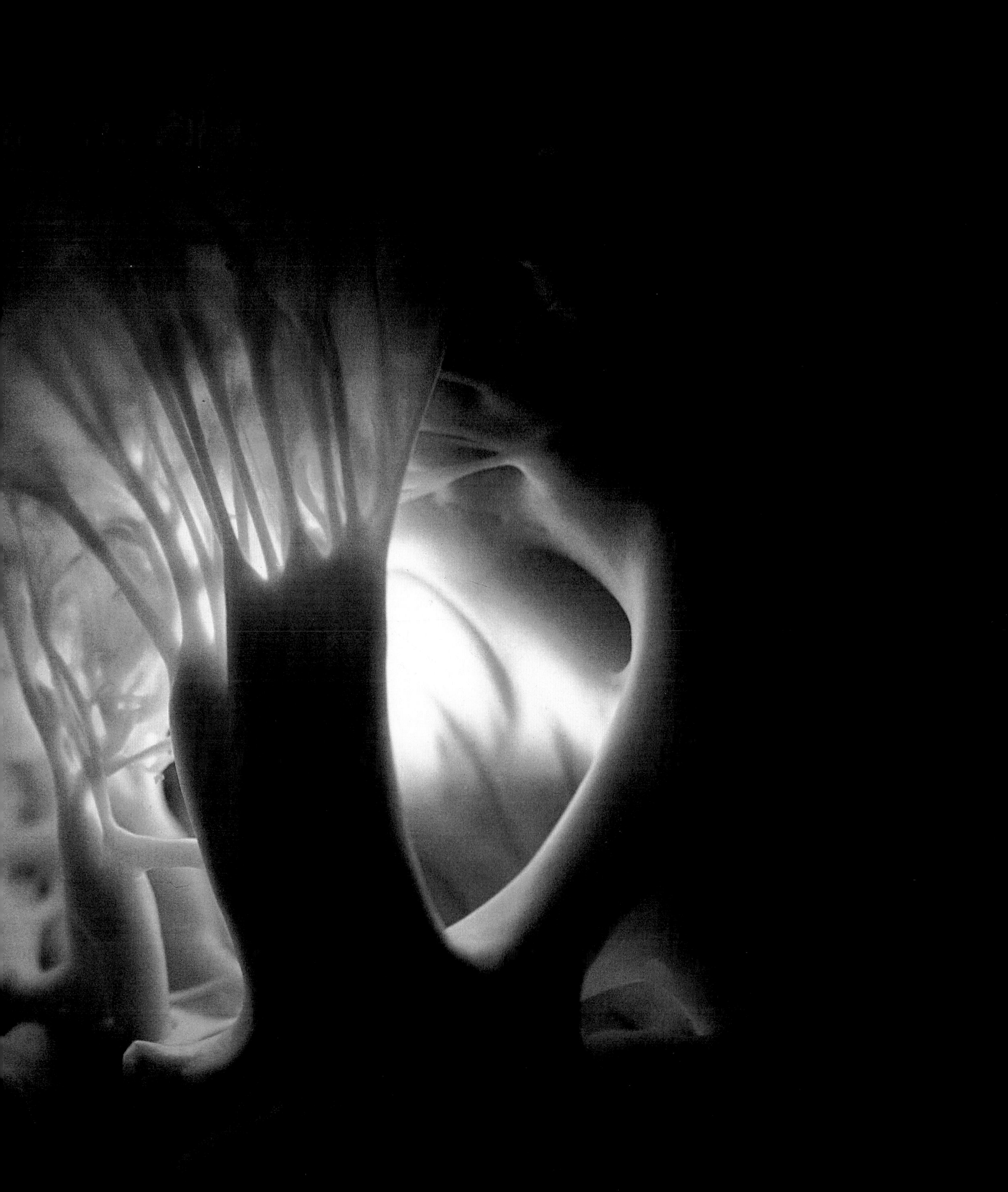

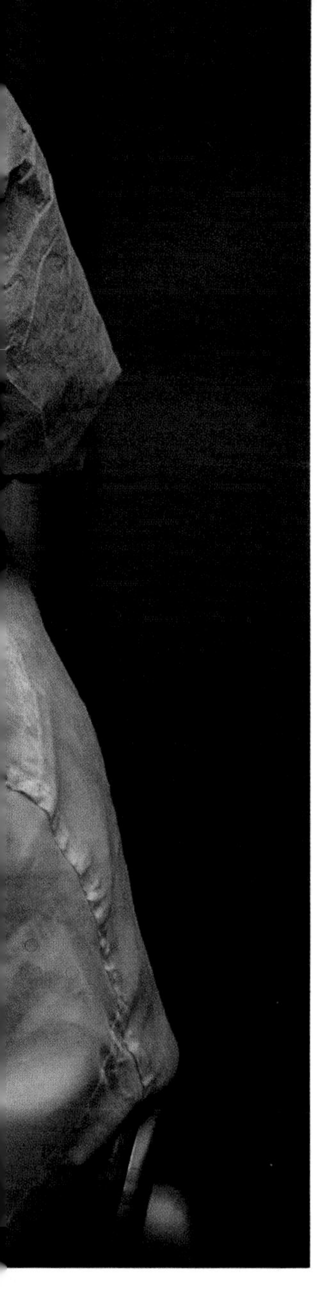

Artificial heart valves can often replace defective ones. These, assembled in California, come in an array of styles and sizes. To make the Bjork-Shiley mechanical valve (left), technicians fit a tilting disk into a precision-polished ring. A cloth cover enables surgeons to suture the valve to the heart. Carefully cut tissue from the outer membrane of a calf's heart is hand sewn to make an Ionescu-Shiley tissue valve (far left). A surgeon will decide on a replacement valve after examining the patient's damaged one. Mechanical valves, tested since the 1950s, have proved most durable; tissue valves seem less likely to trigger clots.

vessel wall is too large for platelets to seal, the damaged tissue sends for reinforcements. Secreting a substance called thromboplastin, the injured wall triggers clotting. Cascading through a complex series of reactions, thromboplastin ultimately causes fibrinogen, a liquid protein normally suspended in plasma, to congeal into long, tough threads called fibrin. Adhering to the torn tissue, fibrin threads trap the spiky platelets in a tangled web, which in turn snares more platelets, eventually damming the breach.

Defenders against disease and infection, the body's white blood cells number only about one for every seven hundred red cells and constitute less than one percent of the blood's volume. Largest in size of all blood cells, white cells also lead the most active lives. Changing form like amoebas, they slither through the bloodstream, stalking and devouring bacteria, viruses, and other microscopic invaders.

Plasma, the third and thickest layer in our test tube, provides the stream through which blood's contents glide. Ninety percent water, this liquid dissolves and distributes the food we eat, carrying salts, minerals, sugars, fats, and proteins to every cell. Plasma also contains its own proteins, including immunoglobulins, or antibodies, used by white cells in the fight against disease, and fibrinogen, the protein essential for clotting.

The Life-Giving Loop

Every 60 seconds, 1,440 times a day, our blood cycles through the body, traveling the double loop—from heart through lungs and from heart through body—known as the cardiovascular system. Fresh, oxygenated blood begins its voyage to the body's tissues by bursting from the left side of the heart into the arching aorta, the body's largest artery. Even the average resting heartbeat hurls about 2 ounces of blood against the aorta walls with great force. These tidal waves of blood smash against the aorta 70 times a minute, delivering their blows 2.5 billion times during the average life span.

Rigid metal pipes could not withstand this battering for long, but the living tissues of our blood vessels have evolved with just this function. Artery walls have three layers: a smooth inner lining, a thick middle layer of elastic membranes and muscle, and an outer layer of fibrous connective tissue. The aorta's elastic membranes stretch with the impact of each surge of blood; its strong muscle fibers then recoil, channeling the intermittent waves into one continuous stream. The resilient muscles also act as a subsidiary pump, propelling blood through our larger arteries at a rate of one foot a second.

As arteries divide and subdivide, the elastic membranes in their walls diminish, and the proportion of muscle grows. A single muscle cell

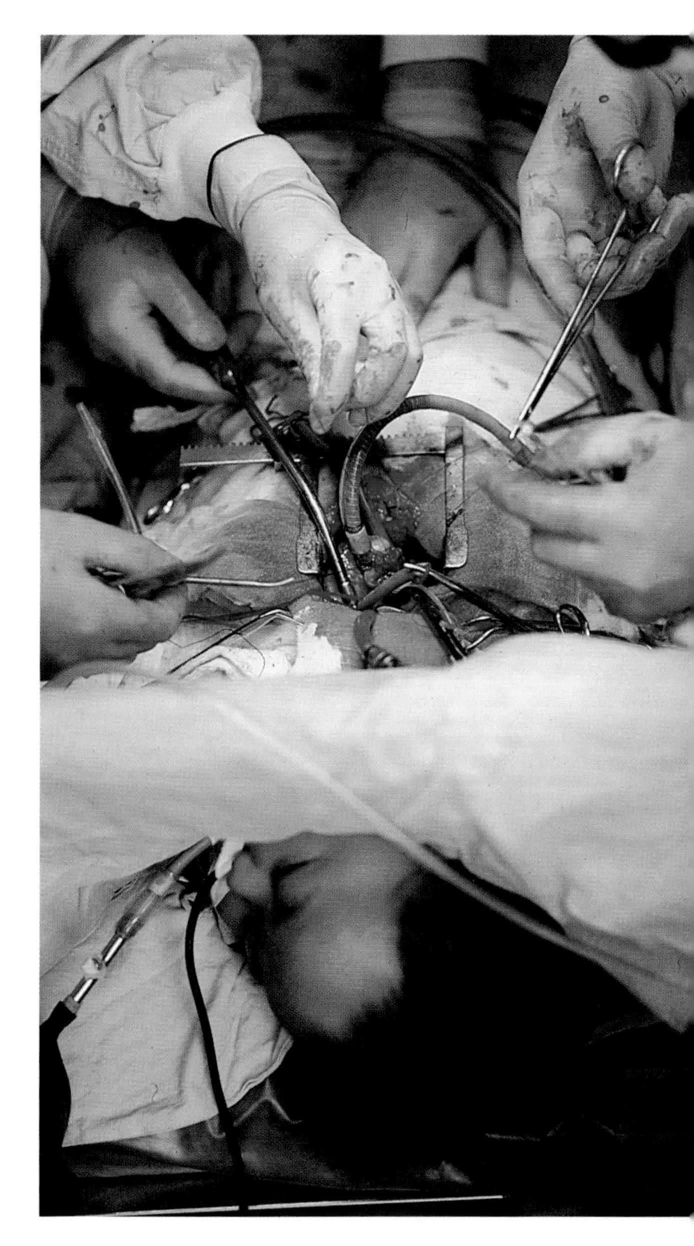

A new heart valve promises prolonged life to a year-old child in China (above). A narrowing, or stenosis, of the pulmonary valve slowed blood flow into the pulmonary artery, giving the child an abnormal heart sound known as a murmur.

The surgeon slices away the faulty flaps and sews the synthetic valve onto the patient's own valve rim. Multiple stitches, tied down for a secure hold, remain in place while the body's healing seals the seam between the natural tissue and artificial valve.

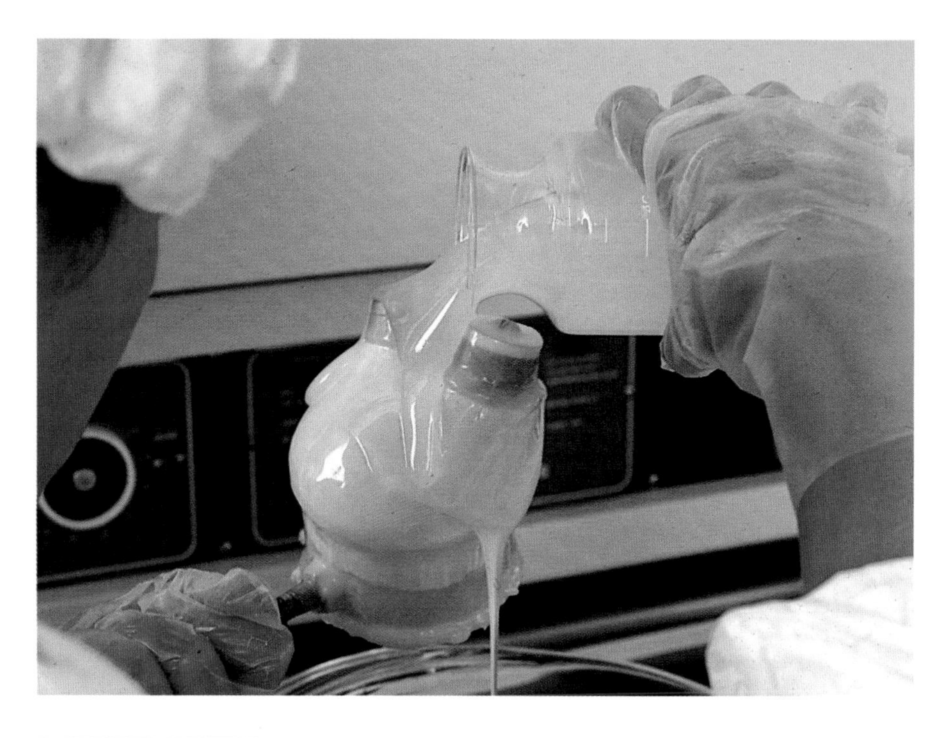

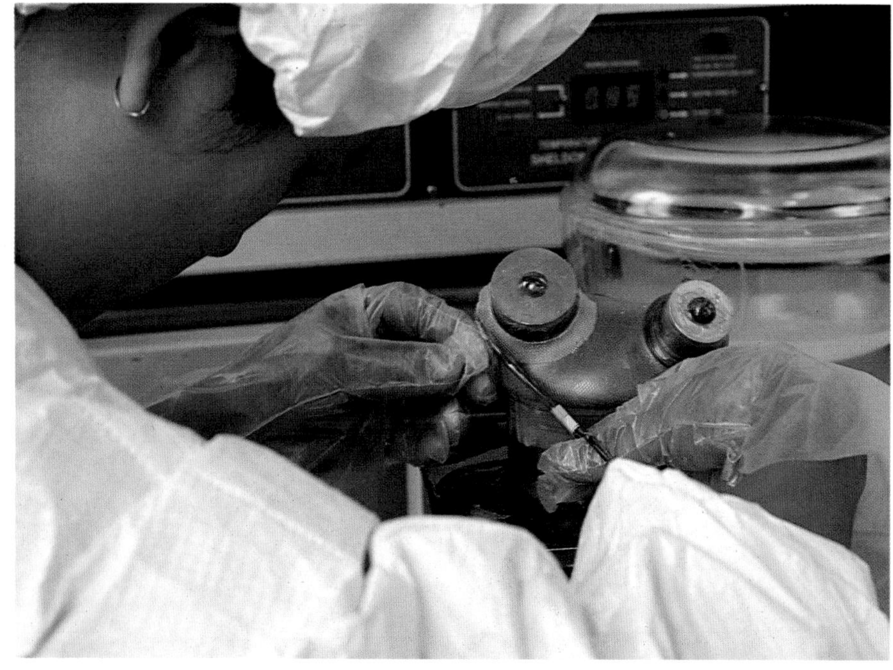

An artificial heart takes shape in a Utah laboratory. The Jarvik-7, named for its designer, contains two ventricles, each with two valves, a diaphragm assembly, and an air line.

To make the ventricles, a gloved technician (upper) pours twenty-four layers of a polyurethane compound over a stainless steel mold. Dacron mesh, added midway through the layers, lends rigidity to the heart and increases its pumping efficiency. To strengthen a point where blood flows in, the technician attaches a Dacron reinforcement ring (above) and trims it. When the polyurethane has set, inspectors look for leaks and measure the thickness of the heart's walls.

To arrive at the best speeds for operating future models, researchers connect the Jarvik-7 to a mock circulation device and monitor its performance (opposite).

may wrap two or three times around each of the smallest arterioles, the last short branches of the arterial tree. Rhythmically squeezing and relaxing, these muscled rings force blood into the ten billion capillaries that fan throughout the body.

Most tissues—brain, intestine, heart, blood vessel—are laced with so dense a network of capillaries that no cell lies more than a millionth of an inch from a blood supply. Capillaries, with gossamer walls only one cell thick, are so fine that even tiny red cells must bend and twist to squeeze, one at a time, through their narrow straits. Yet these vessels perform the cardiovascular system's vital task: They replace waste and carbon dioxide with oxygen and nutrients, delivering life to the cell.

Capillaries deliver their cargo by maintaining a precise balance of pressures between the blood flowing within their walls and the fluid in and around the body's cells. Arterioles pump blood into the capillaries with force sufficient to drive plasma and its dissolved nutrients through porous capillary walls. Cells, busily combusting nutrients to create energy, have a lower concentration of oxygen than the blood in capillaries. Thus tissue fluid hungrily sucks oxygen out of the red cells' iron grip and through the capillary membrane. Inside the capillary, blood cells and plasma proteins too large to pass through the vessel walls travel on in the thickened stream.

A Delicate Balance

Now the process of osmosis takes over. Water laden with carbon dioxide diffuses from the tissues into the capillary's dehydrated plasma. In one minute the water in blood plasma changes place with tissue liquids forty-five times. Still, our blood volume remains constant, thanks to the equilibrium between blood pressure, which forces fluid out of capillaries, and osmotic pressure, which sucks it in. If we lost this balance—if we lost our plasma proteins with their osmotic attraction—the liquid in our bloodstream would quickly and irretrievably flow into body tissues.

By the time blood leaves the capillaries, the swift current propelled by the heart and muscled arteries has slowed to less than a fraction of an inch a second. Bluish blood, laden with carbon dioxide and other wastes, flows sluggishly into the body's veins for the return trip to the heart and lungs. Equipped with fewer muscles than the arteries, veins cannot pump blood as vigorously as arteries do. This is especially true of veins that carry blood uphill from the feet. The thin-walled, elastic veins oppose gravity by relying on the body muscles that surround them. Whenever we move, muscles of the leg, arm, and even stomach press against the veins, sending blood on its way. In addition, one-way valves keep the blood *(Continued on page 119)*

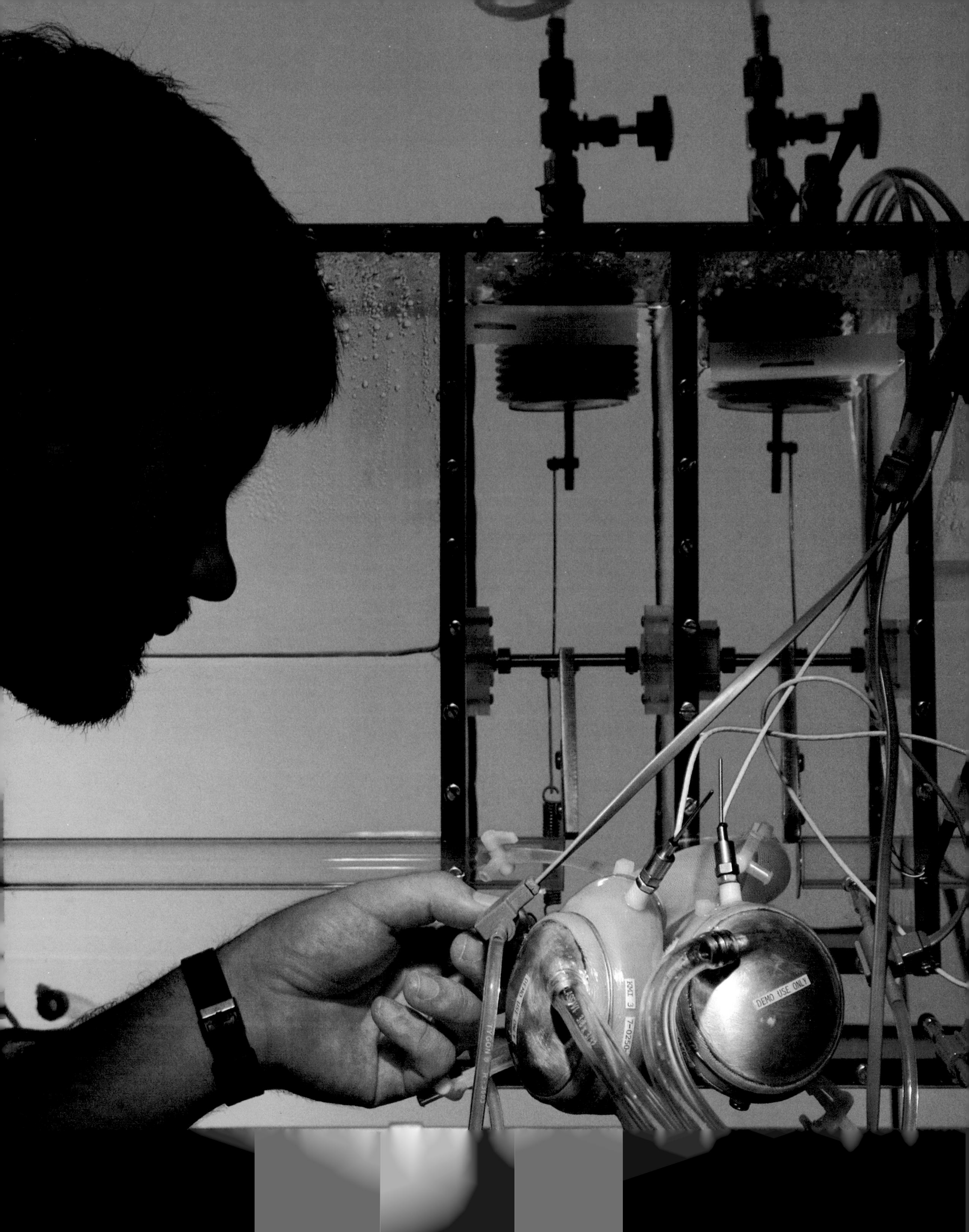
TYGON
KMI 3
DEMO USE ONLY

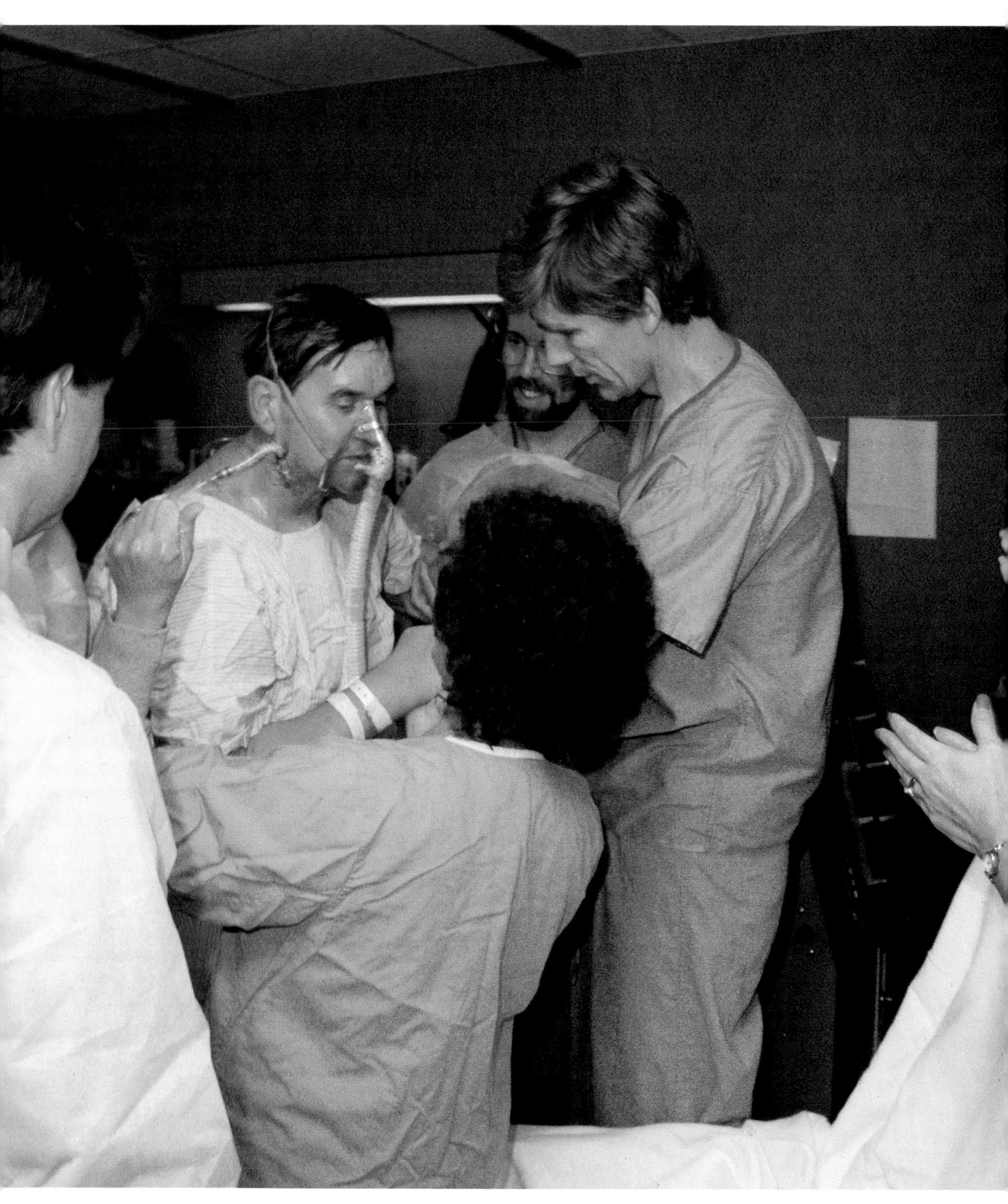

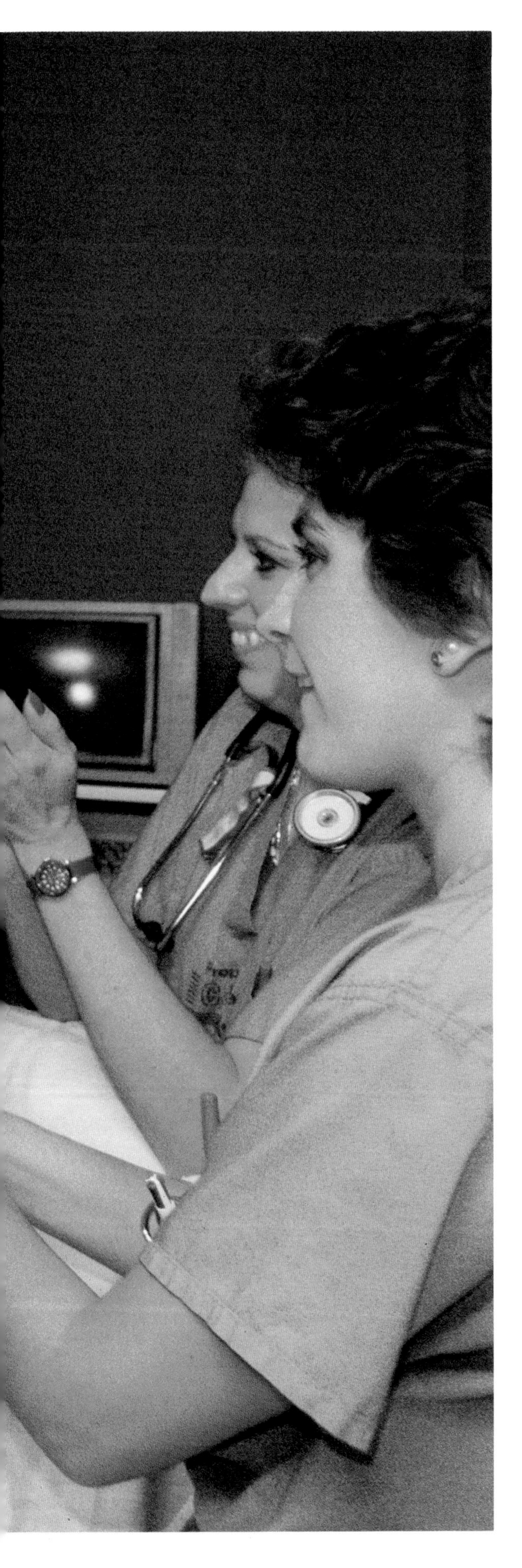

Tethered to his motorized console and bolstered by extra oxygen, William Schroeder (left) takes the first steps following his artificial heart operation. Dr. William DeVries, who put in the heart, assists. Only a week earlier, Schroeder's new heart sat on an operating-room table (right, upper). During the operation, which took more than 6 hours, DeVries removed the patient's diseased ventricles and stitched connective cuffs to the atria, the pulmonary artery, and the aorta. Then he snapped the Jarvik-7 onto the cuffs and sewed it securely in place (right, lower). Finally he connected the heart to a 400-pound drive system, capable of pumping 7 to 12 quarts of blood per minute. The console also has an alarm to warn of sudden drops in blood pressure or heart rate.

In the months following Schroeder's operation, delight with the procedure gave way to controversy and despair as a series of strokes and other problems stalled his recovery. Critics dislike the diversion of resources from more important medical research, such as heart disease prevention, and think that the experimental procedure raises false hopes for patients and family.

Proponents believe that the mechanical heart, though still unrefined, may hold the only hope for critical cardiac patients, especially in the absence of an extensive donor network for human heart transplants.

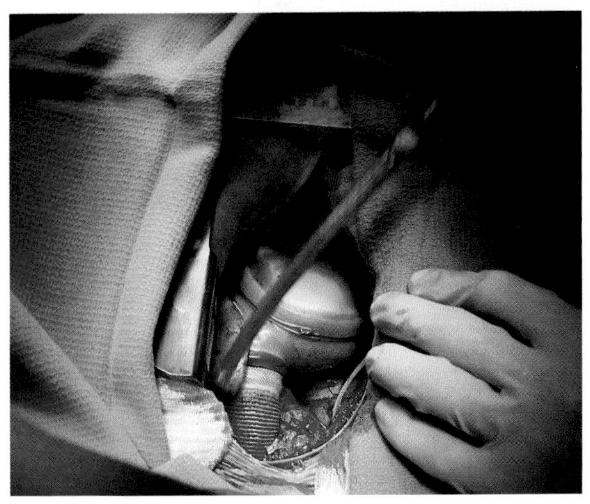

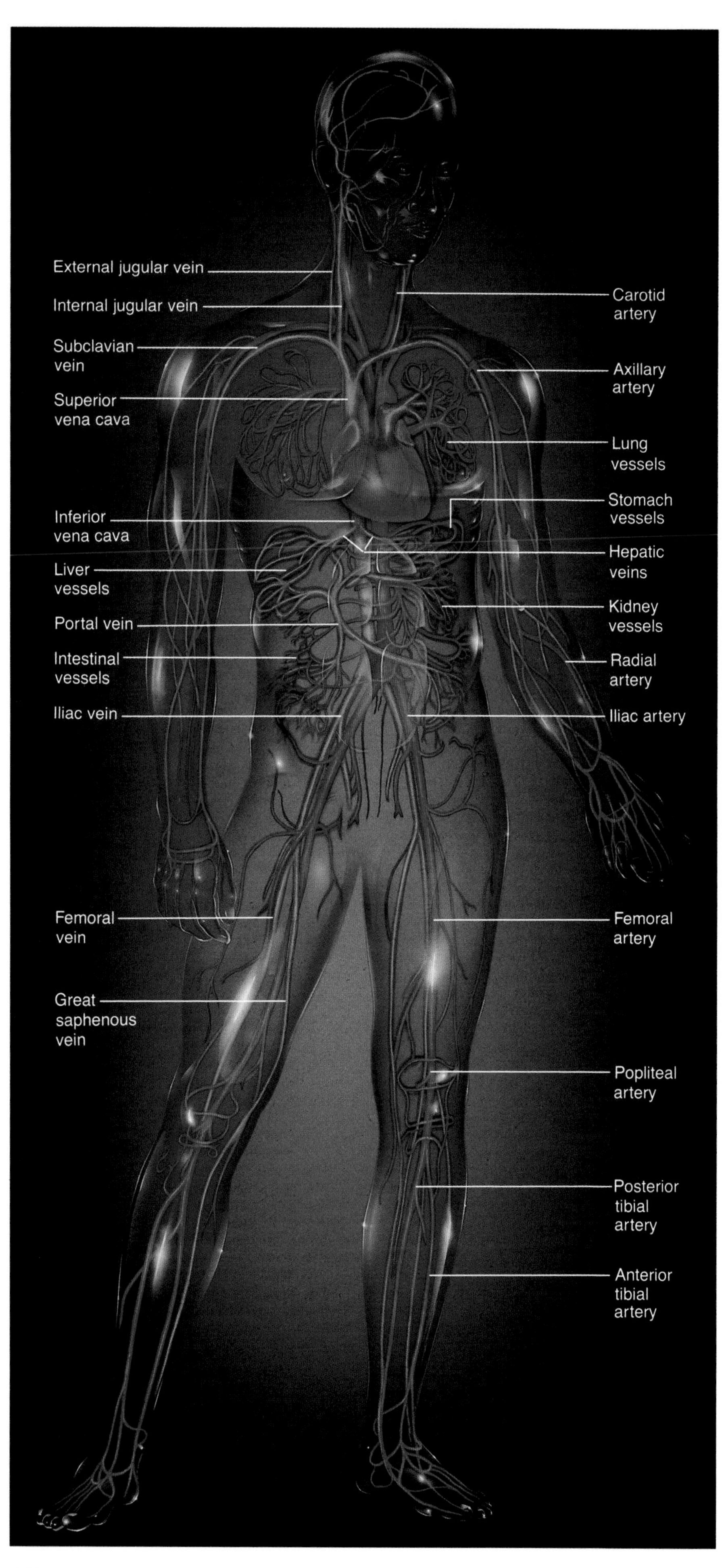

The circular route of our life-giving river begins and ends at the heart (left). Oxygen-rich blood pulses into the yawning aorta, quickly branches into narrower arteries and arterioles, then slows down as it reaches the dense capillary network.

Most capillaries (right), finer than human hairs, force red blood cells to pass through single file. The capillaries do the work of the circulatory system, exchanging nutrients for wastes through their thin walls. Red blood cells, after squeezing through the capillary network, enter venules, which join to veins. These, larger and more thinly walled than arteries, carry red cells back to the heart. One-way valves keep the blood moving along. At any given moment, blood in the venous network makes up almost 75 percent of the blood in the 60,000-mile system of the human body.

In sterile caps and masks, Utah scientists work on a polyurethane blood vessel (far right) that soon will be placed in a human body.

Magnification: 3,000 times

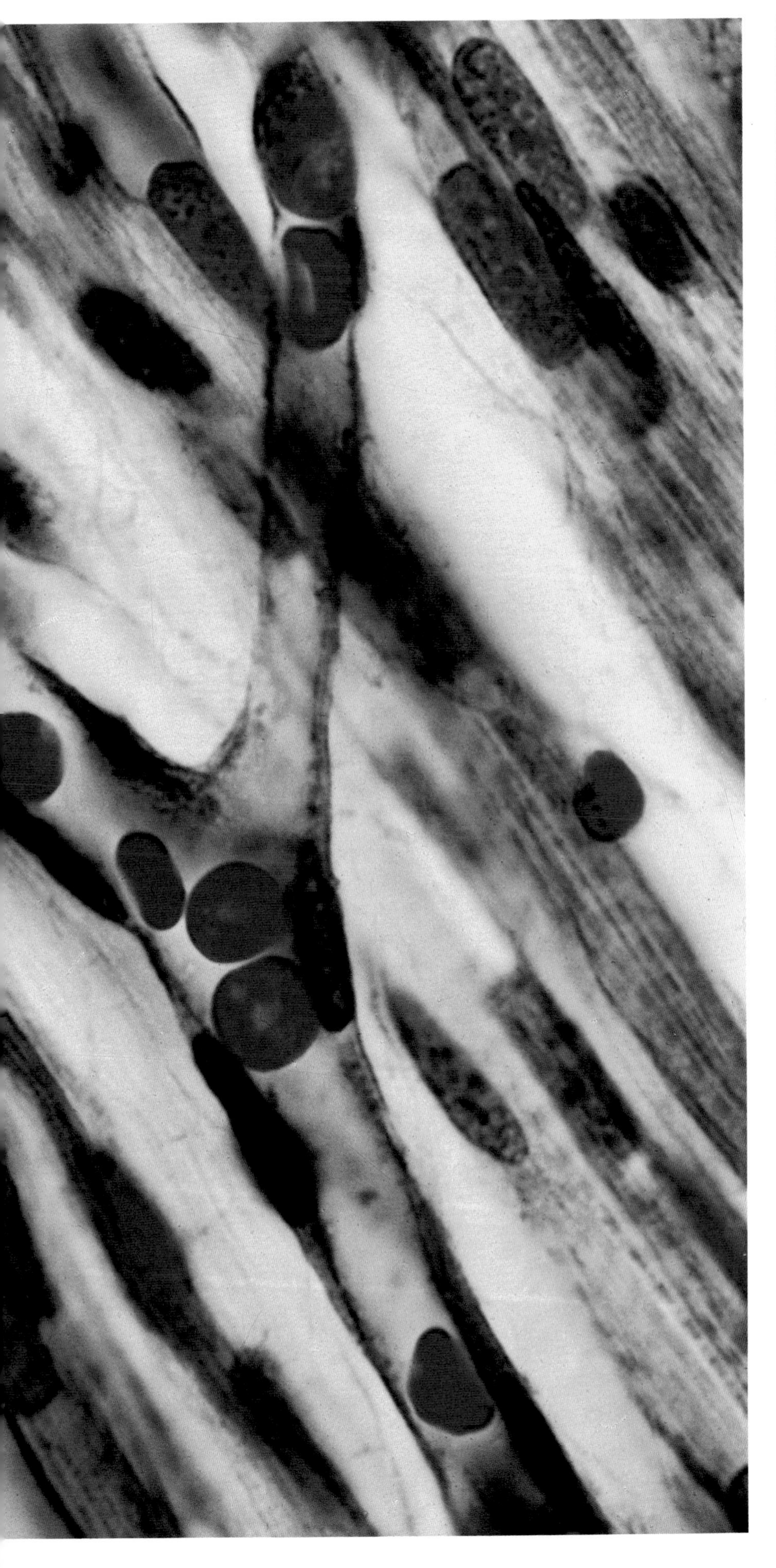

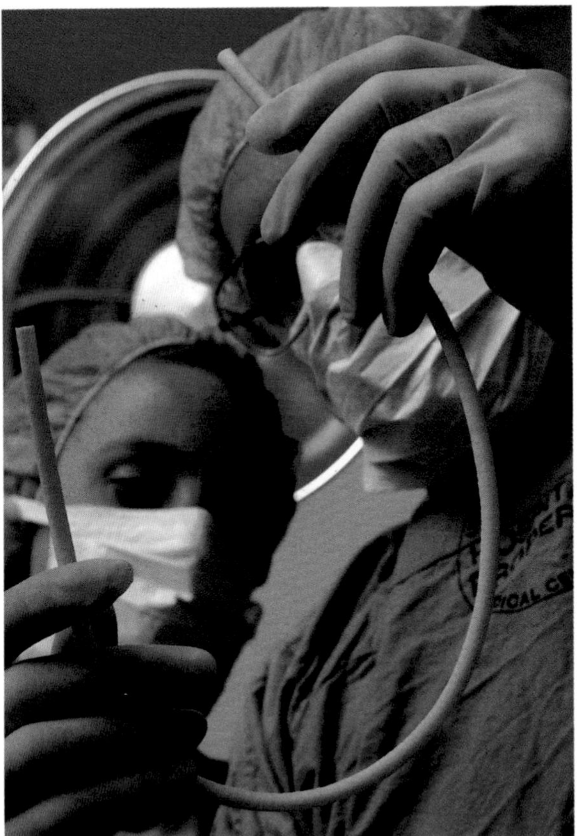

from flowing backward. If the valves in a leg vein become damaged and a reverse flow occurs, the blood will pool, resulting in a varicose vein.

Our veins carry blood to the heart like rivers carry water to the sea, smaller tributaries merging into wider currents. Blood finally flows into the body's major veins, the venae cavae. One leads from the brain and chest, the other from the lower torso. These two great veins empty used blood into the right side of the heart, which immediately pumps the nearly toxic fluid into the lungs.

The capillaries in our lungs serve one essential purpose: They bring blood into such close proximity to fresh air that the carbon dioxide bound to hemoglobin will exchange places with the oxygen from the atmosphere. This apparently simple job becomes a herculean task given one physiological fact: Seventy times a minute, the heart pumps a new supply of blood into the lungs. Its capillaries must complete the gaseous exchange in less than a second.

The short distance air travels in the lungs makes possible this exchange. Air enters the lungs through two large tubes called bronchi, which subdivide many times until they become bronchioles, narrow passageways that end in clusters of air-filled sacs, the alveoli. Some 300 million alveoli, laced with capillaries, line the lungs. Spread flat, the surface area of these tiny sacs would total 750 square feet, about the size of a racquetball court. Through walls only 10 millionths of an inch *(Continued on page 123)*

A sea of red blood cells (right) carries oxygen to body tissues. Each disk-shaped cell consists of a semipermeable membrane that holds water and nearly 300 million hemoglobin molecules. A computer-simulated image of a hemoglobin molecule in cross-section (below) reveals its complexity. A backbone (tinted white) supports the structure. A pocket formed by the globin protein (tinted blue and yellow) holds an oxygen-carrying heme (tinted red). The oxygen must adhere to the heme's iron atom as it is carried through the roiling bloodstream—but it must let go when it reaches its destination in the body tissue.

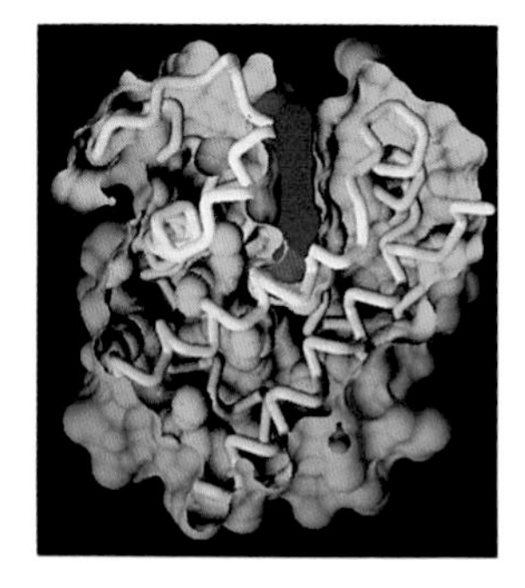

Magnification: 6,000 times

Magnification: 27,000 times

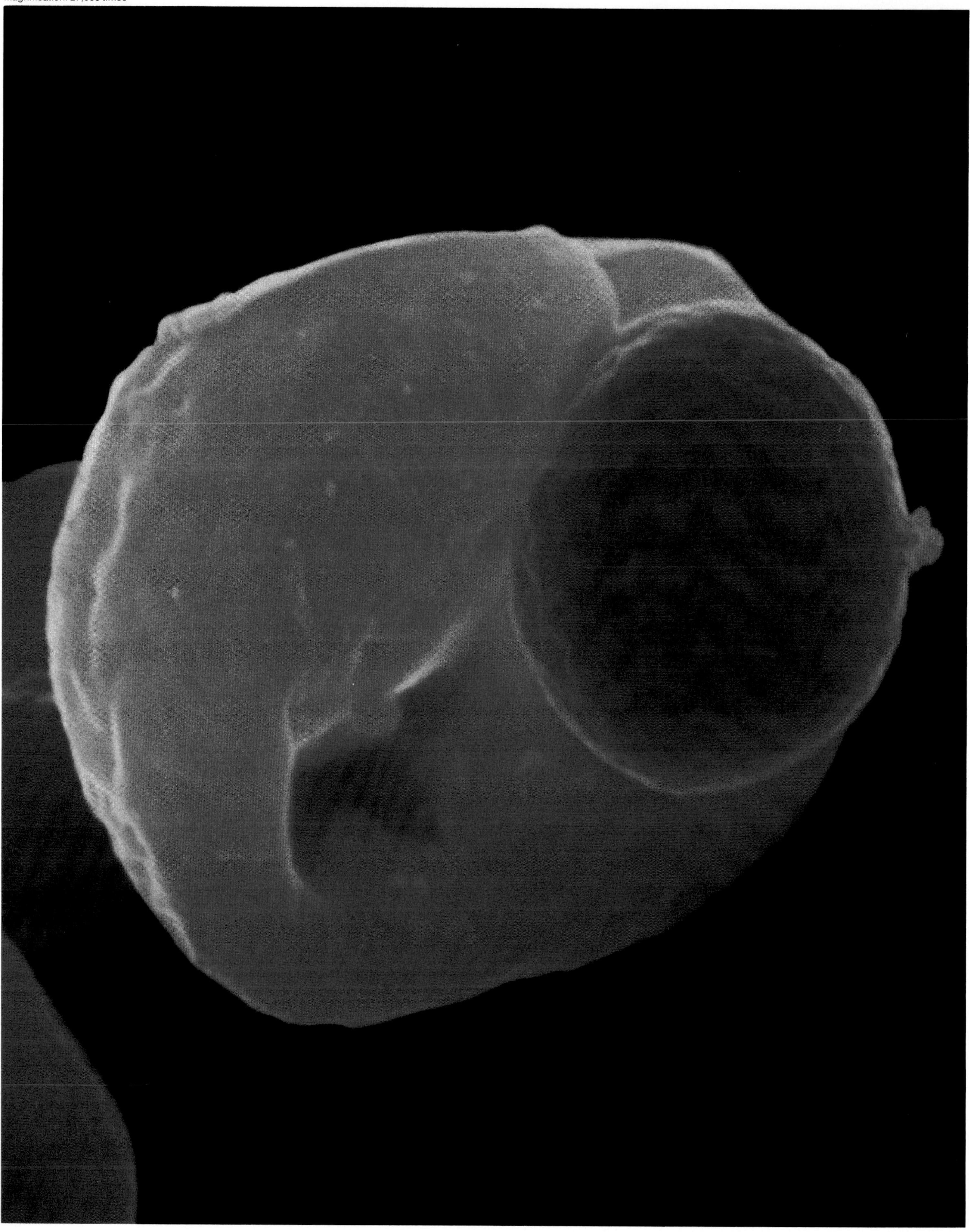

thick, capillaries and alveoli perform their vital trade. Oxygen, under higher pressure in the lungs, slips readily through the filmy membrane and into hemoglobin's tight embrace. Carbon dioxide in turn escapes into the alveoli and is expelled when we breathe. Renewed, the scarlet blood flows into the left side of the heart, ready for its life-giving journey to the cells.

The heart—simple pump and powerful machine—rules over the body's circulation with steadfastness and strength. It begins to beat just 4 weeks after conception. With every beat, the adult heart expels 2 ounces of oxygen-rich blood, 5 quarts a minute, 220 million quarts over 70 years of life. Leg muscles soon tire when we run at top speed, but heart muscle works twice as hard when we relax, even harder when we exercise, never pausing for rest or repair. When surgeons replace heart valves with silicone parts, the hard, man-made materials become battered out of shape after only a few years. Yet the delicate, durable tissues of a healthy heart exert their force, and withstand the beating, for a lifetime.

A Precision Pump

The living dynamo that pumps with such power also pumps with precision. The heart must drive blood through our bodies with enough force to send it surging to the farthest capillary; yet it must pump blood gently to the lungs. If the heart sent blood through lung capillaries and into the air sacs with the same force that it pumps blood through other parts, we would drown in our own plasma. So our one heart, divided by a wall down the middle, has two sides. Each side contains two chambers: an atrium, or receiving tank, at the top, and a ventricle, the pump itself, at the bottom. The left ventricle, which sends blood through the body, has four times the muscle of its counterpart on the right. This is why we feel our heartbeat on the left side even though a third of the heart lies in the right side of the chest.

Though the left and right ventricles flex with different forces, they nonetheless beat at the same time, ensuring a blood flow that is smooth and continuous. Nature ensures this synchrony by wrapping both pumps in one muscle. Spiraled around the organ, this muscle wrings blood out of the heart, simultaneously emptying both pumps.

The contraction of the heart is one of its independent powers. It begins to beat in the embryo, before any nerves connect it to the brain. In transplant surgery it even continues to throb after all nerves have been severed and the diseased organ removed from the chest. Even a single heart cell alone on a microscope slide pulsates as long as it has a fresh supply of blood.

This relentless pulse proves that the heart's beat originates from some power in its tissues.

Magnification: 10,000 times

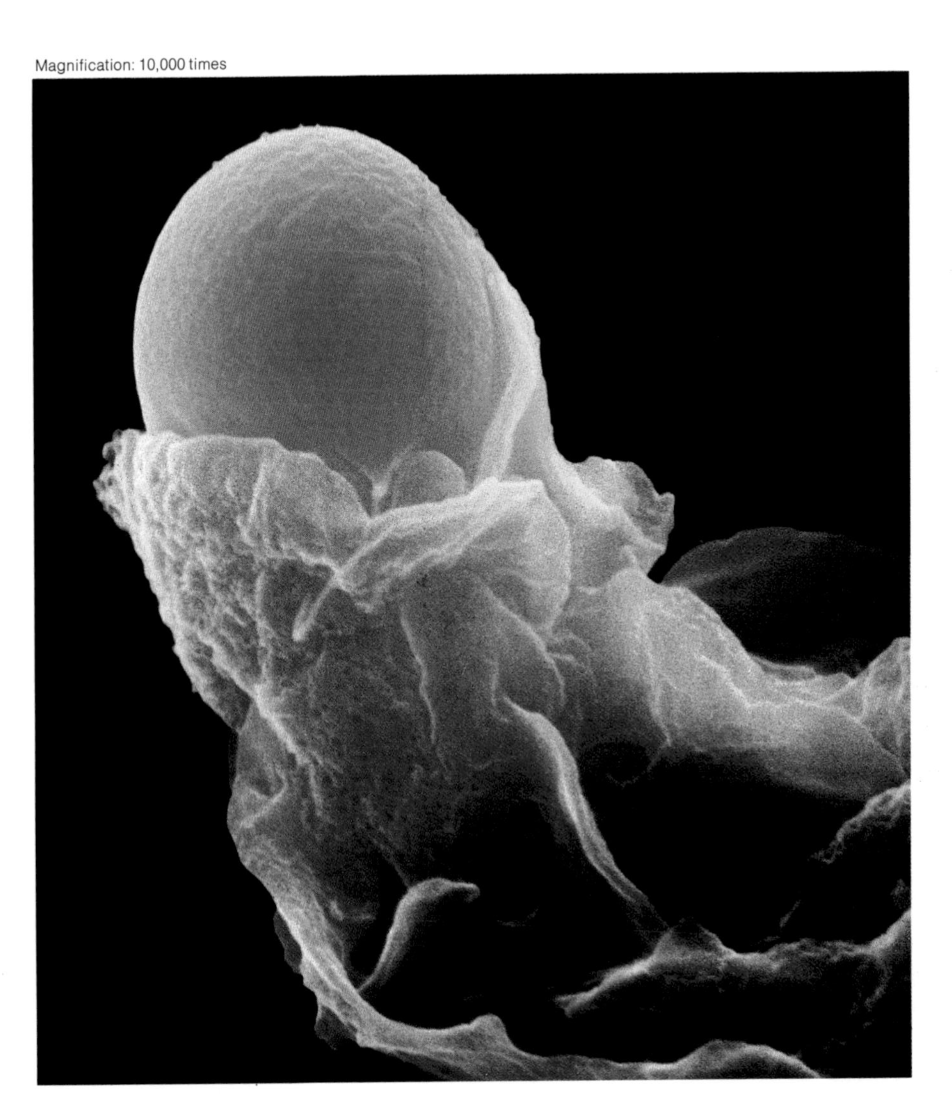

The brief, vigorous life of a red blood cell begins deep in the bone marrow. A sheltered and still primitive marrow cell begins to divide and fill with hemoglobin when signaled by a special hormone that acts as the body's oxygen level drops. In 6 days the cell convulses and expels its spherical nucleus (opposite). Now the fledgling red blood cell slips into the bloodstream and twists into its familiar disk shape.

When it is about 120 days old, the red cell's life draws to a close. One theory holds that the cell, without a nucleus to renew old parts, depletes its inner resources. Its outer membrane may also wear out, like tread on an old tire, from thousands of trips through the bloodstream. A macrophage, the white blood cell responsible for cleanup, senses that the time has come. Catching the aging cell in its embrace, the macrophage engulfs and digests it (above).

Every cardiac cell is, in fact, a living battery, crackling with chemically created energy that stimulates the movement we call a heartbeat. The heart cell generates electricity through two elements plentifully supplied in blood: sodium and potassium. The atoms that make up both elements frequently lose a negatively charged electron, leaving them with an extra proton, or a positive charge. These "charged" atoms are called ions.

Heart cells contain a high concentration of potassium ions, while the liquid surrounding the cells abounds in sodium. The cell membrane constantly pumps sodium out of the cardiac muscles and potassium into them. Because the membrane pumps sodium out faster than it pumps potassium in, an excess positive charge builds outside the cell. When it reaches a certain threshold, the flow suddenly reverses and sodium ions rush back into the cells. This sudden shift sparks an electric charge, and the heart cell flinches in contraction.

When scattered sparsely across a microscope slide, individual cardiac cells beat at different rates, but as they multiply and join, they form a single heaving sheet. Thus do heart cells behave in the human chest: They do not pulsate discordantly, each sparking to its own beat; they explode in rhythmic harmony. Buried high in the right atrium, a minute knot of cells sets the heart's pace. Called the sinus node, its sparks send electrical impulses racing through the heart to other electrical cells woven throughout cardiac tissue. In perfect rhythm each successively explodes. This trail of electricity flashes so rapidly across the heart that all its cells appear to beat as one.

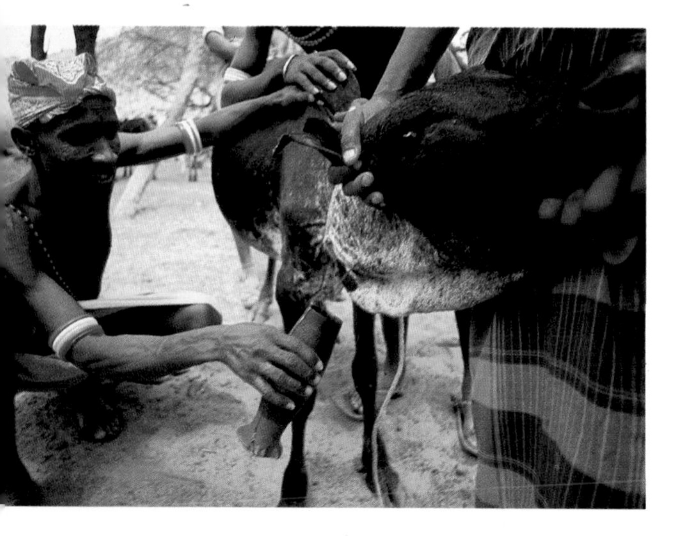

Turkana tribesmen, who live on the barren soils of the Great Rift Valley in Kenya, add iron to their diet by drinking cow's blood. To draw the blood, a tribesman (left, upper) punctures the cow's jugular vein with a sharp arrow, then catches the spurting liquid in a clay jug. The cows, though bled frequently, suffer no ill effect. A youth (left, lower) drinks the fresh blood.

The body can use iron from almost any source, even from the filings of an iron pot. Preparing a dish of protein-rich mopani worms (right), a Venda woman of South Africa benefits from cooking with iron—like millions of other homemakers and their families around the world.

Iron deficiency is brought on by too few red blood cells or too little hemoglobin due to blood loss, poor diet, or pregnancy. It can be corrected by adding iron-rich foods to the diet: Dried beans and peas, red meat, and liver are good sources.

Blood When You Need It

As we each lay in bed this morning, the cells in our bodies burned a minimum amount of oxygen, enough to keep us alive. As we began to stir, our cellular furnaces ignited for the action of the day. When we stood to walk across the room, our bodies doubled their demand. And if we exercised, our cells busily consumed eight to twelve times the oxygen used at night. Some strenuous athletic feats and dire emergencies require such furious energy combustion that our cells burn up to twenty times the oxygen they use at rest.

The heart and blood vessels do more than speed or slow our blood flow to meet these needs. They carry the scarlet stream to different tissues under differing pressures to fuel different actions. Blood rushes to the stomach when we eat, to the lungs and muscles when we swim, to the brain when we read. To satisfy these changing metabolic needs, the cardiovascular system integrates information as well as any computer, then responds as no computer can.

How does it register—and respond—to needs we do not consciously recognize? On constant duty in our brain stems are chemical sensors called carotid bodies that continually "taste" the blood flowing from the heart to tissues, savoring it for the acidic flavor of carbon dioxide. Rising levels of carbon dioxide signal the brain to increase the rate at which our lungs expel it. Other monitors, situated mainly in the aorta and the carotid artery, the major vessel leading to the brain, regulate blood pressure. These stretch receptors activate when a surge of blood, signaling a faster heartbeat, stretches the arteries. The sensors immediately alert the brain, which orders the heart rate to slow.

Nerves Regulate the Heart

The brain presides over the heart and blood vessels by designating authority to the autonomic nervous system, two groups of nerves that oppose and balance each other. The sympathetic nerves send the heart rate soaring: Danger, stress, and exertion signal this system to speed the flow of blood. The parasympathetic system counters by slowing the heart down, conserving energy for the demands of everyday life.

The sympathetic nervous system is part of the fight or flight response that primes us for action. You step off a curb; suddenly a speeding taxicab skids around the corner, heading straight for you. The body responds to such dangers by signaling the sympathetic nerves to pour out adrenaline and noradrenaline (now often called epinephrine and norepinephrine). Both chemicals constrict blood vessels, raise blood pressure, and speed the heart. As the cab bears down, your heart pounds faster and faster, your blood pressure rises, breathing deepens, and muscles tense. You leap for safety and collapse on the sidewalk, sweating and gasping for breath. In extreme emergencies the sympathetic nerves can send the heart rate soaring as high as 200 beats a minute, preparing us for unusual feats of strength or action.

In concert with its control of heartbeat, the sympathetic nervous system regulates the flow of blood. Arterioles—the thickly muscled vessels leading to tissue capillaries—instantly respond to the commands of nerves, squeezing tight to reduce blood flow to certain tissues, opening wide to flood other organs with a copious blood supply. The sympathetic nerves operate the arterioles like living faucets, turning the flow of blood on or off according to our actions. As we eat lunch, sympathetic nerves constrict arterioles to our leg muscles and open the blood vessel gates to the intestines, energetically engaged in digesting food. When we run, the sympathetic system performs its most dramatic role, pouring out enough noradrenaline to close off blood

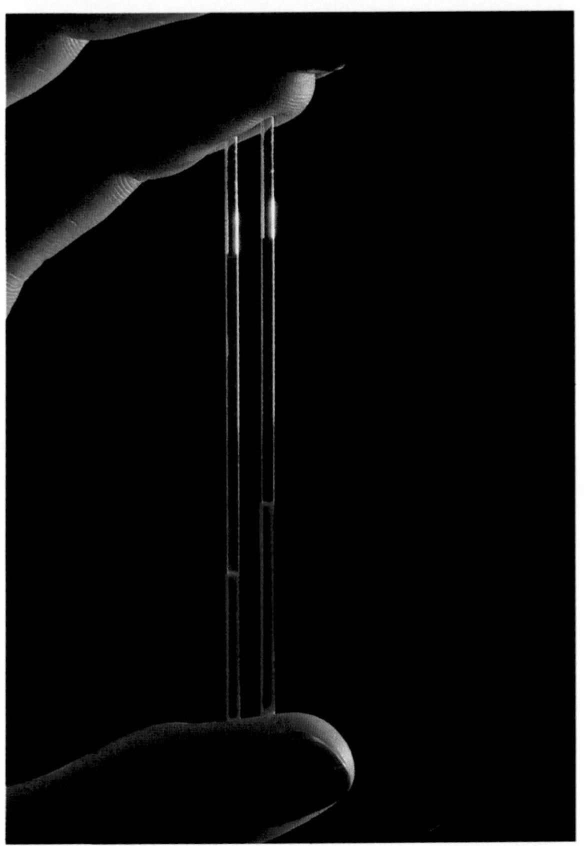

A Moroccan woman living high in the mountains (opposite) has more red blood cells than someone her size living at sea level.

Individuals unaccustomed to thin air get dizzy and short of breath when they climb to altitudes where oxygen is scarce. Within days, the body adjusts by increasing its production of red blood cells and hemoglobin.

A centrifuge (above) monitors red cells by separating the three components of blood: red cells, white cells, and plasma. By examining pipettes of the separated blood (left), a technician can see the difference in the red cell count of persons living at high and low altitudes.

flow almost completely to organs not immediately or urgently in use. In turn the system expands the valves, flooding flexing muscles, the brain, and the heart itself.

Like the accelerator in a car, however, the sympathetic nerves speed the heart; they do not slow it down. That is the job of the parasympathetic system. Rising heart rate can increase cardiac output five times over resting levels; if the arterioles constricted proportionately—increasing pressure as narrowing banks swell a river—our cardiovascular network could explode. Vessels could rupture, spilling blood into brain tissues and causing a stroke. The parasympathetic nerves serve as the cardiovascular brake. When the receptors in the arteries sense the blood pressure rising, they alert the brain, which signals the parasympathetic nerves to release the chemical acetylcholine. Blood vessels dilate; pressure drops. The parasympathetic system can slow heart rhythm to forty beats a minute if necessary, easing the force with which the heart muscle contracts.

A Thermostat in the Brain

Heart rate can rise or fall, but the temperature of the blood must remain constant. A severe drop in body heat can damage cells by inhibiting critical enzyme reactions. Even a mild rise in temperature makes us feverish, and we cannot survive for long if our temperature shoots above 108°F. The body monitors its temperature through a thermostat that measures the heat of blood flowing through the brain. If air temperature drops even a fraction of a degree and our blood cools, the autonomic nervous system responds instantly: Parasympathetic nerves slow the heart; sympathetic nerves constrict vessels in the skin. Blood flows through deeper pathways, away from the cold air at the skin. When weather turns hot or when we exercise—combusting more oxygen and thus generating heat—blood changes course. The sympathetic nerves open arteriole valves, and the blood vessels in our skin act like radiators, cooling the body by casting off heat to the surrounding air.

Continually balancing one another in a state known as homeostasis—nature's balance in humans—the sympathetic and parasympathetic nerves mete out our blood supply, regulate our pressure, and mediate our temperature. Together, their coordinated actions adjust heart rate and blood flow when we stand suddenly, bend over, or even perform acrobatics upside down. If we were to climb from sea level to the thin atmosphere of the Himalayas, sympathetic nerves would quicken the heart pace to send the oxygen we need to our cells. Breathless and dizzy at first, we would quickly adjust; our bone marrow would step up red *(Continued on page 136)*

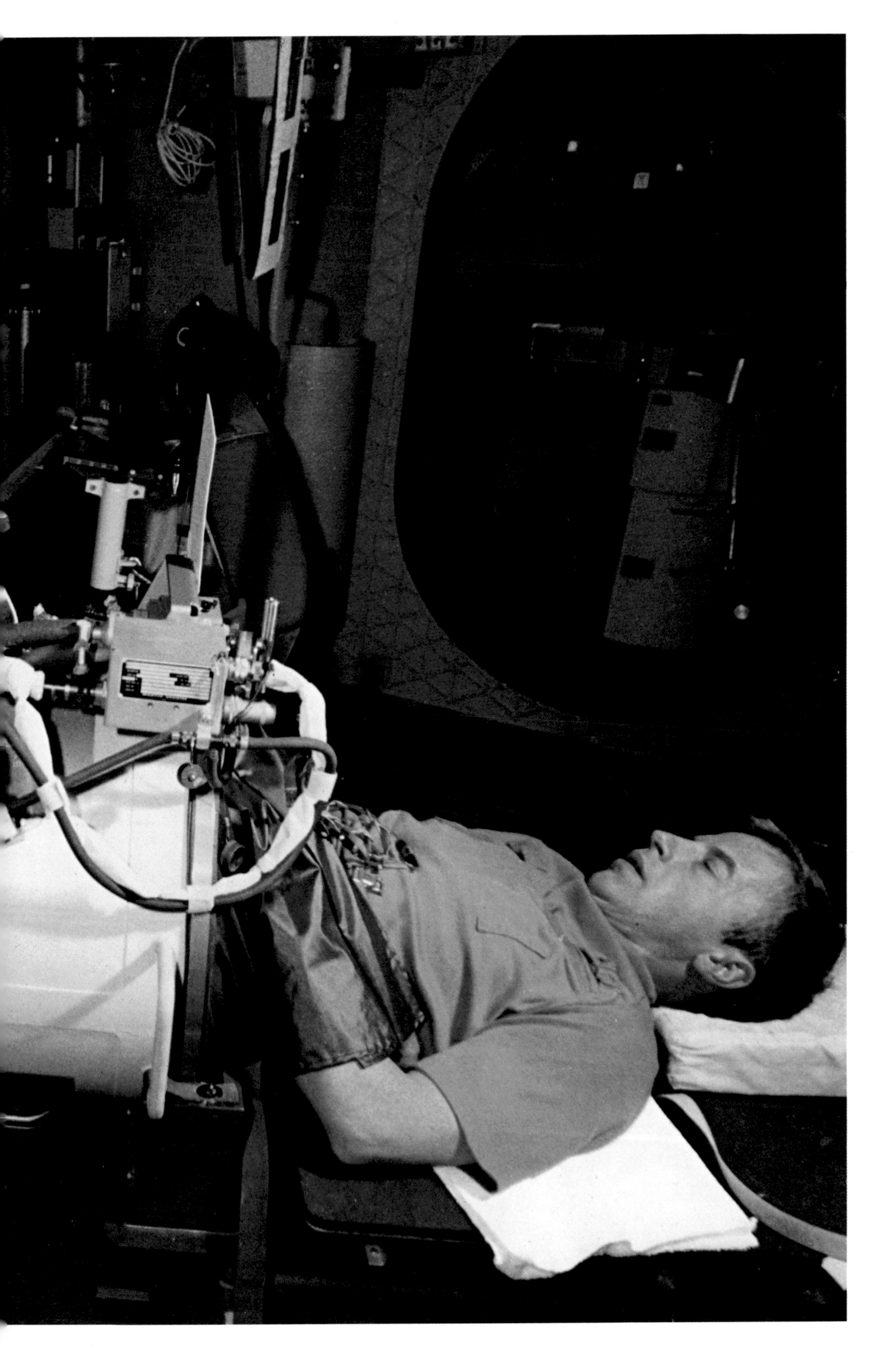

When Skylab orbited Earth for 12 weeks in the early 1970s, astronauts spent time every 3 days in a machine like this to measure their cardiovascular reflexes under weightless conditions. The machine pulls blood to the legs the way gravity would, then measures the heart's ability to pump it back.

Earth's gravity pulls blood so that it makes the trip from heart to lower body in only 10 seconds. In space an astronaut's heart rate adapts to the loss of gravity by slowing down, but once back on Earth the heart is under strain as it works to pump against gravity's force.

The Lungs: Fresh Oxygen on Call

A champion swimmer gulps for life-giving air (right). For an Olympic victory, her lungs must rapidly and repeatedly deliver fresh oxygen to her blood and to every cell in her body. A racing athlete's breathing may jump to 42 breaths per minute, or triple the normal resting rate of 12 to 14 breaths. Depth of breathing increases too: Resting lungs take in about half a quart of air with each breath; during strenuous exercise the lungs can expand to hold 6 or 7 quarts.

The back view of a newborn's lungs (below) reveals the asymmetrical structure of each organ; the three-lobed right lung weighs a few ounces more than the two-lobed left. Adult lungs, whose convoluted surfaces could cover a racquetball court, weigh less than two pounds. A substance lining the walls keeps the lungs from collapsing, lowers surface tension, and reduces the muscular effort required for breathing.

Actual size

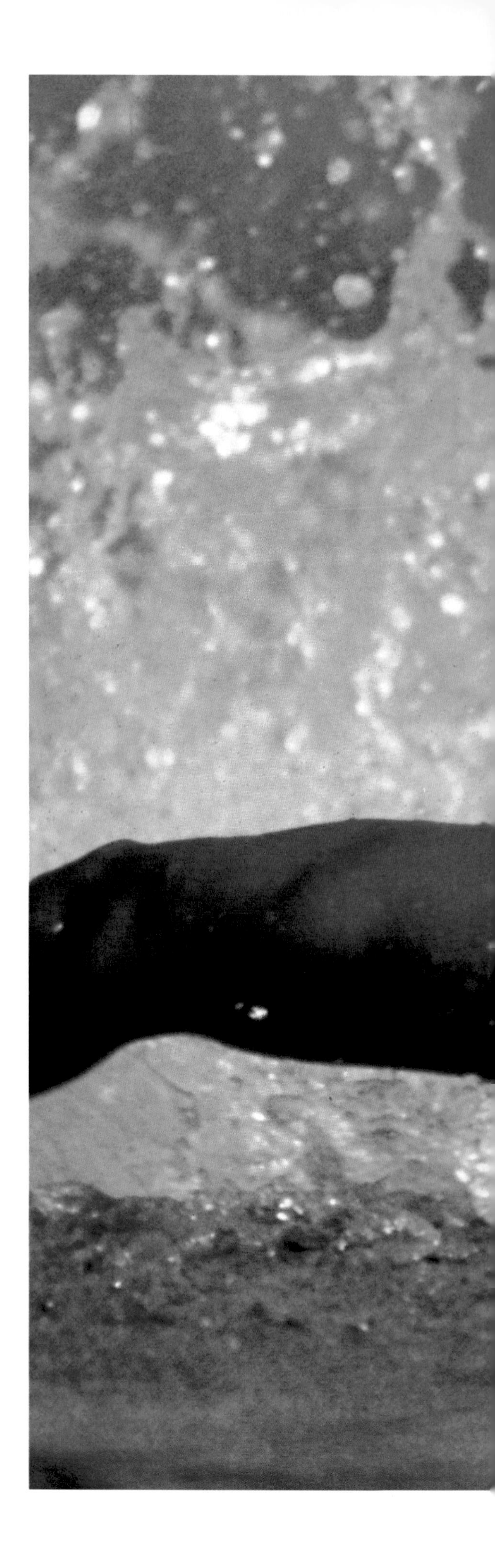

SPEEDO

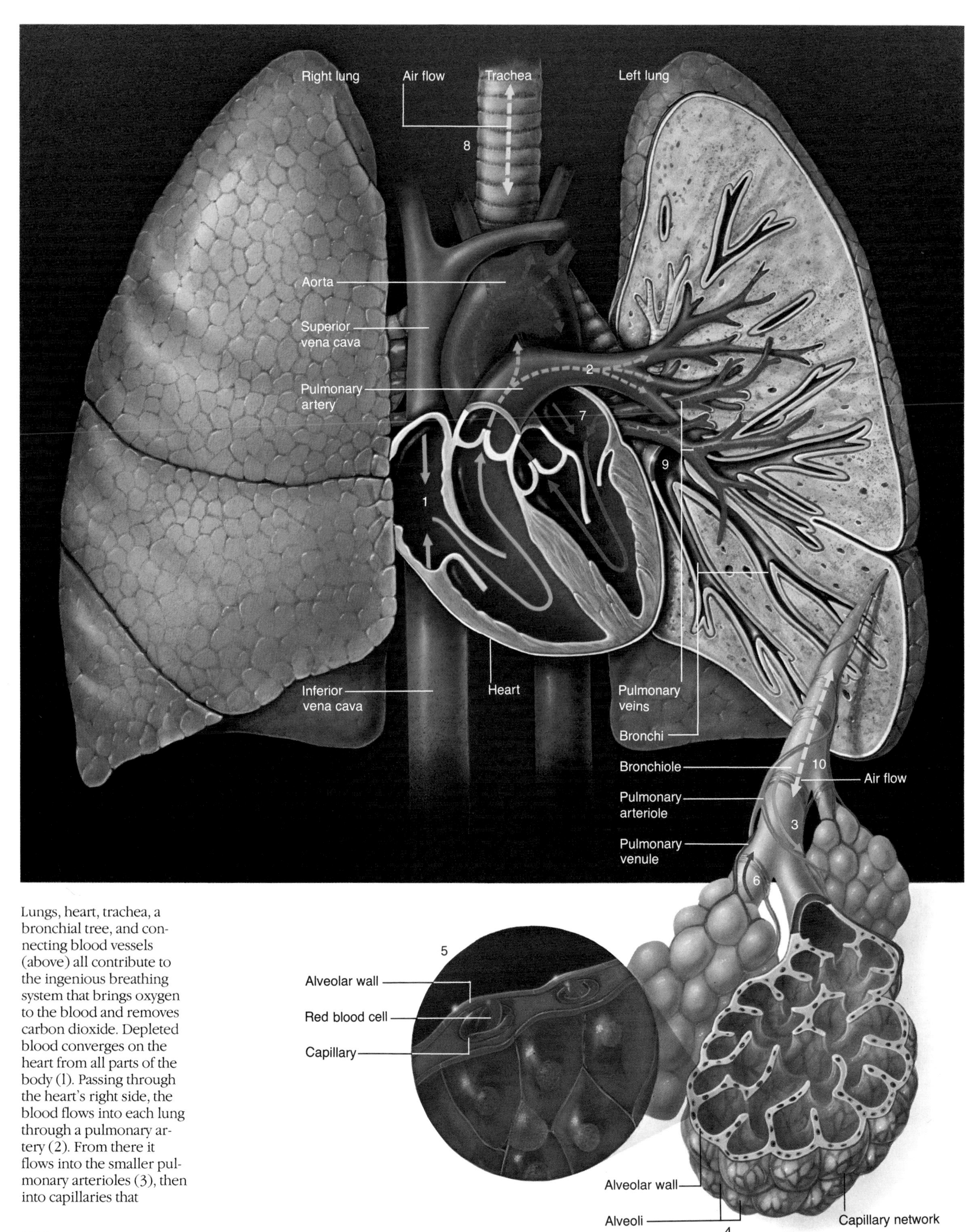

Lungs, heart, trachea, a bronchial tree, and connecting blood vessels (above) all contribute to the ingenious breathing system that brings oxygen to the blood and removes carbon dioxide. Depleted blood converges on the heart from all parts of the body (1). Passing through the heart's right side, the blood flows into each lung through a pulmonary artery (2). From there it flows into the smaller pulmonary arterioles (3), then into capillaries that

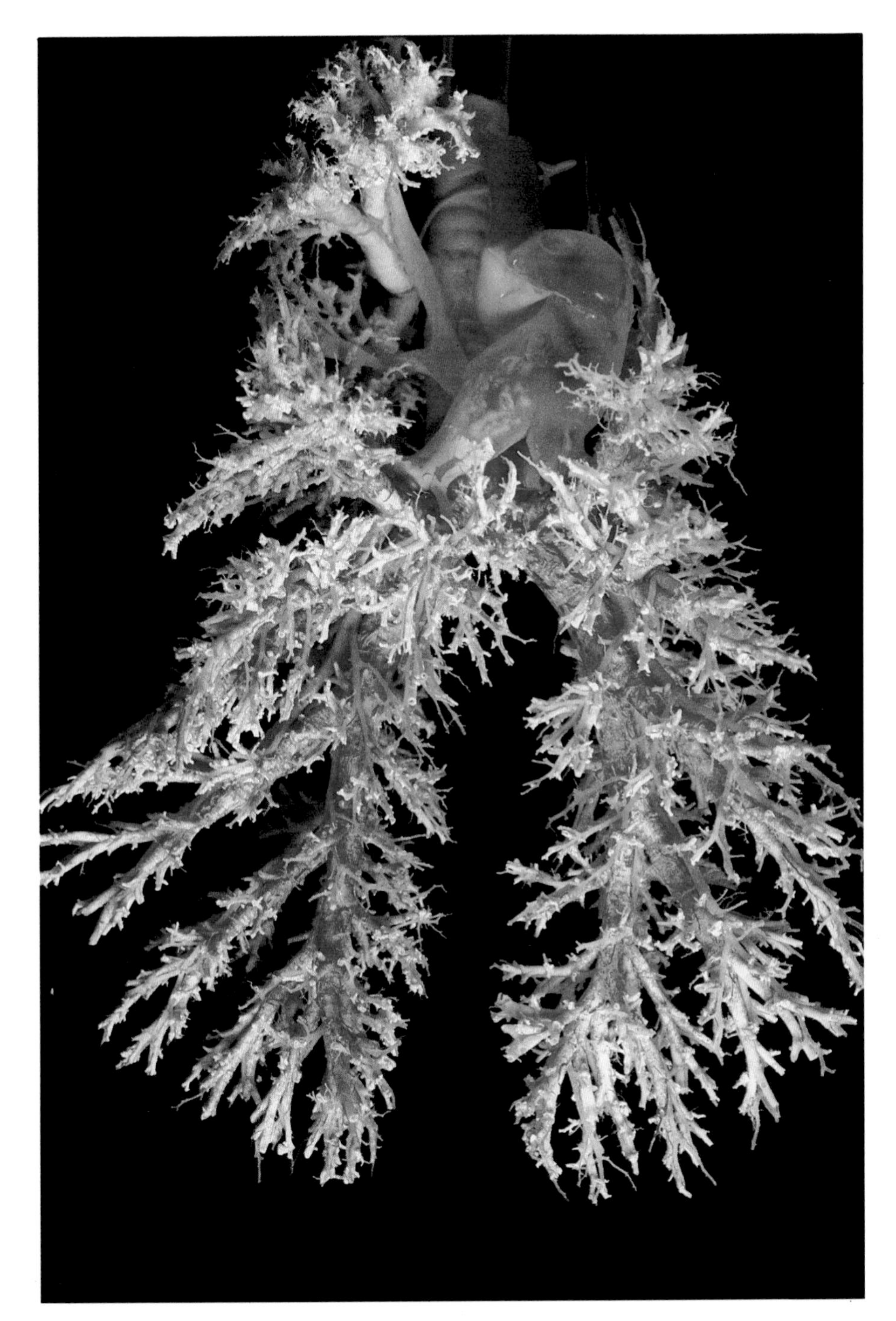

Magnification: 70 times

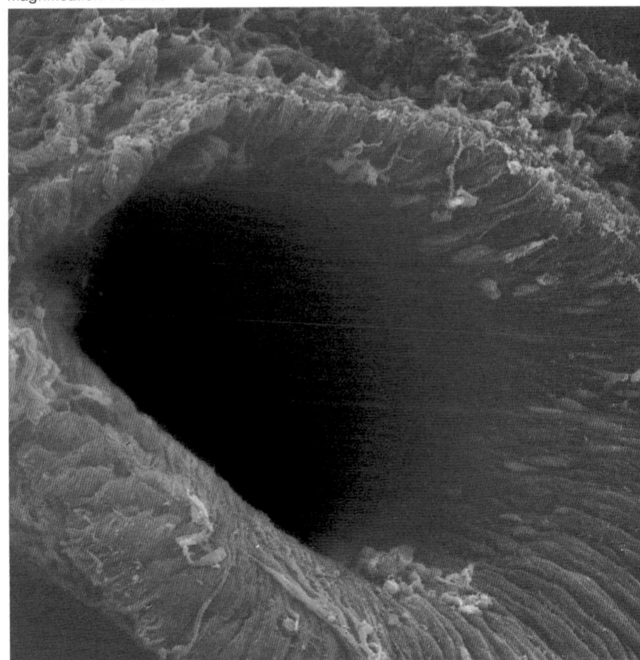

Lennart Nilsson, © Boehringer Ingelheim International GmbH.

surround each of the hundreds of millions of spongy air sacs, or alveoli (4), inside our lungs.

The alveoli conduct the real business of gas exchange, or respiration. Constructed of fibrous strands of connective tissue and surrounded by a fine network of blood capillaries with walls a cell thick (5), the alveoli's moist inner surfaces absorb oxygen molecules from inhaled air. The molecules then diffuse into the surrounding capillaries; carbon dioxide simultaneously passes from the capillaries into the alveoli and is exhaled.

The blood, now rich in oxygen, flows through the pulmonary venules (6) into the pulmonary veins, then through the heart's left side (7) and back out through the aorta into the body's tissues.

To supply enough oxygen, an alveolus fills and empties more than 15,000 times in a day of normal breathing. With every breath, air entering the nose or mouth travels to the lungs through the windpipe, or trachea (8), and a system of breathing tubes called the bronchial tree (above, left). The "tree" follows the branching pattern of the pulmonary vessels, carrying the air from large branches, the bronchi (9), to tiny ones, the bronchioles (above, right and 10), and, finally to the alveoli at the bronchioles' ends.

Magnification: 10,000 times (below); 4,000 times (opposite); Lennart Nilsson, © Boehringer Ingelheim International GmbH. (also opposite)

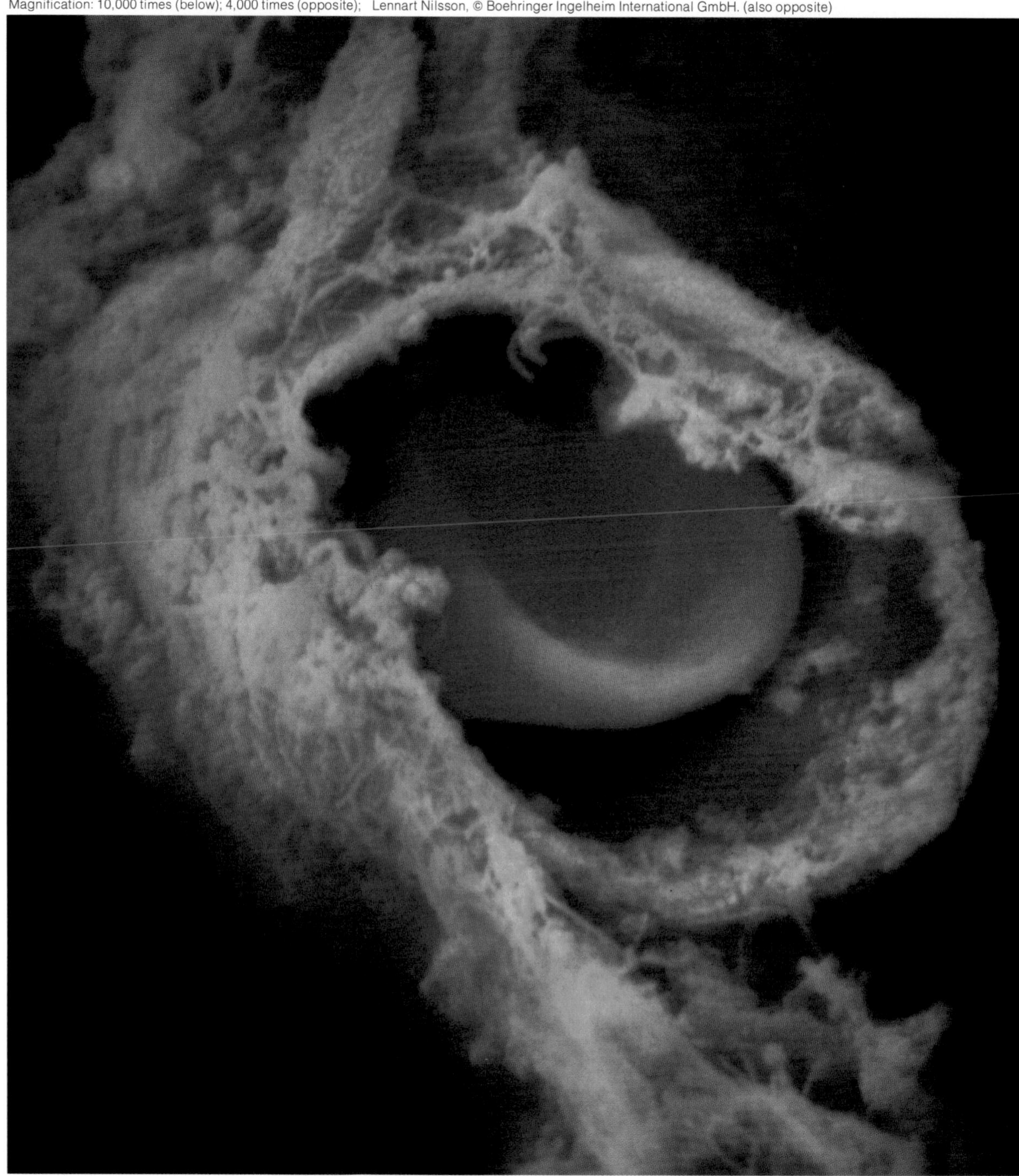

A red blood cell squeezes through a capillary (above) to rendezvous with air inside alveoli less than .00004 inches away.

In a fraction of a second the swap occurs: Oxygen, under higher pressure in the alveolus than in the capillaries that encircle it, diffuses through the alveolar wall and into the blood. Carbon dioxide, under higher pressure in the capillaries than in the alveolus, passes from blood to lungs. By this exchange, the capillaries that lace the alveoli oxygenate blood.

An invading dirt particle (tinted brown, opposite) is caught in bronchial mucus and entangled in cilia before it can reach the lung. Like the steps of an escalator, the hairlike cilia move constantly, transporting the particle up through the windpipe to the throat, where it is swallowed or coughed out.

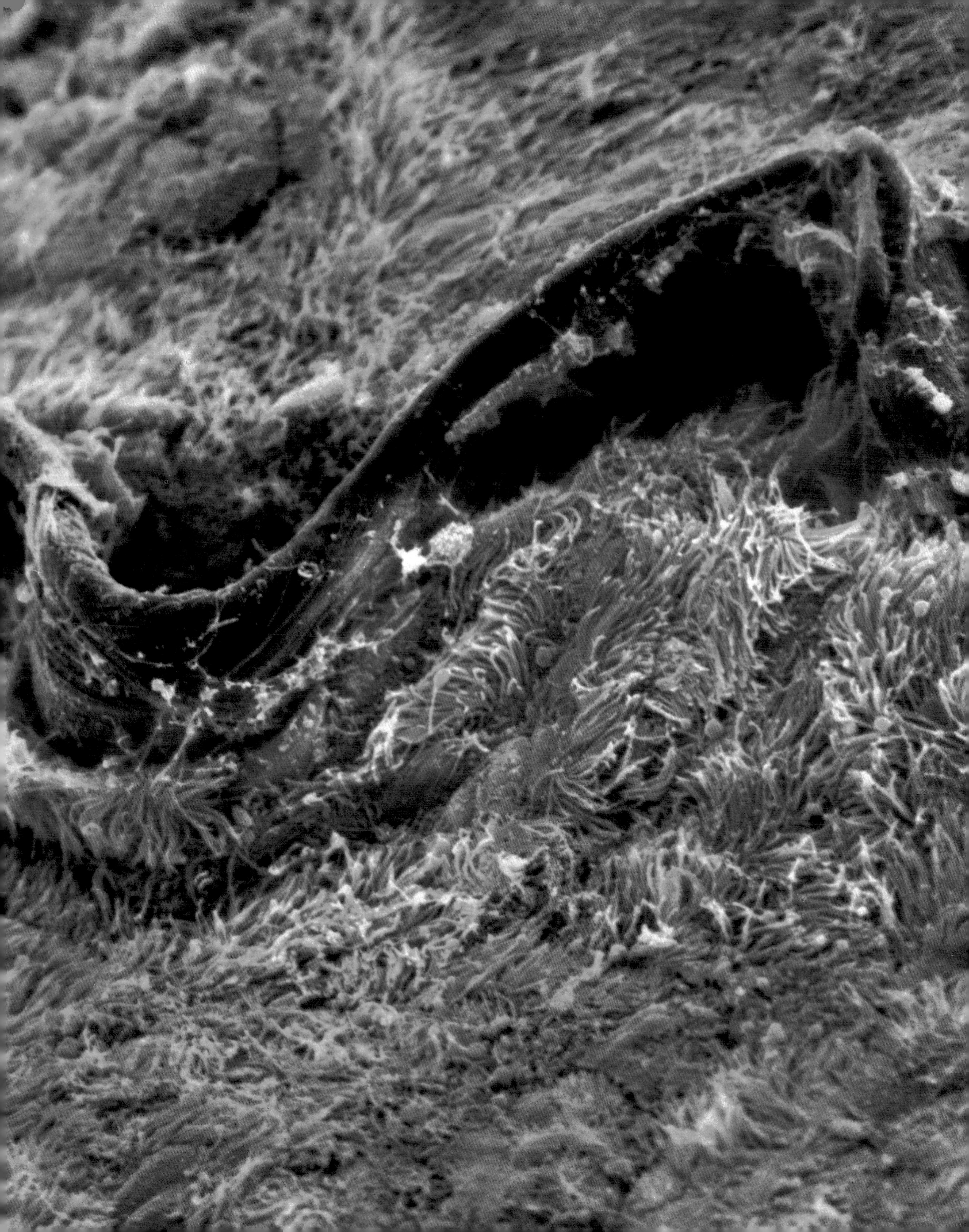

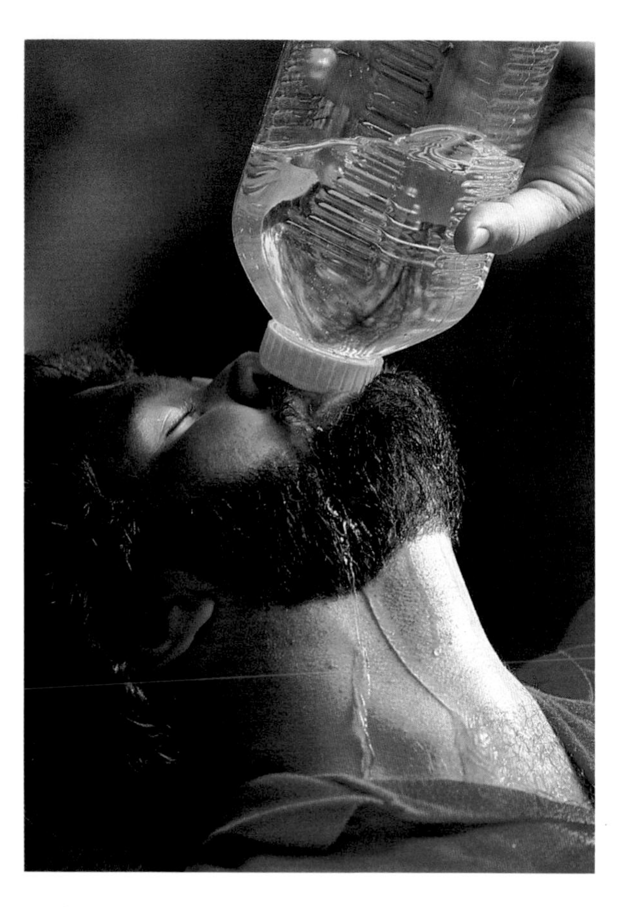

A drink of water quenches a powerful thirst (left) and helps perform the crucial task of washing wastes into the kidneys and out of the body. The kidneys use water to remove toxins that would otherwise poison the blood.

If both kidneys fail, dialysis can take over the job. Portable dialyzers (right) permit patients to stray far from the hospital room for their thrice weekly treatment. From a vein in the arm, blood flows into the unit where it brushes by a semipermeable membrane that allows wastes but not blood cells to pass through it. A second filter removes air bubbles before the machine pumps the cleansed blood back into the body.

cell production, churning out 50 percent more of the oxygen-carrying cells than at ocean level.

Conversely, if we dive underwater, the parasympathetic system slows the heart, conserving our limited oxygen supply. The sympathetic system constricts blood vessels. This shuts off blood to almost all tissues and changes the cardiovascular system into a shortened circuit that cycles mainly from heart to brain. Arteriole sensors register the rising level of carbon dioxide waste in the blood and flash the brain a signal to surface from the water.

In general, the more complicated the capabilities of a computer, the greater the possibility of a breakdown. Yet the heart—center and power of the complex cardiovascular system—is most remarkable because it lasts so long. "Heart attack," an often misused term for a specific type of cardiac malfunction, is rarely caused by the heart itself. Strictly defined, a heart attack, or myocardial infarction, occurs when a patch of heart muscle dies. Most heart attacks stem from cardiovascular disease, primarily atherosclerosis.

Derived from the Greek words for "hard porridge," atherosclerosis begins with an injury to the fragile lining of an artery. As platelets gather to repair the damaged tissue, droplets of cholesterol, a fatty substance floating in plasma, congeal. They stick to the debris and harden into a deposit called plaque. The injured artery wall grows brittle and loses its ability to snap back after each surge of blood. In some cases the plaque builds up until it blocks the flow of blood. In others the pounding blood hammers

at the weakened artery wall, forming a bulge known as an aneurysm, which may one day burst. In still other cases the plaque attracts platelets, which in turn trigger the growth of a thrombus, an abnormal blood clot that reduces blood flow through the vessel. Sometimes a thrombus breaks free and floats through the bloodstream until it lodges in a vital artery, causing an embolism.

If atherosclerosis blocks either of the heart's two arteries—called coronary because they encircle the heart like a crown—heart muscle may die. Each coronary artery feeds a distinct portion of the heart; although many smaller arteries branch from these major vessels, there are few connections between them. Thus, a cholesterol buildup or blood clot in one of the two coronary arteries can block blood flow to crucial sections of heart tissue.

When a coronary artery is partially blocked, it often carries only enough blood to oxygenate the heart at rest. Any stimulus that makes the heart beat faster or more forcefully—and thus causes heart cells to use more oxygen—can drive blood demand above the limited supply. The oxygen-starved tissues then burn with pain. Known as angina pectoris—literally, a "strangling in the chest"—the pain often starts under the breastbone and radiates down the left arm.

Frequently, angina serves only to warn a person to stop activity. But if the pain worsens or increases in frequency, it can deliver a more ominous message. Deprived too long of oxygen, heart cells die, and a heart attack occurs.

Magnification: 2 times; Lennart Nilsson, © Boehringer Ingelheim International GmbH. (also opposite, upper)

Two hard-working kidneys control the amount of water, acids, and salts in the body. A child's kidney (above) is lined with creases that disappear with maturity. In adults, these remarkable structures weigh only about 5 ounces each. They process about 425 gallons of blood a day, of which all is recycled except a fraction that is converted to urine. While ridding us of waste products from every cell, the kidneys also produce at least 3 hormones, including renin which helps control blood pressure.

Three layers protect the fragile kidneys: First comes an inner membrane, then a cushioning layer of fat, then a thin tissue layer that separates the kidney fat from fat around other nearby organs.

The Cholesterol Factor

What makes a person a likely candidate for a heart attack? Studies show that no single factor causes heart disease, but rather a combination of many. However, high blood cholesterol has proved so important a cause that the higher a person's is, the higher the chance of disease, regardless of whether or not the person smokes or has high blood pressure—the next most important causes. By taking medication and changing habits, people can control these risk factors and others such as diabetes and obesity. The contributing factors of age, sex, heredity, and race obviously cannot be controlled.

Because atherosclerosis affects so many Americans and causes a majority of heart attacks, scientists are searching for ways to diagnose and treat it in its earliest stages before the onset of symptoms such as angina. Experiments in Boston have tagged with a radioactive substance certain proteins that carry cholesterol through the bloodstream; then tracked the proteins with a special camera. And researchers have devised ways to clear diseased arteries without major surgery. In a technique *(Continued on page 149)*

Magnification: 400 times

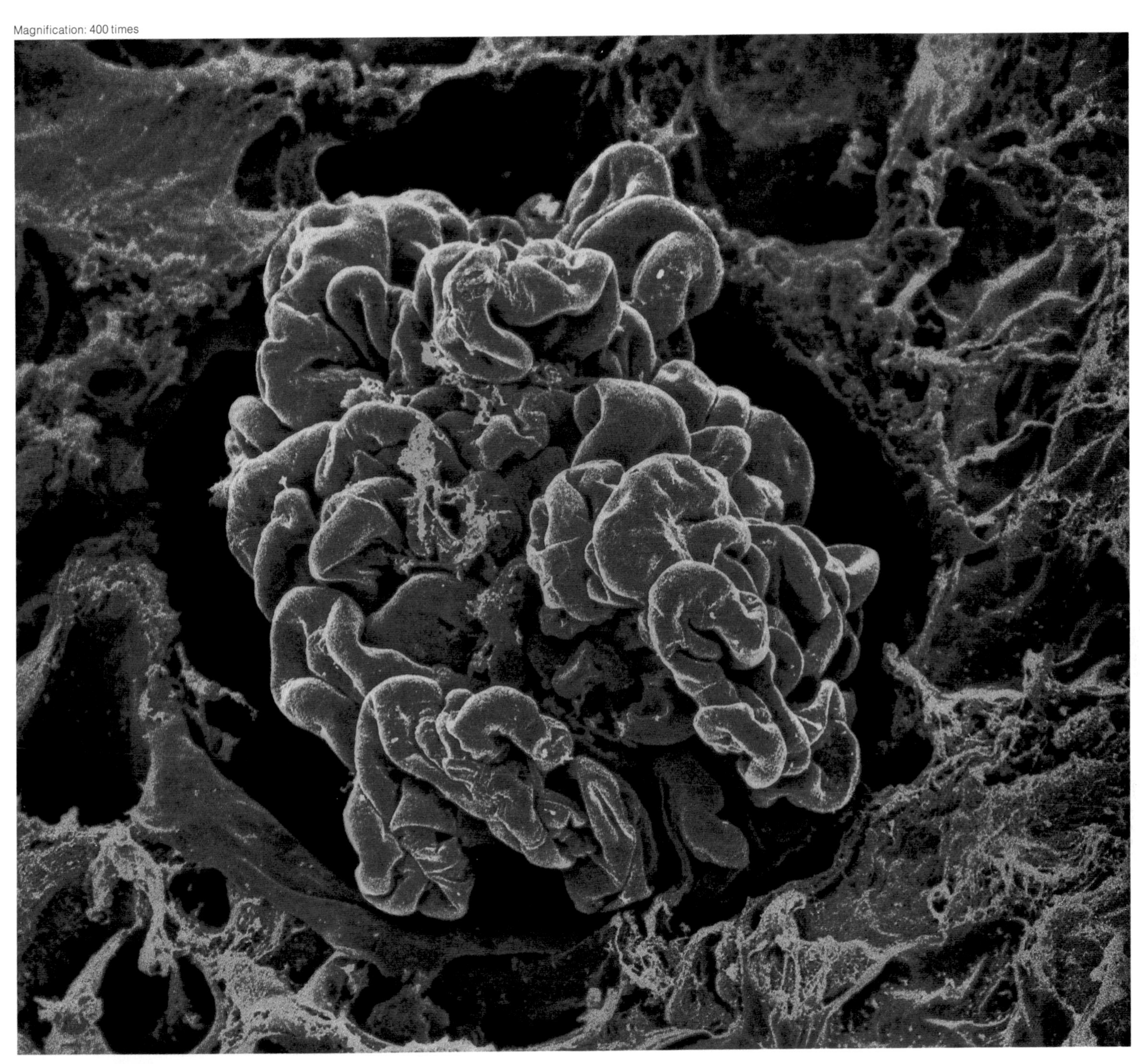

Magnification: 1,250 times

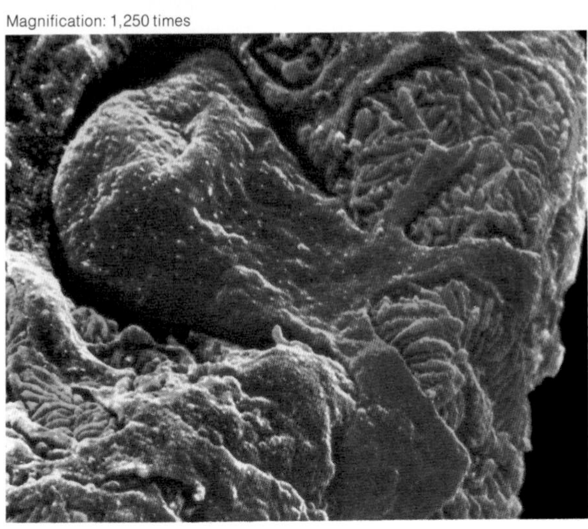

Over a million intricate cleaning units called nephrons riddle the kidney. Blood enters each nephron through a ball of capillaries called a glomerulus (above), which acts as a filter for relatively large particles such as red and white blood cells. The remaining particles, including salts and other chemicals small enough to pass through the walls of the glomerulus, are directed into the nephron by octopus-like cells called *podocytes* (left). As the liquid travels through the nephron, most of it is reabsorbed into the circulatory system. The rest flows into larger and larger ducts and eventually into tubes leading to the bladder, from which it is expelled as urine.

Plugging the Wound

A punctured capillary just under the surface of a fingertip (below) triggers a flurry of activity to clot the blood. The same swift drama unfolds when an artery ruptures (right), and blood cells burst forth.

A break in a blood vessel exposes collagen, a fibrous protein permeating the vessel wall. Platelets, drifting through the bloodstream, adhere to the collagen, then swell up and release adenosine diphosphate (ADP), a chemical that draws more platelets to the site of injury. Within seconds the platelets clump together and plug the cut if it is small.

Platelets, the smallest of the blood cells and the force behind the clotting process, originate in the bone marrow. More than a trillion of them patrol the bloodstream. Each tiny platelet lives for about 10 days; 200 billion new platelets move into the circulatory system each day to replace those that die.

If a simple platelet plug cannot seal the wound, platelets temporarily stem the flow of blood by constricting the broken vessel with another chemical, serotonin. Meanwhile, the wound itself triggers a chain of events leading to a longer lasting clot.

First, the injured vessel wall secretes a substance that combines with chemicals in the blood to produce a product that turns prothrombin, a protein in blood, into thrombin, an enzyme. As an enzyme, thrombin can accelerate chemical change. Encountering a substance called fibrinogen in blood plasma, the newly created thrombin converts it into fibrin threads that adhere to the injured site and gradually spin a loose web around the platelet plug.

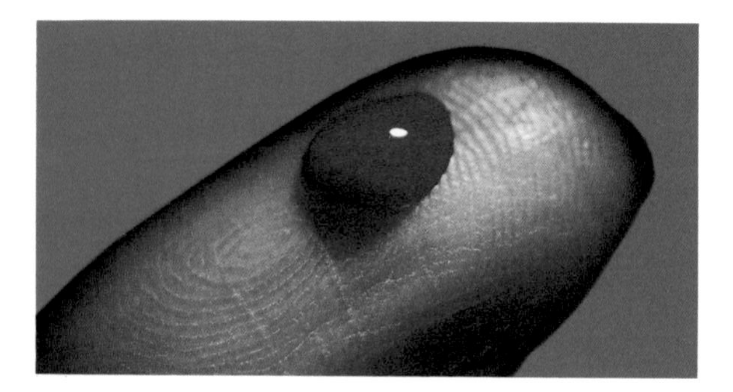

Magnification: 10,000 times

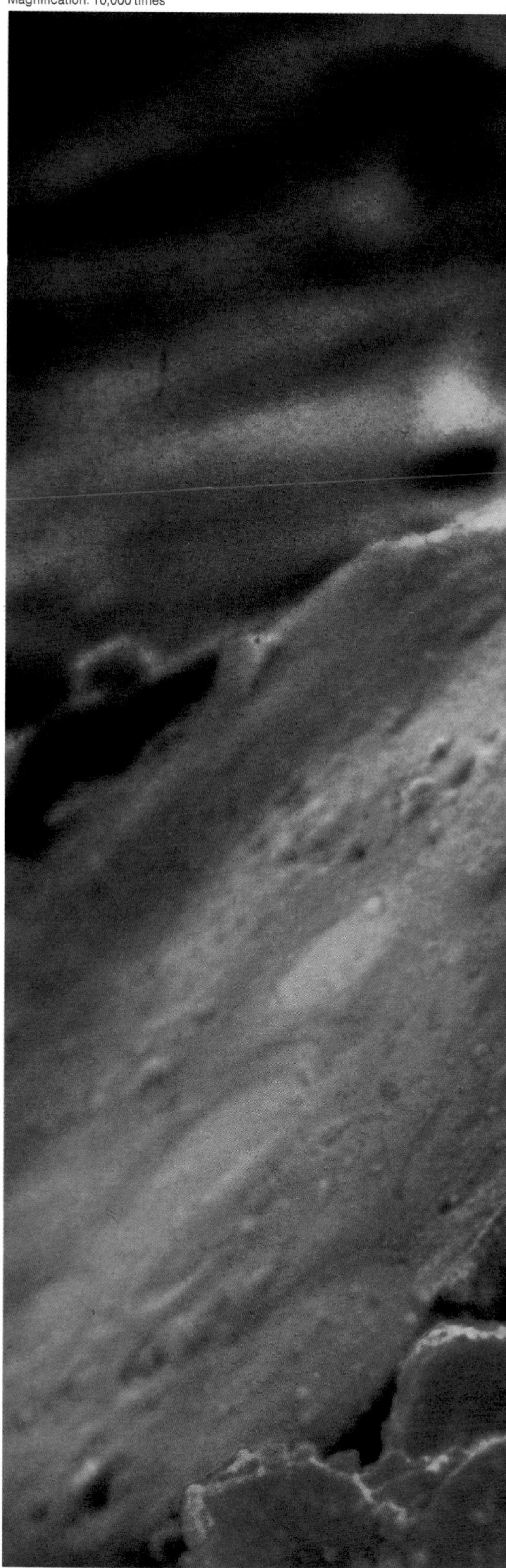

Lennart Nilsson, © Boehringer Ingelheim International GmbH.

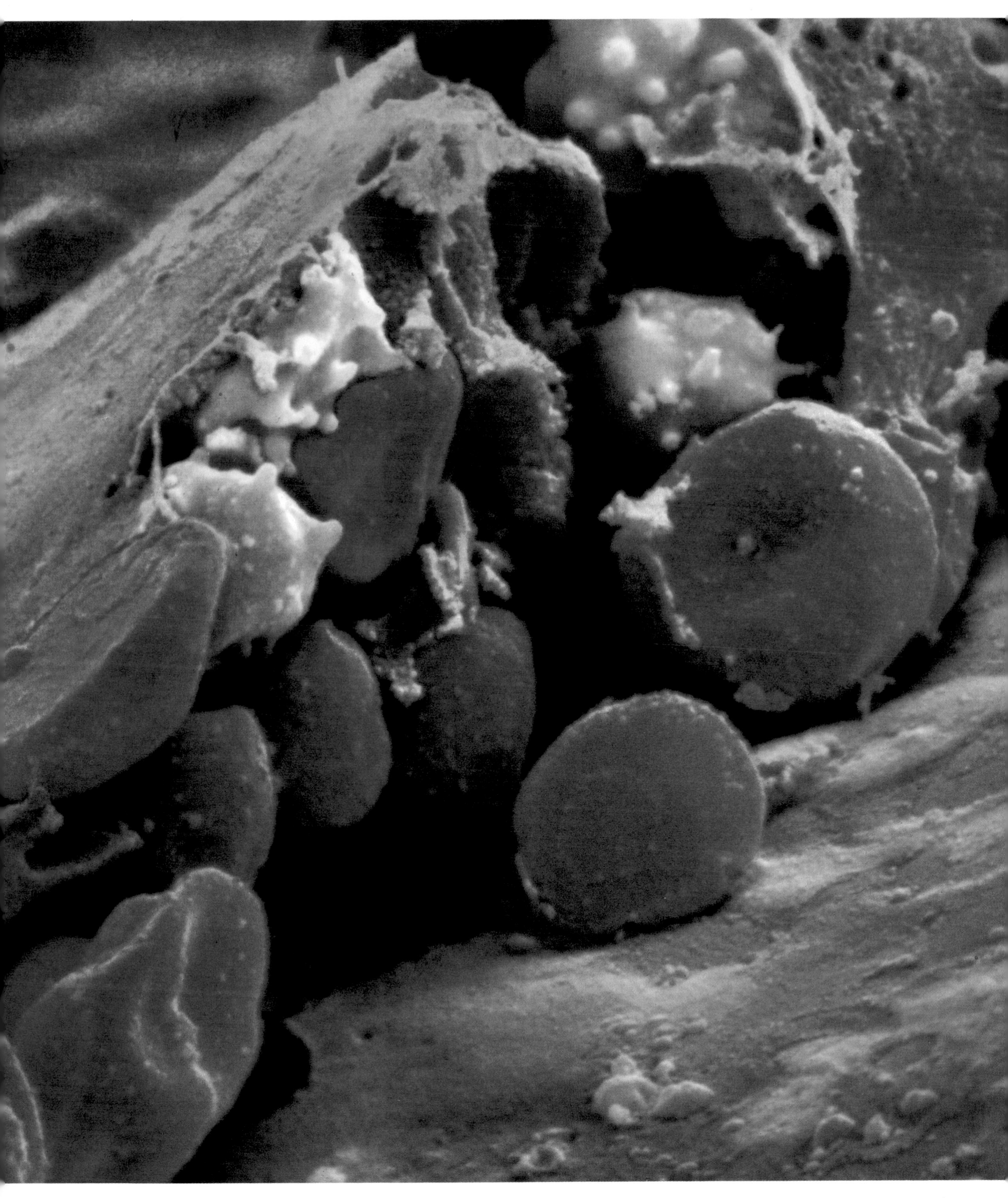

Magnification: 20,000 times (below); 35,000 times (right)

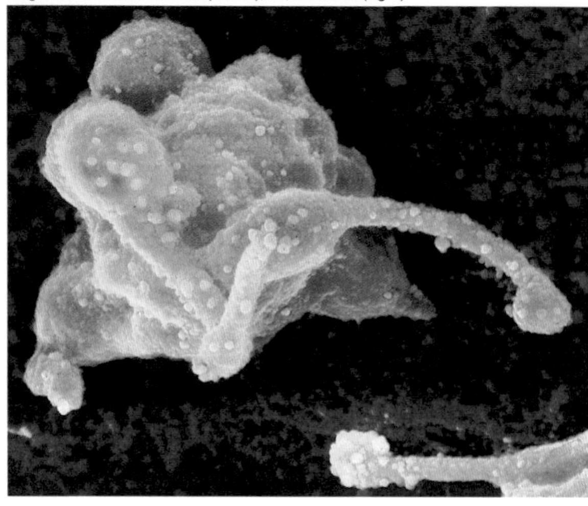

To enlarge the plug, the fibrin net snares passing blood cells (right). Platelets (above) deliver other chemicals through their pseudopods, or feet, including one that links the fibrin threads together and reinforces the growing net (far right). The newly formed clot, which is 99 percent water, dams the gushing blood.

Moments after a blood clot forms, it changes from a gel to a solid. The hardened clot binds the wound until the tissue can heal.

When the clot has served its purpose, one of its proteins changes into an enzyme called plasmin. The plasmin breaks down fibrin threads and disperses the clot debris into the bloodstream. Special white cells devour the waste; the bone marrow, liver, and spleen filter stray blood clots from the circulatory system.

Lennart Nilsson,

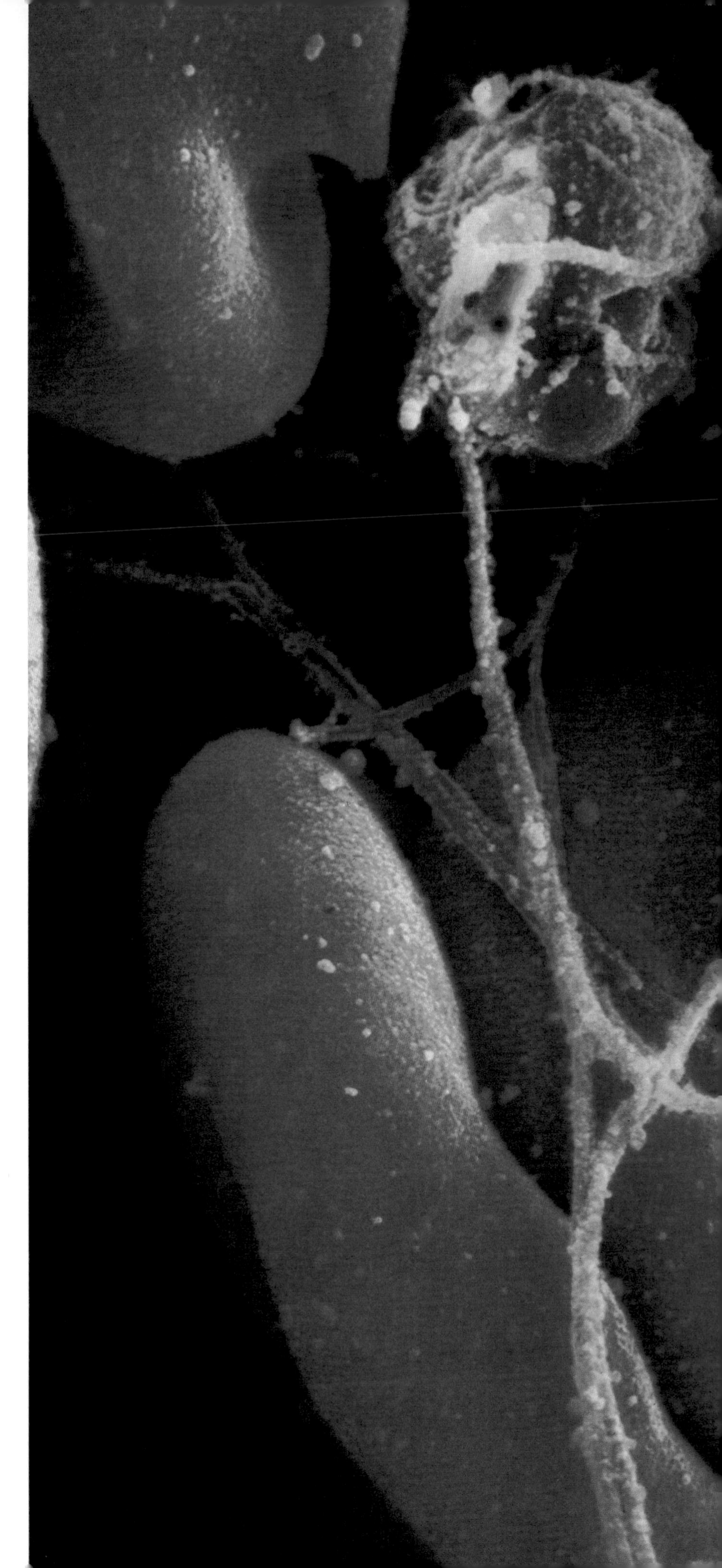

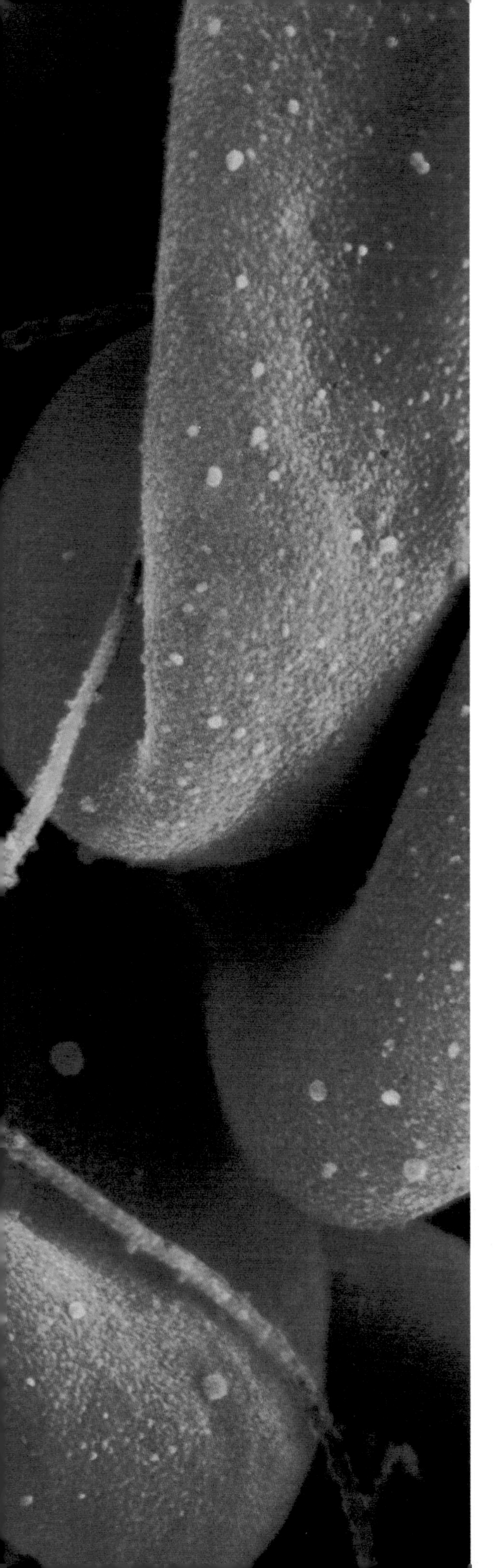

Magnification: 7,500 times

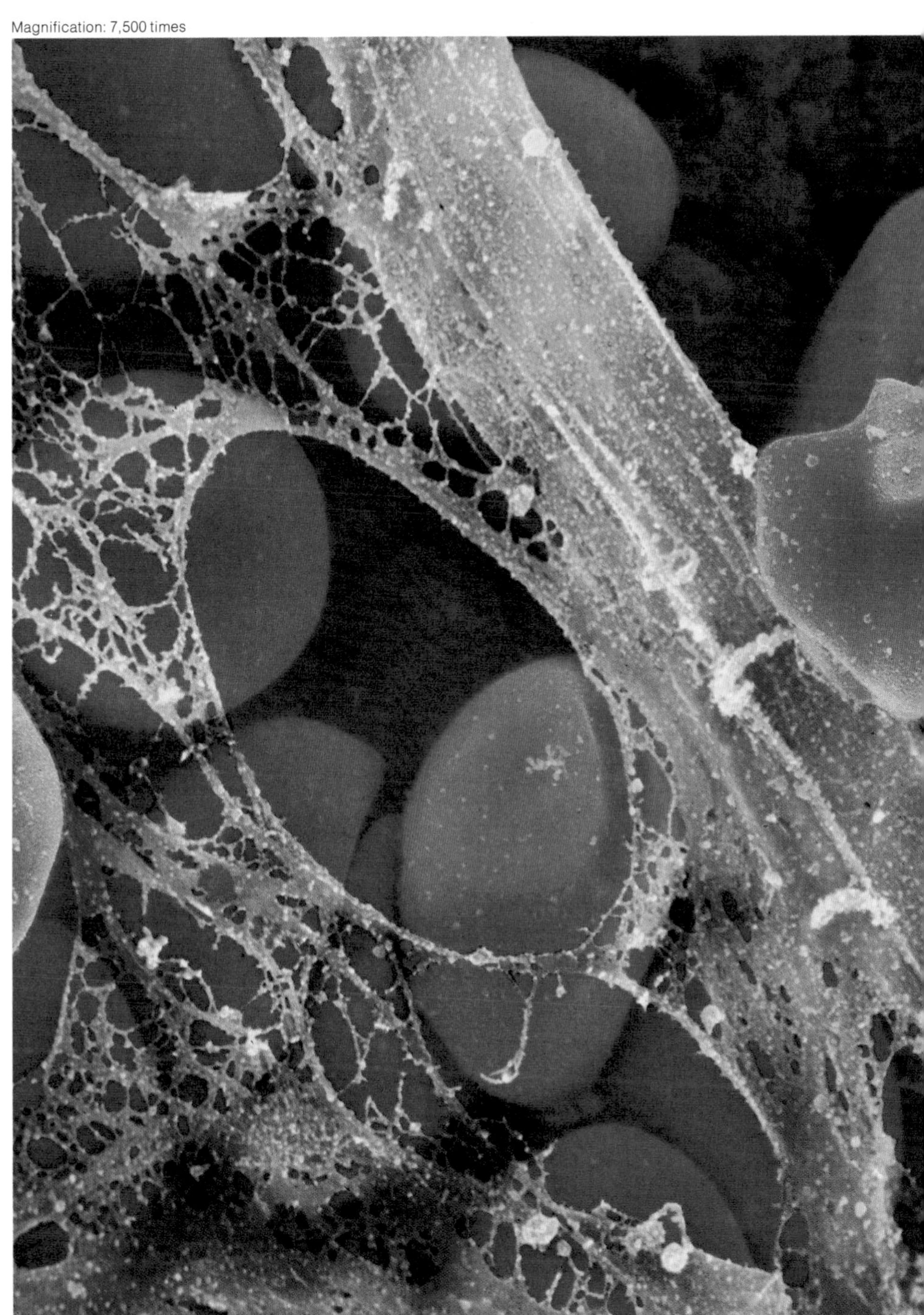

Magnification: 400 times (below); 30 times (right); Lennart Nilsson, © Boehringer Ingelheim International GmbH. (also right)

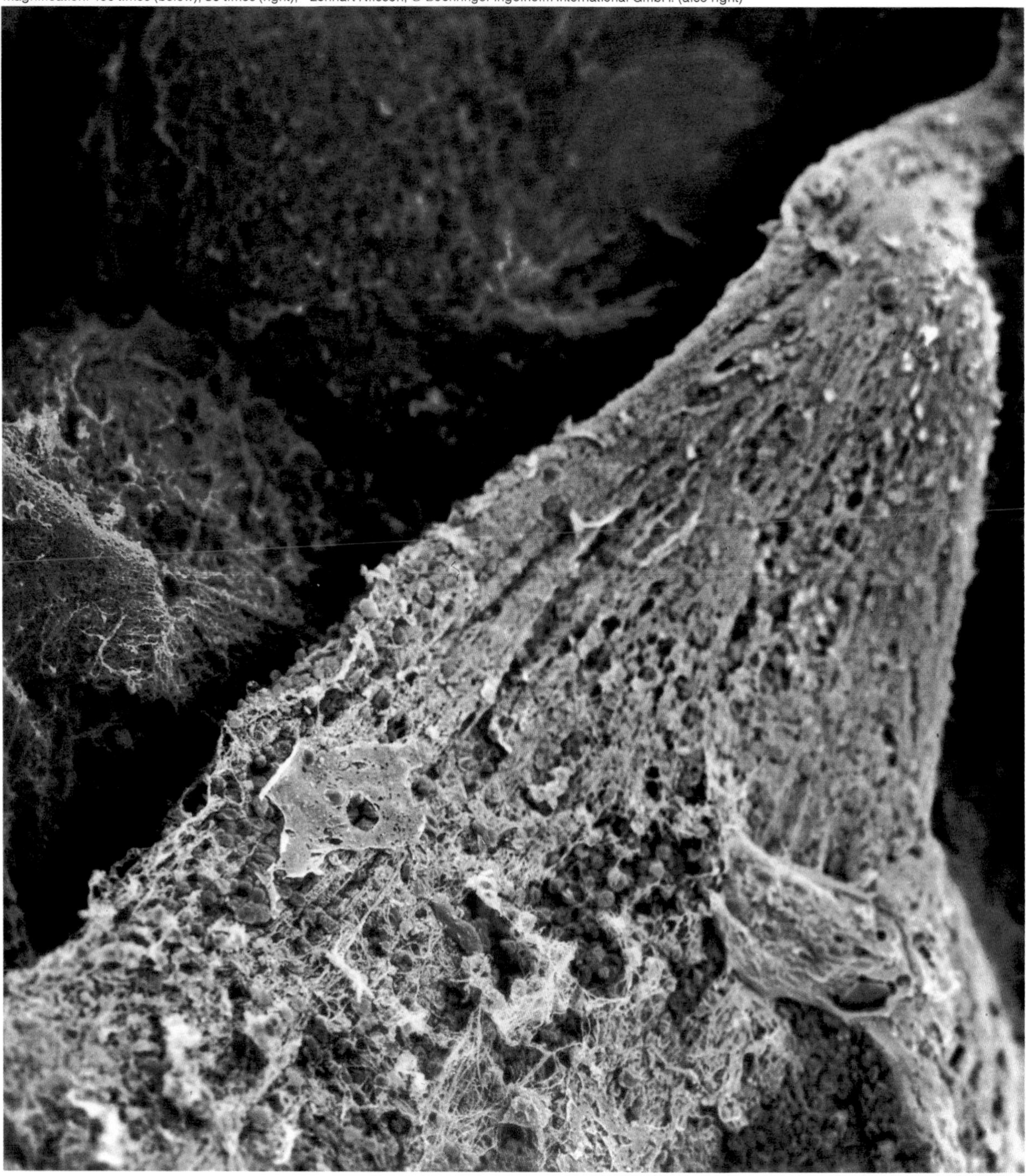

If coagulation proceeds unchecked, it can be dangerous. Normally, anti-clotting agents around the wound keep the process from spreading throughout the circulatory system. As added security, fast-flowing blood carries unneeded clotting agents away from the site. If the blood flow slows, or if anticoagulants fail to perform, a clot may grow to monstrous proportions. The overgrown clot, called a thrombus (above), inhibits blood flow through the vessel. Lodged in a coronary artery (in cross section, right), it can cause death. Sometimes a thrombus breaks free from the vessel wall. Now known as an embolus, it floats through the bloodstream until it suddenly sticks in a smaller vessel, cutting off blood flow.

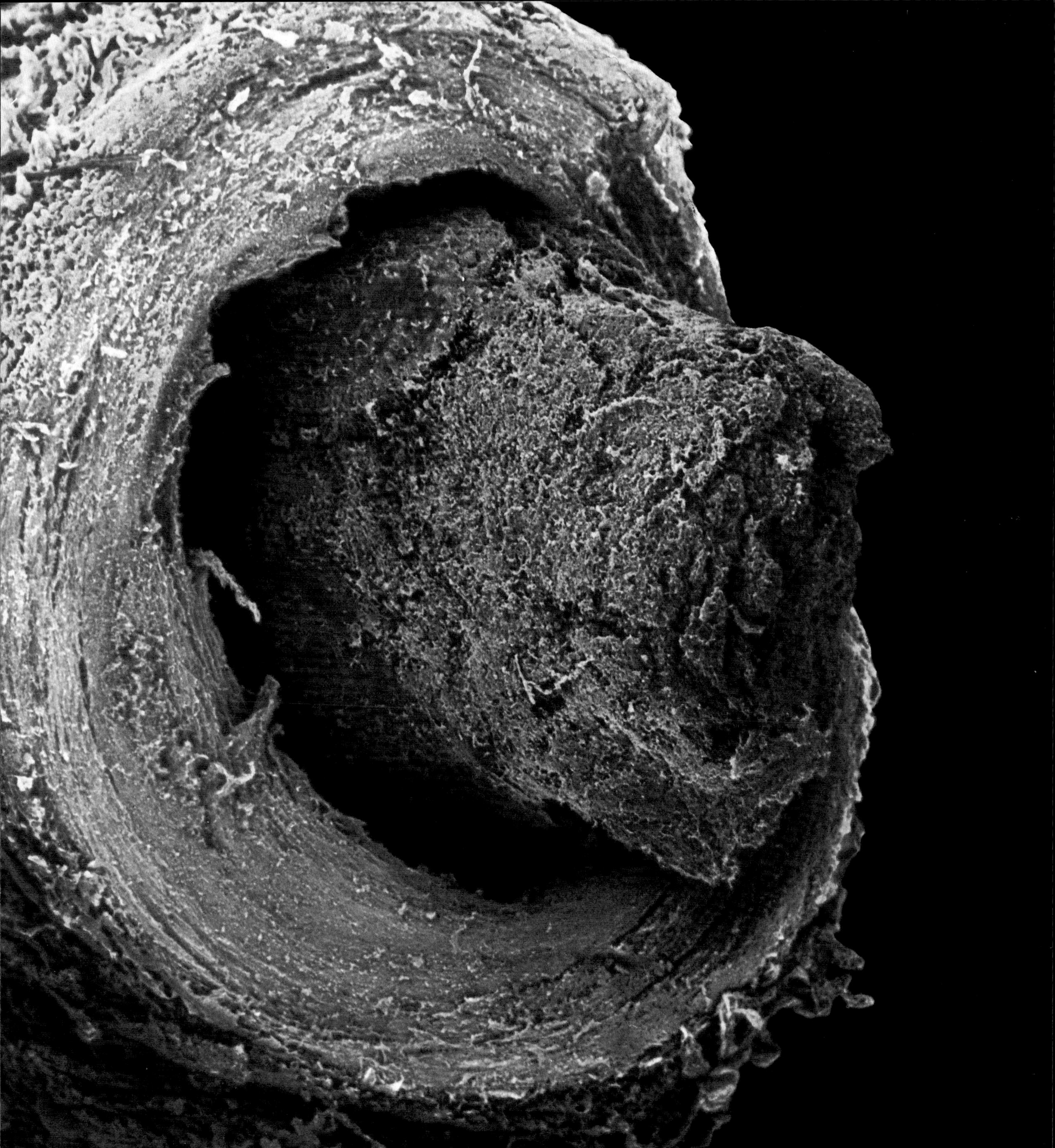

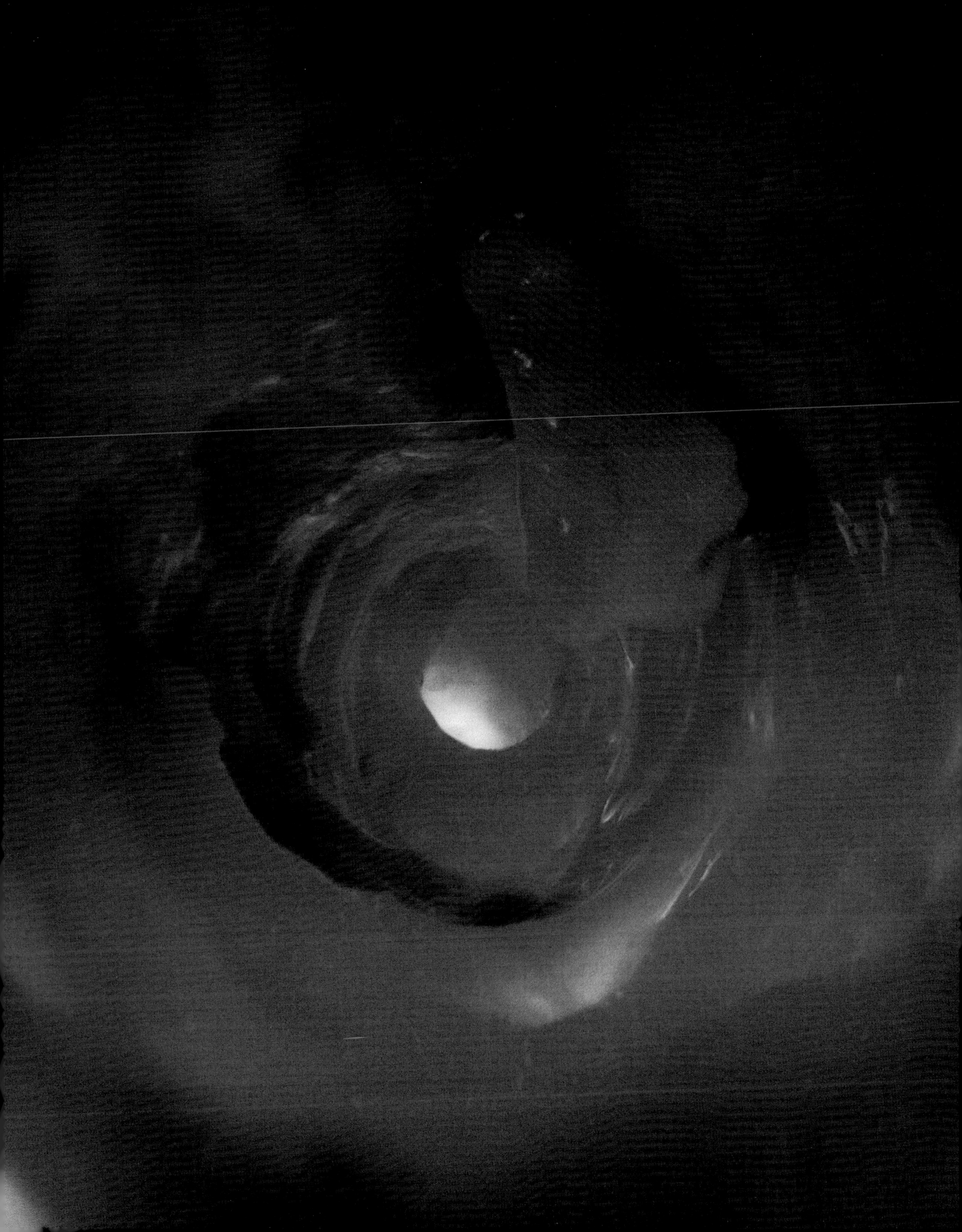

Magnification: 10 times (opposite); Lennart Nilsson, © Boehringer Ingelheim International GmbH.

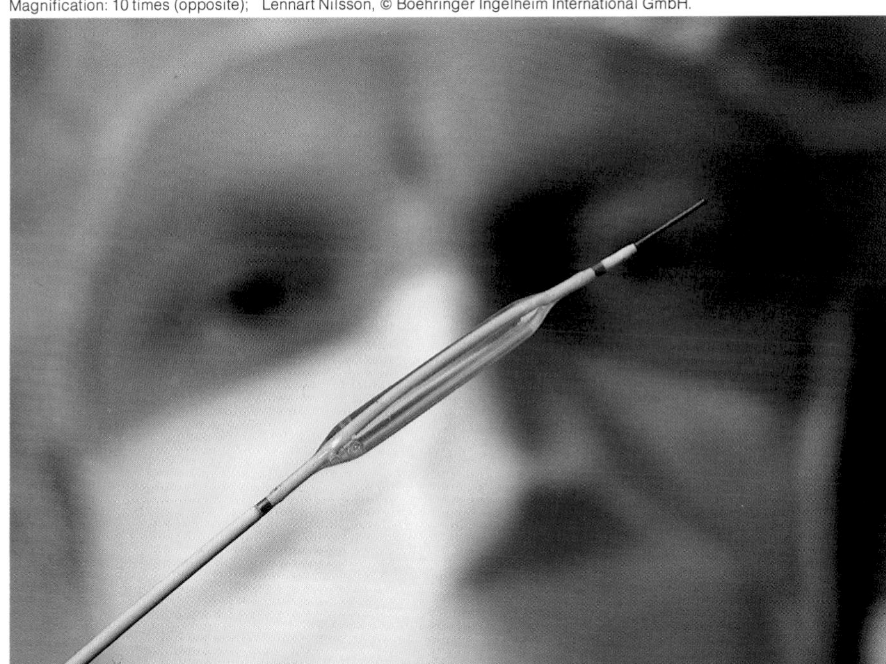

Arteriosclerosis, or hardening of the arteries, threatens blood flow to the heart. It begins with a buildup of minerals, fibrous tissue, or deposits of fat called plaque, which clings like a stalactite inside a normally smooth artery wall (opposite).

If a piece of the plaque breaks away, it may block the artery completely, cutting off the flow of blood. Or it may glide through the bloodstream to distant parts of the body, choking off vessels and damaging tissues and organs.

A procedure called coronary angioplasty clears the blocked artery without surgery. The physician threads a tiny balloon into a catheter (left) that has been guided through vessels leading from the patient's groin to his heart. At the site of the blockage, the balloon is inflated, pushing the plaque against the arterial wall. Doctors watch and work with the aid of live X rays projected on a screen, while the patient lies awake on the table (below). Angioplasty can sometimes replace or delay coronary bypass surgery in the treatment of severe artery disease.

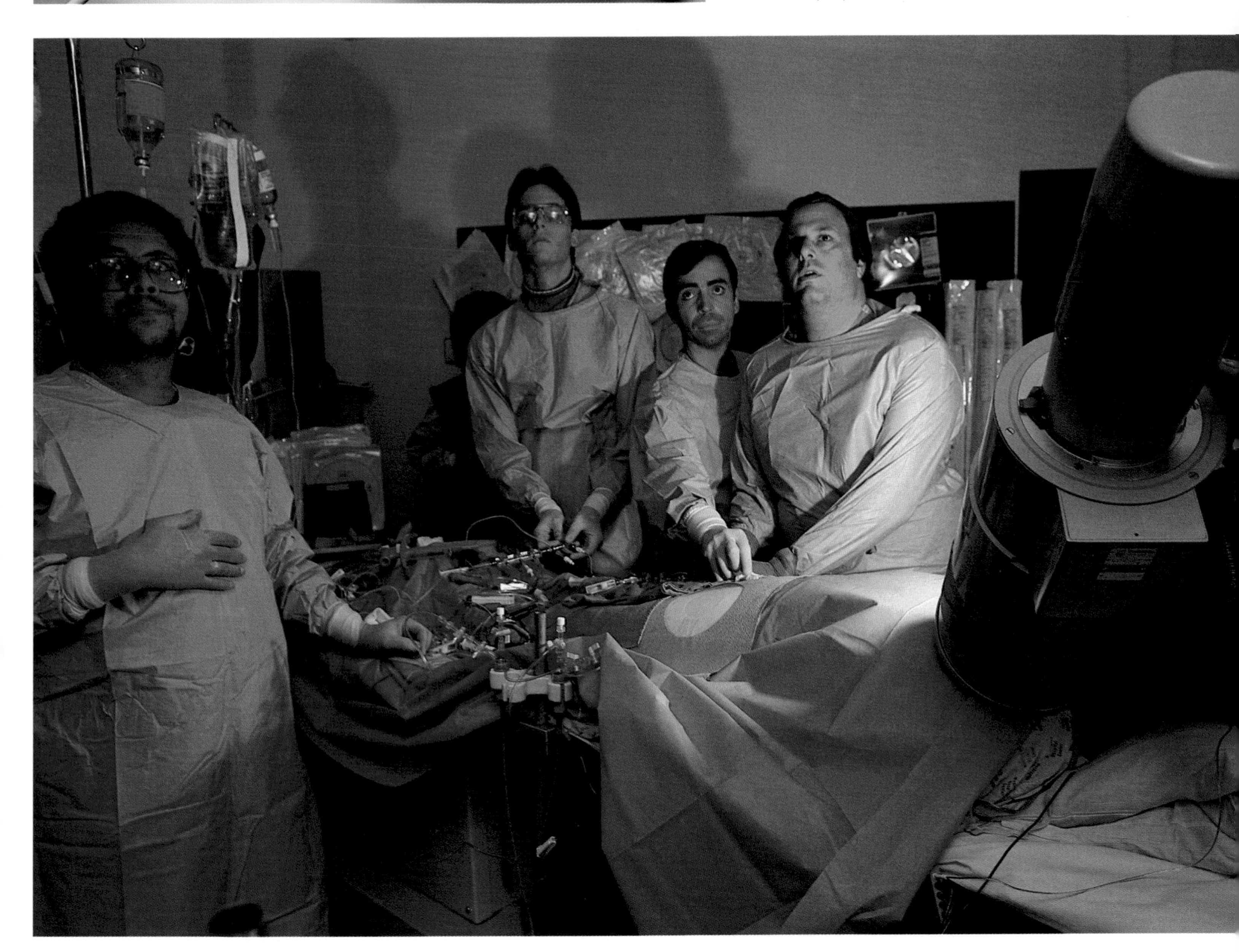

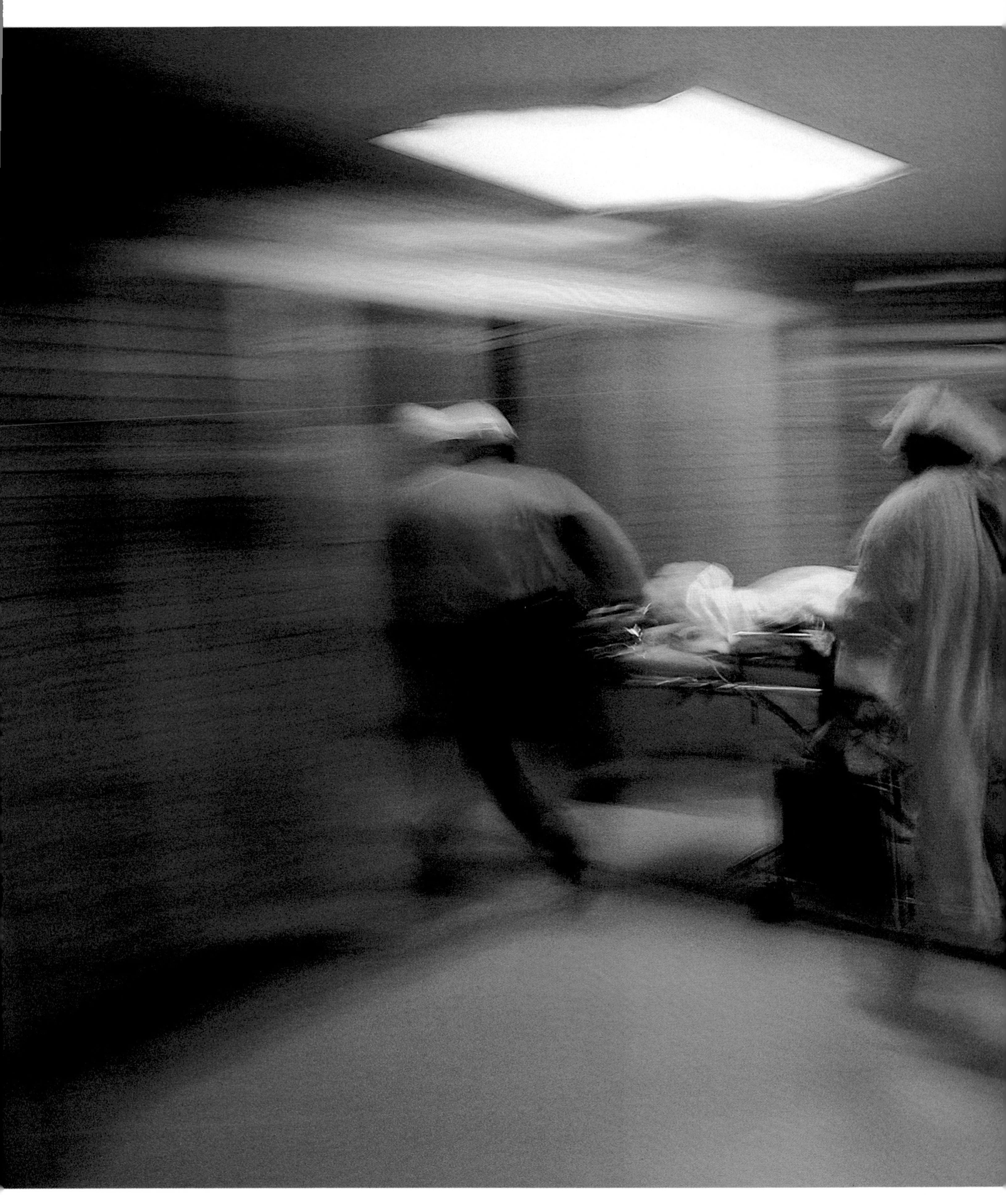

Speed and medical skill count when a heart attack strikes. Delay may be deadly—more than half a million Americans will die of heart attacks this year, 350,000 of them before reaching a hospital. So doctors urge people to heed warning signals, such as acute chest pains that spread down the left arm.

The death of heart muscle—called myocardial infarction or heart attack—occurs when heart cells are starved of oxygenated blood. Following the attack, dead, stalklike muscle cells (tinted brown, right) no longer show the striations of normal muscle (tinted red).

Magnification: 1,000 times

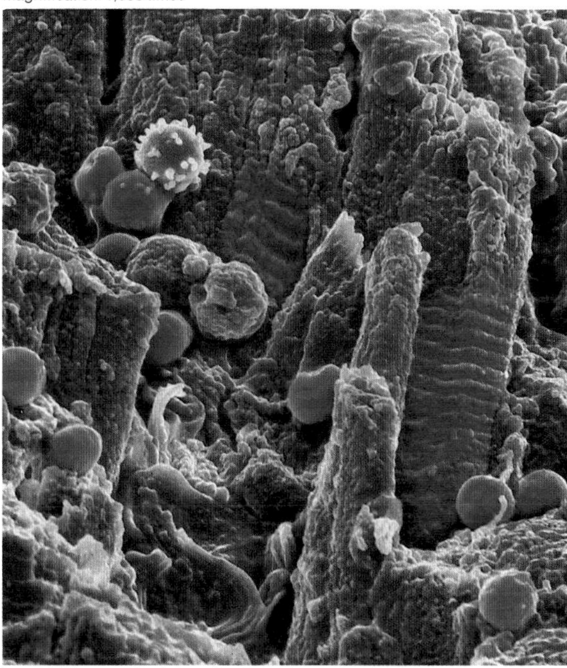

Lennart Nilsson, © Boehringer Ingelheim International GmbH.

called angioplasty, doctors thread a thin tube with a deflated balloon on the end into the circulatory system through a small incision. When the tip of the tube reaches a blocked artery, the balloon is inflated, opening a passage through the blockage. Another, more experimental method puts a tiny blade onto the tip of the tube and takes "shavings" from the cholesterol deposits.

Yet another experiment shows that synthetic glycosides, similar to a natural substance found in alfalfa, will bind to cholesterol in rats and monkeys and carry it through the intestinal tract to be excreted. The Oregon scientists working on this project estimate that a pill might be available in five to ten years that could make eating high-cholesterol foods somewhat less risky.

After a heart attack has occurred, the body swings into action to contain the damage. White blood cells converge on the injured site and clear away debris. Within weeks, scar tissue replaces the dead muscle, patching the heart with stiff, electrically inert fibers that cannot contract like heart tissue does. If the heart attack victim rests completely, the scar will set; after about three weeks, the heart recovers enough to sustain mild activity, pumping with its lifeless but lifesaving patch.

Infarction not only destroys cardiac tissue, it can also disrupt the heart's electrical system. In a healthy heart, individual electric cells frequently fire out of turn, momentarily ignoring the lead of the sinus node, or natural pacemaker. Such random sparks—routinely sensed as skipped heartbeats—rarely present any danger. The sinus node immediately regains control.

During the first few days after a heart attack, however, this erratic firing can trigger a chaotic burst of electrical pulses in the damaged tissue. Even after the scar has set, the danger persists.

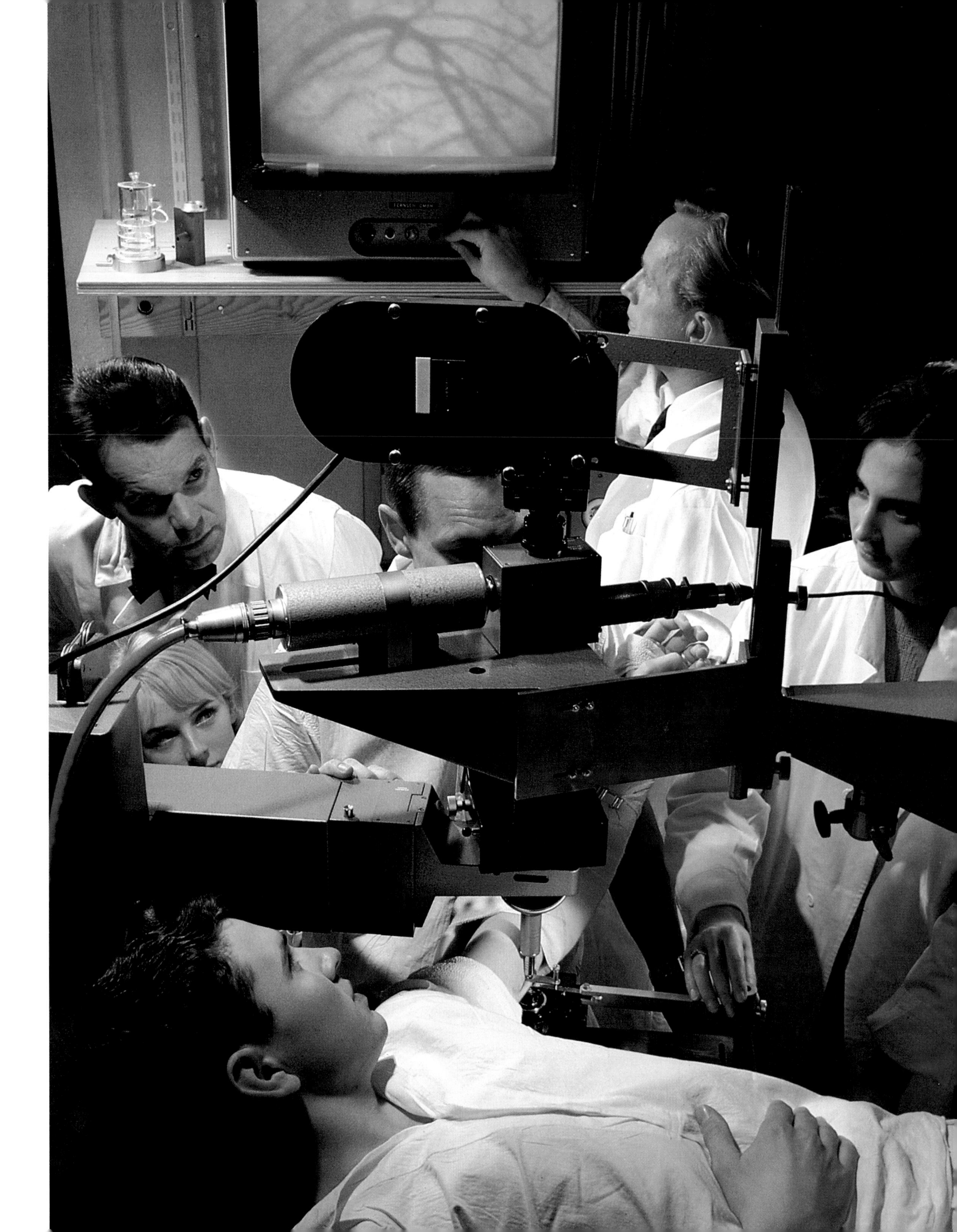

Scientists in a Swedish laboratory (left) study the link between atherosclerosis, the kind of arteriosclerosis in which fat, or plaque, accumulates in the artery wall's inner lining, and a high cholesterol diet. They insert a glass window (below) in a volunteer's arm and expose a bed of capillaries deep within his skin. Poised above the subject's forearm, a camera records and magnifies the capillary landscape and projects it on a screen. Fat particles float in the patient's bloodstream after he has eaten a high-fat meal.

Prompted by a continuously high diet of animal products such as eggs, meat, cheese, and whole milk, globe-shaped fat droplets accumulate and adhere to a coronary artery wall (right). The droplets thicken into a layer of fat that restricts blood flow to the heart.

Magnification: 40 times

Magnification: 1,000 times; Lennart Nilsson, © Boehringer Ingelheim International GmbH.

The pacemaker's normal trail of impulses may hit the inert patch and ricochet. Impulses split and multiply; cells fire haphazardly; the heart's beat falters, then flutters, building to an arrhythmic frenzy called ventricular fibrillation. In a fibrillating heart, cells can fire as rapidly as 500 times a minute. Twitching uncontrollably, the heart collapses into a writhing mass of tissue. Every second counts. If fibrillation is not stopped by cardiopulmonary resuscitation or electric countershock within 4 minutes, irreversible brain damage can occur.

Yet the heart is more resilient than any other organ. Should oxygen deprivation damage the heart's pacemaker, a second, backup node will set a steady pace. And if that node fails, a third bundle of cells, able to spark 15 to 40 times a minute, will start the beat. Meanwhile, the carbon dioxide and other wastes that accumulate when the oxygen level drops, act as potent artery dilators, widening blocked vessels so that blood can nourish threatened tissue. Although platelets, the cells that build clots, pose a danger during infarction by further blocking blood flow, they also perform a lifesaving function. As platelets gather at the damaged tissue, they secrete serotonin, a powerful vessel-constricting chemical that slows bleeding and helps tissues deflect wayward electric sparks.

Though not much bigger than a fist, the heart is wrapped in so much muscle that it can continue pumping even if a third of its muscle mass is destroyed. The left ventricle, the heart's most powerful chamber, can lose up to a quarter of its tissue and still pump effectively. And if the right ventricle—our pulmonary pump—is destroyed, the heart will still send blood to the lungs.

The body protects itself from blocked, hardened arteries by growing bypasses. Small, new arteries, or collaterals, grow from branches above the blocked artery and connect to small arteries below the blockage. Yet the level of protection varies: In some people, collaterals deliver only enough blood to sustain the heart at

FRAMINGHAM HEART STUDY

From a modest clapboard house in the Boston suburb of Framingham, Dr. William Castelli (opposite) directs the Framingham Heart Study. Now in its 37th year, the study is the most comprehensive heart disease survey in history. The Framingham staff has monitored more than 5,000 people for much of their lives, studying their heart rate, blood, weight, and lung capacity and scrutinizing their habits. From the mass of data, scientists have determined that health habits hold the key to heart disease—much more than heredity.

One Framingham patient is checked for blood pressure as part of her biennial examination (right, upper). Sixty-eight years old and in excellent health, she plays tennis two or three times a week (right, lower). Regular exercise and a low-fat diet can keep heart disease at bay, Framingham research proved. Apparently its lessons are being heeded: Between 1963 and 1980 the rate of heart disease deaths fell 30 percent.

resting levels; in others, blood flows normally to the heart even when tests show that the main artery to the left ventricle is completely blocked. Researchers do not understand why some heart attack victims have good collaterals that quickly supply life-saving blood, while others have poor collaterals that do not.

Despite the heart's resilience, a perplexing fact remains: The heart occasionally fails in the absence of any threat at all. One quarter of the people who die of a heart attack show no previous sign of heart disease. Tales of inexplicable heart arrest echo through the ages. Emotions such as fear, anger, and depression can kill as surely as disease. So can stress. Nearly a century ago Sir William Osler, one of the most prominent medical educators of all time, noted that his typical coronary patient was a "keen and ambitious man, the indicator of whose engine is always set at full speed ahead."

The Role of Emotions

Modern studies agree with Osler. Type A personalities, aggressive, achievement-oriented, impatient people suffer heart attacks at double or triple the rates of the Type Bs, people with more placid personalities—a trend that applies across the borders of countries and cultures.

Severe emotional upheavals can also cause heart arrest. Researchers have gathered clues from personalities of victims and events that precipitated fatal heart attacks. People can actually die of broken hearts, or feel they are dying. "My heart aches, and a drowsy numbness pains/My sense, as though of hemlock I had drunk," wrote John Keats in *Ode to a Nightingale.* In the first six months after the loss of a spouse, the coronary death rate among widows and widowers jumps far above the average for their age brackets. Unrelenting depression also seems to cause sudden death. So does personal danger, loss of self-esteem, even overwhelming joy—as evidenced by the man whose heart stopped while celebrating the news that his heart was sound.

While we have known for centuries that emotions affect the heart, until recently we have not understood how. In the 1940s physiologist Walter Cannon, intrigued by the puzzling fatalities of South American Indians who died when condemned by their medicine men, proposed a biological mechanism of anxiety, a voodoo theory of death. He speculated that the flood of adrenaline released during fear and rage prepares the body so powerfully for attack or escape that it overwhelms the heart and blood vessels. Physical action dissipates this energy. In anger or fear the battle rages within.

Every discovery about the cardiovascular system confirms Cannon's theory. Cardiac patients tend to react to stress by releasing far more

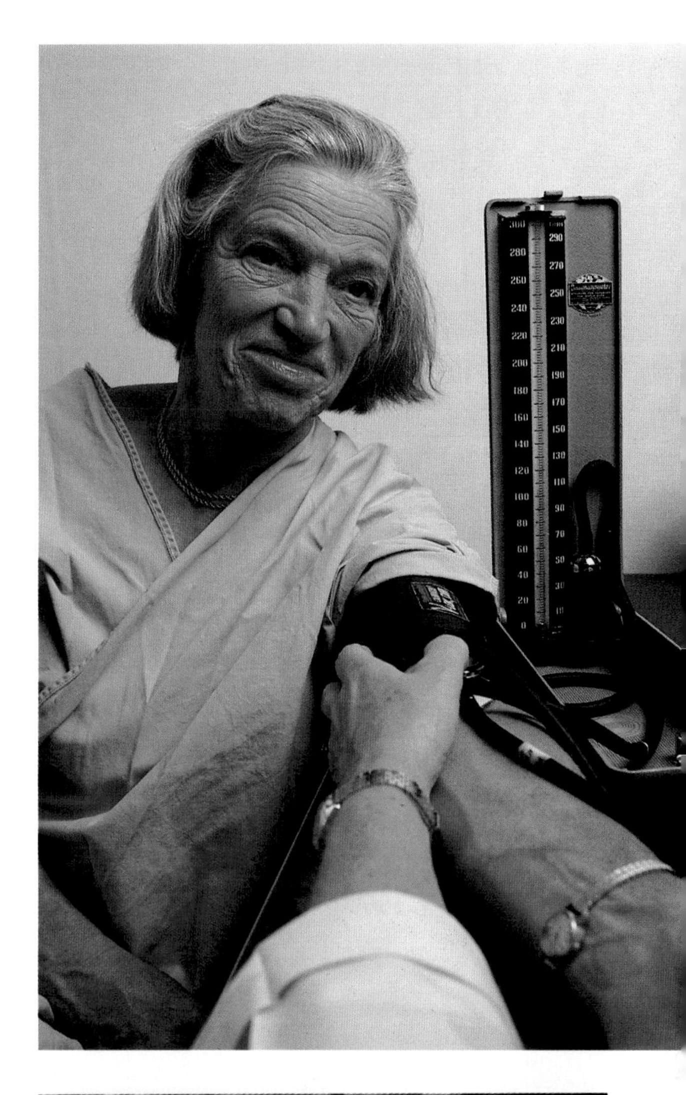

adrenaline and noradrenaline than other people. The sympathetic nerves, which secrete adrenaline, normally use body fat for fuel during an emergency; thus, some people under stress show high levels of fats in their bloodstreams, as well as more severely clogged arteries. The sympathetic nerves also prime platelets for clotting. During fear, anger, or anguish, platelets and blood cells seem to stick together more readily, increasing the chances of a fatal blood clot. But the most devastating evidence linking the sympathetic nerves and adrenaline to sudden cardiac death involves the heart's electrical system. Adrenaline speeds heart rate during stress, requiring heart muscles to burn more oxygen to keep pace. When the heart's racing beat overwhelms its oxygen supply, cells lose their electrical stability and begin to fibrillate. Even the stress of traffic can cause a driver's otherwise healthy heart to pound irregularly. Studies show that heart attacks most often occur in the morning, when physical and mental stress are at a peak.

The sympathetic response, with its galloping heartbeat and skyrocketing blood pressure, accounts for the deadliness of fear and anger. But the parasympathetic system, by slowing the heart, can also wield a fatal power. Rats placed in a water-filled tank with no obvious escape route have drowned when their heartbeats drastically slowed—a parasympathetic response.

Occasionally, the sympathetic and parasympathetic systems compete, each struggling for the upper hand. A person can be torn apart emotionally by outside experience and biologically by events within. Overwhelmed by the urge to run and the urge to give up, gunning the motor and braking it at the same time, the human engine can careen out of control and stop dead.

A Positive Approach

Yet the mind can exercise a positive influence on the heart as well. Many forms of meditation, including biofeedback, can actually lower blood pressure, oxygen consumption, and production of carbon dioxide in 10 to 20 minutes to levels that, usually, only hours of sleep produce. Practitioners of yoga have learned how to slow their hearts to 30 beats a minute or less—through willpower alone. People who develop a more relaxed approach to life have been able to lower their blood pressure. And a patient's heart rate can be slowed by as much as 30 beats a minute by a nurse who takes his hand and comforts him.

In the past few decades we have learned how physical exercise strengthens the heart. Now we find that the exercise of positive emotions—optimism, hope, and love—can strengthen the sturdy organ as well.

SUSAN SCHIEFELBEIN

In a Los Angeles health center, a nutritionist teaches proper diet to a 48-year-old heart bypass patient (above), as other cardiac patients work out on cycles (opposite). Exercise and a low-fat diet can prolong life by raising the level of high-density lipoproteins (HDLs) in the body. These proteins help keep cholesterol from invading an artery's lining and forming the plaque that restricts blood flow.

High- and low-density lipoproteins (LDLs) transport cholesterol through the bloodstream. LDLs carry the most cholesterol and are associated with arteriosclerosis. HDLs take fat from artery walls and deliver it to the liver for removal. Exercise and careful eating habits can lower the level of the bad LDLs and raise the good HDLs.

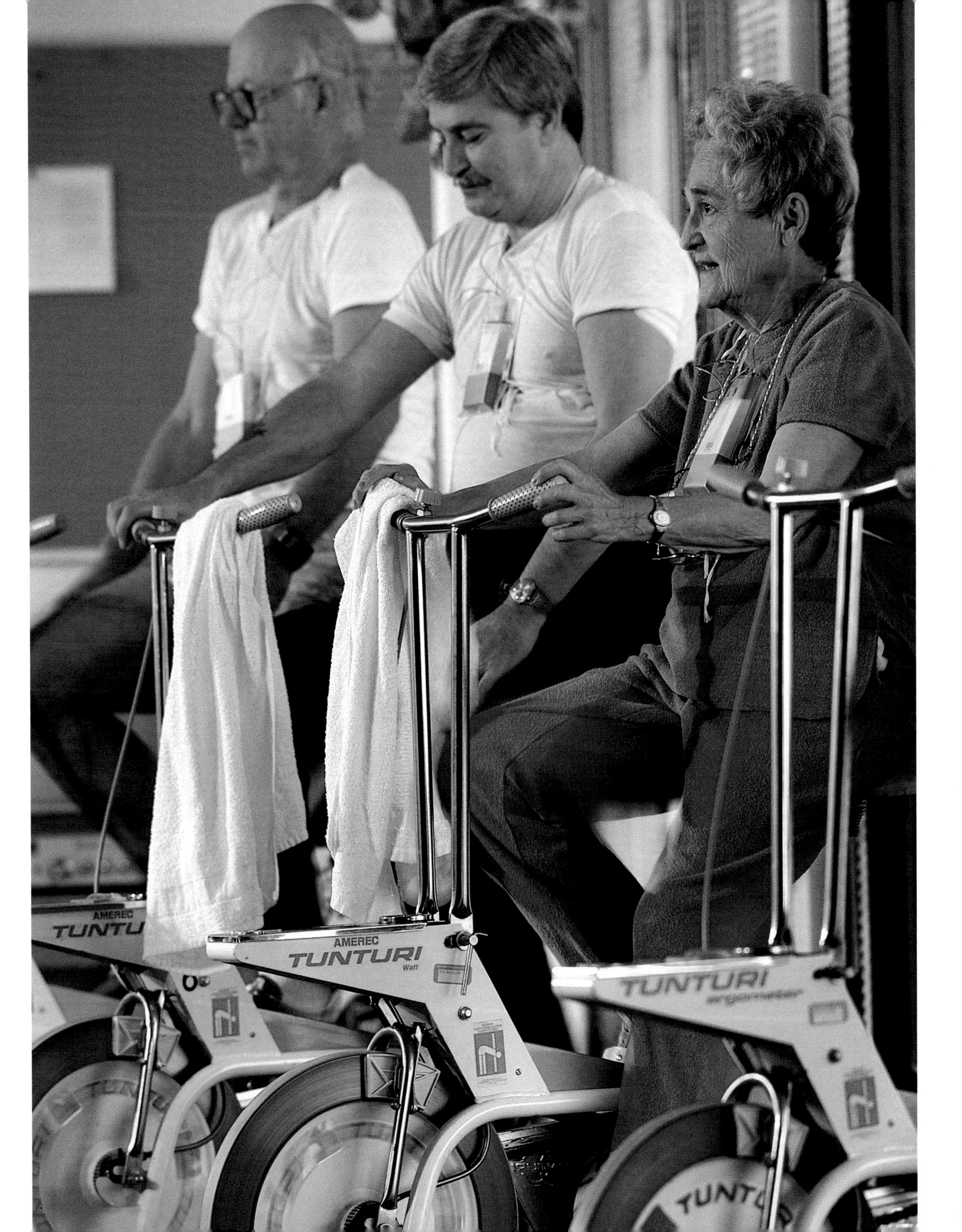
AMEREC
TUNTURI
AMEREC
TUNTURI
Watt
TUNTURI
ergometer
TUNTURI

The Healer Within

We live in a world dense with microbes: Bacteria and viruses, parasites and fungi abound in the air, water, and soil, and on the living things around us. Most of these organisms have little interest in the human species. But a specialized few find the human body an inviting habitat: warm, protected, and well stocked with nutrients. Some settle in the nose and ears, some on the skin and in the intestinal tract.

Usually, we live in harmony with these microscopic residents. Most stay on the body's surfaces. But under certain conditions—when we are malnourished, exhausted, injured, or under severe stress—resident organisms and other microbes may invade and multiply in our tissues, or set forth in the bloodstream, traveling to all parts of the body. If unchallenged, they can cause serious, even fatal, afflictions.

Considering the number of potential interlopers, disease occurs very rarely. This is no accident. Nearly every human possesses a sophisticated and efficient system that works 24 hours a day in every part of the body to ensure good health. Known as the immune system, this network of cells and organs responds almost instantaneously to the presence of any disease-causing intruder, mustering its forces to halt the progress of a poliovirus or to thwart the efforts of a meningococcus bacterium.

We rely on this powerful system not just to repel disease-causing microbes, but to keep house inside the body. Good health depends on order and consistency among body cells, tissues, and organs. The immune system preserves this state of balance by removing dead or damaged cells and by seeking out and eliminating wayward or mutant cells.

Known as the "system of self," the immune system possesses the remarkable ability to tell self from nonself, friend from foe. It recognizes and destroys cancer cells, transplanted tissue cells, and a wide range of organisms, from the minute picornavirus—so small that more than a million lined up would fit in the space of an inch—to some parasites that are visible to the naked eye. At the same time, it usually respects the body's own tissues, varied as they are. Sometimes the difference between self and foreign is slight—a matter of only a molecule or two—as in the distinction between a normal cell and a cancerous one. How does the system tell self from nonself? How does it know what to respect and what to reject?

Nearly every substance known to humankind bears a chemical identity card made of a characteristic pattern of molecules at its surface. Each of the cells that make up our own tissues and organs carries such an identity card. Because genes determine the shape and nature of human self-markers, they are unique to each individual. Lodged in the outer surface of our cells, they stand as flags of our identity.

Magnification: 12 times; Lennart Nilsson, © Boehringer Ingelheim International GmbH.

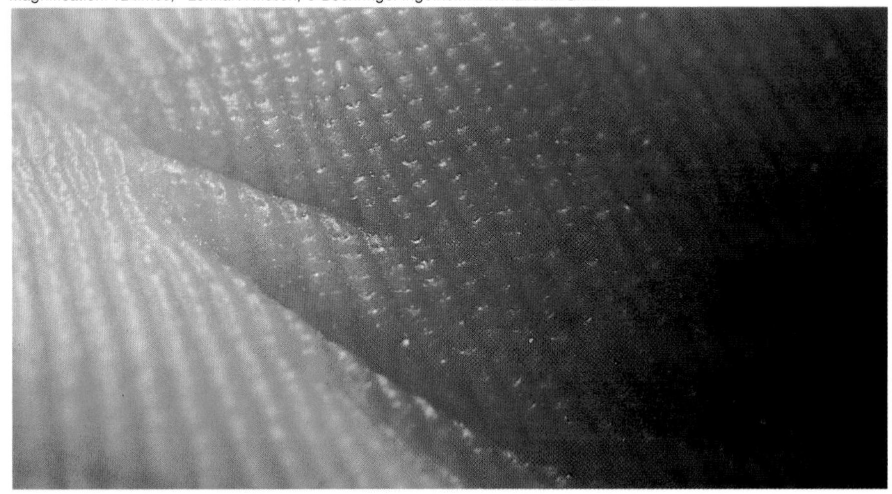

Magnification: 600 times

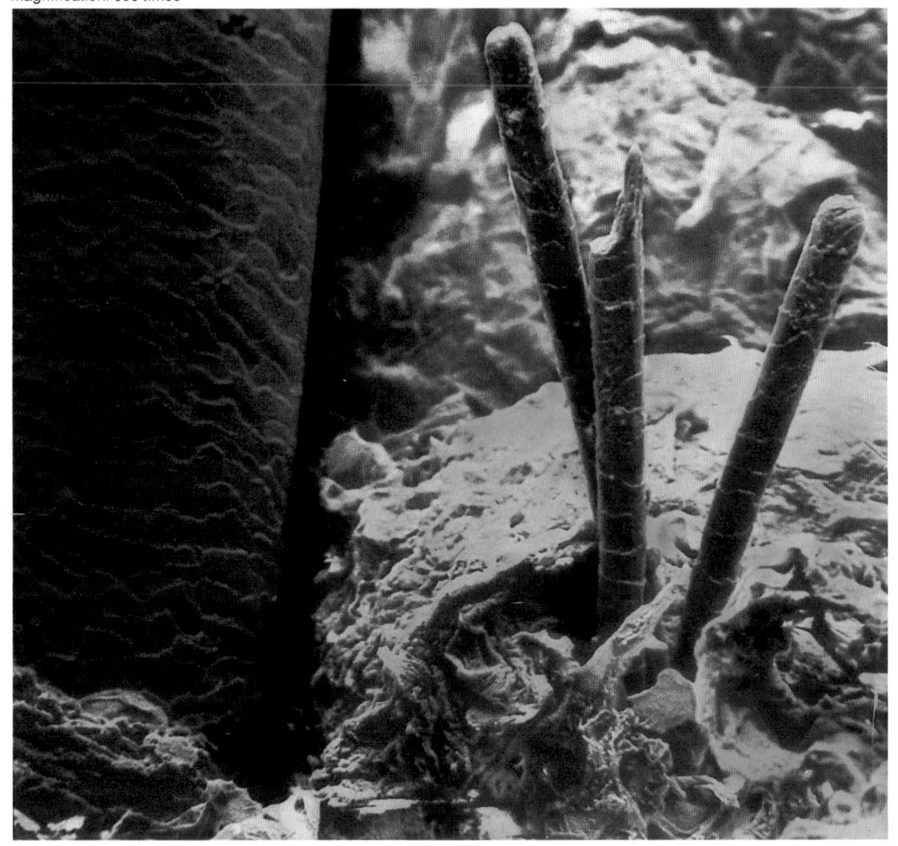

The skin is the body's largest organ. It preserves internal fluids and shields us from an often hostile world. The versatile armor is thinnest on the eyelids, and thickest on the feet and palms. Epidermal cells on the fingertip (upper) form the ridges and grooves necessary for a good grip. Throughout life the same kind of cells may develop instead into tubes that become coarse hair (above, at left) or fine, downy hair (at right).

A tough outside covering of dead cells filled with the protein keratin serves as the first line of defense against the assaults of water (opposite) and a host of invisible attackers. More than a protective shield, skin helps regulate the body's temperature. Acting on signals from the brain, skin adjusts to rid the body of excess heat or keep in needed warmth, enabling humans to survive extremes of desert heat or Arctic cold.

At all times, the immune system surveys the chemical markers of every molecule and cell in the body. Because the marks on our body cells differ from those that brand foreign substances, the immune system can pick out intruders and avoid the reckless execution of friends. If the system recognizes a mark as self, it will usually respect the substance; if it detects a foreign mark, it will launch an attack to destroy the invader.

Any substance that triggers such an attack is called an antigen. Viruses, parasites, fungi, and bacteria can act as antigens. So can blood cells or tissue from another human being and altered self-components, including cancer cells or cells infected by a virus. Even seemingly innocuous substances, such as ragweed pollen, mildew, animal dander, or house dust, can provoke a full-blown attack.

First-Line Defenses

Everything we encounter bears a rich load of foreign material. The air we breathe carries dirt, exhaust fumes, and bits of debris, including pollen grains from hundreds of different plants.

Dust that settles on the furniture and floors of our homes often holds human or pet dander—tiny scales that flake from hair, skin, or feathers—as well as microscopic cousins of the spider called mites. Even the food we eat harbors bacteria, mold, and fungal spores. Fortunately, the immune system does not have to confront the bulk of these substances. A set of first-line defenses keeps most foreign material outside the body, away from its inner tissues. These defenses include the body's tough outer covering of skin and membranes; its protective reflexes, such as coughing and sneezing; and a variety of fluids that wash over its surfaces.

As long as the skin remains uninjured, it holds the body's insides in and keeps the rest of the world safely out. This remarkable organ also has the ability to regenerate. Despite daily scratching, ripping, tanning, burning, and exposure to irritating soaps and drying heat, skin retains its integrity. It constantly relubricates and replenishes its outer surface, the epidermis, and heals ruptures in its deeper layer, the dermis.

A sheet of dead cells at the outer surface of the epidermis acts as the body's primary shield. The cells that form this sheet arise from a layer of living, dividing basal cells, so named because they reside at the base of the epidermis. As the newborn cells move up from the base, pushed from beneath by rapidly dividing cells, they produce greater and greater quantities of a protein called keratin (from the Greek *keras,* or horn). The fibrous keratin accumulates within the cells until it nearly replaces their living cellular machinery. When the cells reach the surface, some four to five weeks after the start of their journey, they

Magnification: 8,000 times (below); 50 times (right)

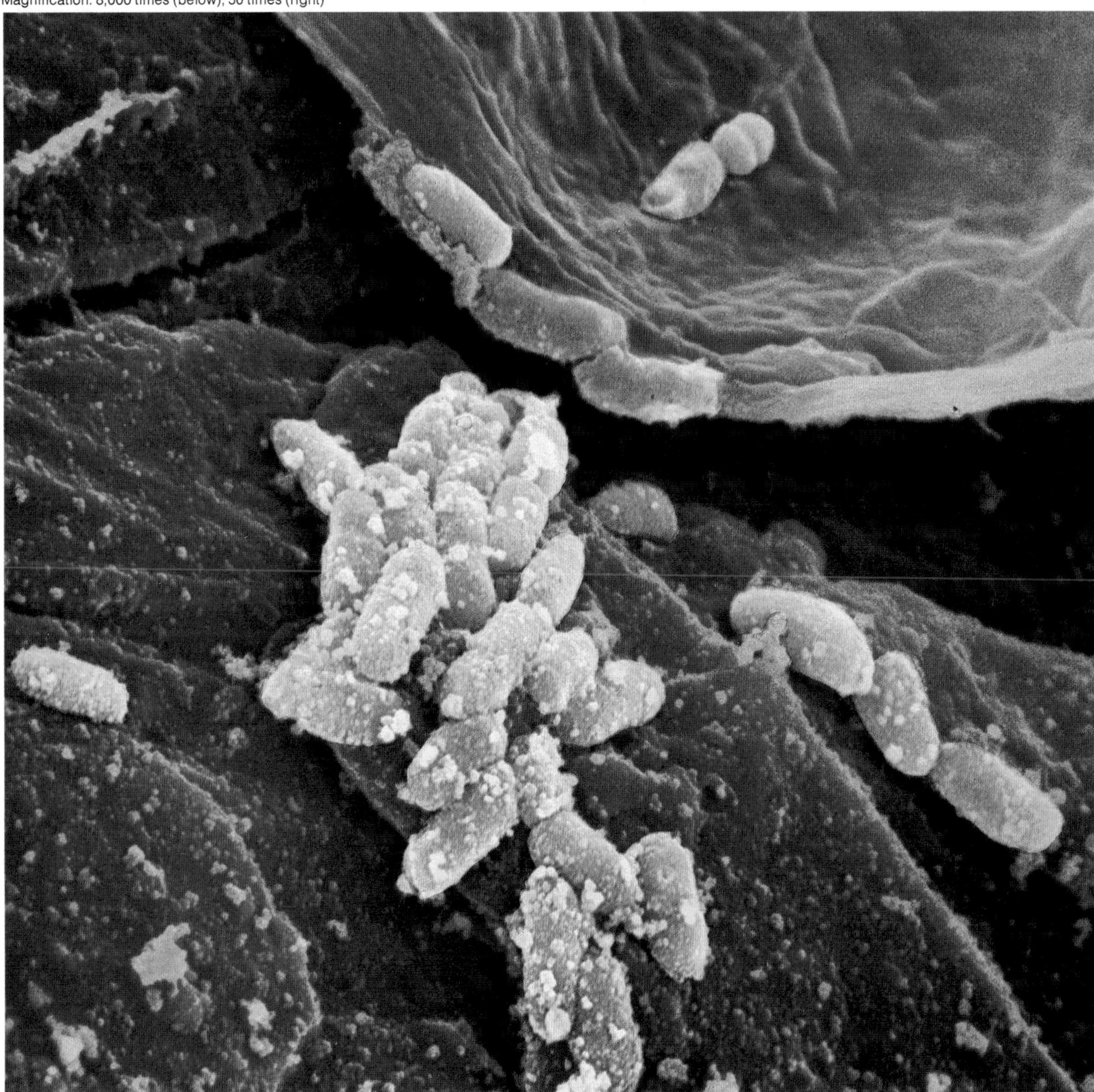

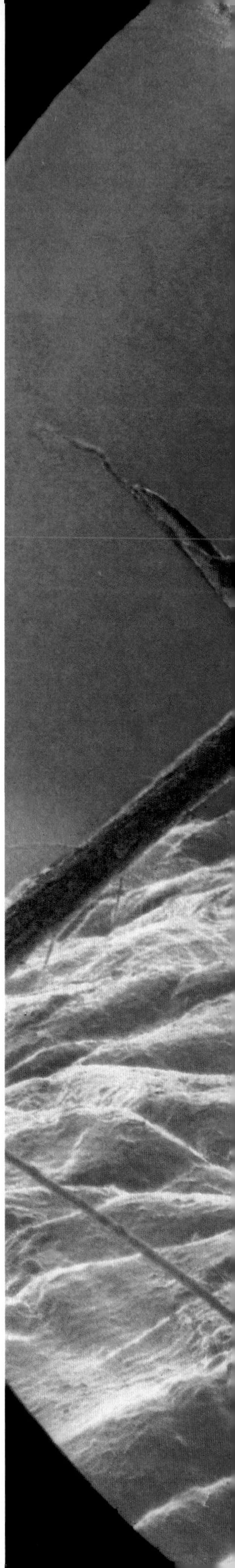

have withered, died, and bound themselves firmly to one another, forming the tough, nearly impermeable part of the epidermis known as the horny layer.

The perpetual shedding of the epidermis prevents many microbes from penetrating the skin. Suppose a bacterium latches onto a surface skin cell. As the epidermis goes about the business of renewing the horny layer, it sheds the desiccated cells at its surface at a rate of about a million every 40 minutes. The old cells curl up and peel off, flinging into space the microorganisms perched upon them. The simple act of rubbing an arm or pulling off a sock wreaks havoc in this microscopic community.

Far more than just a passive covering, the epidermis houses special cells that join the immune system in defense against disease. Cells known as Langerhans cells give agents of the immune system information regarding the nature of foreign substances that enter the body through the skin. When the system receives such a message,

Agent of many ailments, a mosquito pierces the skin to drink blood (right). When a mosquito alights on a human and "bites," it may inject microbes that cause malaria, encephalitis, or yellow fever.

Most microbes encounter millions of resident, friendly bacteria on the rough, irregular surface scales of the skin (above). Even when skin is freshly soaped and rinsed clean, some areas may be home to as many as 20 million bacteria on every square inch (tinted green, above). Most resident bacteria are harmless to their host. They can also work to the host's advantage, by defending the body from dangerous invaders.

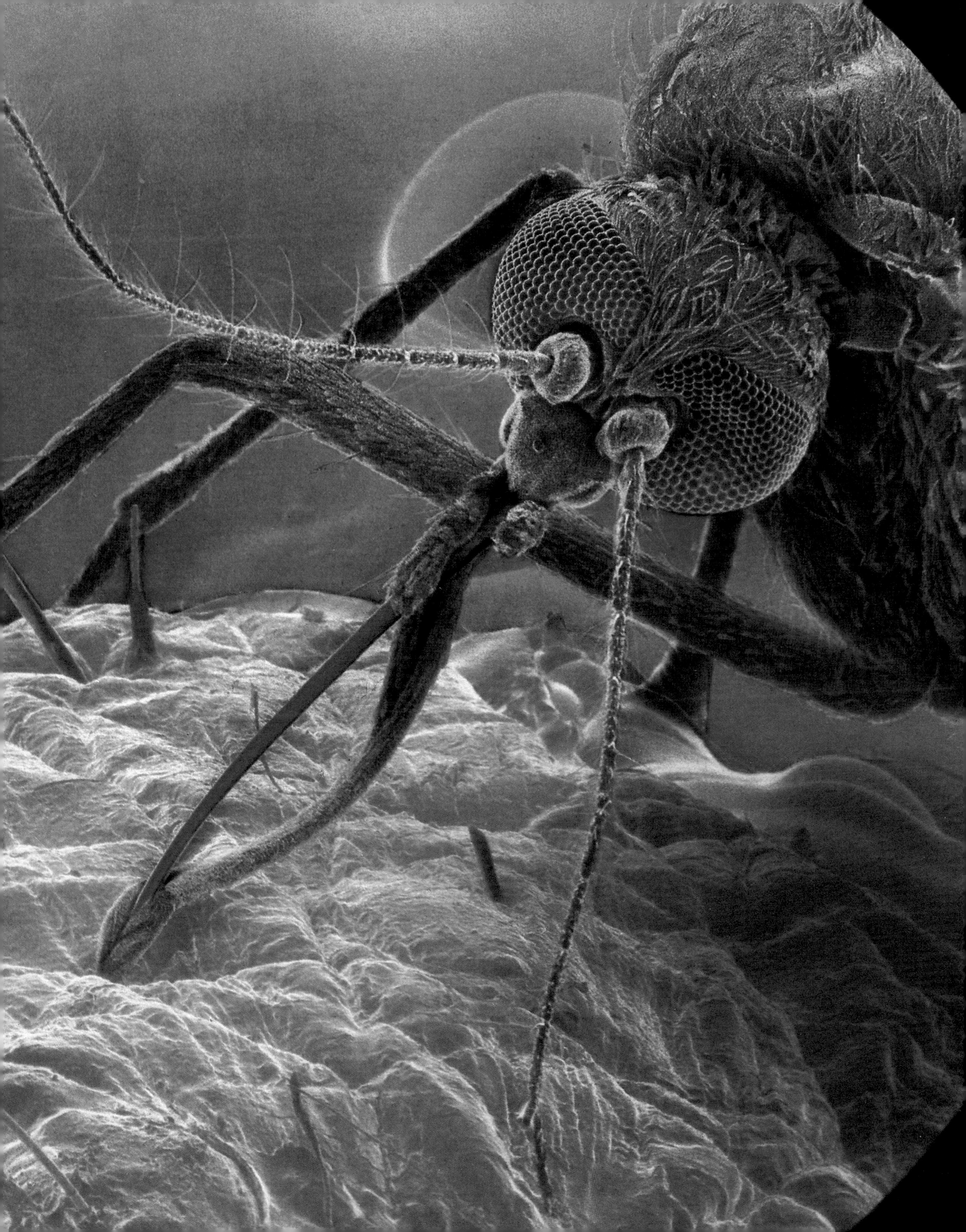

Beneath a forest of hair, the terrain of skin throbs with life. One square inch may hold 650 sweat glands, 20 blood vessels, and more than a thousand nerve endings.

On the top layer, or epidermis, a sheet of dead cells forms a horny shield of keratin. Many microorganisms perish on contact with this surface, which is bathed in salty sweat and acidic, oily sebum. Other microbial invaders fall away as surface cells dry up and flake off. The shed skin is replenished by a living layer of basal cells that divide and move to the surface.

Sebaceous glands pump out the sebum that lubricates skin and hair. To limit the damaging effects of sun, melanocytes inject surface cells with the pigment melanin.

Deeper down, the thick mass of connective tissue called the dermis, and lower-lying fat cells, act as shock absorbers, padding the body's inner tissues from outside blows.

Blood vessels help regulate temperature, widening or constricting to release or conserve heat. On cold days the arrector pili muscle contracts, causing the hair to stand up. In animals this reaction traps an insulating layer of warm air near the skin. In humans it results only in goose bumps.

Snaking through this environment is a warning system of nerve fibers that terminate in endings—either free or enclosed in corpuscles. This network tingles in response to touch, pressure, heat, and cold, alerting the brain to the world outside.

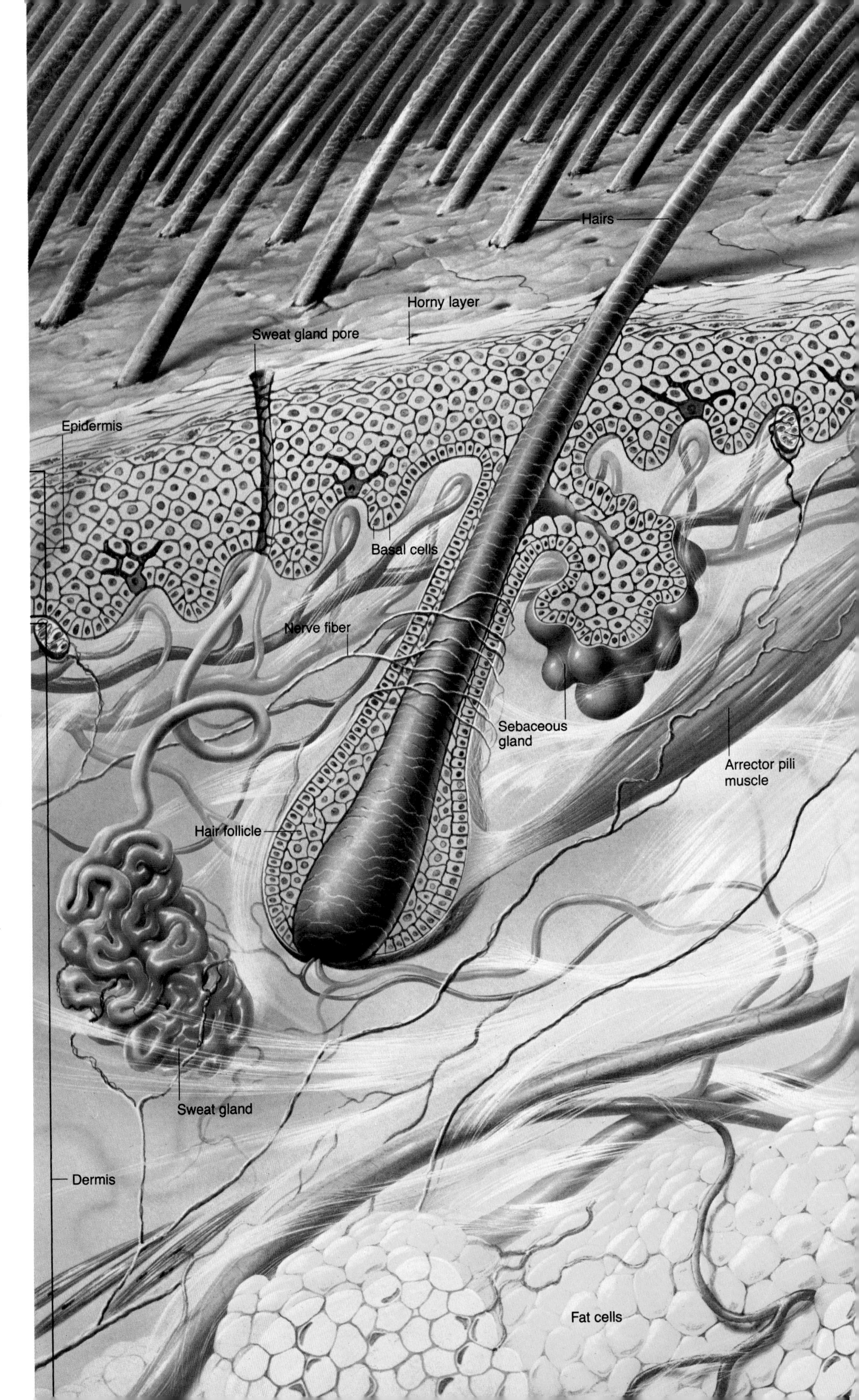

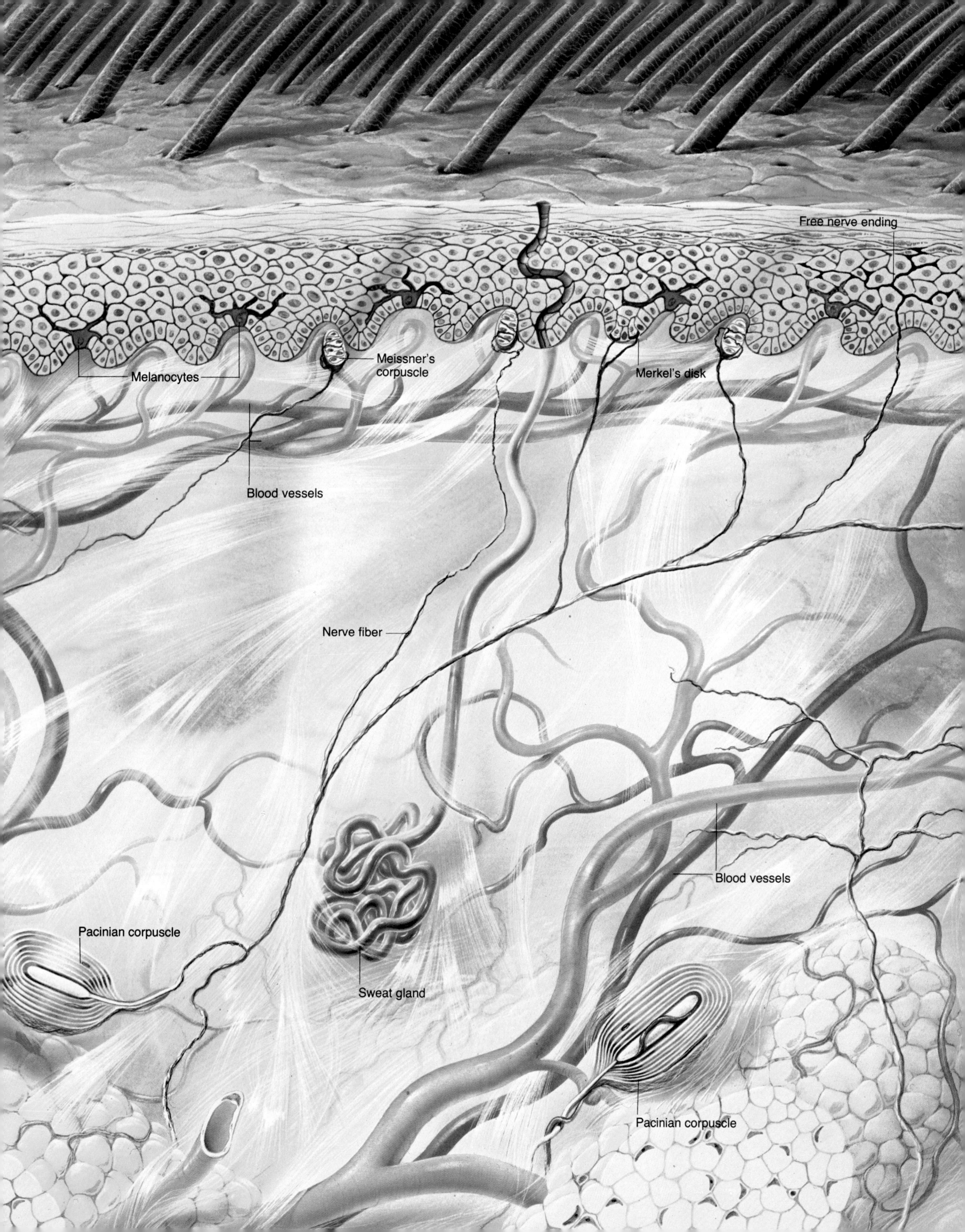

Free nerve ending
Melanocytes
Meissner's corpuscle
Merkel's disk
Blood vessels
Nerve fiber
Blood vessels
Pacinian corpuscle
Sweat gland
Pacinian corpuscle

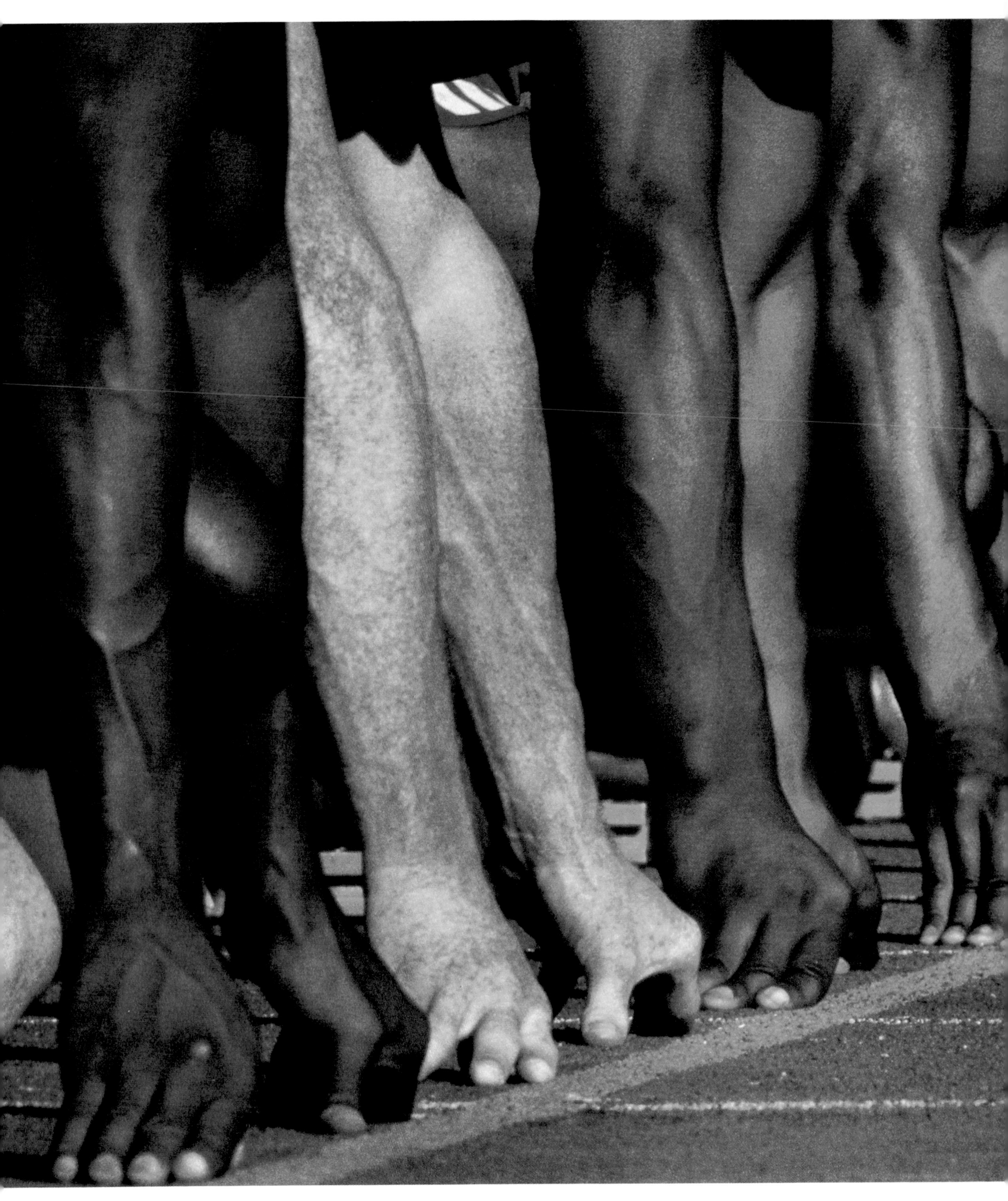

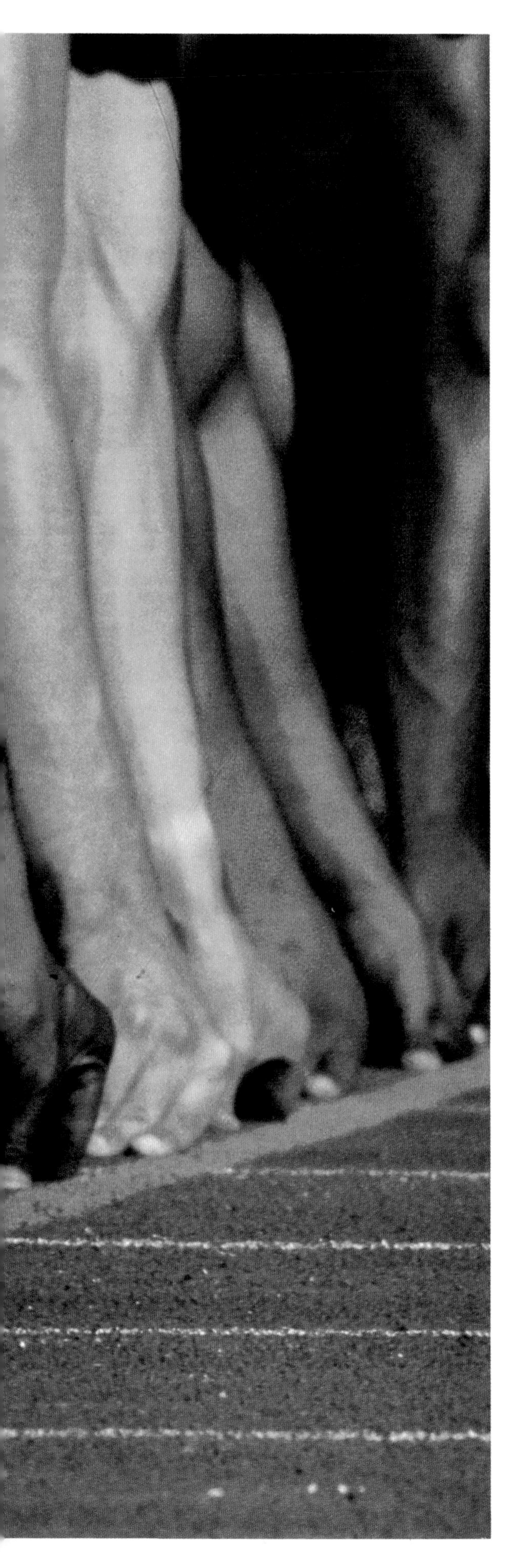

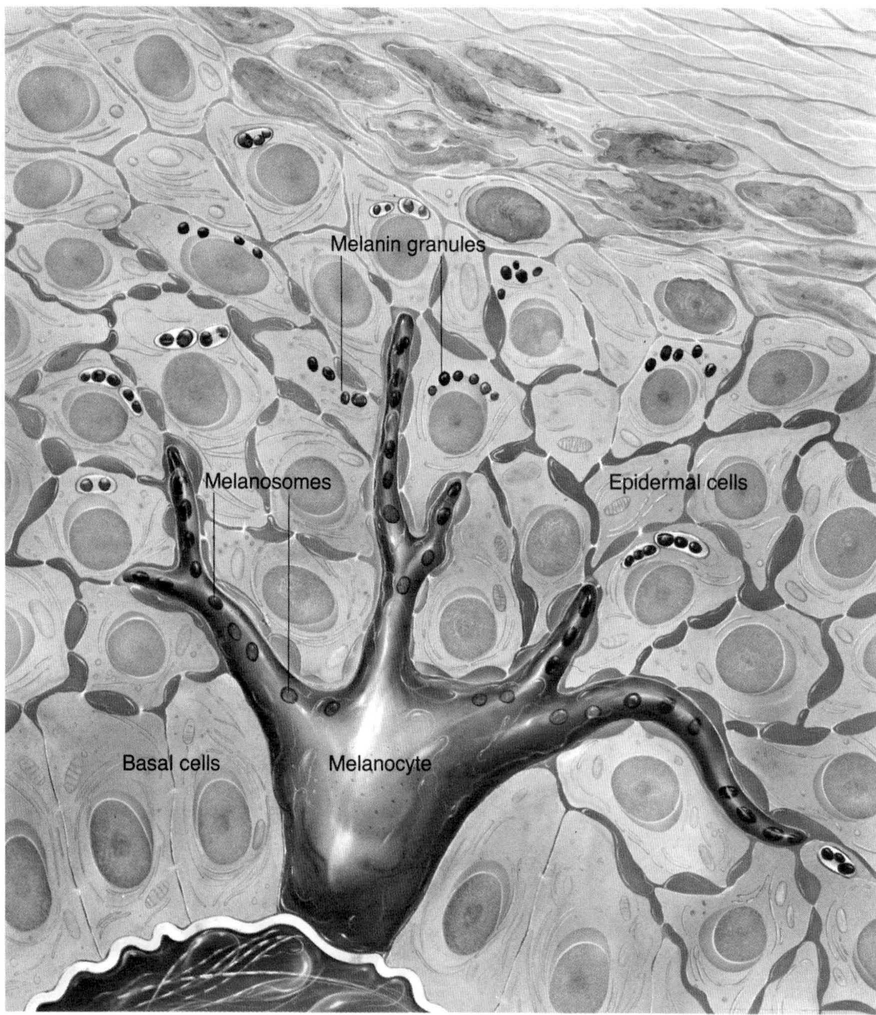

Olympic runners (left) display a range of skin color. Spider-shaped cells called melanocytes (above) produce melanin, the pigment that decides skin color. We possess 60,000 melanocytes in every square inch of skin. Color differences depend on the amount of melanin produced and on its distribution.

Triggered by sunlight, a melanocyte forms small sacs of accumulating melanin—melanosomes. When filled, the sacs migrate as granules into surrounding cells and park between the nucleus and cell membrane. There, melanin acts as an umbrella, protecting the cell's genetic material from ultraviolet rays. This is what we call a tan.

it sets in motion a response to counteract the invader. Extremely high doses of ultraviolet light rays can damage Langerhans cells, preventing them from sending the appropriate warning signals to the immune system—a good reason to avoid overexposing the skin to the sun.

Though we rarely think of them as such, the membranes that line the body's internal surfaces, including the respiratory and digestive tracts, are as much a part of the body's protective covering as the skin is. These membranes encounter microbes and other foreign material in quantity, and so must be armed with a set of defenses equal to the skin's.

In the respiratory tract—the air passages that lead to the lungs, such as the nose, trachea, and bronchial tubes—a collection of highly efficient mechanisms work around-the-clock to ensure that only moist, temperate air, almost free of debris, reaches the lung's air sacs. During the course of a day in a city, we inhale up to 17,000 pints of air. That air contains some 20 billion particles of foreign material, including dirt, dust, and chemicals, most of which never make it to the lungs.

Airborne matter that enters the nose must pass through a trap of stiff nostril hairs that catches many of the larger particles. Just past this trap, the direction of the airstream shifts abruptly with the curve of the bones in the nasal passage, forcing some of the larger particles to collide with the wall of the pharynx. Here, tonsils and adenoids—strategically placed tissues containing agents of the immune system—trap foreign material and see to its destruction.

Smaller particles may make it farther down the tract, landing on the walls of the trachea and bronchial tubes. Special cells and glands in the membranes that line these walls secrete a slightly sticky fluid, mucus, which traps and holds dirt, debris, and microorganisms. Tiny hairlike projections called cilia, which carpet the membrane, then sweep the material away from the surface. With rapid, forceful strokes, the cilia push the mucus and debris out of the passage at the rate of about an inch a minute. This escalator of cilia removes nearly all foreign material to a part of the throat near the mouth called the oropharynx, where it can be coughed out, or swallowed and eventually eliminated from the digestive tract with other wastes. Heavy smoking can paralyze the action of cilia and thus lower the smoker's resistance to respiratory infections.

Sometimes we inhale particles that excite sensitive receptors in the nose, triggering a sneeze, or in the air passages beyond the nose, provoking a cough. The rush of air produced by a cough moves at a speed approaching 600 miles per hour, propelling debris and mucus up and out of the respiratory tract.

Microbes that enter the body by way of the mouth confront waves of saliva loaded with the enzyme lysozyme and other microbe-killing substances. Lysozyme, which also occurs in tears and nasal secretions, destroys bacteria by digesting their cell walls.

Microbes that avoid the protective agents in the mouth find their way to the stomach. There most succumb to the powerful acid secreted by cells in the stomach lining. Others get caught in the sticky mucus that coats stomach and intestines. The wavelike motion known as peristalsis, which moves food through the digestive tract, pushes the mucus and microbes from the body.

Friendly Bacteria

The body's internal and external surfaces support communities of microscopic allies. Populations of resident flora, or friendly bacteria, live on the skin, in the mouth, stomach, and lower intestines. They also inhabit the ears and other parts of the body. As long as these microorganisms do not penetrate the skin or membranes, they do not cause disease. Their presence prevents virulent organisms from multiplying.

Any disease-causing microbe trying to settle on the skin must contend with a well-entrenched colony of bacteria, *(Continued on page 172)*

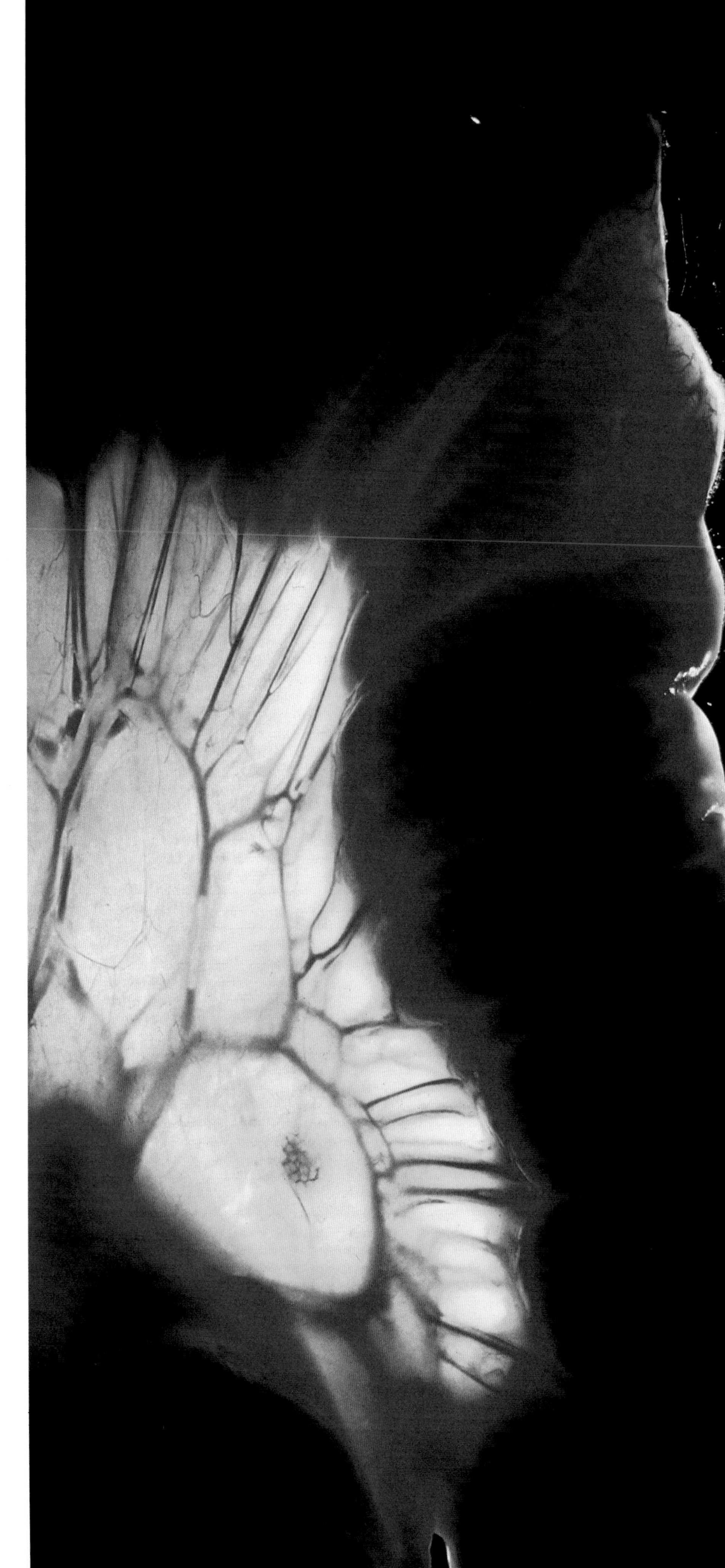

Lymphatic vessels link the scattered organs of the immune system (right). A river of lymph sweeps dead cells and other debris through the channels, and dispatches white blood cells to battle infection. Lymph nodes cluster along the vessels, with major groups in the groin, abdomen, armpits, and neck. They filter the body's detritus. Abdominal lymph nodes (reddish spots, left) trap bacteria carried there, via the lymphatic vessels, from the small intestine.

Connective tissue encapsulates a lymph node (below) and its densely packed nodules where lymphocytes and macrophages are stored. These agents of the immune system identify, track down, and destroy bacteria, viruses, and cancerous cells before they can seriously harm the body.

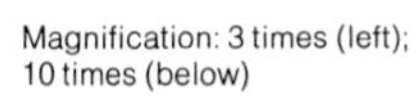
Magnification: 3 times (left); 10 times (below)

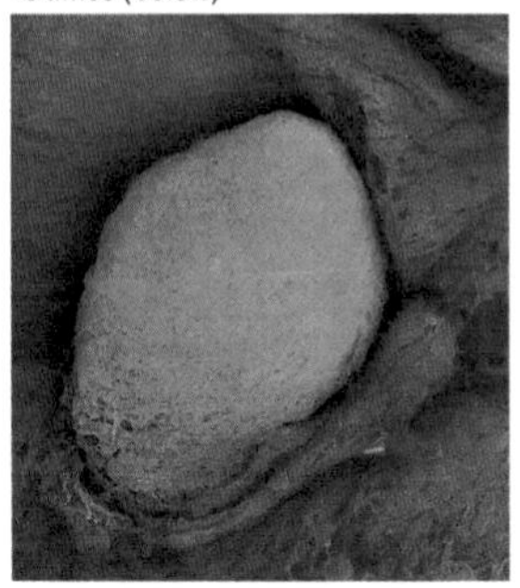

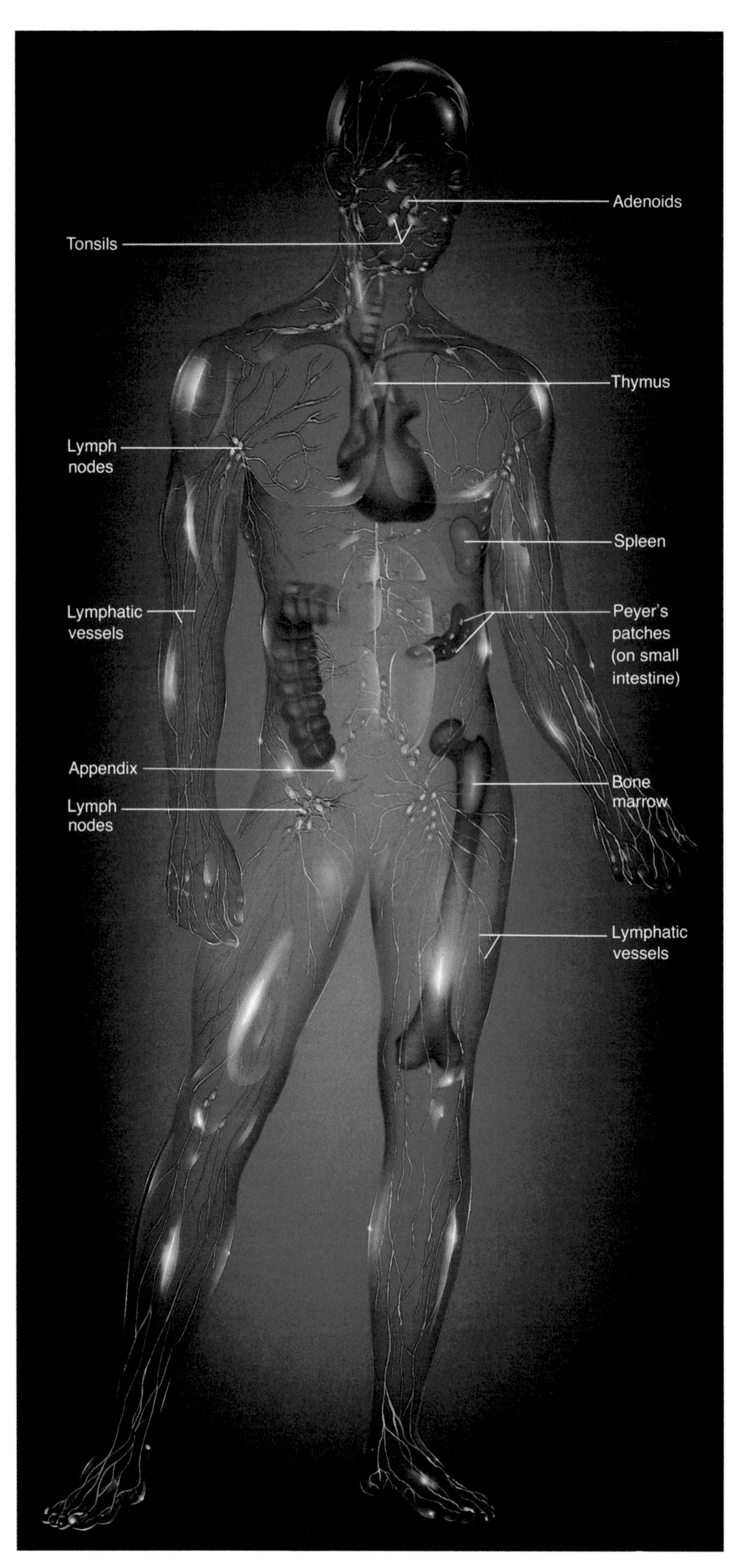

The Inflammatory Response— Healing a Wound

A splinter breaking the skin sets off a battle within the body—the inflammatory response (1). Mast cells at the site of injury release chemicals that affect nearby capillaries (2). The tiny blood vessels expand their walls and become more porous. As additional blood flows into the area, the skin reddens and heats up. Blood serum seeping from the leaky capillaries makes the wounded tissue swell and grow tender (3).

Within an hour, white blood cells called neutrophils mobilize to fight invading microbes that rode in on the splinter. These small cells speed to the battle site, slither through capillary walls, and gobble up bacteria (4).

Later, larger white blood cells, the macrophages, begin to arrive (5). These scavengers sweep over the site, wrapping their fingerlike pseudopods around the bacteria and eating them. Macrophages also devour dead neutrophils and other debris.

For hours, often days, the white blood cells muster forces to defeat the invading horde and prepare the way for the process of healing and repair.

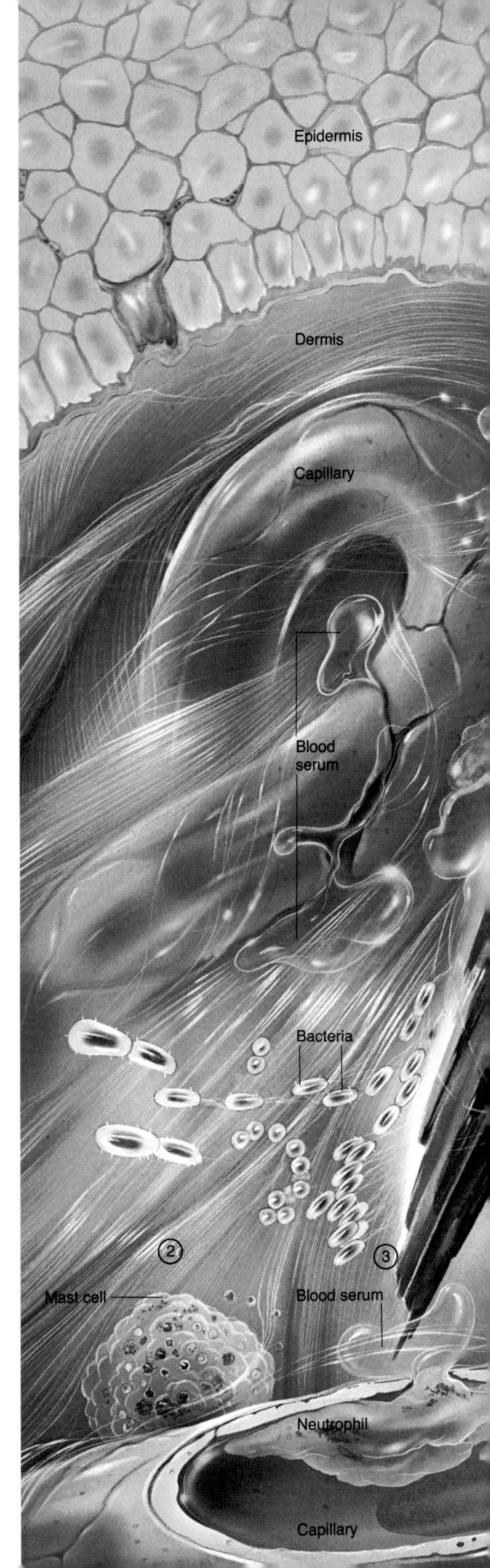

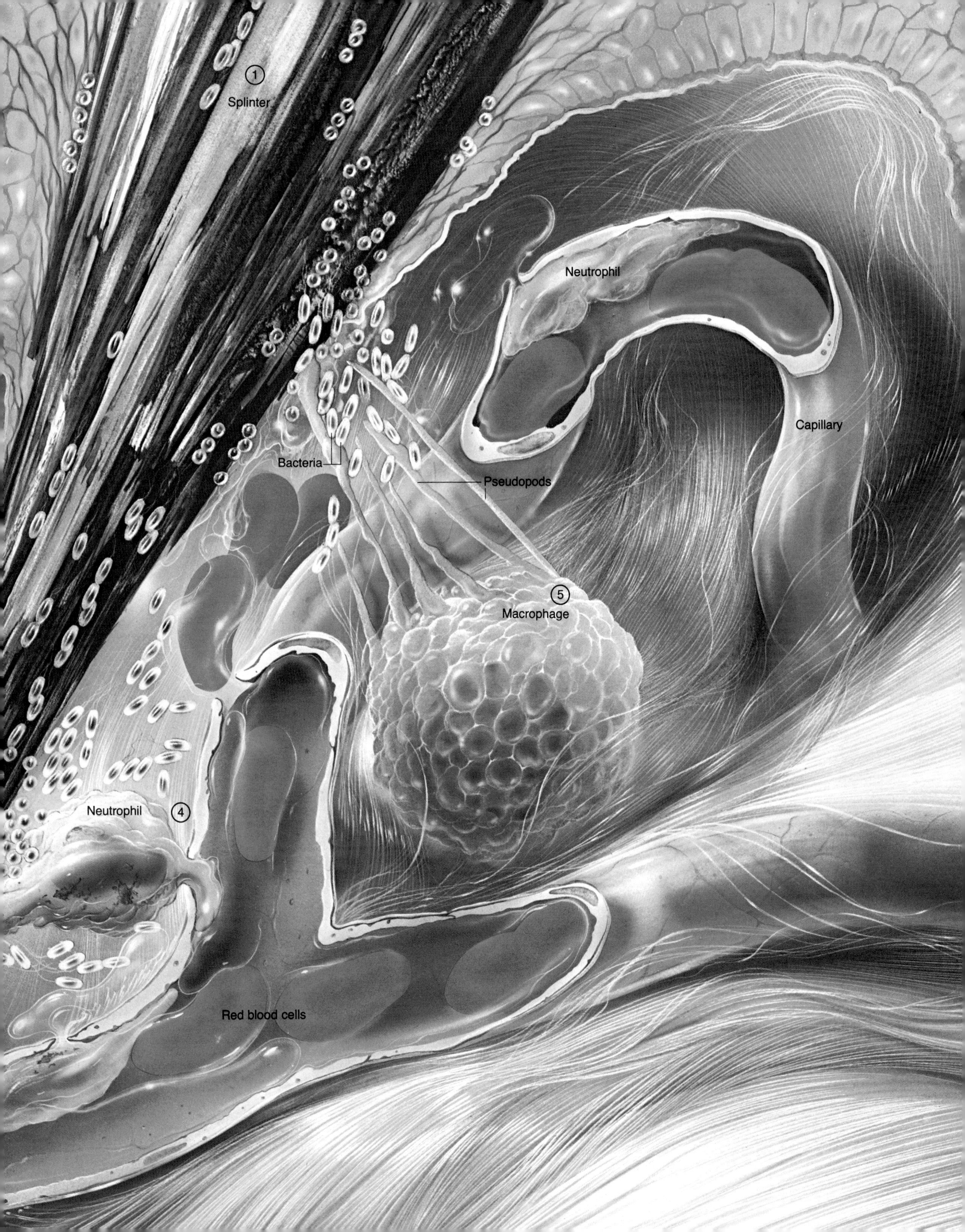
1
Splinter
Neutrophil
Capillary
Bacteria
Pseudopods
5
Macrophage
Neutrophil
4
Red blood cells

Magnification: 25,000 times

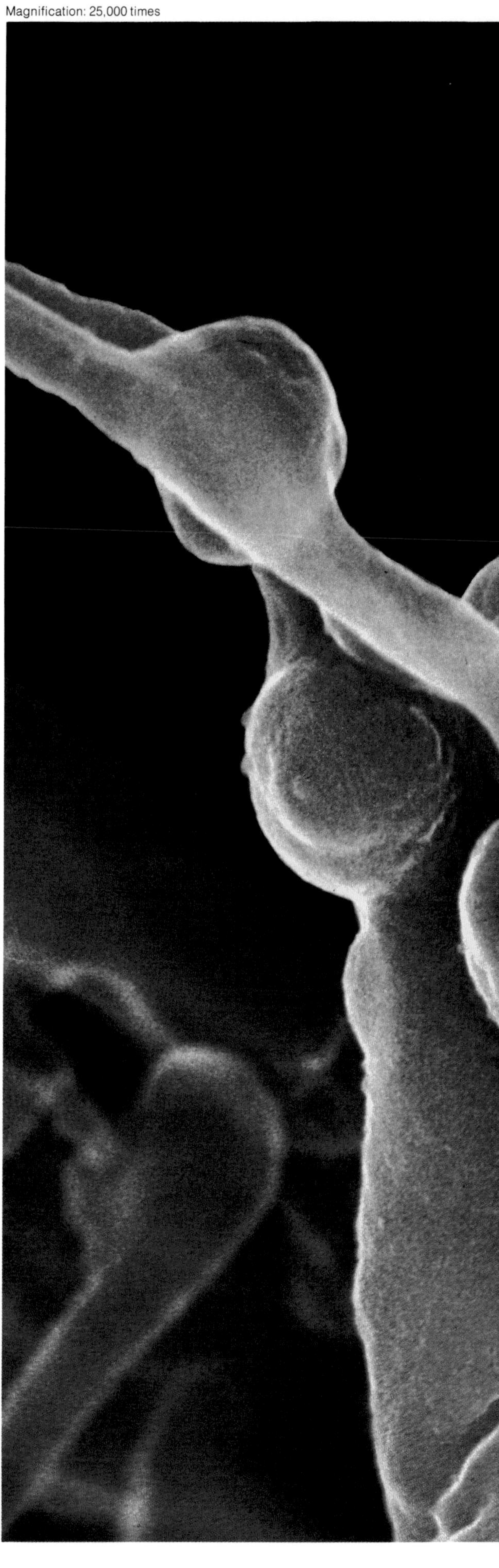

Magnification: 4,000 times; Lennart Nilsson, © Boehringer Ingelheim International GmbH. (all)

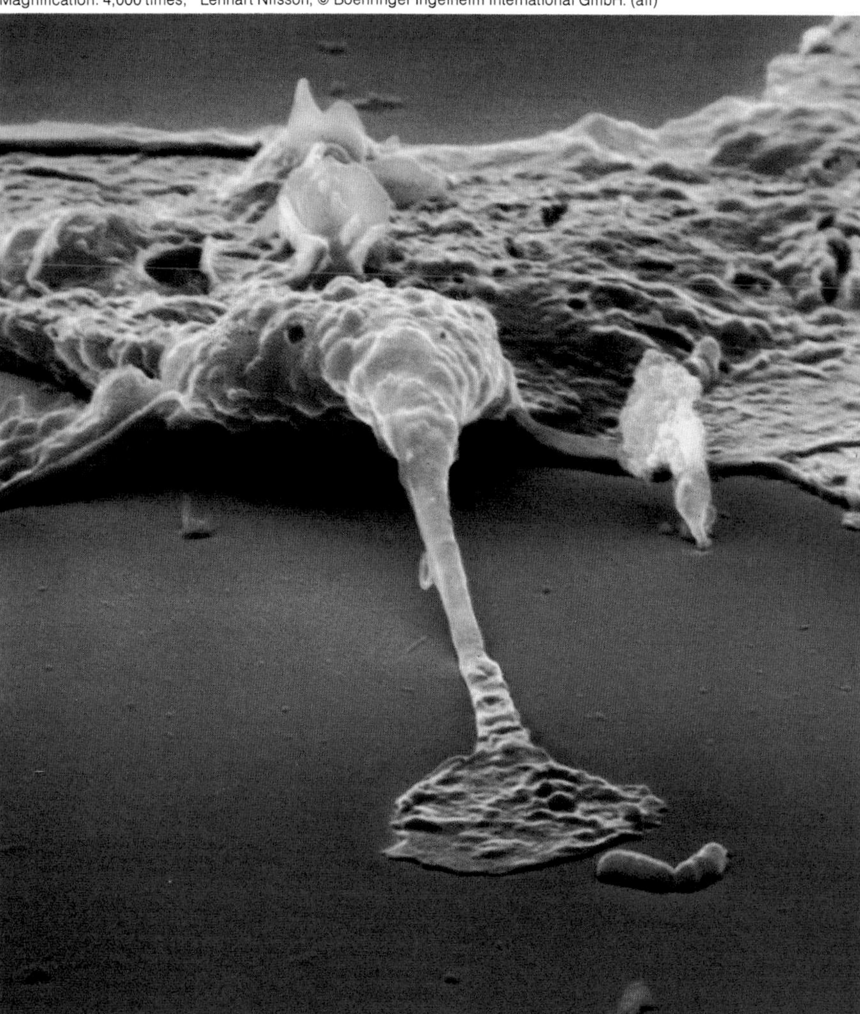

Like a hunter stalking its prey, a macrophage patrols the body in pursuit of bacteria, parasites, and other agents of disease.

This hunter, a class of white blood cell, creeps along the chemical trail of an *Escherichia coli* bacterium. With sure aim, the macrophage (tinted gray, above) extends a pseudopod toward the enemy, which is already multiplying just beyond reach (tinted green, above). Then the macrophage ensnares the bacteria in a deadly embrace (right).

Now the macrophage changes, sucking its own cell membrane and all attached bacteria into a gaping pit (opposite, upper). The unwelcome *E. coli* will vanish inside the macrophage as the defender's cell membrane closes over them, sealing their doom. Inside the macrophage, digestive enzymes destroy the bacteria. One meal finished, the insatiable macrophage starts another, engulfing one more *E. coli* (opposite, lower).

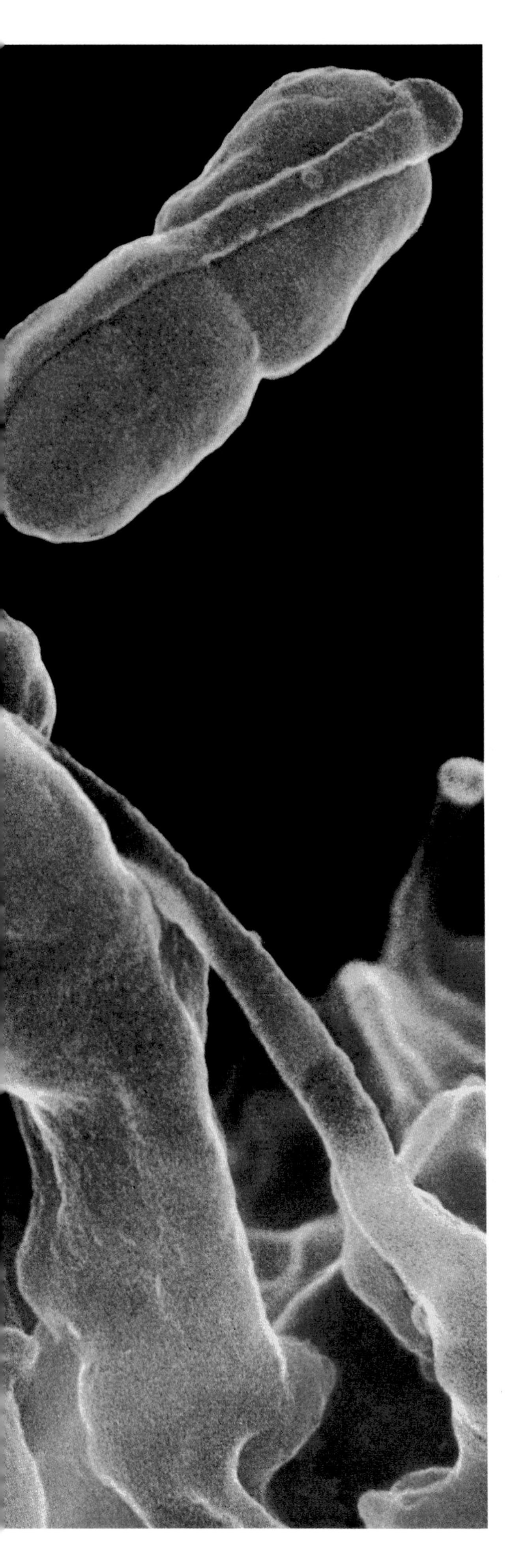

Magnification: 8,000 times

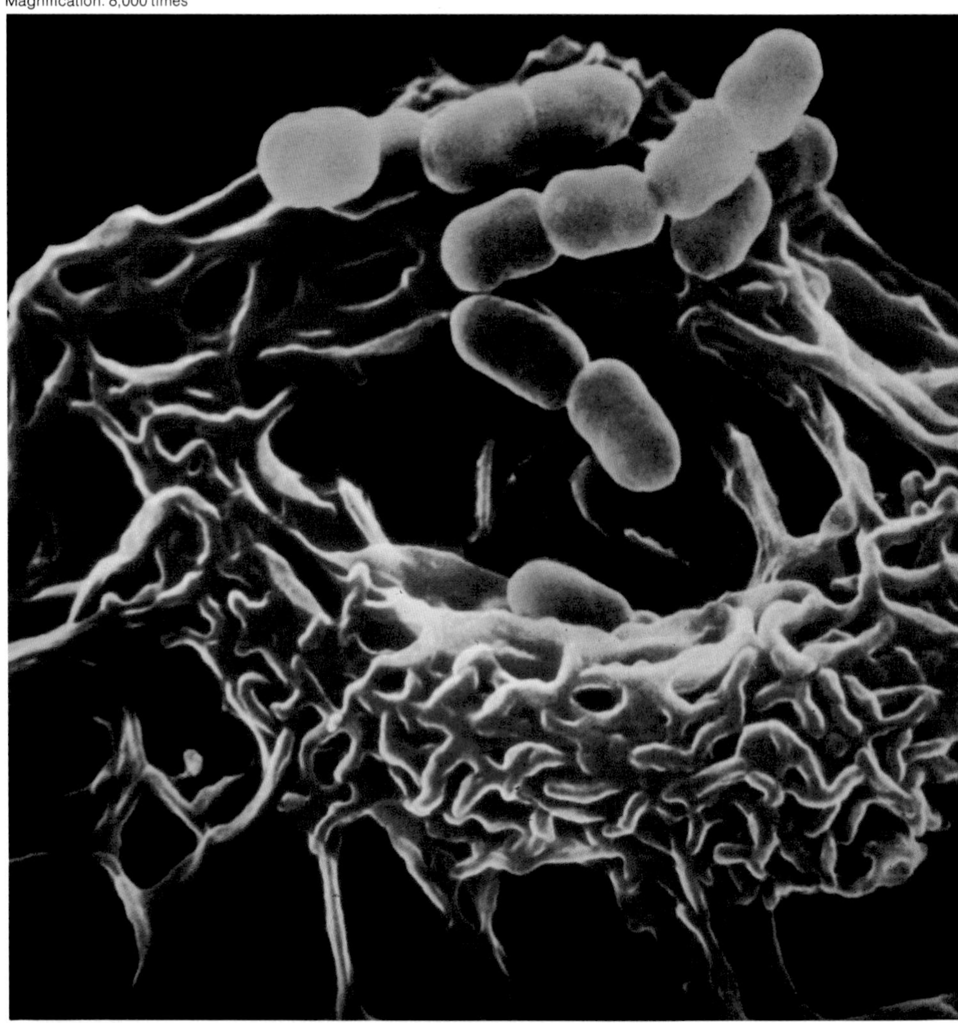

Magnification: 25,000 times

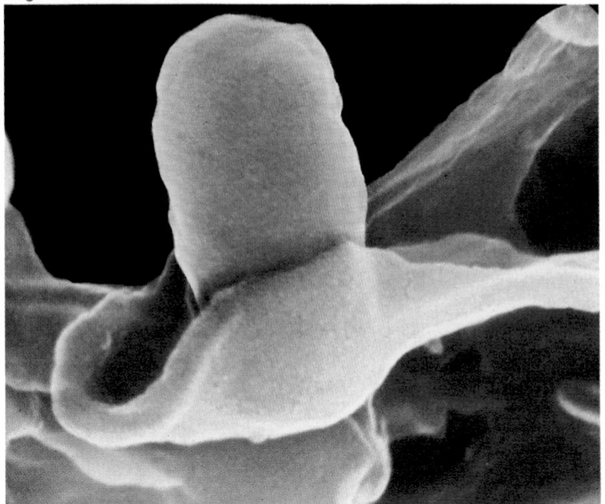

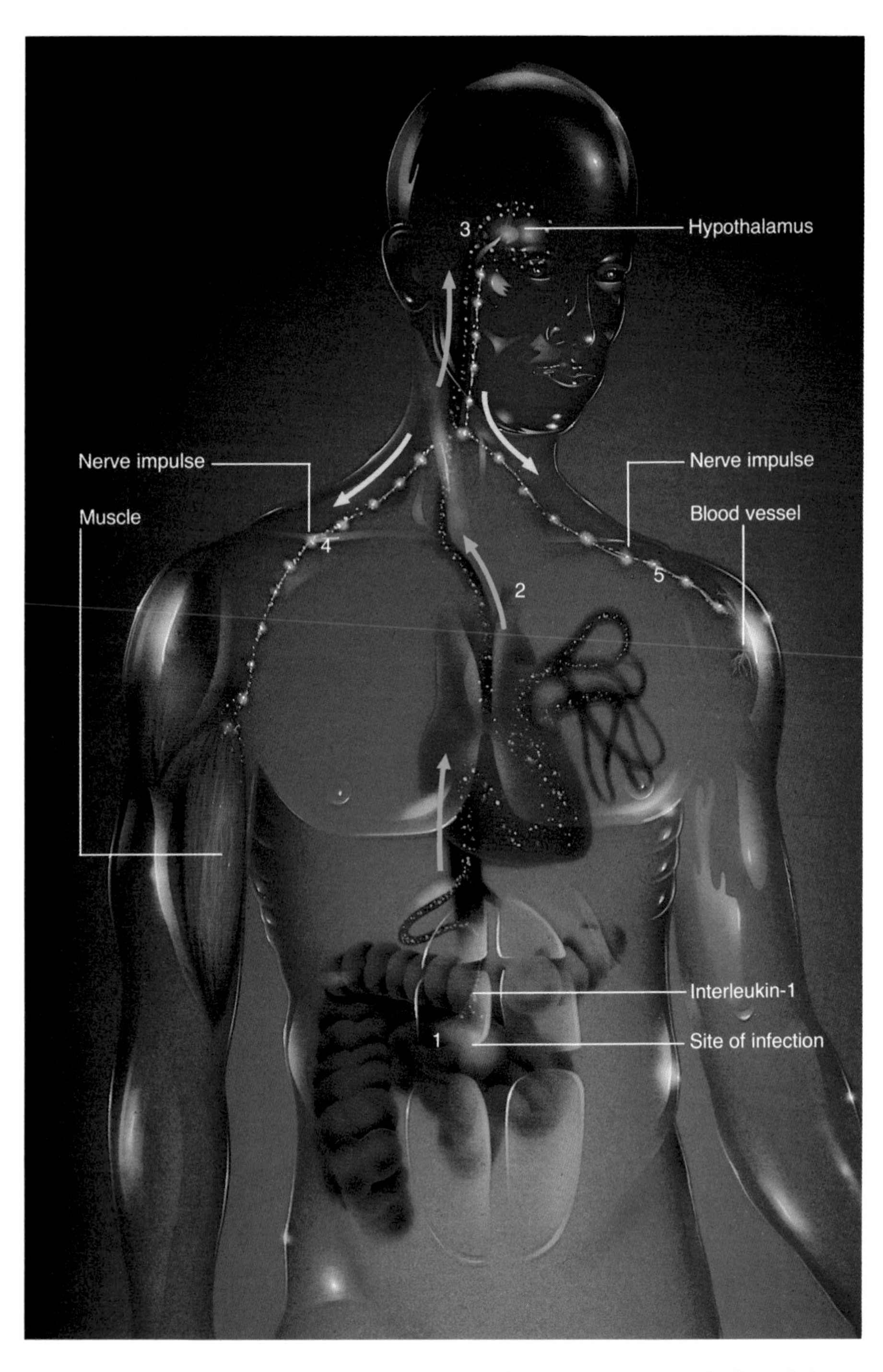

How does fever work? It usually begins when agents of disease invade the body. At the site of infection, white blood cells called macrophages engulf the invaders and release a chemical messenger (1) known as interleukin-1. The chemical journeys through the bloodstream to the brain (2). There it sparks a tiny part of the hypothalamus that acts as the body's thermostat, triggering it to set a higher temperature (3). To help the body reach the new temperature, the hypothalamus sends nerve messages to muscles, telling them to contract (4). We shiver, producing heat. Other signals order blood vessels near the skin to constrict, minimizing heat loss (5). Stored fat breaks down to further warm us.

The raised temperature makes some of the body's defenders operate more efficiently and may help to slow or stop the growth of invading organisms.

in places nearly 20 million microorganisms per square inch, some of which make life unpleasant for newcomers. Certain friendly bacteria produce fatty acids that hinder the growth of other strains of bacteria and several kinds of fungi. The bacteria *Escherichia coli,* which live in the intestinal tract and work as part of our nutritional system, simply use up the nutrients that other, less favorable types of bacteria require to live and reproduce, in effect, starving out their competitors.

Experience with antibiotics has demonstrated the dangers of disturbing the microbial life that normally inhabits the body. Long-term use of these substances can wipe out friendly and neutral germs as well as hostile ones—with disastrous results. Once rid of their competitors, disease-causing microbes quickly establish themselves, sometimes causing intestinal ailments, fungal infections such as thrush, and other maladies.

The body's physical barriers and other first-line defenses most often prevent microbes and other foreign matter from entering the body's inner tissues—but not always. Bacteria sometimes enter the deeper layers of the skin through a cut in the finger. Viruses slip through the lining of the respiratory and digestive tracts, penetrating the lungs or intestines. Such invading microorganisms usually encounter the full force of the immune system.

Immune Cells Permeate the Body

Unlike the digestive or circulatory systems, the immune system is not contained within a set of organs or network of vessels: Its elements permeate nearly every part of the body. Imagine the components of the system glowing from within. Its key operators, a class of white blood cells known as lymphocytes, appear from head to toe. Like minute twinkling lights, a trillion or more lymphocytes illuminate the blood, lungs, liver, stomach, and nearly every other body tissue. So do scavenging white blood cells called phagocytes (from the Greek, *phagein,* to eat).

Against the dark silhouette of a human form, two of the system's organs glow bright: the thymus, a small two-lobed organ just behind the breastbone, and the soft, gelatinous tissue of the marrow deep within our long bones. In these primary lymphoid organs, lymphocytes grow and develop.

Also glowing are the secondary lymphoid organs, the sites where lymphocytes are stored and where some immune responses take place. These include the spleen—an organ in the upper abdomen that filters blood—and the lymph nodes, pulpy clumps of tissue, as well as the tonsils, adenoids, appendix, and the Peyer's patches, bits of lymphoid tissue embedded in the wall of the small intestine.

A network of lymphatic vessels connects these widely dispersed organs. The vessels carry lymph, a colorless fluid that leaks from the bloodstream, collects between our cells, and then seeps into the small lymphatic capillaries, whose walls allow fluid in but prevent it from escaping again. Lymph, like blood, transports the cells of the immune system, as well as foreign substances that find their way into body tissues.

The network of vessels begins in a multitude of thin-walled capillaries that branch throughout the tissues. Like small tributaries that feed into major waterways, these tiny tubes drain into larger and larger vessels. From the top of the scalp, they run down through the neck; from the hands and feet, they flow up through the limbs to the torso. In the lower part of the neck, the biggest vessels pour their contents into two large lymphatic channels. These channels, in turn, ultimately converge with veins that lead to the heart.

Unlike the blood circulatory system, the lymphatic system has no pump to keep its vital fluid in continual motion. Instead, body movements and muscle contractions squeeze the vessels, propelling lymph along its course.

Most of us have felt the glands in our necks enlarge and grow tender when we have the flu, or those under the arm or near the elbow swell when a finger becomes infected. These glands are actually lymph nodes.

Usually one or more of these nodes lie in the pathway of the lymphatic vessels and filter the lymph on its way to the bloodstream. In each node, a labyrinth of channels weaves through a dense webbing of tissue divided into compartments. Each compartment houses a distinct population of white blood cells. As the incoming lymph trickles through the channels of the node, some particles get caught in the webbing or fall prey to white blood cells. In this way, the nodes filter out foreign chemicals, particles, and microorganisms before they enter the bloodstream. This function was discovered during an autopsy performed on a heavily tattooed sailor: His lymph nodes showed traces of ink.

When a disease-causing organism arrives at a node from a site of infection, the node swells as the white blood cells within divide and multiply in response to the invader.

Even though aches and chills are no fun, there is often good reason for fever: If mild and short-lived, it often helps the body fight disease-causing invaders. We drink hot tea and burrow under the blankets while our bodies launch a full-scale counterattack.

The Body's Eating Cells

Patrolling phagocytes, the body's "eating cells," often intercept foreign substances that find their way into the body's lymph, blood, or tissues. The two most common types of phagocytes—the smaller, short-lived neutrophils, and the larger, tougher macrophages—seek out viruses, bacteria, fungi, protozoa, and other invaders.

Each day, some 100 billion neutrophils leave the bone marrow and enter the bloodstream.

Inside cell
Lipids
Protein molecule
Protein molecule
Carbohydrate chains
Outside cell

The surface of a cell (opposite) ripples with strange spikes and beads—molecules that are its identity card. Imagine them as protein icebergs floating in a sea of lipids, or fatty beads. Some of the proteins carry treelike antennae composed of carbohydrate chains.

Every living cell bears a set of identity molecules—usually proteins and carbohydrates—which announce it as a member of a specific cell species.

When an unwelcome guest—a parasite or bacterium—invades the body, white blood cells, such as the B-cell (below), quickly read the pattern of molecules on the invader's surface. Recognizing it as foreign and unfriendly, they launch an attack. Any substance bearing a set of molecules that can trigger an attack by immune cells is called an antigen.

B-cells make antibodies to fight invaders. These Y-shaped proteins, such as the one simulated in the computer-generated image (right), bind with particular antigens that they match. Bound to an antibody, an antigen becomes an easy target for white blood cells that engulf and destroy the invader.

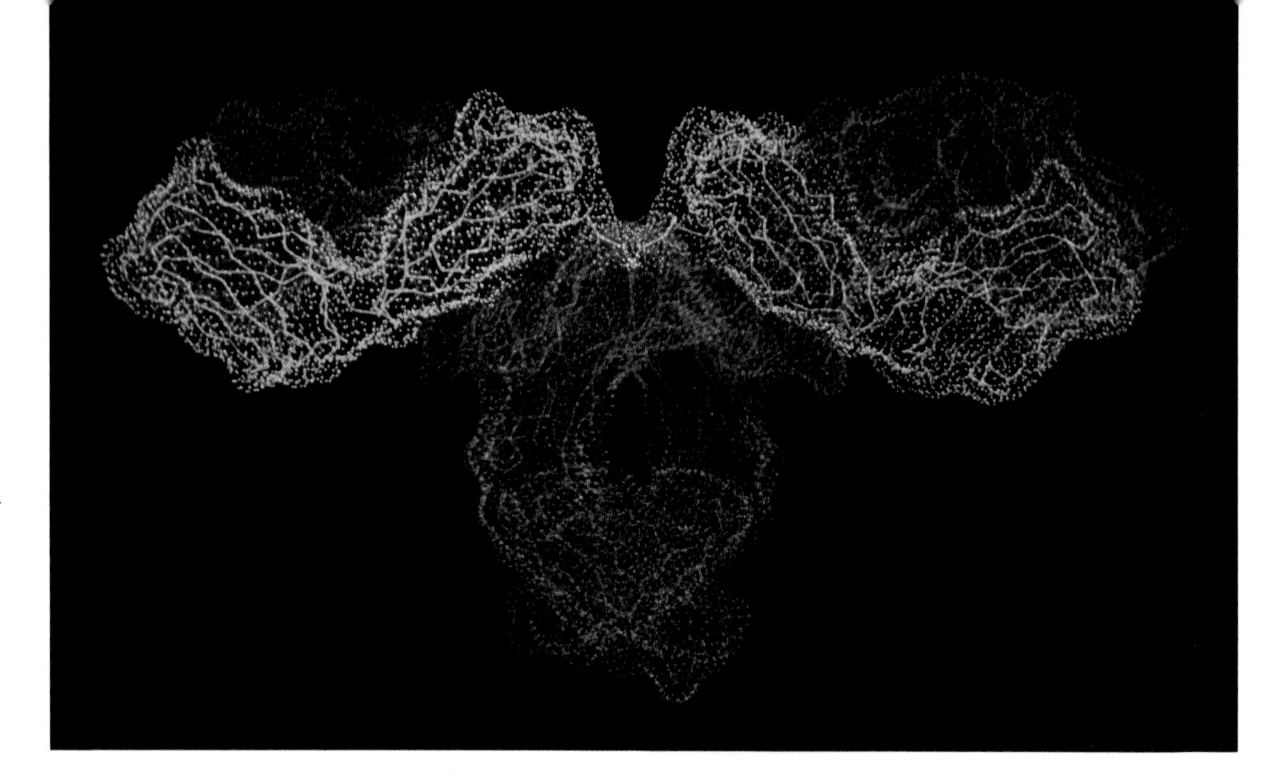

Magnification: 4,000 times

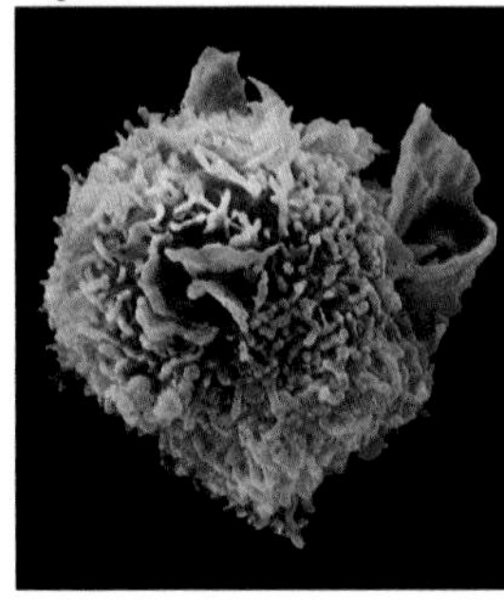

Nearly half of them circulate with the blood; the rest, known as the marginal pool, cling to the walls of the blood vessels. Almost any kind of stress to tissues in the body triggers a rise in the number of blood-borne neutrophils. During severe infection, their number may increase more than fivefold—some originating from the marginal pool, others arising fresh from the bone marrow. At the appropriate signal, the neutrophils leave the bloodstream and migrate into the tissues to pursue invading microorganisms.

The macrophage also originates in the bone marrow. Known as monocytes in their immature form, these cells leave the marrow, travel through the bloodstream for a few days, and then migrate into the tissues.

Here they mature into macrophages, or "big eaters," professional scavengers with a nearly insatiable appetite for undesirable cells. Many macrophages settle at the common sites of entry for disease-causing microorganisms, including the tissues of the lungs, digestive system, and circulatory system. When a macrophage receives a warning signal from infected cells or tissues, it develops still further, acquiring even more sophisticated cellular machinery.

Unlike neutrophils, which live for only a few days, mature macrophages live in the body's tissues for months, perhaps even years. Some act as janitors, sweeping up dirt, damaged tissue, and aged cells. Each day, the macrophages in a single human body consume more than 300 billion dead or dying red blood cells. In the lungs, macrophages continuously clean the surfaces of the air sacs, mopping up the bits and pieces of matter that find their way past the nostril hairs and cilia of the respiratory tract. They can even clear lung tissues darkened by the tar from tobacco smoke. As long as they do not have to cope with additional smoke pollution, macrophages can eventually restore the lung's normal appearance.

The materials that a macrophage cannot digest, such as asbestos or silica, it stores away in internal sacs, sometimes only a temporary remedy. The inert material can eventually destroy the lining of the sacs and kill the macrophage. The dead macrophage may release a substance that stimulates the growth of fibrous tissue in the lung, inhibiting its function and causing diseases such as asbestosis or silicosis.

The Inflammatory Response

Both macrophages and neutrophils participate in one of the body's most primitive reactions to injury or invasion by a foreign substance—the inflammatory response. Injury of any sort, within the body or on its surface, triggers this response. A wound to the skin reveals the outward signs of the reaction.

Suppose that you cut your finger with a knife, and bacteria hovering on the horny layer find their way into the deeper layers of your skin. A macrophage wandering in the tissues near the wound may happen across the invaders and engulf them. More often, though, the bacteria begin to divide and multiply, crowding and injuring nearby body cells. The damaged body cells secrete a medley of chemicals into the surrounding tissues, including an ammonia-like substance called histamine. In response to the chemicals, the blood vessels near the wound swell. More blood flows into the affected area, making the wound look red and feel warm, or inflamed. Gaps form between the cells that line the blood vessel walls. Through these tiny rifts and fissures oozes blood serum loaded with proteins

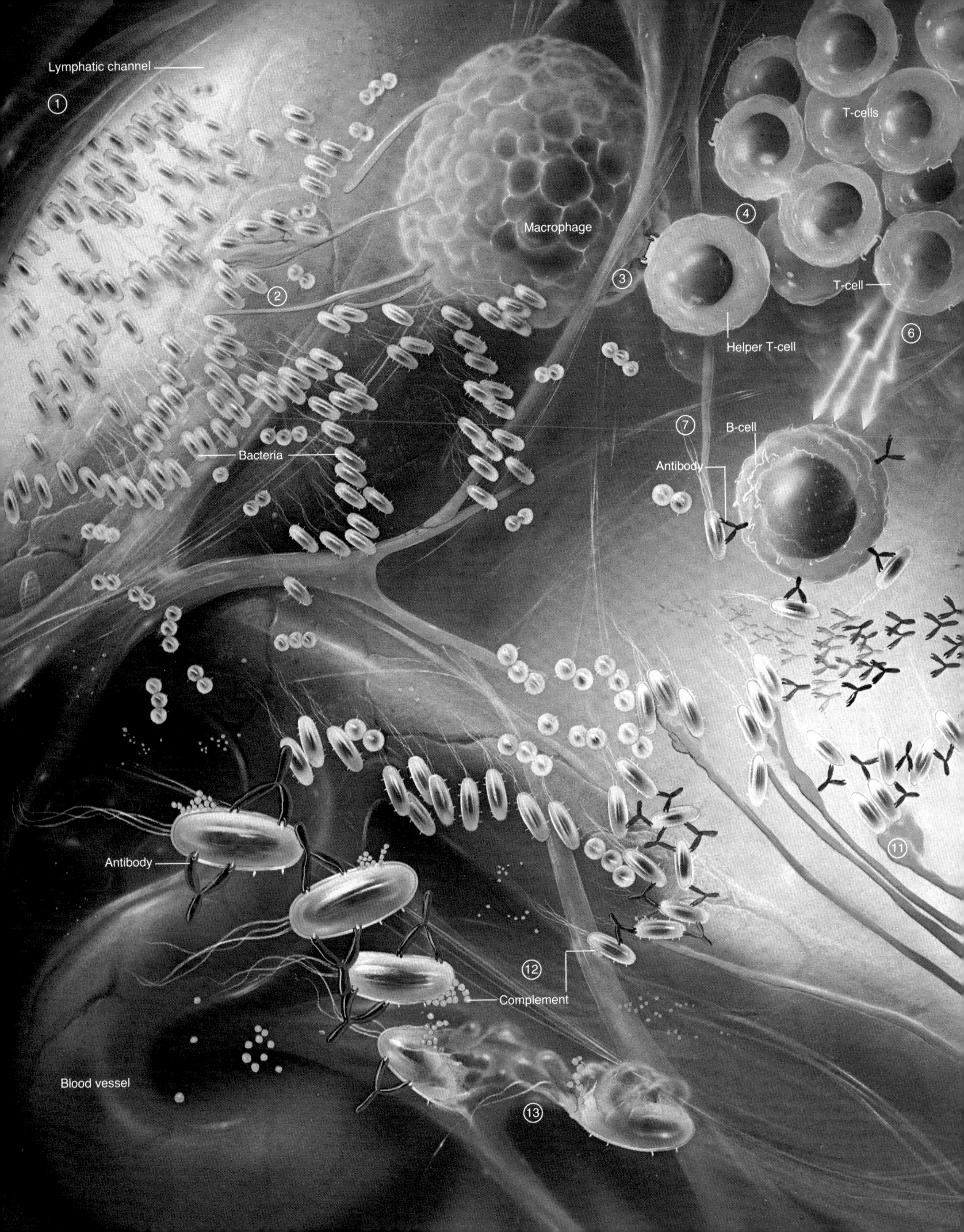

Lymphatic channel
1
2
Macrophage
3
4
T-cells
T-cell
Helper T-cell
6
7
B-cell
Antibody
Bacteria
11
Antibody
12
Complement
Blood vessel
13

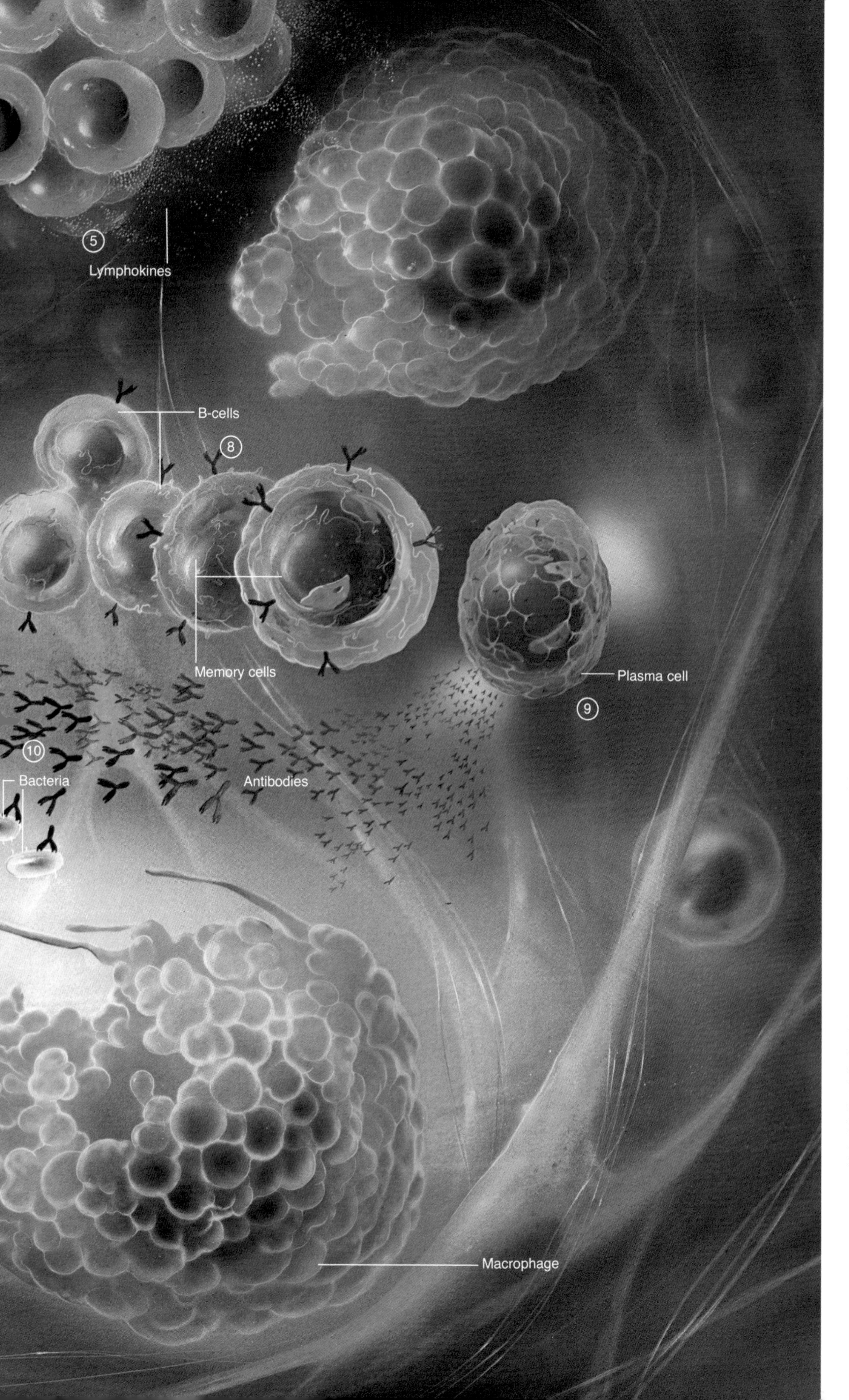

Within a lymph node, a battle rages between the body's attackers and defenders. Bacteria pour in through a lymphatic channel (1). A macrophage engulfs the invaders (2), digests them, and displays their identity markers on its own surface. The macrophage presents this chemical message to a white blood cell known as a helper T-cell (3). The T-cell responds to the urgent message by multiplying (4). Its progeny release lymphokines, chemical messages that call more defenders to arms (5). Some T-cells send a signal to new fighters—known as B-cells—telling them to join the battle (6). Only B-cells specially programmed to recognize this specific kind of bacterium are put on alert. The B-cells link up with the bacteria and begin to reproduce (7). Some new B-cells become memory cells that store information to help the body fight the same kind of bacterium on another day (8). Other B-cells become plasma cells and join the battle, spewing out thousands of antibodies each second (9). Like guided missiles, the antibodies home in on the bacteria, forcing them to clump together (10). Macrophages sweep through, swallowing the clumped bacteria (11). A group of protein molecules—called complement—helps antibodies make bacteria into palatable morsels for macrophages (12). Or complement may kill bacteria directly by puncturing their cell walls (13). Scavenging macrophages will clean the lymph node of battle debris, engulfing scattered antibodies, complement, and dead bacteria until the infection finally subsides.

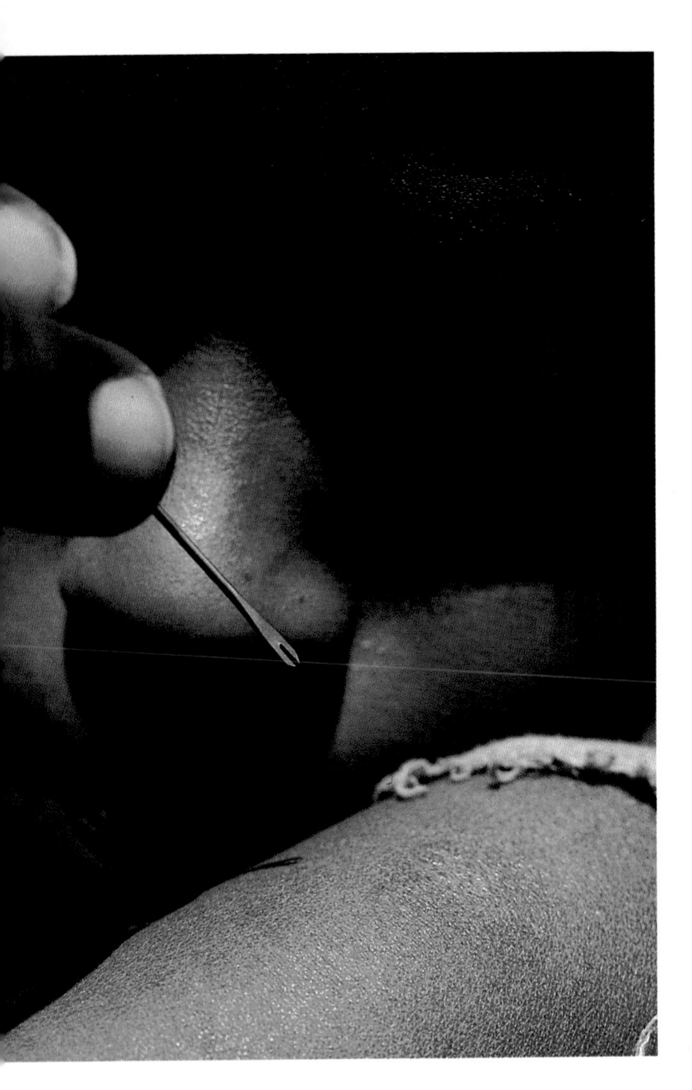

A tiny, two-pronged needle (above) delivers the vaccine that rid the world of smallpox. Other vaccines—for polio, measles, tetanus, whooping cough—help reduce suffering today.

Scientists are now preparing what they hope will be the world's first antiparasite vaccine. Its target is the parasite that causes malaria, a disease that kills more than two million people a year. In Africa, where malaria is common, a doctor (right) examines children for enlarged spleens, a symptom of the disease.

and other large molecules. As the fluid accumulates in the tissues, it causes swelling and exerts pressure on the nerve endings in the area. The pain and tenderness warn the rest of the body to spare the injured area further abuse.

Bacteria-Eating Cells

Within minutes of infection, a wave of neutrophils arrives at the site—the advance guard of the body's professional "eating cells." Each neutrophil, convulsed by the chemical signals, thrusts a portion of its cell body between the crevices in the blood vessel walls, squeezes through, and slithers toward the microbes. A struggle ensues.

The bacteria, dodging their attackers, spew powerful toxins that can disable or kill the neutrophils and surrounding cells. A persistent neutrophil seizes a bacterium and wraps a portion of its own cell membrane around the microbe. By sucking the membrane inward, deep into its body, the neutrophil creates a miniature sac for its prey. Once imprisoned within the white blood cell, the bacterium may writhe and twist and disgorge its poisons in a last-ditch effort to escape its captor. But now the neutrophil dispatches its own weapons—small bags, each filled with a load of digestive juices and microbe-killing agents, which release their contents on the bacterium. The powerful juices quickly digest the prey. Each neutrophil may engulf and destroy up to 25 bacteria, but its efforts take a heavy toll. At the end of its bout, the neutrophil dies from the accumulation of its own digestive juices and the poisons released by the bacteria.

Neutrophils last for only a short time, but the body sends in a steady stream of reinforcements. More neutrophils arrive to join the struggle. So do macrophages, which can manufacture as many as 50 different types of enzymes and antimicrobial agents.

Macrophages use some of the enzymes they produce to cut their way through the thick tangle of fibers, proteins, and debris at the site of infection as they move toward the microbes. Unlike the neutrophils, which can consume only one big meal, activated macrophages engulf numerous intruders, digest them, and move on with relentless energy to pursue more prey, sometimes destroying up to a hundred bacteria before they expire. As the struggle continues, dead tissue, digested microorganisms, spent phagocytes, and debris may ooze from the wound as pus.

The action of phagocytes alone or together with the inflammatory response sometimes results in the removal of invaders from our tissues, clearing the way for healing and repair. But these mechanisms—part of the *nonspecific* branch of the body's defenses—do not always succeed. Phagocytes often exhaust themselves by trying to

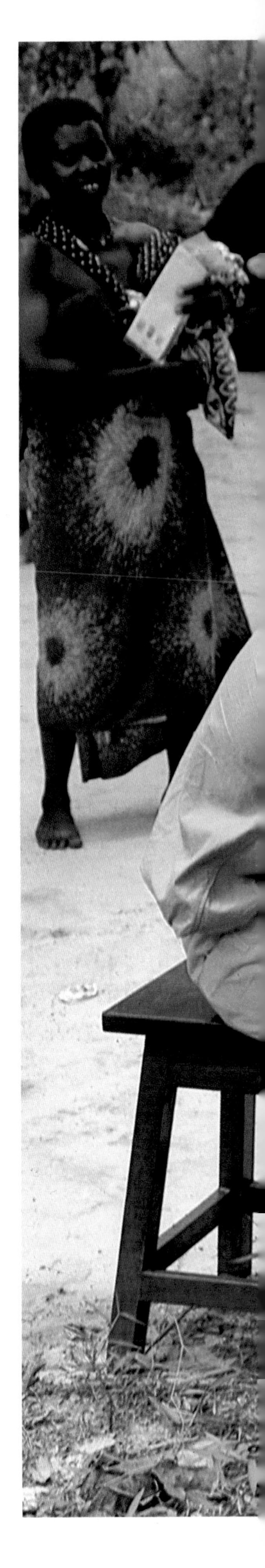

consume all the bits and pieces of miscellaneous debris at a site of infection. Though they can distinguish self from nonself, they do not recognize the particular nature of an invader and so, do not adjust their response accordingly. Should an infectious agent reappear, the phagocytes will respond in the same indiscriminating fashion—unless their activities are enhanced by lymphocytes, the major players in the *specific* branch of the immune system.

The Specific Immune Response

In the fifth century B.C., the Greeks observed that people who recovered from the plague were forever protected from the deadly disease. Somehow, their bodies remembered the agent that caused the plague and could respond to it more effectively at second meeting. We now know that lymphocytes endow the body with this ability to build immunities to infection.

Unlike phagocytes, lymphocytes have the ability to recognize the precise identity of virtually any antigen, or foreign substance—millions of different molecules—some not even found in nature, but made in a laboratory. Some lymphocytes can tell an infected liver cell from its

Around the world, practitioners of traditional medicine attempt to help the body heal itself.

A young asthma patient in a Shanghai hospital is treated with moxibustion (left). Leaves of mugwort burn on her chest, providing curative heat, a traditional treatment still practiced in China.

In a western Morocco marketplace (above), a healer uses a cupping device to draw blood from a patient who could be suffering from high blood pressure. The idea is to remove "excess" or "bad" blood—perhaps a pint or so. Modern versions of bloodletting, such as plasmapheresis, can remove harmful plasma from the patient and replace it with good plasma.

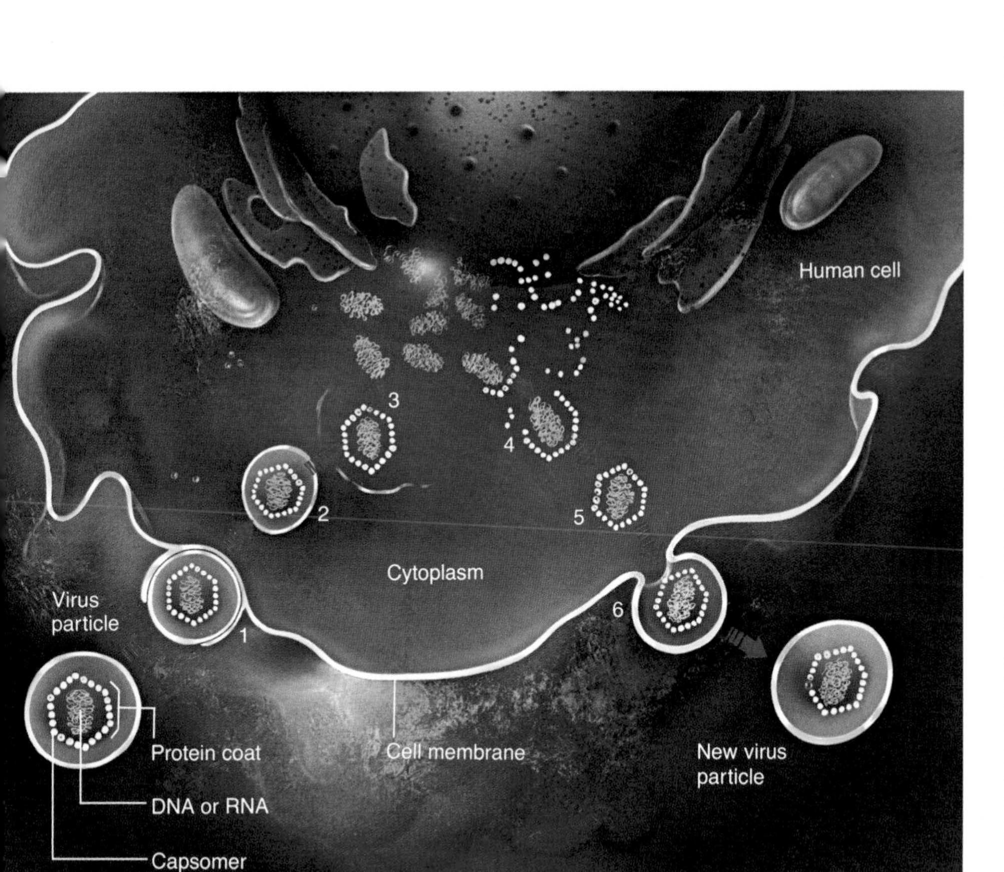

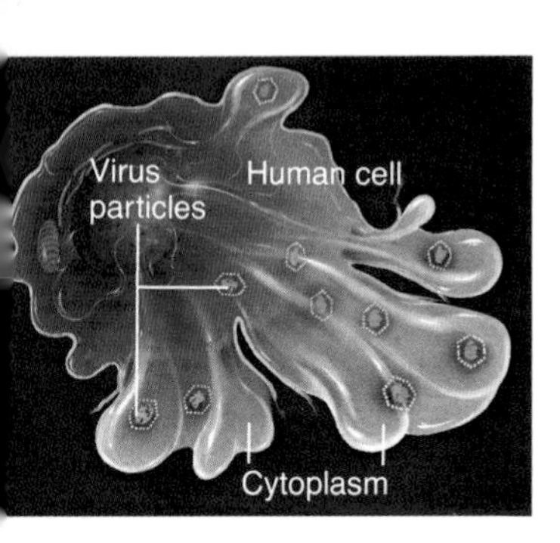

Viruses invade a cell. Unlike bacteria, a virus lacks the materials necessary for its own growth and reproduction. It must take over living cells to survive. A virus is simply a core of DNA or RNA surrounded by a protein coat; some viruses also have an outer envelope.

A virus particle encounters a body cell that carries a chemical receptor for that virus. It attaches to the surface of the cell (1), enters the cell's cytoplasm (2), and there sheds its protein coating of kernel-like capsomers (3). Now the particle seizes its host's raw materials and machinery to replicate its own DNA or RNA and its protein coat (4). New viruses quickly form within the hijacked cell (5). The new viruses may exit particle by particle, through a budding process (6). A portion of the host cell membrane wraps around the viral particle and seals off from the host cell. The viral particle floats free, carrying with it the host cell's membrane as its own outer wrapping. Or the invaders may burst out all at once, destroying the cell (left) that gave them life. By commandeering their hosts, viruses have ravaged humans with diseases as devastating as polio, influenza, rabies, and smallpox.

The tiny dots and strands of a *togavirus* cluster like deadly pearls (right) on epithelial cells; such cells line the mouth and throat. After an attack, a host cell (right, at center) has been stripped naked of its cell membrane by thousands of viral particles departing the cell.

Most of the *togavirus* family are transmitted to humans by insects. These invaders cause yellow fever and encephalitis.

Magnification: 20,000

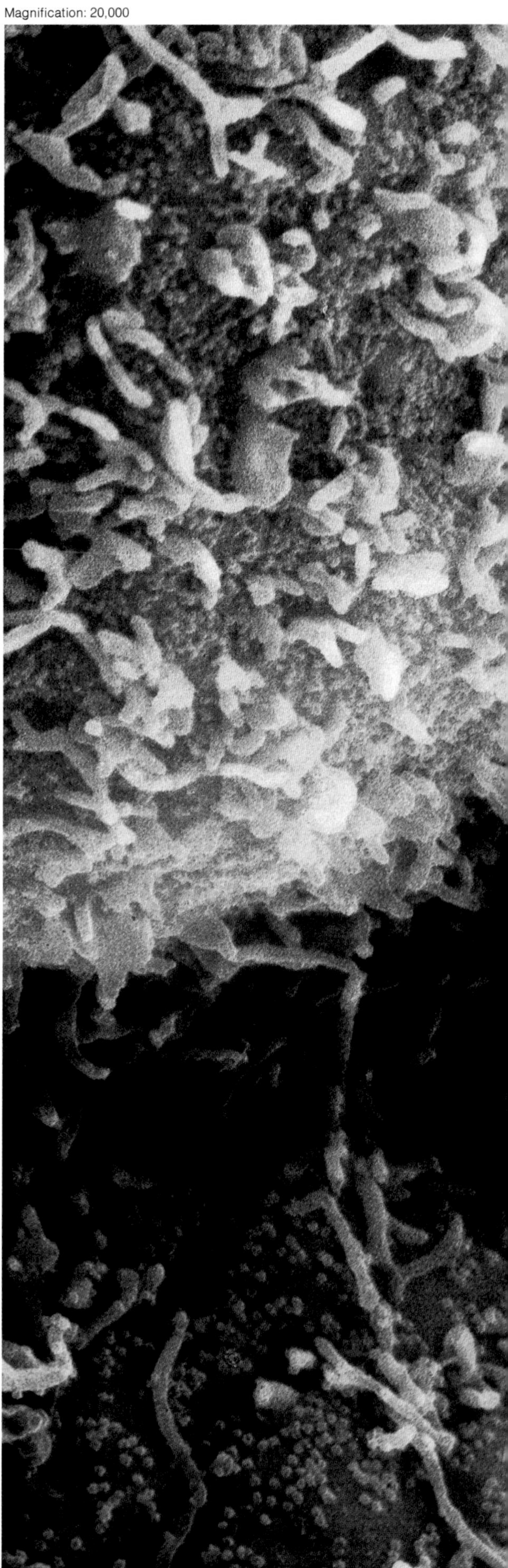

Lennart Nilsson, © Boehringer Ingelheim International GmbH.

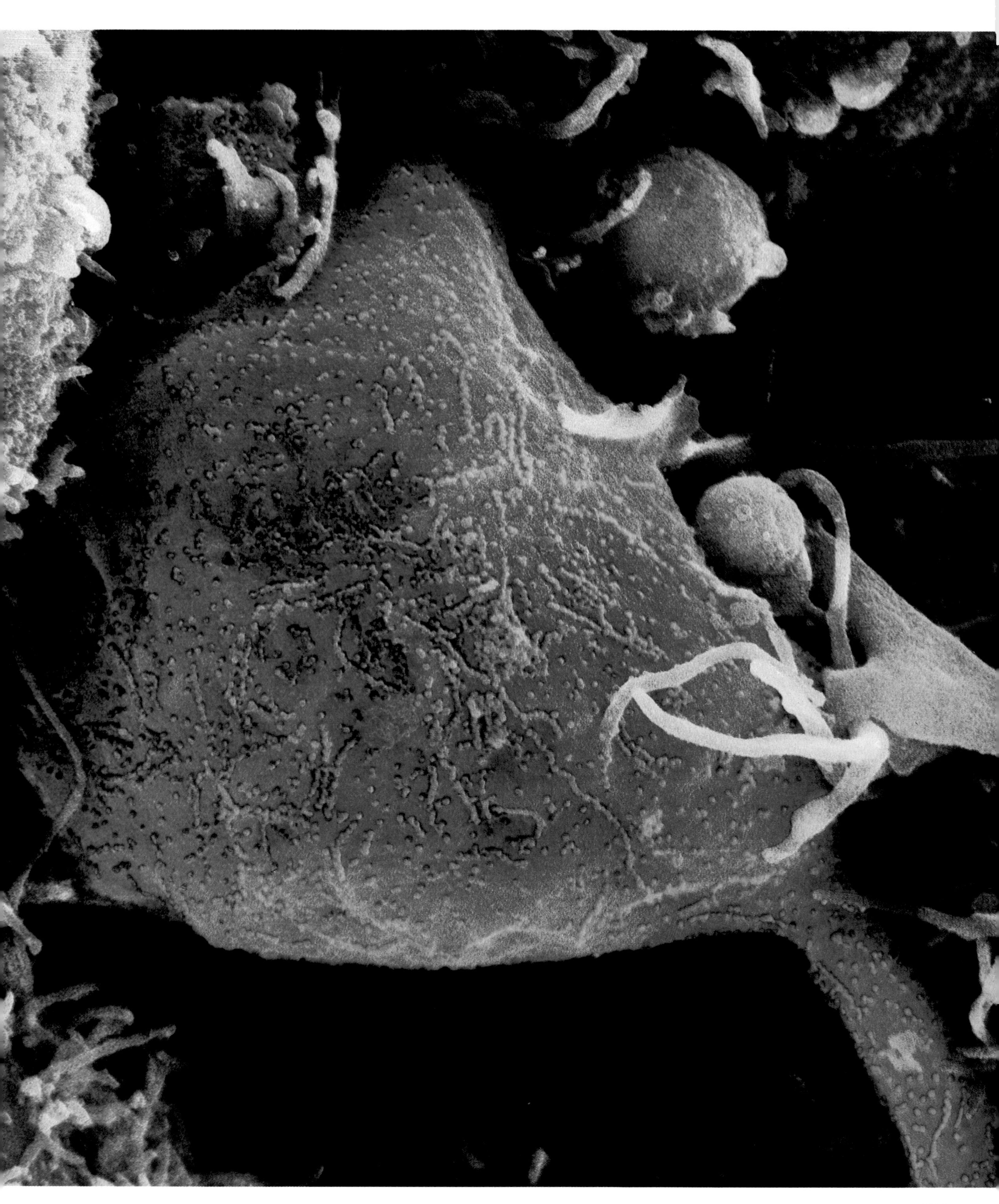

Magnification: 80,000 times; Lennart Nilsson, © Boehringer Ingelheim International GmbH.

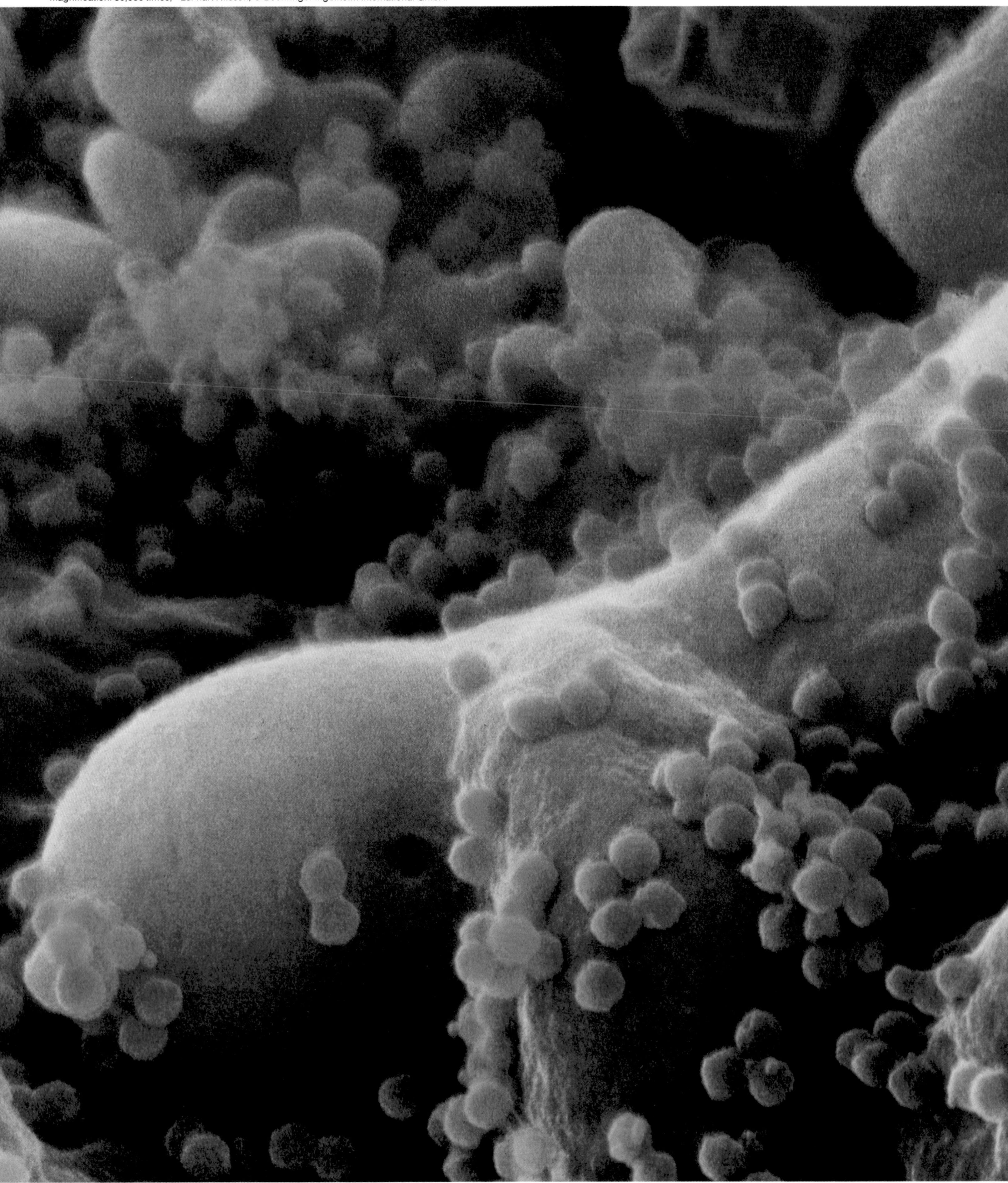

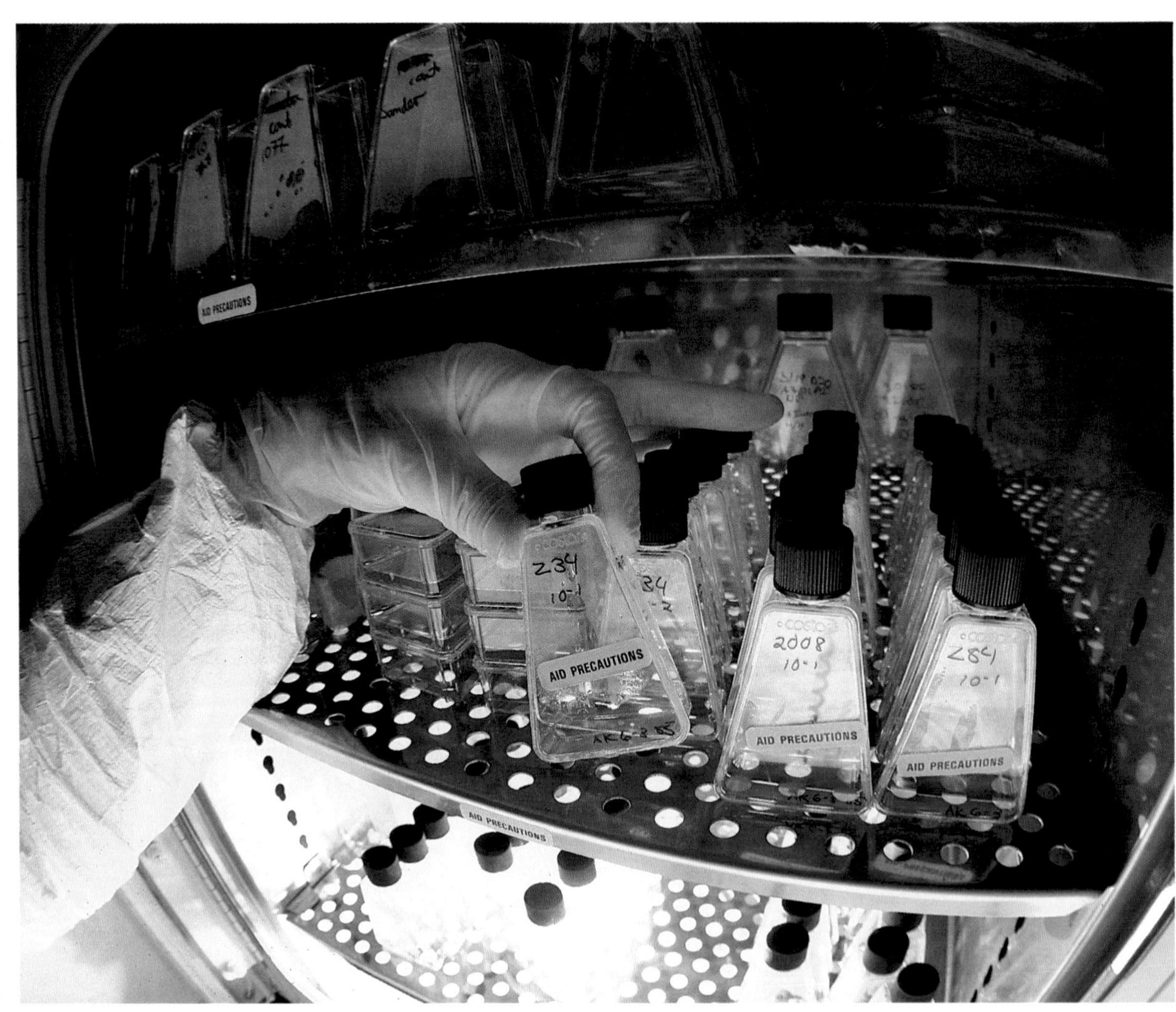

healthy counterpart or a cancerous body cell from a normal one by recognizing small differences in the cells' chemical markers. Others can discriminate between an influenza virus and a smallpox virus or a *Staphylococcus* bacterium and an *Escherichia coli*. They do so by means of specific recognition sites called receptors that each lymphocyte carries on its surface.

A single lymphocyte recognizes only one antigen: Its genetically determined receptor fits that antigen almost like a key fits a lock. Billions of lymphocytes wander through the bloodstream and tissues, each one hunting for the specific antigen that matches its receptor. The body can afford to store only a few cells programmed to react to any single antigen. So, many times the invaders outnumber the lymphocytes designed to confront them. To overcome this disadvantage, nature has provided a unique solution. When a lymphocyte encounters its matching antigen, the lymphocyte undergoes an internal explosion: It swells up and promptly begins to divide and multiply, producing a clone of identical cells, each capable of recognizing and reacting to the same substance that spurred its parent.

Born in the same place from identical parents, all young lymphocytes look alike. In the marrow deep within the long bones of the skeleton,

The deadly, invading AIDS virus creeps across a T-cell—a white blood cell (left). The virus particles (tinted blue) wreck the immune system by attacking white blood cells, especially helper T-cells, which normally activate the body's fight against infection.

AIDS, which stands for acquired immunodeficiency syndrome, took more than 7,000 lives in the United States between 1981 and late 1985.

Scientists, searching for a cure, grow the virus (above) at the National Institutes of Health in Bethesda, Maryland. Researchers hope that experimental drugs will soon provide an answer, either by limiting the virus's ability to reproduce or by stepping up the body's immune response to AIDS.

Magnification: 84,000 times

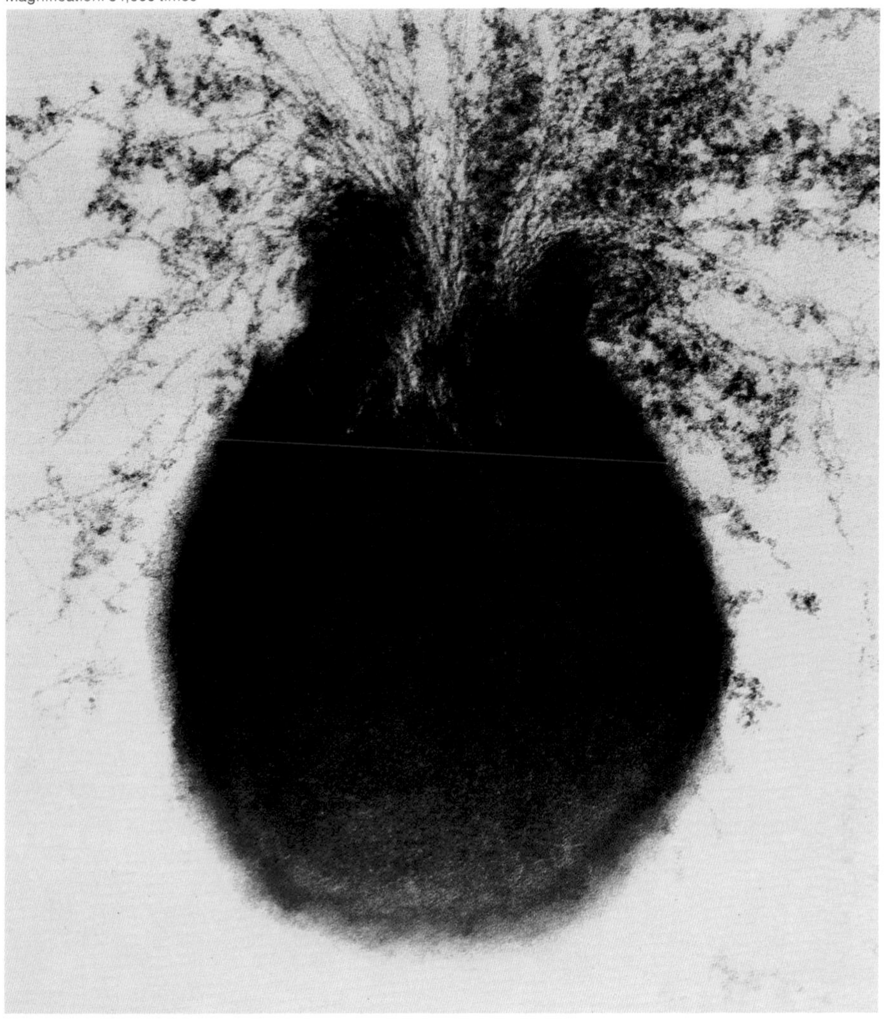

Exploding from its cell wall, a bacterium dies (above), an experimental victim of an antibiotic called fosfomycin. Although this bacterium, *Staphylococcus aureus,* often lives harmlessly on the skin of healthy humans, it can cause boils, abscesses, pneumonia, and a bone disease called osteomyelitis.

Antibiotics have cured millions of people of pneumonia, tuberculosis, meningitis, and other life-threatening infections. Although antibiotics first seemed to be miracle drugs, researchers soon discovered that the drugs are not foolproof. Bacteria can build a resistance to an antibiotic and spawn generations that will not succumb to standard treatment.

Today, scientists grapple with the problem of bacterial resistance as they search for new, safe drugs. Antibiotics take shape on the glass sketchboard of scientists Noel Jones and Michael Chaney (opposite). They examine the molecular structure of an experimental drug that could someday help the body fight new strains of harmful bacteria.

primitive stem cells give rise to these white blood cells at a rate of about a million every second. Young lymphocytes are capable of responding in a highly targeted fashion to both foreign and self-substances. Sometime early in their development, those cells programmed to attack self-molecules are eliminated or suppressed. Lymphocytes, then, come to accept self-substances and reject what is foreign to the body. Mature lymphocytes are divided into two classes, distinct but cooperating: B-cells and T-cells.

B-Cells and Antibodies

B-cells travel from the bone marrow directly to the spleen and lymph nodes, where they settle until summoned to fight an infection. When called to action, they produce antibodies (also known as immunoglobulins), Y-shaped protein molecules that circulate in the bloodstream and lymphatic system. Each B-cell makes one particular antibody, which is specifically tailored to match one antigen. Some of the antibody molecules sit on the surface of the B-cell, acting as receptors on the lookout for the appropriate antigen. Even though antibodies are tiny—only about a thousandth the size of a bacterium—they counteract a wide range of antigens. They are ineffective against most microorganisms that reside within body cells, but they can disable free-floating agents, including viruses, bacteria, and many types of bacterial toxins.

When an antibody encounters its matching antigen, the two arms of the Y attach and bind to it. In this way, antibodies neutralize toxins and prevent viruses from attaching to body cells. When antibodies bind to the surfaces of bacteria, fungi, or protozoa, they slow down the invaders, clump them, and "butter them up" so that they're more palatable to phagocytes. Sometimes antibodies activate a phenomenon known as the complement sequence—a complex chemical reaction that blasts a deadly hole in the enemy's cell wall.

The body makes millions of antibodies. All antibodies are divided into classes according to their structure and the defensive tasks they perform. One group fights bacteria with great efficiency. These antibodies, because of their large size, are restricted to working almost entirely within the blood vessels. The members of another group, built in a way that allows them to cross the placenta, provide the fetus and newborn baby with protection until the baby's own immune system becomes fully developed.

B-cells make up one class of lymphocytes. The second class, known as T-cells, mature in the thymus (*T* stands for thymus). One subgroup of these T-cells, known as killer T-cells, attack and destroy certain cancer cells, body cells infected by viral agents, *(Continued on page 196)*

Fighting the Dread Cancer

A CAT scanner explores for cancer (right). Every day, an adult produces some 300 billion cells. Each cell replicates according to its regular, controlled pattern. But sometimes this cycle breaks down: One cell, carrying the seeds of hereditary damage, divides and proliferates into a tumor. A cancer has begun.

To locate a brain tumor, specialists place a patient in a CAT scanner (for computerized axial tomography). The device rotates around the patient's head and emits X-ray beams, from which the machine's computer makes a three-dimensional map. On the map (below), from Huntingdon Medical Research Institutes, Pasadena, California, a tumor is a red zone; yellow and black areas are parts of the skull; blue depicts the hole where the spinal cord connects with the brain.

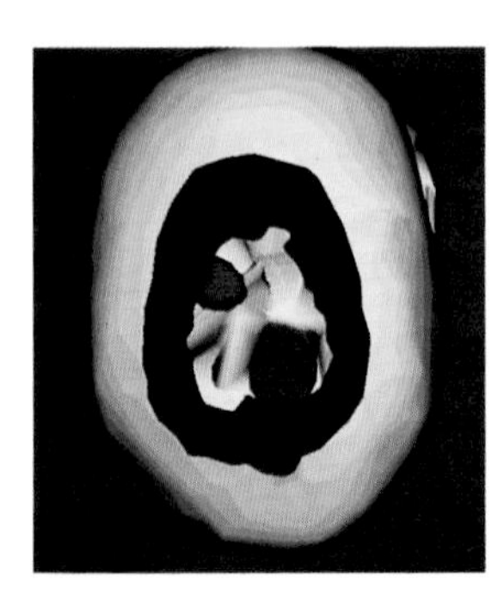

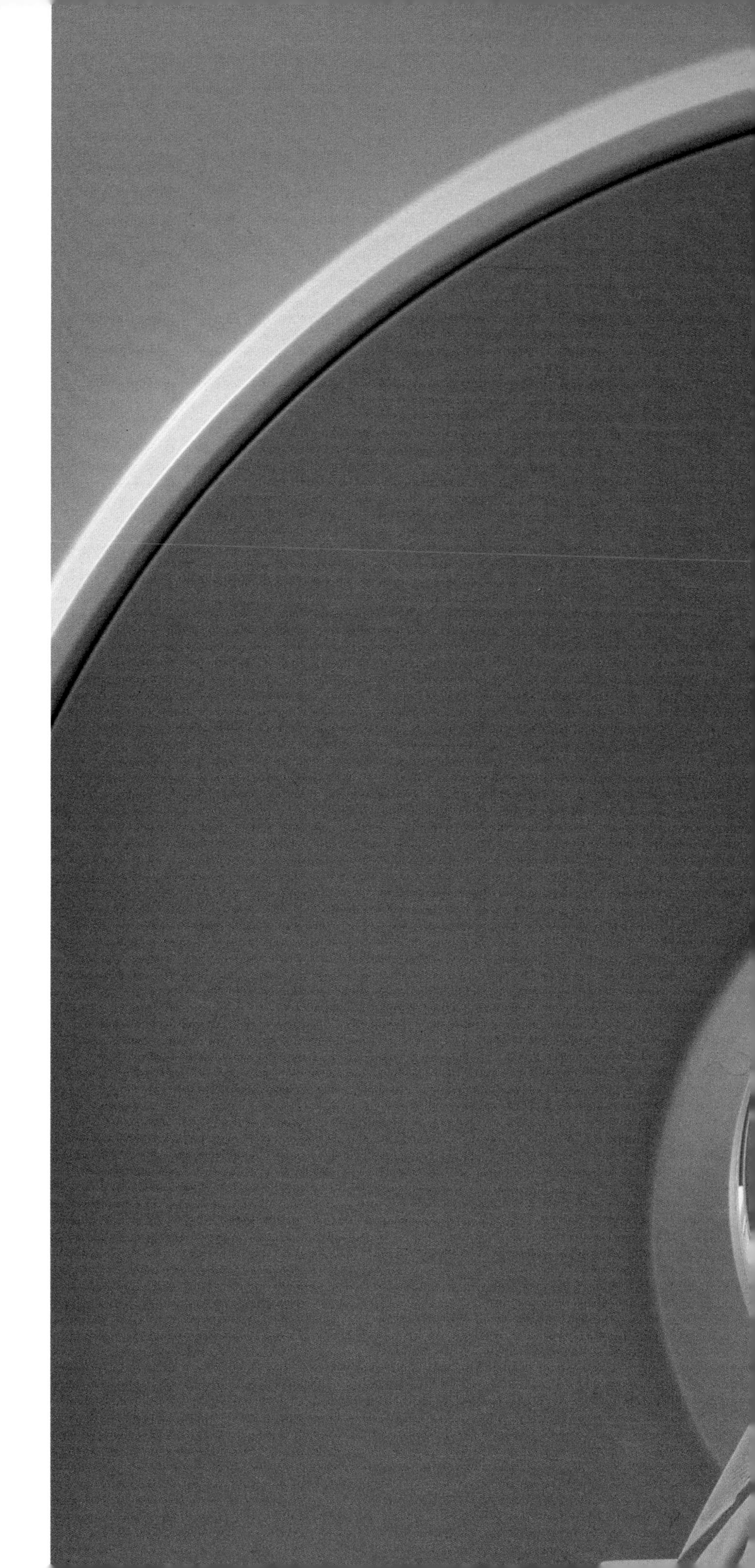

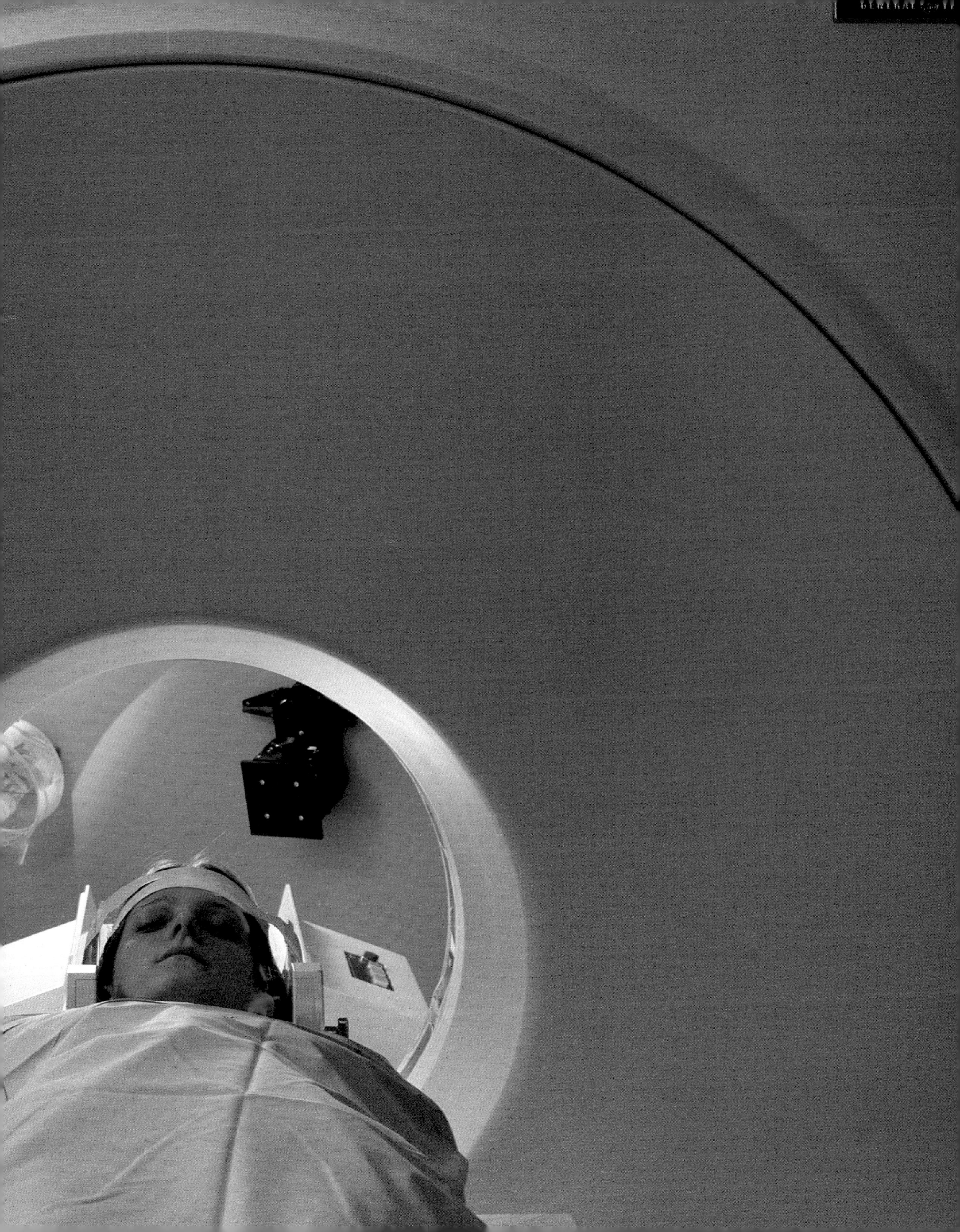

Magnification: 4,000 times

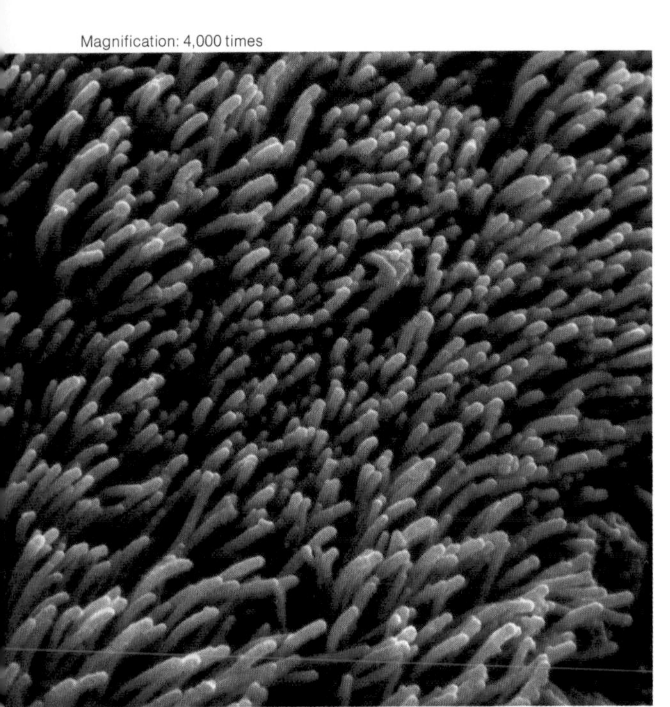

Magnification: 3,000 times

Lennart Nilsson, © Boehringer Ingelheim International GmbH. (all)

Cancer cells invade the bronchial tract (left, lower). Hairlike cilia cover the healthy cells; the irregular, lumpy shape distinguishes the cancer cells. This cancer crowds out normal cells, whose cilia form a carpet of living fibers that normally sweep dust and foreign particles away from the lungs. A healthy bronchial wall is covered with countless cilia (left, upper), which can be killed or paralyzed by cigarette smoke and other pollutants.

When cancer cells invade healthy cells, our immune system can counterattack. A T-cell (tinted purple), one of the body's defenders, tries to destroy a much larger cancer cell; the presence of numerous fingerlike projections, or microvilli, on the cancer cell indicate that it is active. A reserve guard of inactive T-cells stays close to the battle, and may gear up to fight the cancer cell.

Magnification: 40,000 times

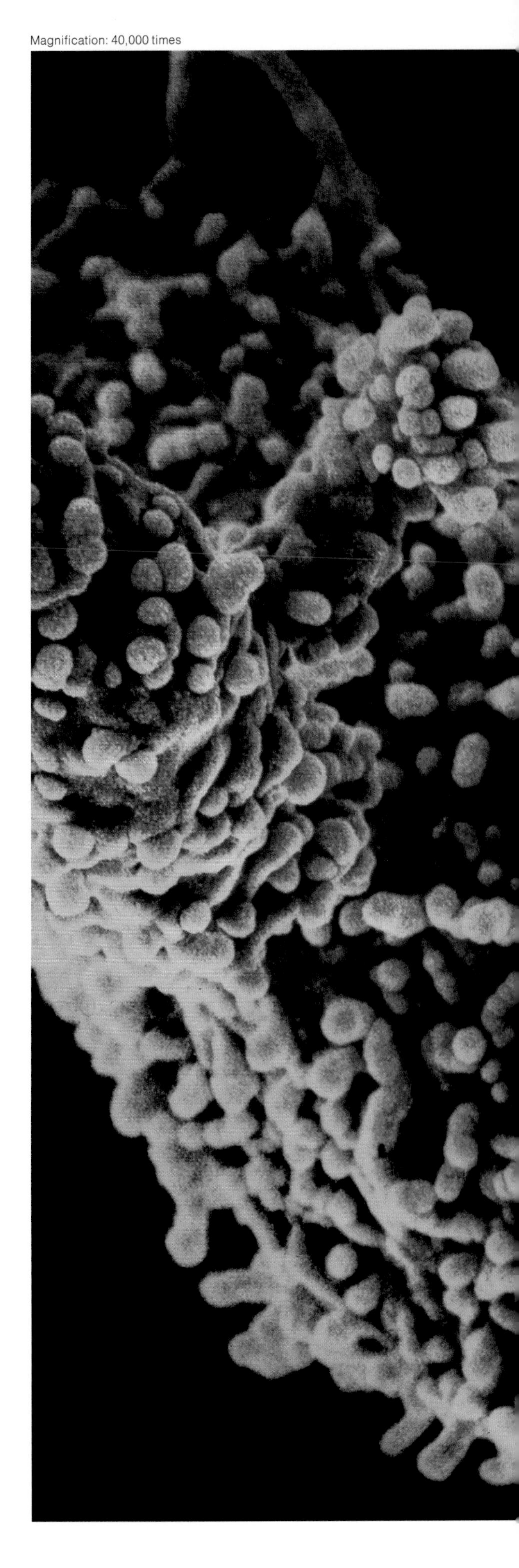

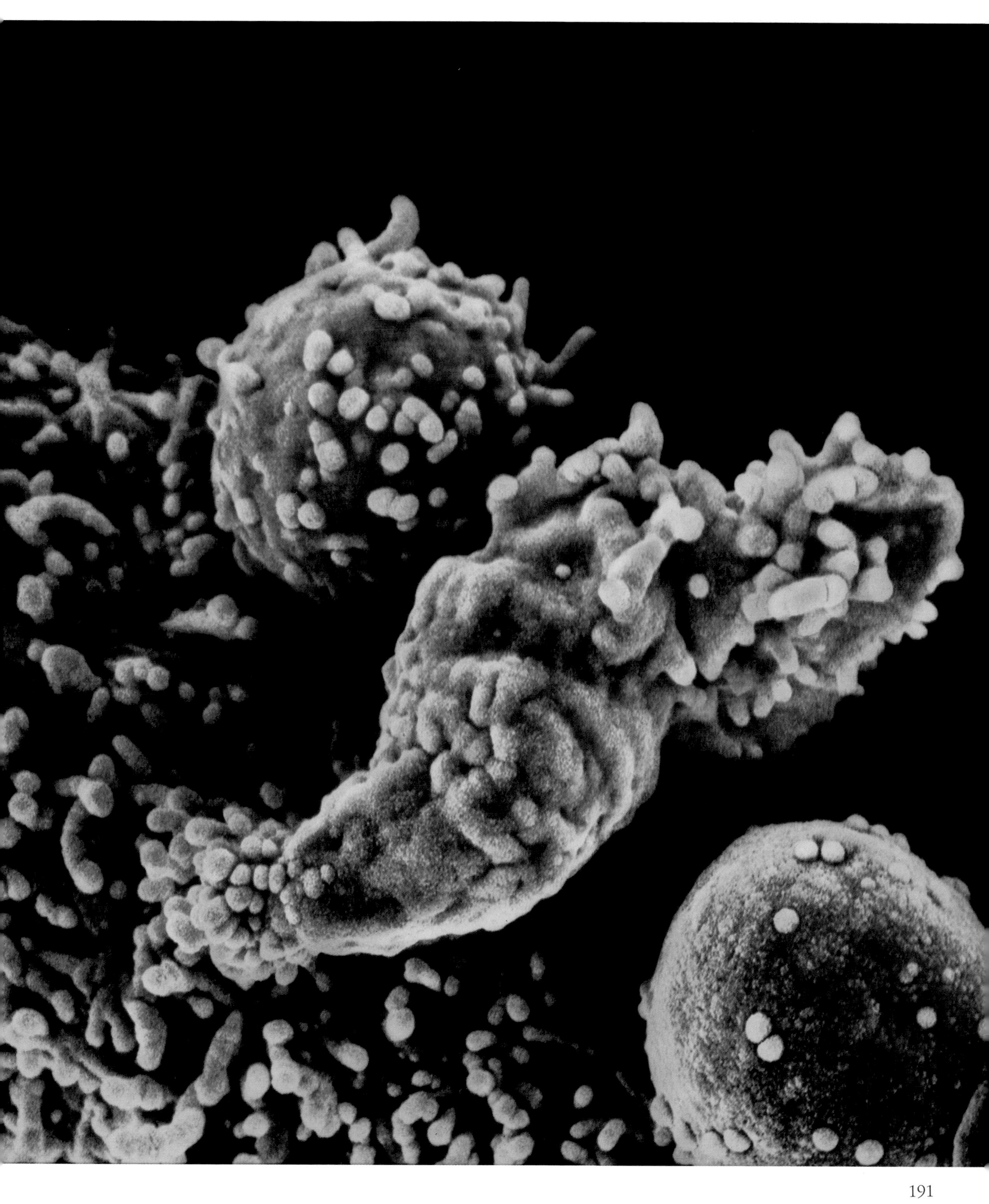

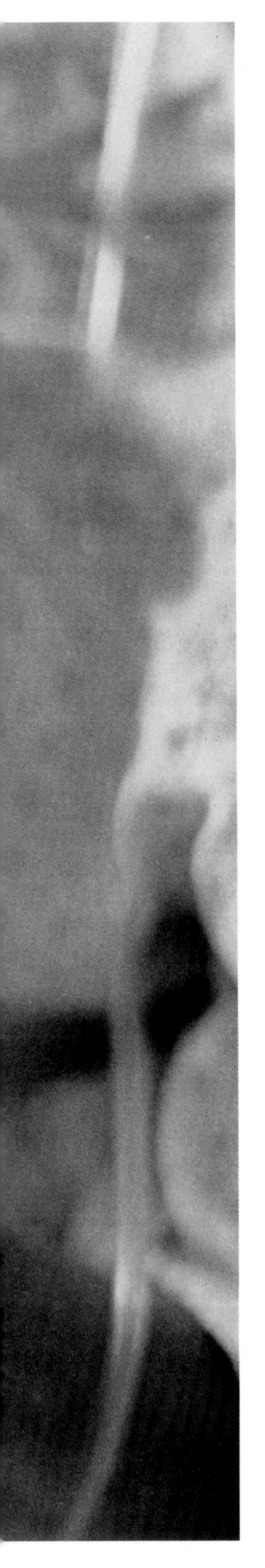

Cancer, once an almost certain death sentence, now meets the resistance of modern medical treatment and innovative therapy—one reason, perhaps, that survival rates have doubled since the 1960s.

At The Wellness Community in Santa Monica, California, patients try to battle cancer into submission through mutual support. Here patients confront the psychological and emotional crisis of the disease in group therapy, in nutrition seminars, and in classes on how to deal with pain. One participant (left) is comforted by a visitor while receiving treatment at a nearby hospital.

At one meeting, a bone cancer patient shares her thoughts with another patient (right, upper). Her experience in The Wellness Community supports, but does not replace, regular medical treatment, which includes daily doses of radiation (right, lower).

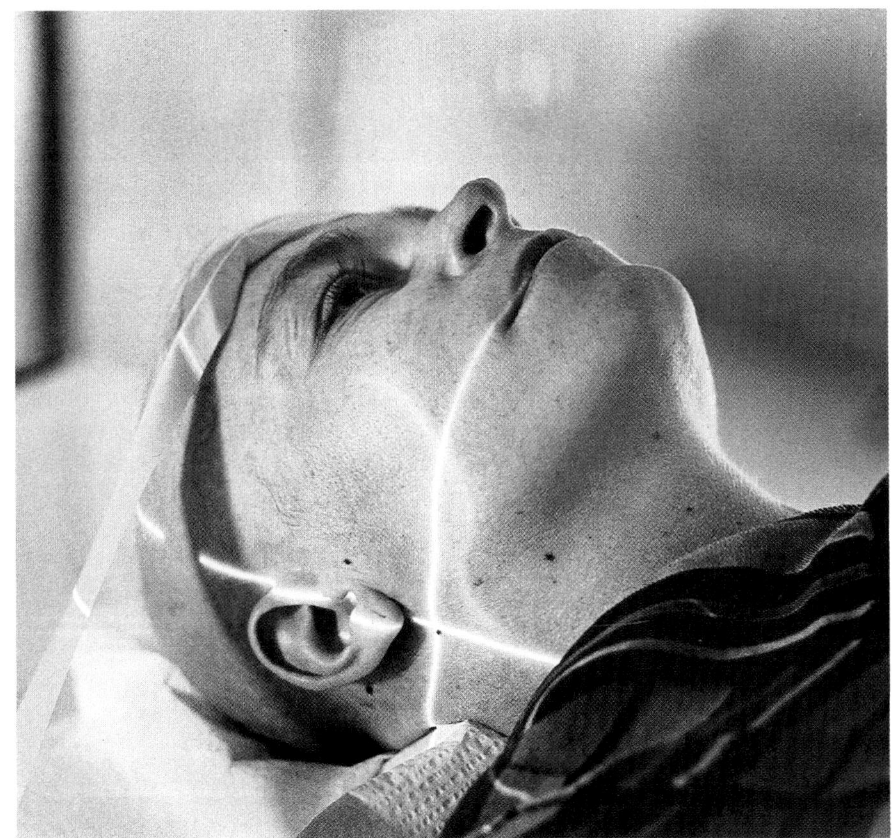

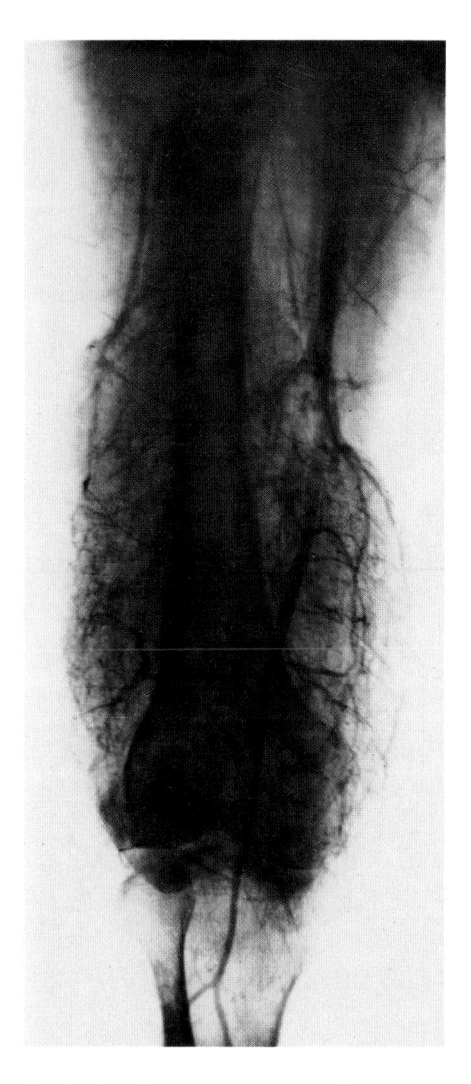
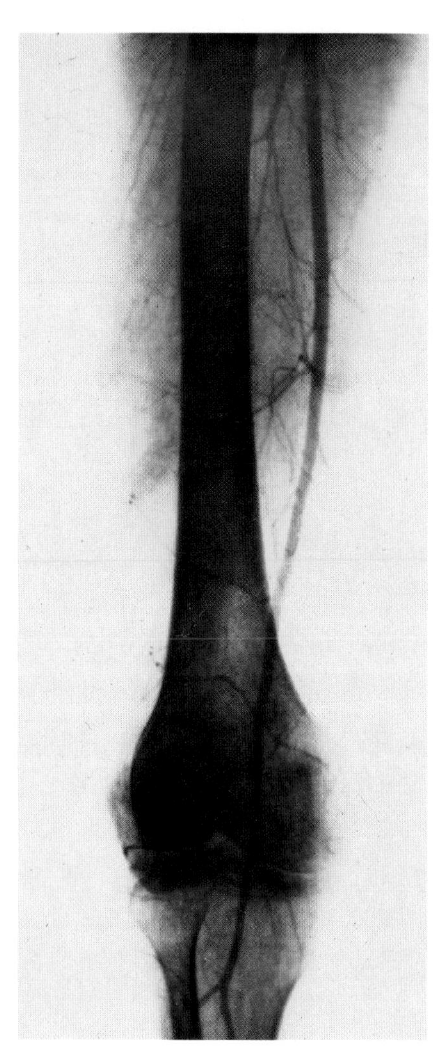

In what has become a controversial procedure, German scientists use a siege strategy against cancer.

The treatment begins as doctors selectively raise acid levels within the cancer tissue, making it more vulnerable to heat. Then the patient is placed in isolation (right) and bombarded with radio waves from a Selectotherm machine, which targets the heat. It raises the body's temperature to 104°F, the cancer's to 108.5°. Doctors simultaneously restrict blood flow to the cancerous area.

Researchers remain uncertain about exactly how the process works, but one theory suggests the following: As the body's temperature increases, circulation in healthy tissue steps up to dissipate the excess heat. But around the cancerous tissue, blood circulation shuts down. The result: Cancer cells, starved for blood, oxygen, and nutrients, die and break apart.

Before treatment, an X ray of a patient's thigh reveals a network of cancerous tissue surrounding the bone and the large femoral vein (above, left). Three weeks after treatment, the cancer disappears, with no damage to the surrounding tissue (above, right).

Pulse/min
°C
°C
°C

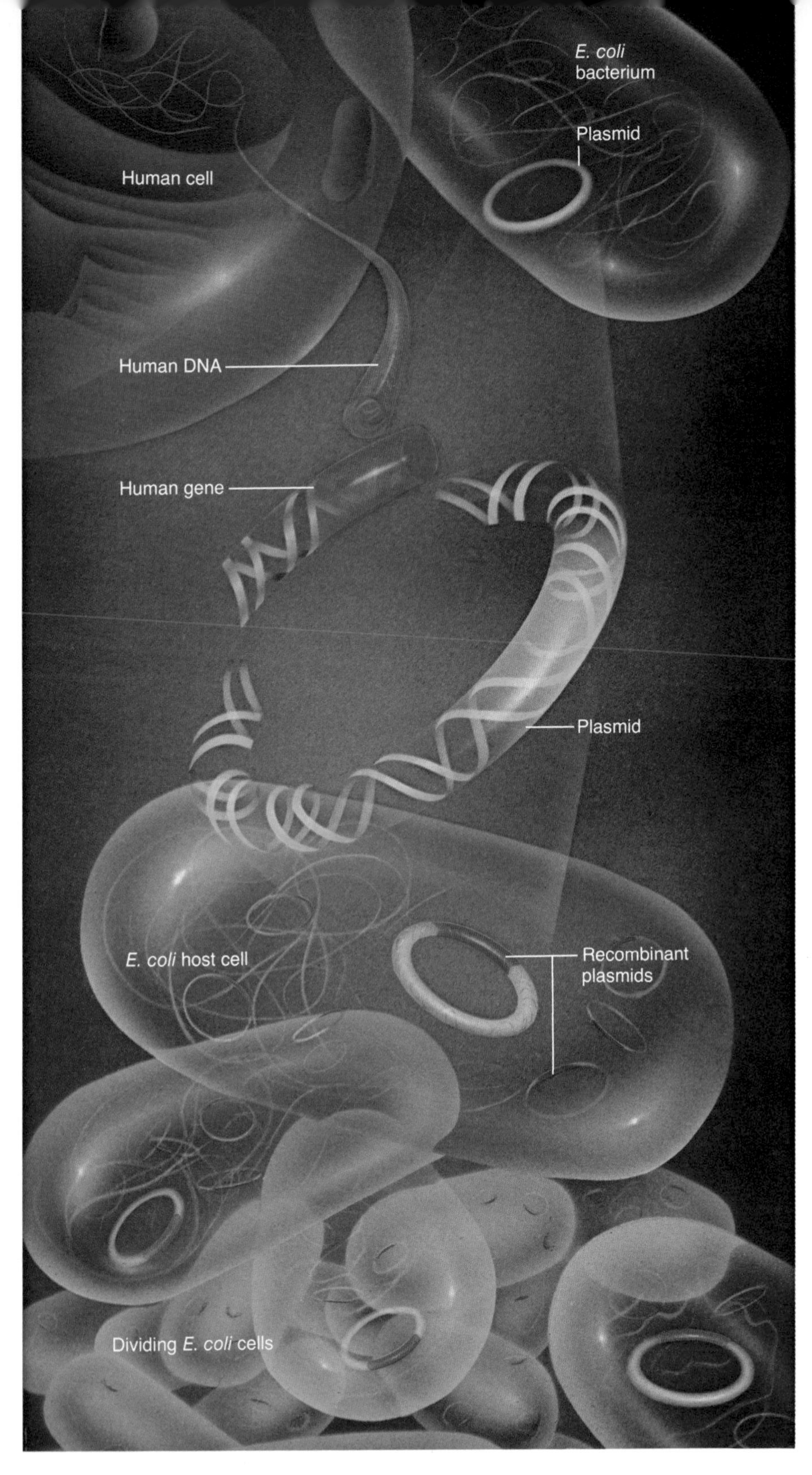

Scientists study a gel slide (opposite) as the first step toward mass-producing a protein by recombinant DNA technology. Specialists produce large quantities of insulin, the essential protein that diabetics lack, and interferon, a protein useful in the treatment of certain cancers.

After a protein such as insulin is targeted, scientists separate a ring of DNA, called a plasmid, from an *Escherichia coli* bacterium (above). Chemical scissors snip the plasmid at a spot where a human gene can attach; chemical glue binds the ends of the gene to the broken plasmid ring. Inserted into a host, a new *E. coli,* the recombinant plasmid tells the bacterium to produce the human protein. The recombinant plasmid multiplies, and the newly created insulin factory is duplicated every 20 minutes, as the bacterium divides. Soon billions of bacteria are manufacturing a protein that can help save a human life.

and the cells of transplanted tissue.

Other T-cells regulate the strength of the immune response. Those known as helper cells secrete substances that turn on antibody production and stimulate phagocytes and other T-cells in times of need; those known as suppressor cells produce chemicals that turn off antibody production and suppress the action of other T-cells. These regulator cells ensure an appropriate response to any single invader.

In people with the disease known as AIDS (acquired immunodeficiency syndrome), the normal ratio of helper to suppressor T-cells is disturbed. The AIDS virus attacks helper T-cells, preventing them from carrying out one of their regular duties—to activate the immune system when a threat arises. This breakdown in normal communication between immune cells leaves the body virtually undefended. Those who suffer the disease become the victims of a rare skin cancer called Kaposi's sarcoma, life-threatening pneumonia, and other serious infections.

In most cases, the various parts of the immune system—T-cells, B-cells, and phagocytes—work together. No single organ or set of organs orchestrates their defensive operations. Instead, the cells of the immune system talk to each other in a language of chemical signals—a language with a large vocabulary and a complex grammar. Each cell sends and receives several different messages. Each message, well timed and precisely directed, stimulates or inhibits other cells or regulates their activities. As the body molds its defense against a particular invader, the pattern of signals shifts slightly to suit moment-to-moment needs. The result is a delicately balanced, sensitive system of defense powerful enough to destroy or neutralize the effects of nearly any foreign intruder.

Invasion by a Cold Virus

Consider the cooperative efforts of the body's immune cells and molecules in an encounter with a rhinovirus, the agent that causes 30 percent of all colds. Like other viruses, the rhinovirus consists of a strand of genetic material—a piece of "bad news"—encased in a protein coat. It enters the body through the mucous membranes in our nose, throat, and eyes. Latching onto the surface of its host cell, the viral agent penetrates the membrane of its target and injects its own genetic material into the cell body.

Once inside its host, this minute fraction of genetic information swiftly diverts the cell's machinery to the production of new virus particles, or virions. As many as a thousand new virions may come bursting out of the now dead and wasted body cell, ready like their parent to attack healthy cells. Left to their own devices, these submicroscopic invaders might multiply

A scientist observes the progress of a gene he is assembling from bottled nucleotides, the basic units of DNA (above). Such research has led to the creation of valuable human proteins, such as interferon, which shows promise in the fight against cancer. Synthetic genes have also helped to provide large quantities of insulin for diabetics—as well as a hormone for the treatment of growth disorders.

The process begins at the keyboard of a computer, where a researcher types in information about the gene's identity: its unique sequence of nucleotides. With those instructions, the "gene machine" begins to make a chemical chain of nucleotides in the proper order and number. The researcher then splices the synthesized gene into a plasmid ring (see page 196). The recombinant plasmid then goes into a bacterium or a yeast cell, now engineered to produce a specific protein. Thus modified, the

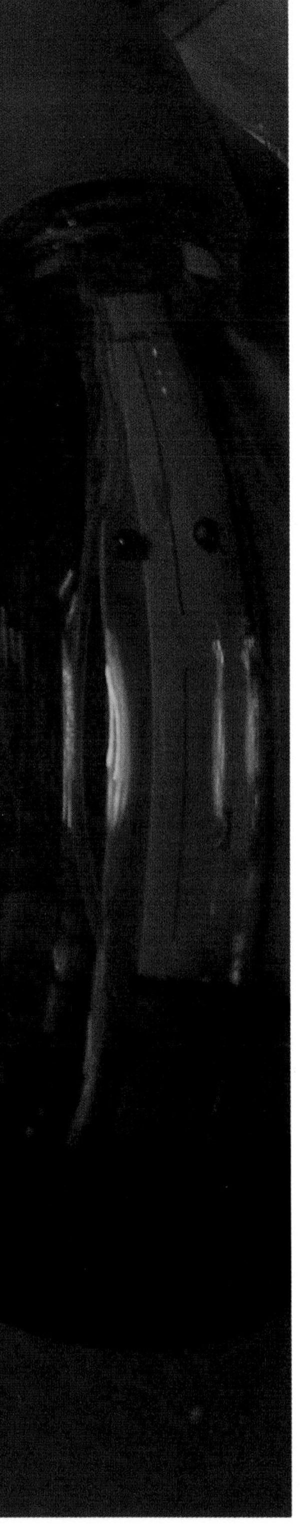

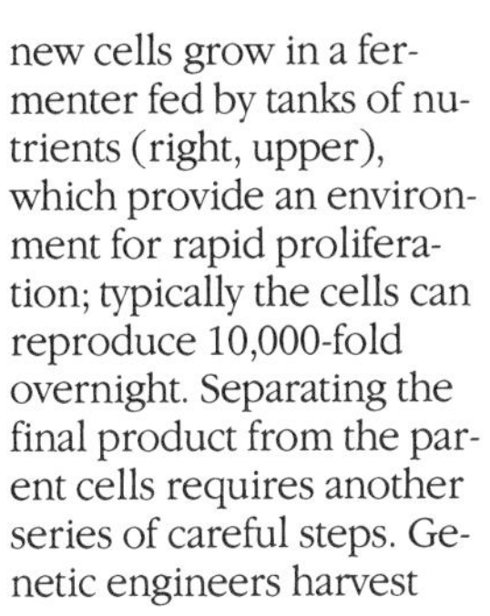

new cells grow in a fermenter fed by tanks of nutrients (right, upper), which provide an environment for rapid proliferation; typically the cells can reproduce 10,000-fold overnight. Separating the final product from the parent cells requires another series of careful steps. Genetic engineers harvest their crop in the glass bottles of a fraction collector (right, lower), which removes nutrients and bacterial impurities and collects the purified protein product for human use.

exponentially and cause serious damage. But in most cases, the body responds immediately to the destruction of that first infected cell.

Before the body's beleaguered cells succumb to their viral attackers, they release a substance called interferon. The powerful chemical alerts nearby cells to the presence of the virus. These neighboring cells then produce a protein that prevents the viruses from multiplying within them, thus limiting the spread of the infection.

Up to this point, the body housing this infection has probably felt little discomfort. The symptoms of a cold do not result directly from the destruction of cells by viruses, but from the body's response to this destruction—specifically, the inflammation that accompanies the immune system's attack on the virus.

Under viral attack, cells in the mucous membranes of the nose release histamine, blood capillaries in the area dilate, and blood serum leaks into the tissues of the membrane. The membrane swells. Some of its cells secrete more mucus, not all of which can easily move up through the constricted nasal passages. The excess simply runs out. An overload of mucus in the throat excites nerve endings that stimulate a cough, which clears the passage before the virus-laden mucus can move down to the lungs. The swelling and cell damage stimulate receptors in the nose, triggering a sneeze. Your head feels stuffy and congested, your nose runs, your throat is sore. Each sneeze and cough releases into the environment a spray of droplets loaded with virus particles that can now spread to nearby victims.

The Body Counterattacks

Meanwhile, the body's white blood cells have begun to converse. As it happens, the macrophage initiates the conversation. The "big eater" engulfs the virus and somehow incorporates a fragment of its prey—a foreign marker—into its surface membrane. Like every other body cell, the macrophage also bears a self-marker on its surface. The macrophage then presents both markers—self and foreign—to nearby T-cells. The self-marker says, in effect, "I belong to your team: You can trust my information," and the foreign marker, "So here is the nature of the invader"—in this case, the rhinovirus.

A T-cell with a receptor designed to recognize this particular kind of virus responds to the signal by multiplying. Newly formed T-cells secrete a chemical that attracts more macrophages to the site of infection and holds them there.

Some of the T-cells travel through the bloodstream to nearby lymph nodes or other lymphoid tissues to spread word of the invasion. There they contact B-cells and killer T-cells genetically programmed to react to the rhinovirus.

The killer T-cells leave the node and migrate

Hybridoma cells frozen at -94°F are removed from cold storage (opposite), thawed, and put to the service of medicine. They produce monoclonal antibodies, substances which have been used to diagnose ailments such as Legionnaires' disease and hepatitis. The antibodies can also mark tumor cells for attack by the immune system, while leaving healthy cells untouched. Someday they may even be rigged with toxic drugs to kill cancer cells directly.

Manufacturing these antibodies is, at least in principle, a simple process: First, a mouse is injected with an antigen (right, upper). The animal's immune system responds as a human's would—by producing antibodies to combat the injected threat; cells that make the antibodies are then withdrawn from the mouse's spleen. To mass-produce the antibodies, scientists fuse these cells with cancer cells, making a virtue from a vice. Like cancer cells, the new hybrid cells reproduce endlessly, but they now produce a continuous supply of a beneficial antibody. The new cells, called hybridomas, clone themselves (right, lower), making antibodies for medical treatment and research.

Monoclonal antibodies contribute to research on malaria, diabetes, allergies, infertility, cancer, and many other maladies.

to the site of infection. Using their specialized receptors, they attach to the surface of the infected cells. Less than a minute after contact, the T-cell delivers a chemical signal to the target cell, which results in its destruction hours later. Meanwhile the T-cell moves on to destroy other infected cells.

At the same time, the activated B-cells in the lymph nodes have begun to multiply and divide, generating a line of cells committed to respond to the rhinovirus. Some of the progeny become plasma cells—protein-producing factories that make antibodies specially tailored to bind with the rhinovirus. It may take several days for plasma cells to achieve full working capacity. Once they do, the new antibodies pour out of the cells at a rate of thousands per second, flood into the bloodstream, and home in on the invaders.

Some bind with the viruses themselves, preventing them from attaching to body cells; others mark the infected cells for destruction by phagocytes. The activated phagocytes sweep over the site, engulfing free-floating virus particles and infected cells and clearing away the battle wastes. As the response mounts, a volley of chemical signals flies back and forth between the lymphocytes and phagocytes. The coordinated efforts of only a few specialized cells becomes a massive, yet sensitive, attack, with all of its force aimed directly at the intruder.

The Assistance of Fever

When confronted by some organisms, macrophages produce a substance called interleukin-1, which triggers fever. Once released by the macrophage, the interleukin-1 travels through the bloodstream to the tiny portion of the brain that controls body temperature—a small cluster of neurons deep within the hypothalamus. Somehow, the chemical prompts the temperature regulator to set a new body temperature. Nerve impulses from the hypothalamus then spark the body's heat-conserving mechanisms—blood vessels in the skin constrict, preventing heat loss; muscles contract, causing shivering. We cover up with sweaters and blankets until the body temperature reaches its new set point. As long as the fever remains mild and does not persist for more than a few days, the raised temperature appears to increase the efficiency of the body's infection-fighting agents. The antiviral substance interferon operates more effectively. Phagocytes attack their prey with more speed and vigor. Fever may even increase production of T-cells.

As the lymphocytes, phagocytes, and antibodies begin to overcome the virus, suppressor T-cells signal the defenders to bring their efforts to a halt. Your head clears, your runny nose dries up, you can swallow easily again.

Out of every such *(Continued on page 207)*

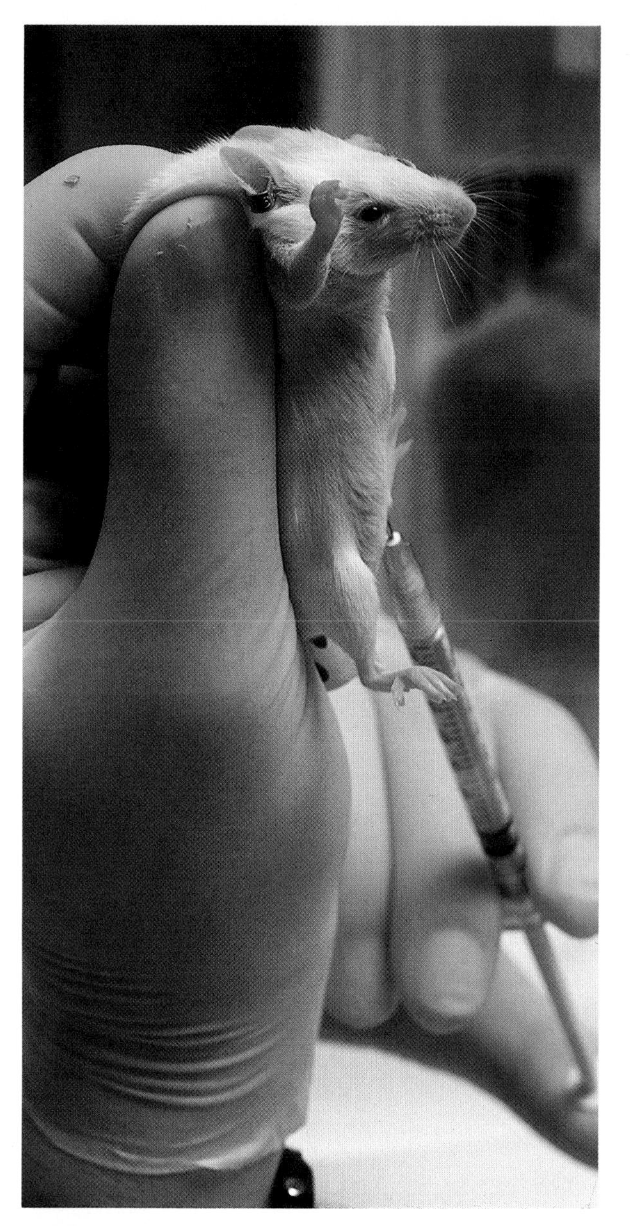

Magnification: 4,000 times

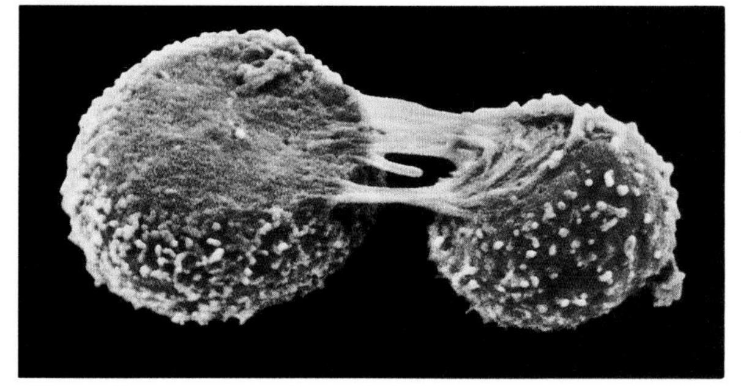

Sorting Blood to Save Lives

Mismatched blood transfusions can be life-threatening. Millions of people receive donated blood safely today, thanks to elaborate testing procedures. The automatic analyzer (right) helps identify blood types. It does this by combining chemical agents with samples of red blood cells and plasma, two of the major components in blood.

Like other human tissue, red blood cells carry chemical markers—their identity cards—on their surface. When one person donates blood to another, the blood types will not mix safely if the markers of donor and recipient do not match.

If a person with type A blood receives a transfusion of type B blood, for instance, antibodies in the type A plasma recognize the type B red cells as alien. These antibodies, known as agglutinins, will cause the donor cells to clump (below) and perhaps clog the body's important blood vessels.

Magnification: 2,000 times

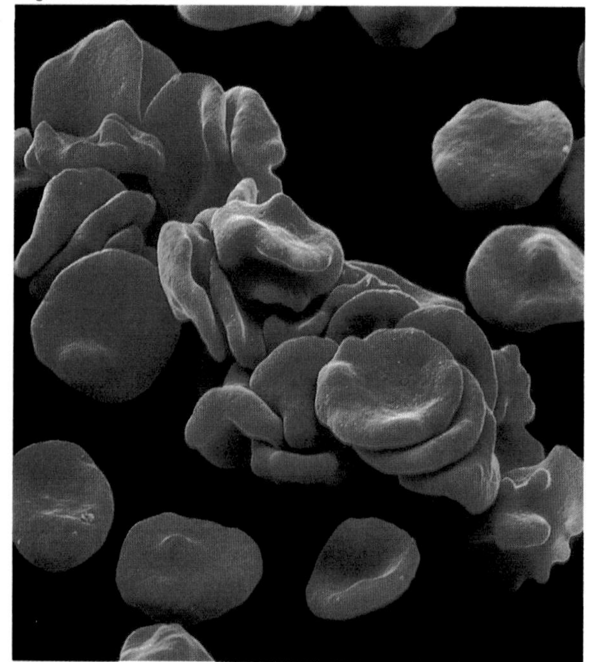

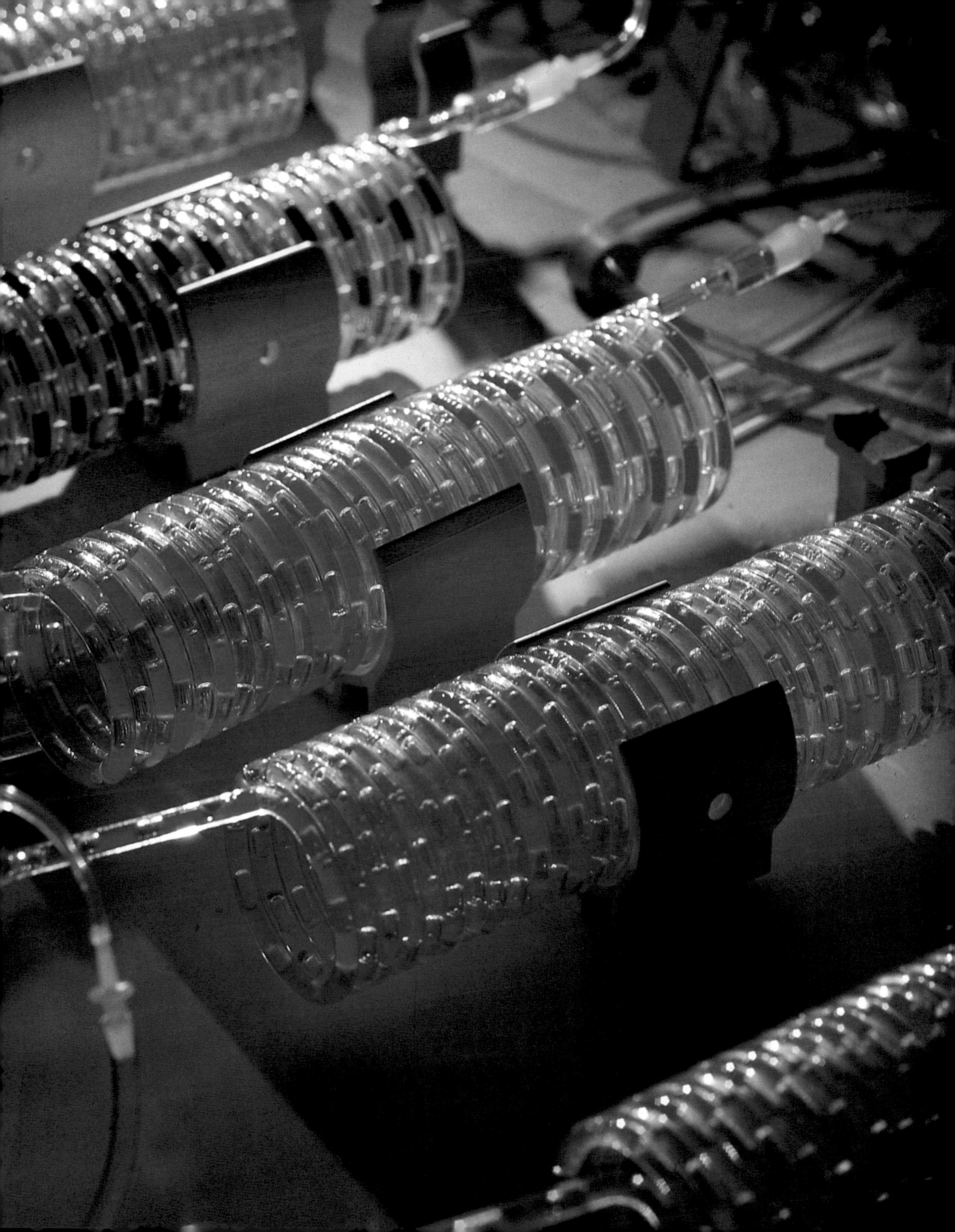

Fresh-frozen blood plasma lies thawing in a national blood bank (left). As the plasma liquefies, a necessary clotting component called factor VIII precipitates out. Extracted from the plasma, it is used to treat hemophiliacs. By the efficient storage of such blood components at strategic locations, thousands of lives are saved each year. The American Red Cross dispatches helicopters (above) to beat the clock in times of need, rushing the vital fluid to a distant emergency.

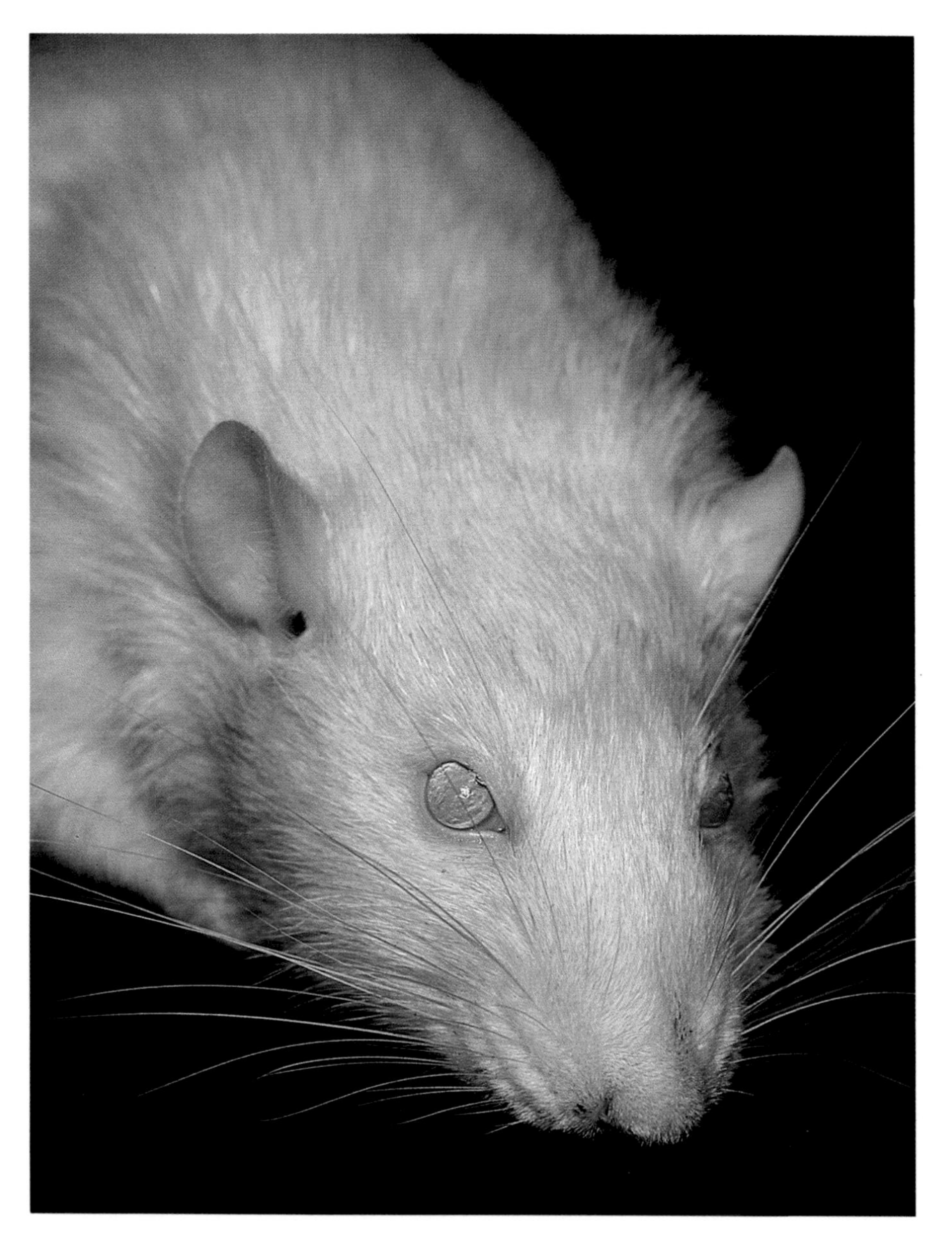

It looks like milk but works like blood. Developed by a Harvard biologist, this "artificial blood" substitutes for hemoglobin, the protein in red blood cells that binds with oxygen and transports it through the body.

The solution, one of a variety of substances known as perfluorochemicals (PFCs), goes by transfusion to a laboratory rat (opposite). The rat's eyes are their normal red color. As the liquid fills its blood vessels, the rat's eyes turn white (above).

Though still experimental, the blood substitute holds promise for humans: In the future, it may be used to treat sickle cell anemia, tetanus, tumors, and other blood-related disorders. If perfected, large quantities could be stored for use when war or natural disaster place heavy demands on a regular blood supply. For the present, however, its use in the United States is limited to emergencies, most of which involve members of the Jehovah's Witnesses. Their religion bars transfusion of real blood, but they accept substitutes.

encounter arises a special group of lymphocytes called memory cells. Like the other B- and T-cells created to fight an infection, these cells are equipped to respond to the antigen that triggered their production. However, they are also able to survive and carry the memory of the encounter for years, even decades, after the fight is over. Before the first exposure to an antigen, only one lymphocyte in a million might recognize and react to that antigen. After an infection, thousands of primed B-cells and T-cells circulate in the bloodstream and pool in the lymph nodes and spleen. If the same antigen invades again, these memory cells will react quickly and forcefully. They usually destroy the invader before it can multiply and cause disease.

Americans catch almost a billion colds a year—up to 6 per person. If every viral infection results in the production of memory cells, why are we not immune to cold viruses? In fact, we are. But immunity is specific to only one particular virus. At least 110 rhinoviruses have been identified—and some 90 other types of cold viruses. While we do not catch the same cold twice, we constantly encounter new viruses for the first time.

Vaccinations depend on the ability of the immune system to remember an encounter. By introducing a dummy enemy, vaccines prepare the immune system for future attack by real disease-causing agents. In a vaccine, this dummy is usually a weakened or killed virus, which triggers the appropriate lymphocytes to reproduce and make antibodies, but does not cause disease. When the full-strength threat comes along later, the body is prepared.

Cancer and the Immune System

The cells and chemicals of the immune system work together to protect the body from outside threats. How does the system handle threats that arise from within?

Every day an adult produces some 300 billion new cells. They usually divide as they should, but sometimes, chemical, physical, or viral agents can sabotage and rearrange the genes that regulate normal cell growth and differentiation. When this occurs, that single cell may begin to divide uncontrollably, multiplying and joining to form a colony of mutant cells, a malignant tumor.

When a body cell becomes cancerous, its membrane may change slightly, so that it bears markers somewhat different from the body's own. Ordinarily, agents of the immune system will recognize and react to the new markers, eliminating the mutant cell.

Some scientists believe that the cellular part of the immune system may have developed originally as a surveillance mechanism for cancer cells. As animals evolved, they became more

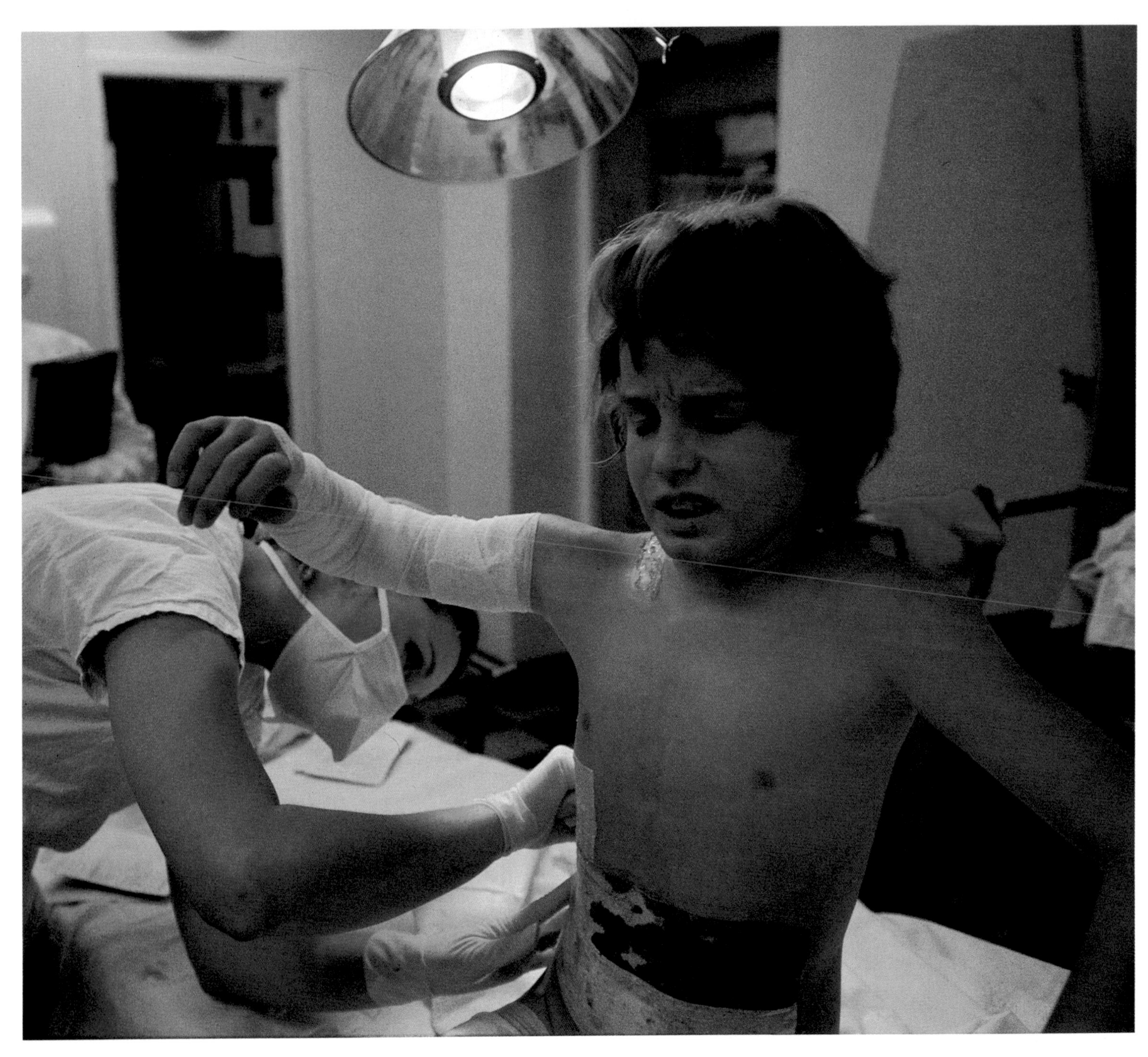

Probably the most torturous of all wounds, burns strip victims of their natural armor, the skin. Burn victims suffer numerous infections until new skin covers their wounds.

To minimize the trauma for a 10-year-old, doctors grafted pigskin to his abdomen (above). The temporary skin maintains body temperature, reduces fluid loss, and helps relieve the pain caused by exposed nerve endings until the human skin grows back.

A potential complication in this and all transplants is the body's immune system, which can attack donor tissue as an intruder. The drug cyclosporine represses that natural reaction in humans and in laboratory animals: With continuous drug treatment, a white rat (right) will readily accept skin from a black donor rat.

complicated in structure. Abnormal variants could occur as new, specialized cells developed at a rapid rate and established intricate connections in forming tissue. If these mutant cells were not discovered and eradicated, they might multiply wildly and destroy the animal. So in the course of evolution, animals developed a highly refined system for recognizing subtle distinctions between self and nonself.

The rejection of transplants may be the price we pay for possessing such an efficient system of surveillance. If a surgeon transplants a patch of skin from one part of a patient's body to another part, the graft is usually accepted as self. But if the surgeon attempts to transplant skin from brother to sister, the borrowed tissue grows puffy, inflamed, and angry looking. Eventually it drops off. Even though the donor and recipient are blood relations, the graft is rejected—as foreign—by the immune system.

When the Immune System Malfunctions

The very characteristics that enable the immune system to act so effectively on our behalf also make us vulnerable to mishaps. Sometimes our immune responses are a little too strong or too weak, and not always comfortable for the individual experiencing them. In some cases, the immune system does serious damage.

Even one of the oldest defenses against infection, the inflammatory response, works to our detriment from time to time. If an infection proves intractable, a chronic state of inflammation may develop. Inflammation flares up, retreats, and flares up again, never accomplishing its healing purpose. The attempt to repair an injury becomes a disease itself.

White blood cells and antibodies sometimes respond to a harmless intruder as if it were a dangerous adversary. An allergic reaction results. We wheeze and sneeze and break into hives as the immune system rallies its huge forces against a simple grain of pollen, a dust particle, a morsel of crab.

Occasionally, the agents that monitor our insides for damaged self-components and foreign substances become overzealous and autoimmune disease can occur. Our defenders attack the body's healthy cells and tissues as though they were foreign. In the disease known as rheumatoid arthritis, misled agents of the immune system brand the tissues of our joints for destruction; in multiple sclerosis, they attack the tissues of the central nervous system; and in myasthenia gravis, they inhibit the substances that regulate our voluntary muscles.

The system sometimes underreacts, too, or is tricked and fails to react. A cancer cell may shed its antigenic markers or camouflage itself behind a coat of protein. *(Continued on page 217)*

Eight-year-old Stormie Dawn Jones (above), the world's first recipient of a simultaneous heart and liver transplant, owes much of her recovery to a daily dose of cyclosporine with her milk.

Cyclosporine, a drug made from a soil fungus, was discovered in 1970 and has revolutionized transplant surgery. It reduces the hazard of tissue rejection, largely by suppressing helper T-cells, part of the immune system's defense network. Because the drug works by inhibiting the body's natural defenses, Stormie and other cyclosporine users face the ever present danger of infection, along with the possibility of side effects, such as kidney damage.

Living with Allergies

The simple act of mowing a lawn can spell misery for some 15 million Americans who suffer from hay fever. For some, a bubble helmet with hose and filter (right) may be the last resort. The need for such devices illustrates how little we know about controlling allergies—the immune system's overreaction to such otherwise harmless particles as pollen, dust, and animal dander.

For the hay fever victim, the first breath of pollen-laden air spurs development of an antibody known as immunoglobulin E (IgE). This antibody attaches to mast cells lining the nose, eyes, and throat. With a second breath of pollen, IgE triggers an explosion in the mast cell (opposite, upper). The bursting cell sheds granules of histamine and leukotrienes (tinted blue), powerful chemicals that inflame surrounding tissue. The familiar symptoms of sniffles, tears, and itches result.

Pollen can also set off an asthma attack. Muscles around the inflamed bronchial tubes tighten, the lining of the airways swells, and excess mucus bubbles from the bronchial walls (opposite, lower).

Magnification: 5,000 times

Magnification: 3 times

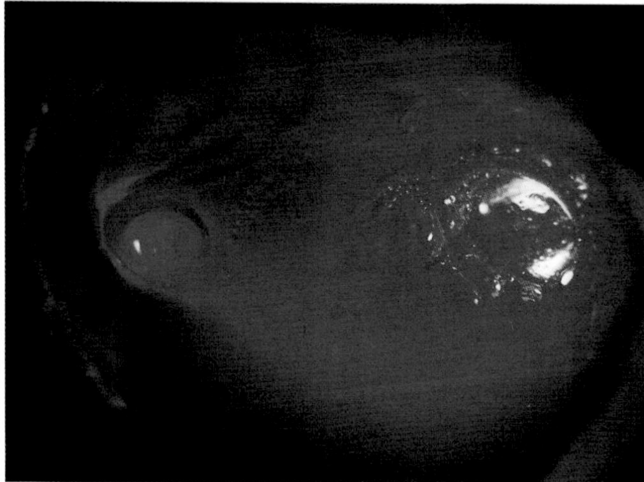

Lennart Nilsson, © Boehringer Ingelheim International GmbH. (also above)

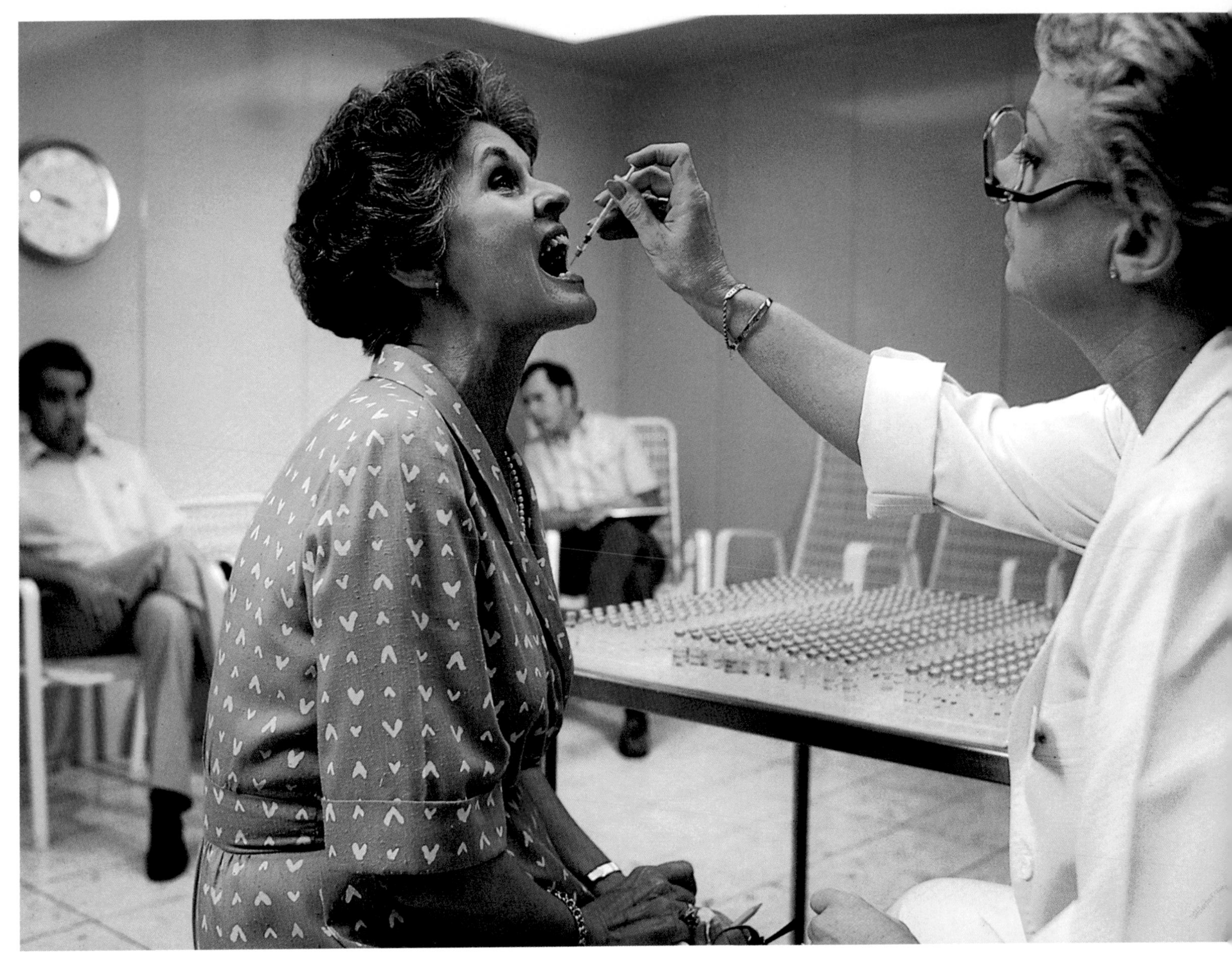

For the youth in the isolation chamber (opposite), the whole world threatens with possible allergies. For 15 minutes, she inhales a vapor that may help pinpoint the source of her drowsiness, headaches, and nausea.

Such "universal reactors" are people who appear to be sensitive to chemical pollution from plastics, pesticides, adulterated foods, cleaning fluids, and a bewildering array of other substances. The culprits behind the attack vary, and the range of reactions is wide: rashes, chills, fatigue, dizziness, depression, sometimes even the loss of voice.

To treat these patients, physicians known as clinical ecologists examine possible allergens by trial and error. In a room specially constructed of porcelain steel, a nurse injects a patient to test her reaction (above).

Not all experts agree that these reactions are allergies. Conventional allergists stress that one classical sign of allergy—an increase in the antibody IgE—often does not occur.

While the controversy continues, people seek answers. Discarded vials of possible allergens (right) attest to the time-consuming routine of periodic exams and treatment.

A trailer park in Seagoville, Texas, offers a temporary refuge for people seeking a decontaminated world. Here, in a rural area, with the help of carefully controlled housekeeping, they look for relief from synthetic pollutants and chemical irritants.

Instead of hanging out laundry, one visitor clips postcards to a clothesline (left, upper) so the sun can dry chemicals out of the paper. Her meals consist of specially prepared foods. The walls and ceiling of her trailer (right) have been lined with foil in an attempt to seal out any chemicals that might remain on its surfaces after repeated washings. Bare floors eliminate the possibility of allergens lurking in rugs.

Outside their sanitized trailers, in air free of a city's pollution, two visitors hold hands and pray (left, lower).

Magnification: 25 times

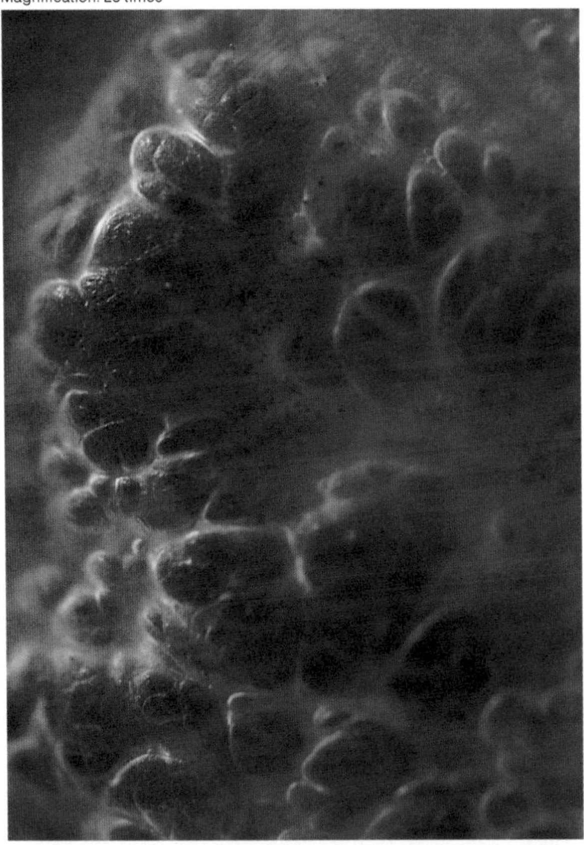

Magnification: 25 times

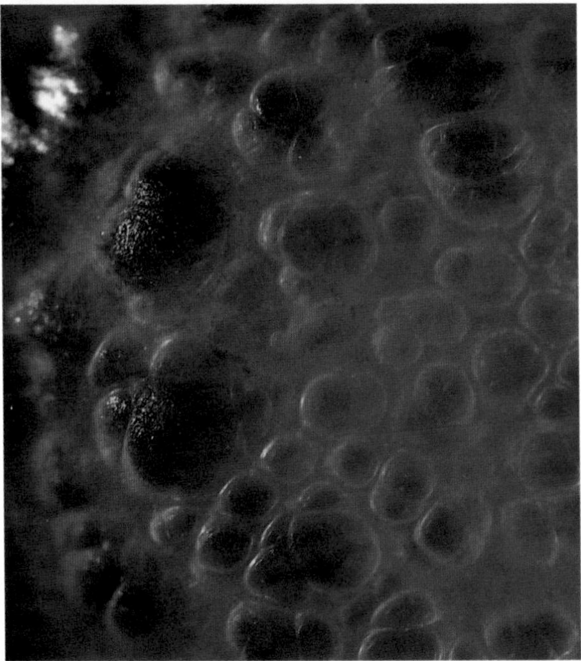

Lennart Nilsson, © Boehringer Ingelheim International GmbH. (also above)

Pollutants spew from a power plant, a dark reminder of the health threats created by an industrial environment.

Harmful gases and particles in the air can slip by the body's respiratory filters and penetrate to the lung's air sacs, or alveoli, where oxygen is delivered to the blood.

Healthy lung tissue (left, upper) shows no evidence of the long-term exposure to dirty air that has left deposits in the bubble-shaped alveoli (left, lower)—deposits that may cause emphysema or other respiratory ailments.

At such times, a single mutant cell may escape detection, divide and multiply, and form a mass of mutant cells that invades and destroys surrounding healthy tissue. If a cell breaks off from the original mass, it may metastasize, traveling through the blood or lymph to distant sites of the body, causing further destruction.

Factors Affecting the Immune System

The immune system works remarkably well for most of us most of the time. Just how it works differs from person to person. The immune system of every human being reflects the life history, the individuality of its owner. How any one person will meet a given challenge depends on many things, including age, genetic makeup, physical fitness, and perhaps, mental well-being.

In general, the very young and the very old more often succumb to infection—the young because they have not yet built up immunity to the host of microbes in the environment; the old, because the elements that make up their defenses—the T-cells, in particular—do not perform as well as they once did.

Genes play perhaps the most important role in shaping our response to a challenge. Fragments of DNA dictate the self-labeling for all body cells, as well as the specialized receptors on the cells of the immune system. Genes also determine the makeup of antibodies. Genetic differences may explain, in part, how one person can resist an attack by an infectious agent, while someone else contracts the disease caused by that agent; why some people are particularly susceptible to certain immune disorders, such as allergy, autoimmune disease, and even cancer, and how this susceptibility may persist over time, from one generation to the next.

Though quite constant in its workings, the immune system is subject to the perpetual cycles of change within the body. It is a system of balance, of dynamic equilibrium, intimately connected to the body's other systems, and thus, prone to their influence.

Diet can affect immune system function. Malnutrition, a deficiency of protein or the trace element zinc, or dietary excess may all have a profound effect on the development of disease. A severe deficiency of protein at an early age can slow or stunt the growth of the thymus, impairing the normal functioning of T-cells.

Perhaps most provocative and least understood is the complex connection between the immune system and the brain. Mental stress, both moderate and severe, can affect the workings of our white blood cells. In one study, it was found that patients hospitalized for severe depression had suppressed or underactive immune systems, making them more vulnerable to infections and disease. Another study showed

Magnification: 20,000 times; Lennart Nilsson, © Boehringer Ingelheim International GmbH. (also opposite)

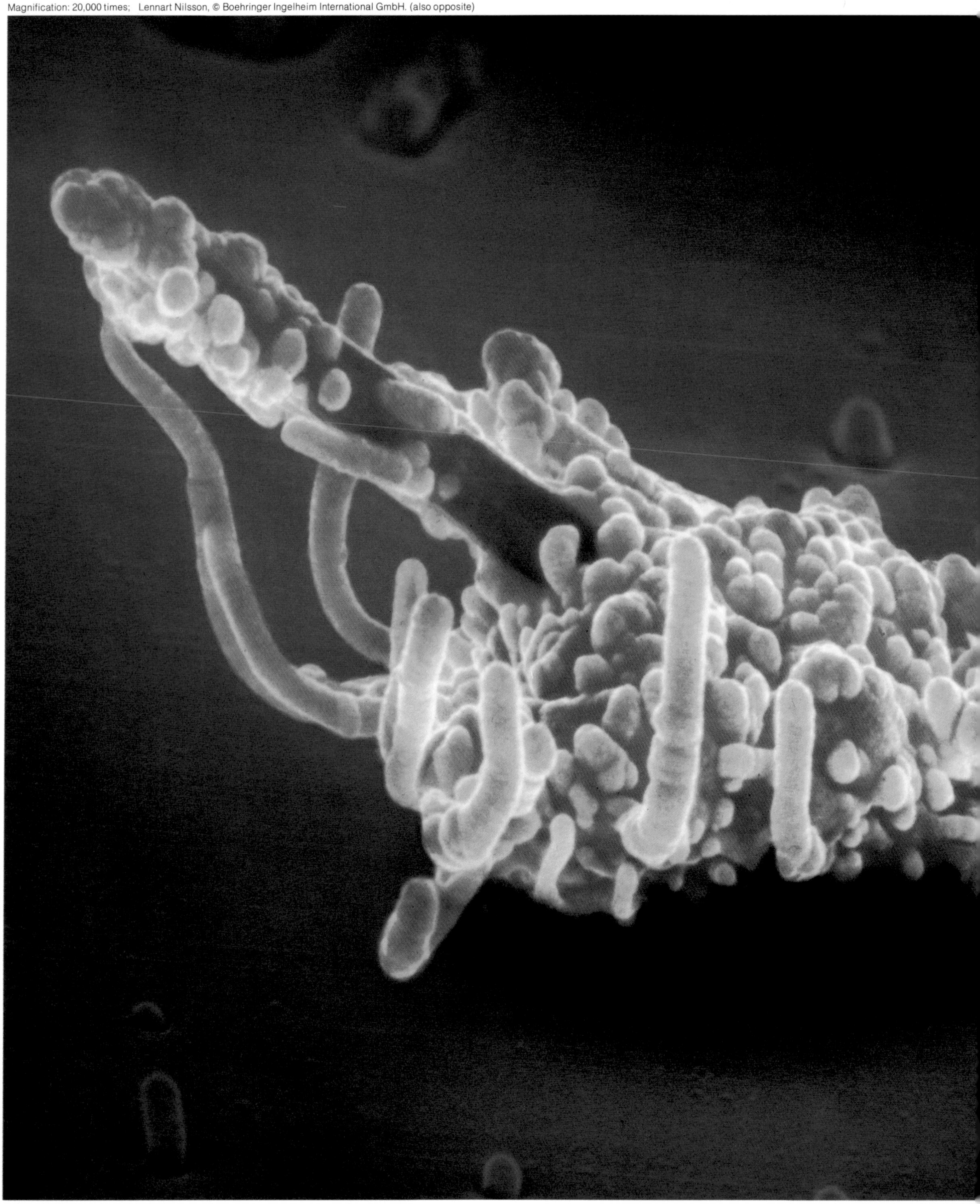

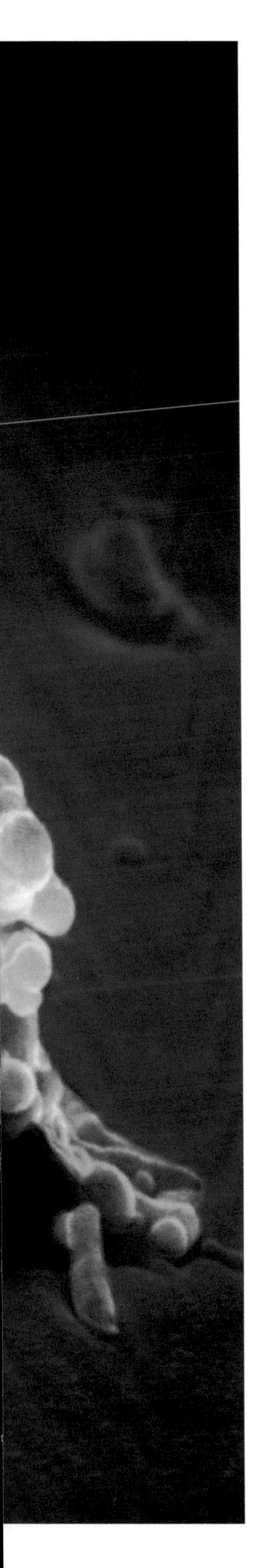

Magnification: 7,000 times

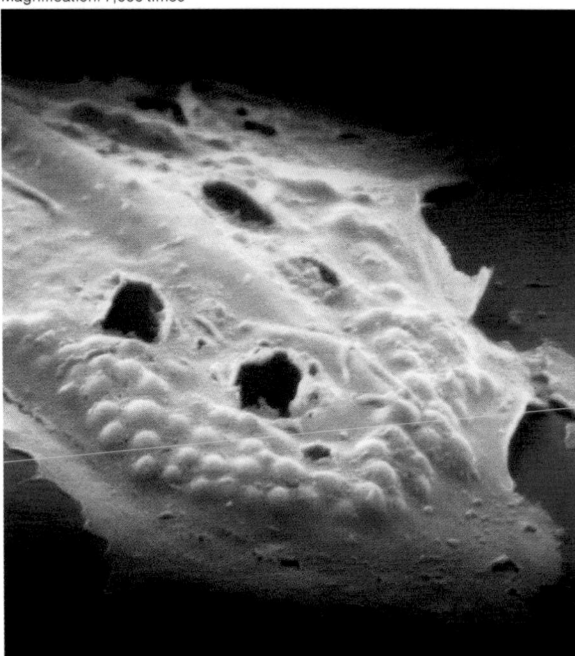

Like an octopus trapping its prey, a macrophage surrounds an asbestos fiber (tinted brown, left).

Macrophages clean microbes, dust, and pollution from the lungs by trapping and then eating the unwanted material.

When a macrophage encounters a pollutant, it springs into action, waving its tendrils and stepping up production of enzymes that enable it to digest the material it has in its grasp.

Asbestos is a deadly exception. Though macrophages can engulf the fine mineral fibers, they cannot digest asbestos. They collapse and die (above) in the attempt, releasing the toxic fiber back into the lungs. Prolonged inhalation of asbestos fibers can cause asbestosis. Victims of this disease suffer from inflamed and scarred lung tissue and can develop lung cancer.

that the suffering of a widower over the loss of his wife may harm his health: Bereavement seems to prevent lymphocytes from reacting appropriately. Even the much milder strain of taking an exam appears to reduce T-cell activity.

In 1886, physicians reported that an artificial rose could provoke an attack in an asthmatic patient who was allergic to roses. Since that time, several studies have indicated that learning and conditioning can affect the activity of the immune system. In one study, researchers found that they could condition the immunological response of guinea pigs. First they injected the guinea pigs with an allergen. In response, the animals produced histamine—a key agent in allergic reactions. Over the course of several weeks, the animals were exposed to the allergen along with an unrelated odor. Soon the researchers determined that the guinea pigs had "learned" to react to the odor alone: When exposed to it, they produced histamine, just as they had when injected with the allergen. The brain—and not an allergen—triggered the allergic reaction.

The biological basis of these interconnections remains much of a mystery. It is clear, however, that the nervous and immune systems are inextricably linked, anatomically and chemically. There is some evidence that lesions in certain areas of the brain can affect the immune system. Damage to parts of the hypothalamus may cause changes in lymphocyte activity.

Nerve cells and immune cells seem to engage in two-way conversations. Some immune cells have receptors on their membranes for neuropeptides, chemicals produced by the brain. One group of neuropeptides, known as endorphins, may at times reduce our resistance to disease. Endorphins produced in response to mental stress seem to suppress the activity of T-cells, perhaps robbing us of help in fighting cancer cells. In some cases, endorphins appear to have the reverse effect: When we are injured, they may boost our immune response by helping to attract macrophages to the wound site.

Evidence suggests that immune cells themselves send messages to the brain. The rate at which brain cells fire changes during an immune response. Some researchers believe that agents released by activated lymphocytes, such as interferon and interleukin, cause this change in the brain's electrical activity.

Whether the immune system evolved originally to protect us from infectious organisms, to monitor our internal environment for mutant cells, or to modulate the body's other cells and tissues, it has brought with it enormous advantage: For the most part, the system works with the body's other cells and tissues to maintain stability and balance within, preserving a most precious possession—good health.

Jennifer Gorham Ackerman

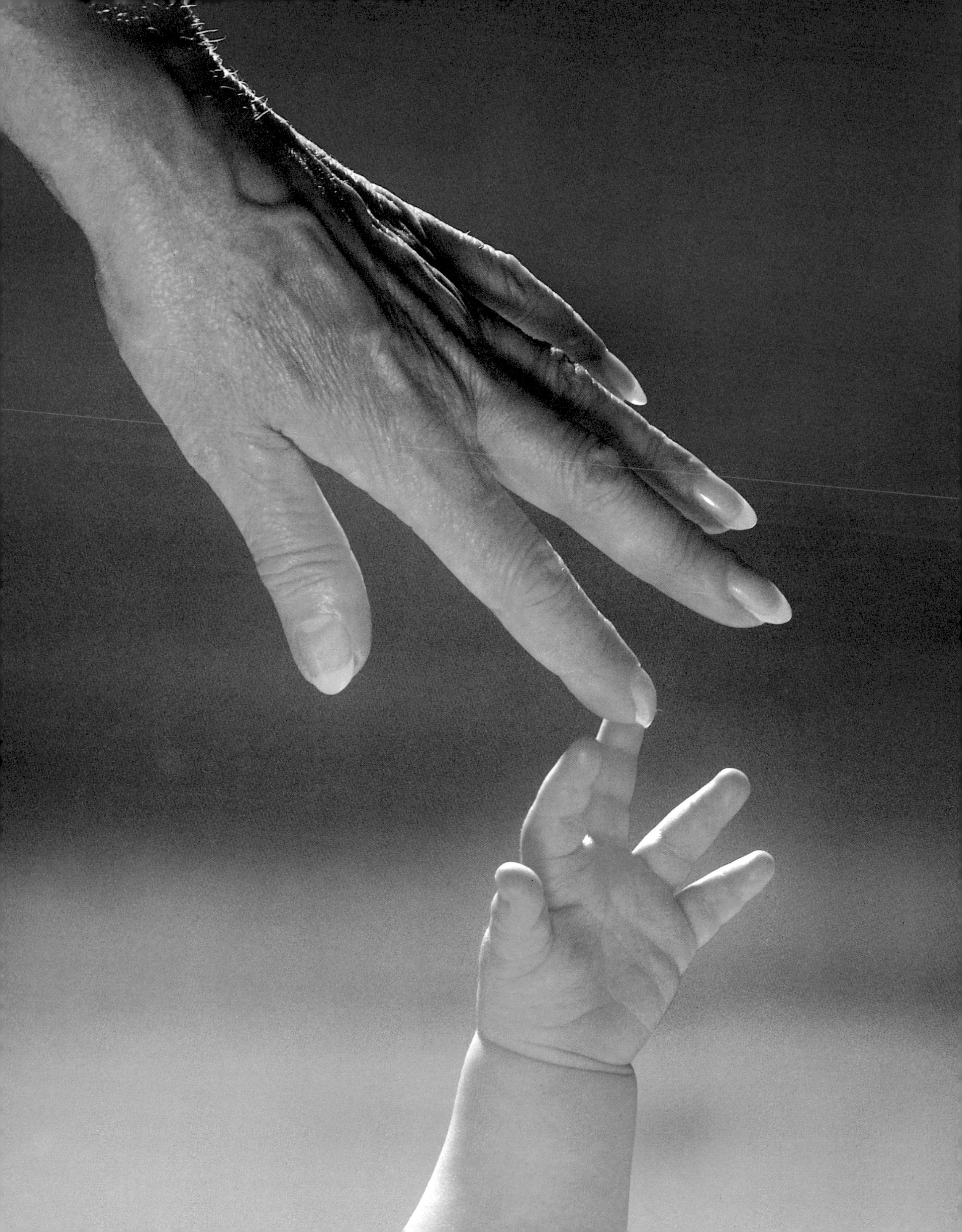

Stages of Life

By rights we humans should not live as long as we do. In the natural order of things, our life expectancy should be somewhere between that of a goat and a horse—that is, between 10 and 30 years of age. All mammals are warm blooded, and their bodies function in much the same way. Generally, the maximum length of life of a mammal can be correlated to its body size. The smallest mammal, the shrew, might see the spring of one year and the fall of the next. A rat will live perhaps 2 or 3 years. A rabbit may survive up to 10 years, while an average-size dog will no longer be man's best friend after 15 or 20 years. An elephant lives 60 years at best, almost matching our current life expectancy—74.7 years in the United States, 69.6 in the Soviet Union—but, of course, the elephant is an exception.

Why do we live more than 50 years beyond sexual maturity and maximum physical strength? Survival of the species does not require a long life span; we need only 20 years at most to care for our offspring. One reason for our longevity is the power and skill of our brains. Far larger than that of any animal of equivalent body size, our brains have helped us master our environment. We can delay starting our families. We have developed weapons to protect ourselves from wild animals (protecting ourselves from the follies of one another is another matter). By taming the plague, polio, tuberculosis, smallpox, and other pestilences that once threatened our survival, we have extended the time we can expect to live.

Life span is the maximum age a person does attain. It is—for the moment at least—finite. Humans may not be able to live beyond 120, the oldest authenticated age: In 1985 the record holder was a Japanese man born June 29, 1865.

Life expectancy, on the other hand, is the number of years a person can expect to live, based on statistical probability. The Bible speaks of living threescore years and ten, but most people 2,000 years ago were fortunate to reach their early 30s. By the end of the Middle Ages, life expectancy had reached 38. It has edged up slowly across the centuries. A baby born in the United States in 1900 could expect to live 47.3 years. Dramatic changes have occurred this century, due mainly to reductions in infant and childhood deaths. Nowadays the life expectancy at birth in the U. S. has shot up to 71 years for men and 78.3 years for women. It is lower in developing countries, but there too longevity has increased.

How do we grow and develop? Nursery rhymes talk of sugar and spice and all things nice, but little girls, just like puppy dog tails and little boys for that matter, are made of cells and of the tissues and organs that cells form. The body of an adult consists of about a hundred trillion cells, all derived from just one cell, the fertilized egg. Growth involves either an increase in the size of existing cells or the creation of new ones by cell division. Both processes are at work throughout life, but one tends to dominate at any given stage of development.

Growth is not a simple story of cells getting bigger and bigger. They cannot. There is a physical limit to their size. Many cells are ball shaped. As a cell enlarges, its volume increases at a rate greater than its surface area. Since all materials needed for the cell to carry out its activity must cross the surface membrane, the surface area of the cell will ultimately limit how much it can absorb. Some cells overcome this size restriction either by altering their shape to an elongated form, like a nerve cell, or to a flattened shape, like a skin cell, or by using hairlike projections to increase absorption the way an intestinal cell does. These adaptations enable a cell to increase its surface area without increasing its volume.

From Simple Beginnings

Without cell division, further tissue growth would be prevented by the size limitation of single cells. In the process of cell division, called mitosis, each new daughter cell grows to the size of the parent. These new cells—more than 200 million are created in your body every minute—replace injured and worn-out cells. Old, damaged cells self-destruct by releasing a powerful enzyme that digests the cell from within. Normal human cells cultured in the laboratory divide up to 50 times before dying.

The time it takes a cell to move through the complete cell cycle—from growth to division—varies enormously. It may take as little as a few hours, or it may last as long as the body lives. Some skin cells live about 8 hours, cells that line the intestine about 1 1/2 days, while undamaged muscle and nerve cells last a lifetime.

Division alone would create a mass of cells that all look the same. But some cells need to turn into skin, some into liver, some into brain, and still others into myriad other tissues that make up the body. To differentiate, they change their structure and appearance and assume specialized functions. This process begins with the embryo and, in some cases, continues throughout life. Nerve cells, for example, develop thin strands up to three feet long that transmit stimuli to and from the brain. Certain skin cells make a pigment called melanin, which tans the skin to protect it from the sun's harmful ultraviolet rays.

Differentiation results when certain genes within a cell are activated, while others are repressed to prevent the formation of unwanted proteins. Differentiated cells no longer perform many of the functions of other cells, although their nuclei retain all the genes necessary to do so. Nerve cells, for example, have a gene for making melanin but do not use it. Scientists are just beginning to understand the way a cell uses only some of its genes while repressing others.

Ready, set, go! Proud parents point their crawlers toward the finish line in a Diaper Derby at the Monterey County Fair in California. Babies of the same age do not always respond to a new situation in similar ways. Some speed toward their destination, while others dawdle along the way.

Researchers disagree about why various talents emerge. Does heredity or environment determine why some children develop athletic prowess while others develop intellectual or artistic abilities? Who finishes the race first seems to depend on an unknown combination of genes and parental encouragement.

4
11

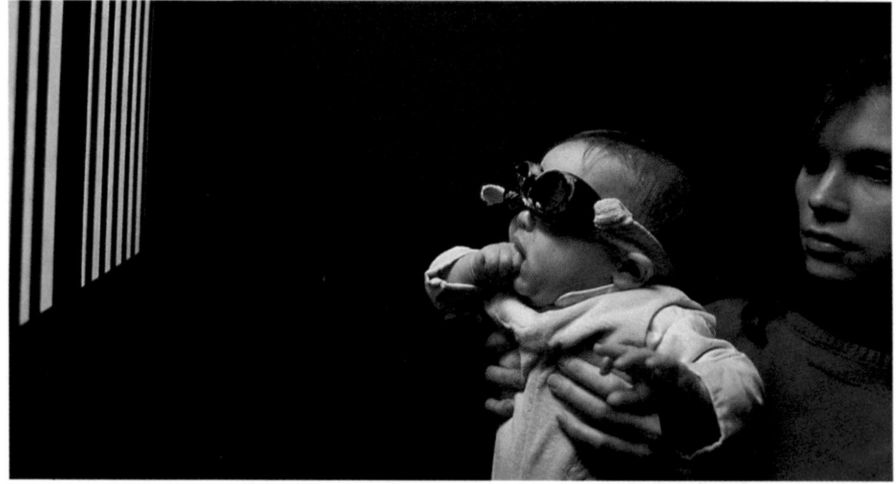

A baby creeps toward its mother's outstretched hands across a glass-topped "visual cliff" (opposite) in a study of depth perception. Infants who crawl at a very early age will creep across the apparent abyss without hesitation, while late bloomers will refuse to cross.

Other tests find answers from babies too young to speak. A researcher flashes a pattern of lights (upper) and measures an infant's heart rate, which increases with mental effort. When a baby views lines through 3-D goggles (lower), head movements indicate that the child sees the design in three dimensions. Some infants even try to grab it.

Differentiated tissues and organs behave and respond to disease and accident in markedly different ways. The skin, the lining of the intestines, and the progenitors of blood cells, for example, divide actively throughout life. The total mass remains constant because lost or destroyed cells are continually and rapidly replaced. The liver, by contrast, maintains a much slower rate of cell turnover, but its cells can divide quickly if challenged by disease or too much alcohol. The cells of the lungs and kidneys, on the other hand, divide during childhood but stop when the body reaches maturity. When lung or kidney cells are lost through disease or accident, the remaining cells simply enlarge and increase their capacity to perform—a process called hypertrophy.

Yardsticks of Growth

Human growth proceeds in four phases. The first, before birth, results mainly from cell division. All nerve cells, for example, are present by the sixth month of gestation; the nervous system continues to grow as these cells enlarge. During the second phase, from birth to maturity, the enlargement of existing cells tends to dominate. A baby's heart, only one-sixteenth the size of an adult's heart, contains the same number of cells as the larger organ. It grows only by enlargement. During the third phase, maturity, emphasis switches to the maintenance of existing functions and the repair of damage from injury or wear and tear. When old age—the final phase—sets in, slowed growth can no longer replace lost cells, and the efficiency of our organs and tissues declines.

Some parts of the body grow faster than others, which explains why a baby's proportions are very different from those of an adult. A newborn's head accounts for one-quarter of its body length; the brain is relatively large and well developed. By contrast, the head of an adult is less than one-seventh of total body length. A baby's legs are about one-third its length, while an adult's legs are half the body's length.

We do not grow at a constant rate. The most rapid rate occurs before birth, when, in the space of nine months, the fetus increases its weight about 2.4 *billion* times. After birth, two spurts of growth, in the first two years and again at puberty, are separated by a slower, more steady rate, where height increases by two to three inches a year and weight increases by five to six pounds. By their first birthdays, babies usually weigh three times their birth weight and have grown in height by 50 percent. Parents of small children can estimate the final height of their offspring. By the time a boy is two and a girl eighteen months, they will have reached about half their mature height. Adolescents may grow four to six inches a year.

Can we boost a baby's intelligence by early training and enrichment? Raised in a musical household, a ten-month-old infant seems to prove we can as she taps a rhythm on the piano in time to music on the stereo. Young children, exposed to the fine arts, can learn to recognize the works of masters (opposite). By age three, other youngsters have learned to read, play the violin, and identify pictures of U. S. Presidents.

The proponents of early learning believe that training children intensively at a young age produces ultra-intelligent superbabies. But skeptics contend that superbabies can turn into little computers; too much stimulation may smother their initiative to explore and learn on their own. Teaching takes time away from cuddling and other basic interactions that give a child a sense of security, so superbabies can become insecure and overdependent on their parents.

Until adulthood the long bones—elongated bones in the fingers, arms, legs, and hips—grow quickly by expanding at each end. These growth centers contain gristlelike cartilage cells that create layer upon layer of new bone tissue. Once the cartilage cells stop dividing, the growth centers harden into bone, marking the end of growth in that region. Most growth centers, such as those in the femur and tibia of the leg, have ossified by the age of 17 to 20 years. The breastbone is one of the last bones to stop growing, around age 25.

By the time you grow from infancy to adulthood, you will have about 144 fewer bones: The 350 or so bones in a newborn's body gradually fuse into the approximately 206 bones in the adult skeleton. The number of bones varies because some people have an extra pair of ribs or fewer vertebrae in the spine.

Sometimes disease stunts the growth of the skeleton by preventing the bones from ossifying completely. Henri de Toulouse-Lautrec, who died at 36, may have suffered from pyknodysostosis, a rare bone disorder that often afflicts the offspring of blood relatives. Toulouse-Lautrec's parents were first cousins. He grew only about 5 feet tall. Though his head was large, the plates of his skull never knitted together fully.

Why do we stop growing when we have reached less than a third of our life expectancy? Aquatic species such as mollusks, crustaceans, and some fish grow indefinitely. One giant clam weighed 600 pounds and may have been 100 years old; giant squid can grow to 50 feet; a 2,800-pound turtle has been reported. One reason: The water helps support their great weight. But land-dwelling creatures have to support their body weight themselves, so they have evolved ways of limiting their size.

The interplay of heredity and environment controls growth. The instructions encoded in the DNA (deoxyribonucleic acid) of our genes determine the hormonal and neural control of growth. But this control is modified by the environment—by climate, disease, nutrition, and pollutants. Genes determine the amount of growth hormone secreted by the pituitary gland, but malnutrition may prevent the pituitary from functioning at its full potential.

The instructions encoded in our DNA determine, to a large extent, our final height and weight. Tall parents tend to have tall children, while short parents are likely to have short children. Genes also regulate the variation in growth between the sexes. There is little difference in the size of boys and *(Continued on page 235)*

COVENTRY

Growing Up around the World

Whatever the culture, close family ties and affection freely given foster healthy emotional growth. In Namibia a grandmother tenderly kisses her granddaughter (below). These San, or Bushmen, live in extended family bands, with relaxed attitudes toward their offspring: Loving relatives help parents care for children, so family friction rarely has a chance to build up. Unmarried boys sleep in a separate house; unwed girls also live together in their own hut.

A Brazilian boy trundles his baby brother in a wheelbarrow (right). Studies confirm the importance of sibling ties in families, where older children protect, teach, and influence their younger brothers and sisters. The very closeness of siblings fosters rivalry in some, but not all, cultures. Sibling rivalry tends to be strong in societies that encourage self-assertion, competitiveness, and strong bonds between parent and child.

Children share common needs for independence, self-realization, and acceptance by their peers. Smearing more food on herself than she gets inside (opposite, upper), a baby learns to eat without her parents' help. In the second year of life, toddlers can help dress themselves, brush their teeth, and comb their hair.

Fantasy fuels the imagination of an eight-year-old as he frolics at Paper Bag Players (opposite, lower). This New York City group makes fanciful creatures out of colorful bags and boxes, and allows children to touch, play, explore, and romp in a whimsical playground filled with cardboard dinosaurs.

Like schoolgirls around the world, young Uruguayans waiting for a bus flash toothy, conspiratorial grins (above). Peer relationships become increasingly important as children leave the strong influence of families and develop their abilities to share, take turns, and handle conflict.

Formal education prepares youngsters for life in today's world. At a grammar school in Dallas, Texas (opposite), students take for granted computer operations that often baffle their parents, and teachers are finding a new "problem"—some students do not want to leave at the end of the school day. No daydreaming in today's classrooms: Computers provide instant feedback, so students must stay alert.

A new striving for excellence in education means more testing. At a New York high school, teachers pore over exam papers (above). Teachers find that high expectations, more demanding courses, and greater stress on the fundamentals pay off for students in higher test scores.

Teenage boys fling themselves and their skateboards above the sunset at Venice Beach, California. Sometimes called the tempestuous years, the teens mean spurts of physical growth, wild energy interspersed with lethargy, and bewildering emotional upheavals as the body's glands pour hormones into the bloodstream, producing the most dramatic changes since birth. The pea-size pituitary gland (cross section, below) performs a giant task, even though it weighs little more than a small paper clip. Located in a protected cavity beneath the brain, the pituitary gland produces the hormone that regulates growth.

Magnification: 10 times

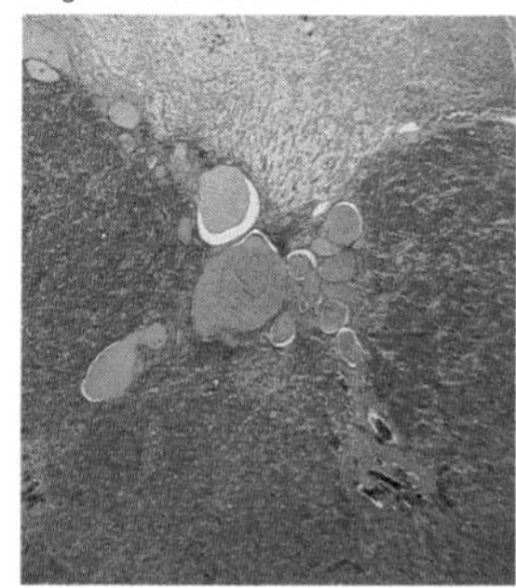

girls until the age of ten. The growth spurt at puberty starts earlier in girls but lasts longer in boys. On average, boys gain more height and, later, more weight.

Genes also control extremes in height. The Tutsi from Rwanda and Burundi in central Africa are among the tallest people in the world, with the men averaging six feet. The Mbuti Pygmies of the Congo River basin are the smallest; the men average only four feet six inches.

Factors in the environment, especially nutrition, modify the genetic blueprint for growth we inherit from our parents. Millions of children in various parts of the world are near starvation. Their growth will be retarded and their final height reduced by lack of nutrients.

To grow normally, we must eat adequate amounts of certain compounds. Proteins provide the amino acids essential for growth. Bones need calcium and phosphorus; hemoglobin, the protein that transports oxygen in the blood, requires iron to function properly. Absence of zinc retards growth and sexual maturation. An iodine deficiency also stunts growth and can lead to hypothyroidism, a condition that causes metabolism to slow and the body to lose vigor. Vitamins A, B complex, C, and D promote the development of healthy bones and other tissues.

Illness during childhood delays growth, but after recovery there is often a spurt of growth to catch up. Children in higher socioeconomic classes benefit from better nutrition, so they tend to be taller than those from lower classes.

Emotional upsets and lack of exercise may also interfere with growth patterns. Even the seasons play a role. The phenomenon could be summarized as "spring up, fall out." Children grow twice as fast in the spring as they do in the fall, but they gain more weight in the fall.

The Body's Chemical Messengers

Hormones secreted by our endocrine glands control how fast we grow. Scattered through the body like tiny islands, these glands affect every aspect of our lives—growth, physical and mental development, reproduction, and cell repair. Endocrine glands act on organs or certain types of tissues located in other parts of the body by releasing hormones, or chemical regulators, into the bloodstream. These hormones come in contact with every cell, but only certain ones, called target cells, will respond to any given hormone. Once the hormone molecules bind to receptor proteins in the target cells, the hormones set off a cascade of reactions, causing specific chemical reactions to speed up or slow down.

Two different mechanisms relay the information brought by the hormone to the target cell. Some hormones enter the cell and bind to a receptor protein in the cytoplasm, the gelatinlike

Throughout the world, different cultures celebrate puberty rites to mark sexual maturity and acceptance of adult responsibilities and status. A 14-year-old Apache girl (left) holds her head proudly as fellow tribe members shower her with cattail pollen, regarded as holy by her people. The blessing is part of a four-day ceremony that signifies the passage from childhood to womanhood.

Puberty rites in Bali, Indonesia, set teeth on edge—literally (right). Girls and boys formally come of age when a priest files the six upper front teeth until all are even. The Balinese believe this almost painless ritual protects the newly initiated adults against the evil in human nature and qualifies them for eventual cremation. Otherwise the gods might mistake them for fanged demons and bar their entrance to the spirit world.

substance encasing the nucleus. Together, the hormone and receptor move to the nucleus, bind to the chromosome, and cause the cell to synthesize certain proteins. Other hormones do not enter the cell at all. They bind to receptor proteins on the cell's surface and trigger the release of a second messenger in the cytoplasm. It is this compound that then initiates the cell's response to the hormone.

Before puberty begins, hormones play a major part in regulating growth. The growth hormone, somatotropin, is the main controller of height. It is one of several hormones secreted by the pituitary gland, which dangles from the base of the brain, just above the roof of the mouth. Somatotropin stimulates bone and muscle growth, maintains the normal rate of protein synthesis in all body cells, and speeds the release of fats as an energy source for growth. The pituitary also releases thyroid-stimulating hormone. This chemical causes the thyroid gland, set like a pinkish bow tie on the windpipe, to secrete hormones that influence general metabolism, especially the growth of bones, teeth, and the brain.

It is tempting to refer to the pituitary as the master gland because it regulates the release of hormones from other glands. But the pituitary is actually controlled by a region in the middle underside of the brain known as the hypothalamus (Greek for "under the inner room"). A special set of blood vessels connects these two glands and carries messages from one to the other. Chemicals released by the hypothalamus tumble half an inch to the pituitary and tell it to secrete

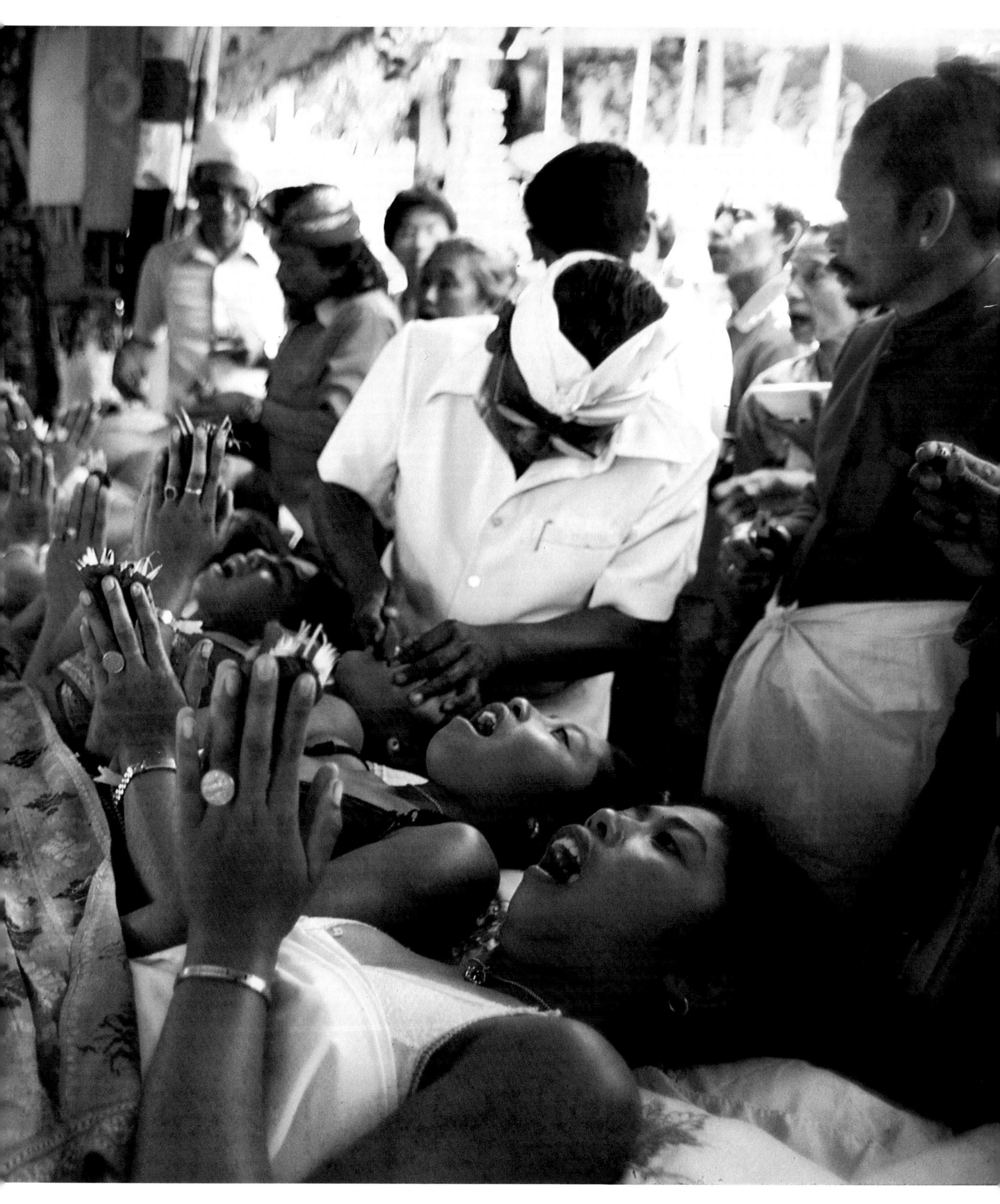

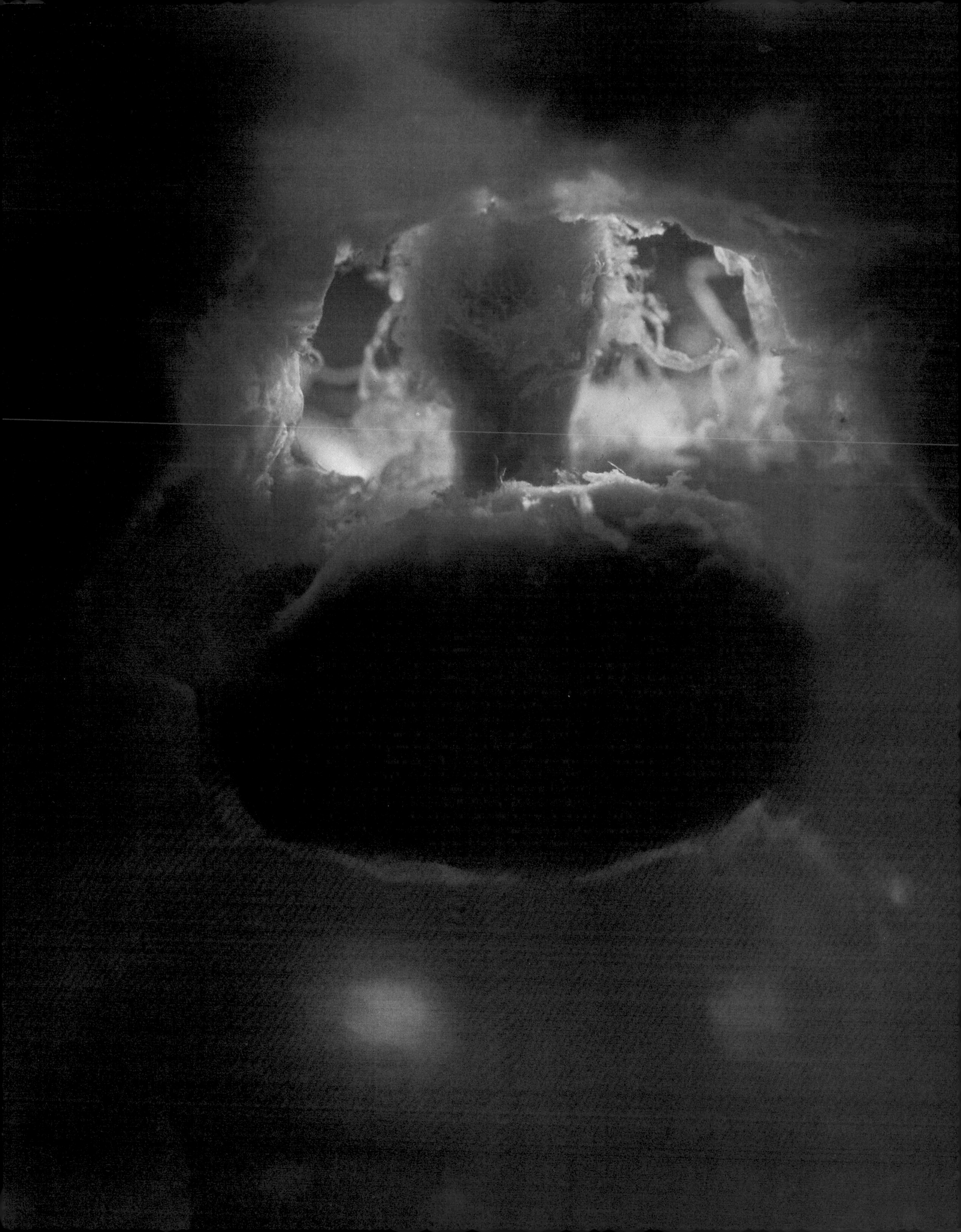

its hormones. When hormones from other glands reach high levels in the bloodstream, they send a message to the hypothalamus to stop releasing chemicals. This in turn slows the release of hormones from the pituitary.

Passage from Childhood

At the bidding of the hypothalamus, the pituitary gland initiates the dramatic changes that take place at puberty. Production of growth hormone slows, and the sex hormones take over. The hypothalamus signals the pituitary gland to secrete hormones known as gonadotropins, which stimulate the gonads—the ovaries and testes—to make the sex hormones estrogen and progesterone in females and testosterone in males. (While the word "testosterone" simply identifies a hormone made in the testes, "estrogen" harks back to the Greek for "producing frenzy," reflecting the biases of the male researchers who coined the term in 1927.) The hypothalamus controls this process by releasing a hormone that migrates to the pituitary gland and triggers the release of the gonadotropins. The pyramid-shaped adrenal glands, perched atop the kidneys, suddenly become active and increase their secretion of sex hormones.

The first signs of puberty occur around age 9 or 10 in girls but closer to 12 in boys. They appear only after a child has passed the peak of the height spurt. This rapid growth starts with the hands and feet, making them look out of proportion to the rest of the body. The arms and legs grow next, followed by the hips and the chest and shoulders. Secondary sexual characteristics then begin to appear. Girls develop breasts and grow pubic and underarm hair. The uterus and vagina enlarge, and menstruation starts around age 12 or 13. Wider hips will help in carrying and bearing children. For boys an early sign of puberty is enlargement of the testes and scrotum, followed by growth of the penis about age 13. The development of the reproductive organs triggers two other signs of manhood: a deepened voice and hair on the face.

Because girls enter puberty earlier, there is a period during which they are taller and heavier than boys of the same age. This phase is temporary, for girls usually reach their full height by the time they are 18, while boys continue growing until nearer their 21st birthday.

Over the last hundred years, children have been getting taller and reaching maturity earlier. Surveys from several European countries have shown that the height of children between five and seven has increased about half an inch every decade since 1900. Better nutrition and disease control probably play a major role.

Scientists are still unable to answer the question of why growth stops. The amount of growth

Magnification: 15 times (opposite)

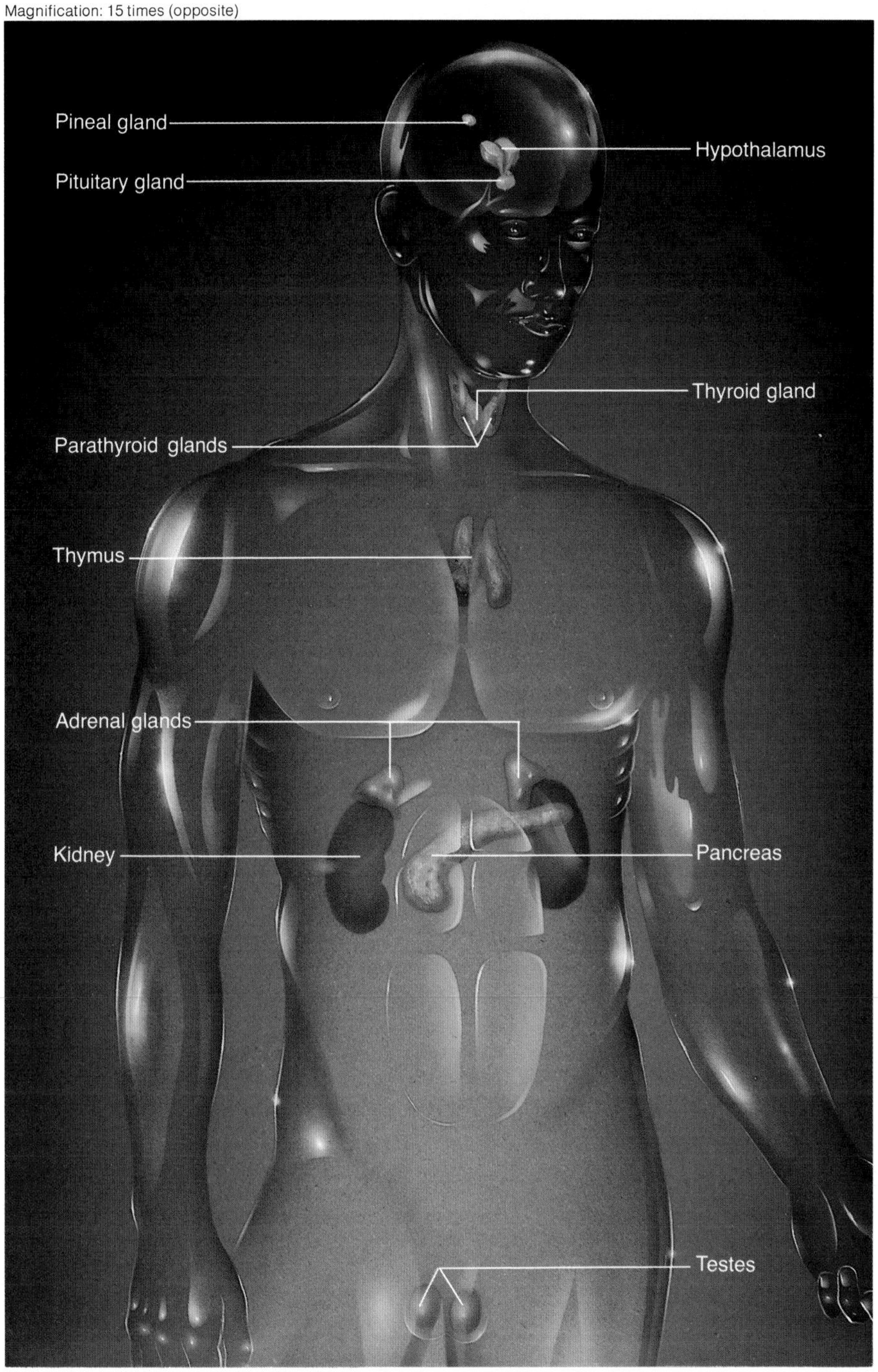

Scattered through the body, the endocrine glands (above) regulate many functions. Their chemical messengers, or hormones, travel through the bloodstream to all parts of the body.

The hypothalamus coordinates the activities of the nervous and endocrine systems from its control center atop the brain stem. A short stalk leads from the hypothalamus to the pituitary (opposite), which regulates how much hormone the other glands release.

The tiny pineal gland acts as the body's clock, signaling the onset of maturation and regulating the menstrual cycle. Hormones secreted by the bilobed thymus stimulate the production of white blood cells.

The thyroid produces hormones that make sure calcium is deposited in the bones and reduced in the blood. Embedded in the thyroid are four tiny parathyroids, in charge of removing calcium from the bones and adding it to the blood. The pancreas secretes insulin and glucagon to control the level of blood sugar.

Perched atop each kidney are the adrenal glands. They influence metabolism, maintain normal blood pressure, and help the body adjust to stress.

Male sex hormones are made in the testes and stimulate sexual development. The female's ovaries secrete the hormones estrogen and progesterone, which enable women to bear children.

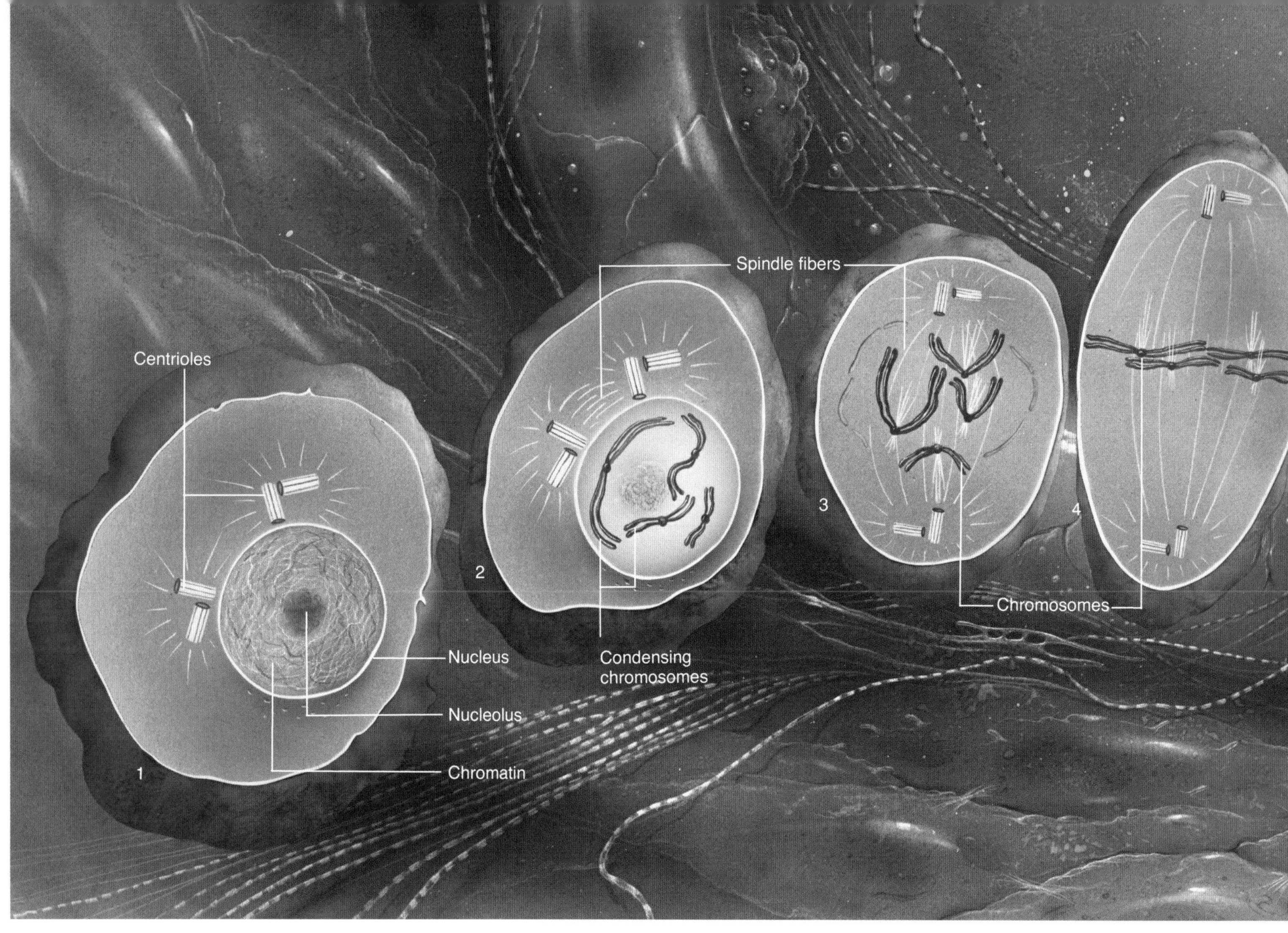

Mitosis—the continuous process by which body cells divide to create identical copies—reproduces the genetic blueprint and transfers it in an orderly fashion to each of two new daughter cells. For clarity, this painting (above) shows only 4 chromosomes, rather than the usual 46. Cell division in humans takes from 10 minutes to a few hours.

Before mitosis begins (1), thin threads of gene-bearing chromatin are dispersed throughout the cell's nucleus. The filaments radiating from the two pairs of centrioles give the cell its shape.

When mitosis begins (2), the chromatin condenses into individual chromosomes, each made of duplicate strands called chromatids. As the centrioles drift apart, threadlike spindle fibers form between them. The nucleus breaks apart (3), and the centriole pairs migrate to opposite ends of the cell, creating a fibrous structure called a spindle.

The chromosomes line up along the middle of the spindle (4), then the chromatids separate (5) and move toward opposite poles. As the chromatids head toward the poles (6), the spindle fibers elongate. The two genetically identical cells pull apart (7), and a nucleus reforms in each. Then they separate. After each new cell doubles in size, mitosis begins again.

As a cell undergoes mitosis (right), the two daughter cells pull apart, their spindle fibers not yet broken. Human cells cultured in the lab divide up to 50 times, then die. With one exception: Cancer cells retrieved in 1952 continue to grow through countless doublings.

Magnification: 8,000 times

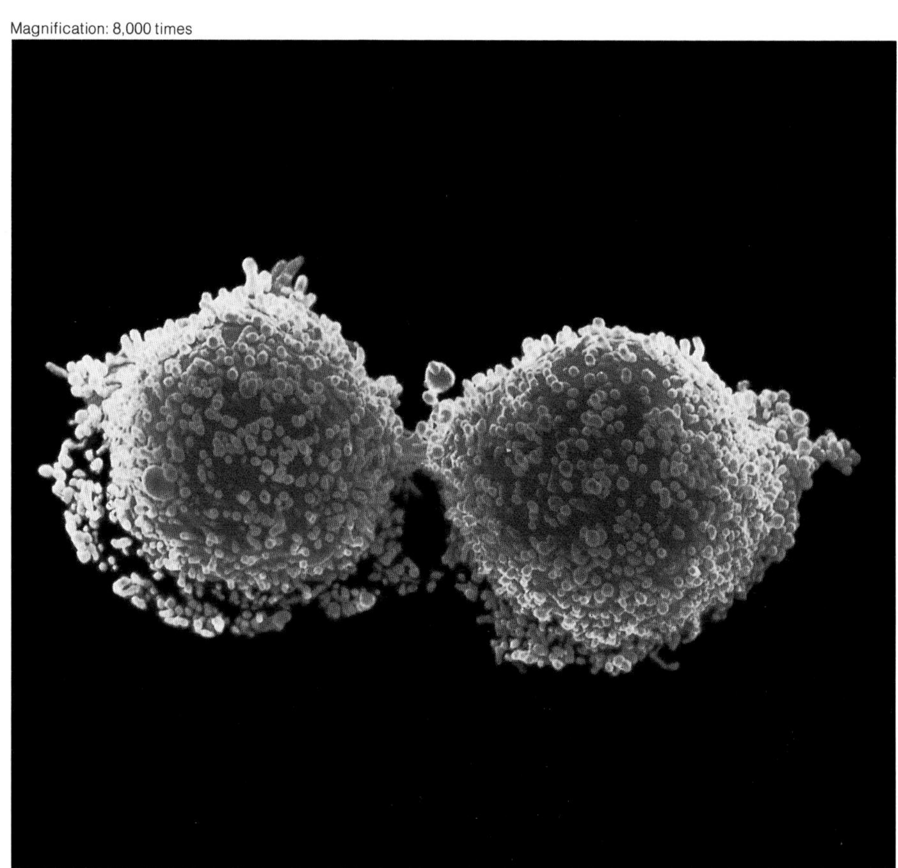

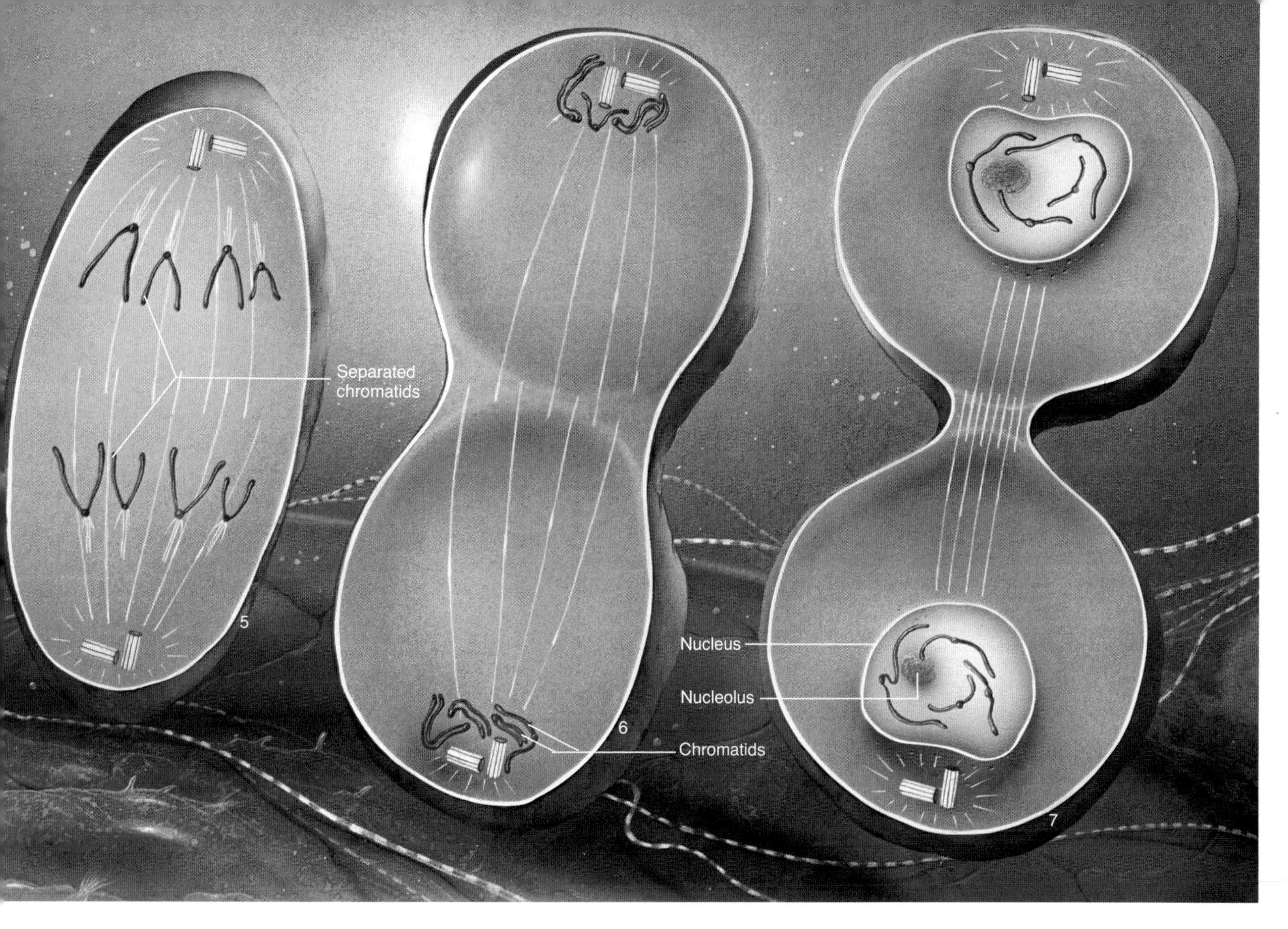

hormone secreted by the pituitary does not change. Sex hormones do inhibit the effects of growth hormone to some degree, but this is not the whole story. Castrated animals produce far fewer sex hormones yet reach normal size before they stop growing.

Sometimes the finely tuned control of growth breaks down. A few children grow too quickly. Robert Wadlow from Illinois weighed 8 1/2 pounds at birth in 1918 and grew normally until age 2. But by age 14 he had grown 7 feet 5 inches, and at 22, when he died of blood poisoning after his ankle became inflamed from an ill-fitting brace, he was 8 feet 11 inches tall. Wadlow's pituitary gland secreted too much growth hormone. A child with this condition, known as pituitary gigantism, grows with reasonably normal body proportions, but the legs tend to be longer and the head larger than normal.

If the pituitary becomes overactive after normal growth has ceased, an otherwise normal adult continues to grow in those few regions of the body that can still grow. The bones increase in thickness, causing the hands, feet, and jaw to enlarge. The tongue, liver, and kidneys also continue to grow. This condition is known as acromegaly. Somatostatin, a hormone found in the hypothalamus, inhibits the secretion of growth hormone and is now being used with some success to treat both gigantism and acromegaly.

Undersecretion of growth hormone causes problems at the other extreme. Pituitary dwarfism may result. The shortest human for whom we have well-documented evidence was Pauline Musters, a Dutch girl who measured 12 inches at birth in 1876. She died of pneumonia 19 years later—only 23 inches high and weighing less than 9 pounds.

Boys are three times as likely as girls to suffer from a deficiency of growth hormone. Without treatment, many of these children do not grow more than four feet tall. Injections of synthetic hormone produced by genetic engineering techniques can help them attain normal height if treatment begins before puberty.

Not all cases of dwarfism are caused by pituitary malfunction. A congenital heart condition or malnutrition, especially a vitamin D deficiency, will slow growth. Failure of the thyroid gland to develop causes cretinism. Cretins have thickened facial features and a skeletal structure with proportions like those of a newborn baby. Achondroplastic dwarfs have a somewhat similar skeletal structure, resulting from a hereditary defect in bone formation.

Yet another tragic abnormality of growth is

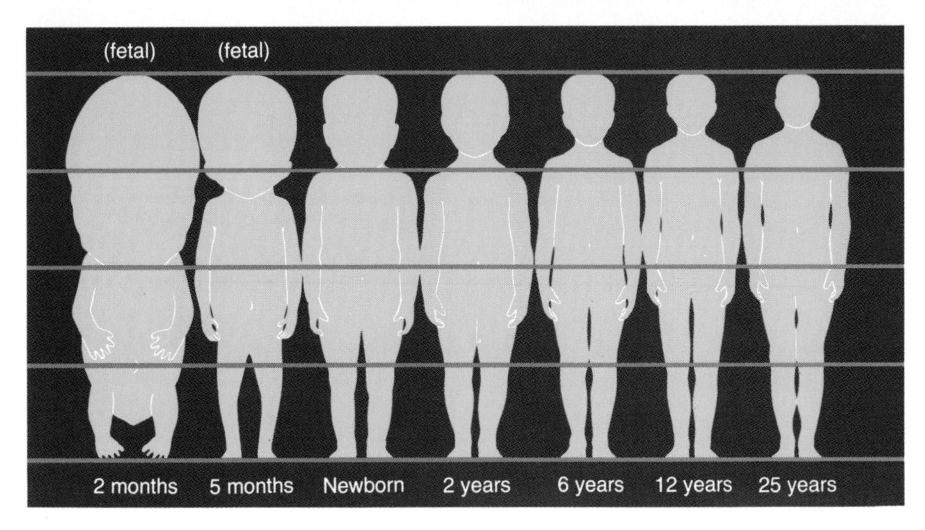

progeria, or precocious aging, which may be caused by a genetic defect that upsets the timing mechanisms of metabolism. The cycle of life speeds up. Balding and severely stunted growth occur within the first few years of life. Signs of old age—sterility, wrinkled skin, premature senility, and bone and muscle degeneration—soon follow. Most victims die in their teens, often as a result of cardiovascular complications caused by atherosclerosis. Progeria, fortunately, is very rare. Fewer than a hundred cases have ever been reported worldwide.

Aging is normally a gradual biological process that begins at birth and ends with death. The healing of wounds is often cited as an example of this process: It is most efficient in the very young and declines thereafter. But by and large, aging takes over when growth ceases around age 20. We become increasingly susceptible to accident and disease as the structure and function of our tissues deteriorate.

We can think of aging as a ladder with, say, eight rungs, each representing a decade. Various bodily functions decline as we climb the ladder. But aging varies markedly between societies and between individuals. We may all climb the ladder at the same pace as the years tick away, but our organs do not wear out at the same rate. The Baltimore Longitudinal Study of Aging, a project begun in 1958, has shown that some older people perform just as well as younger ones on intensive physical and mental tests.

Moreover, the broad terms we use in the West to denote age may represent quite different norms in other parts of the world. Our term "middle age" refers to the years 45 to 65, "old age" to the years beyond the usual retirement age of 65. But such a simple correlation between the number of years lived and a measurement of aging is misleading. In regions where nutrition is poor or backbreaking toil is needed to sustain a living, the organs deteriorate more rapidly. The same bodily changes occur, but in a compressed time frame. In much of India, Southeast Asia, and Africa, a baby born between 1975 and

Our body proportions change dramatically from before birth to maturity (above). In a newborn the head accounts for one-quarter of body length, but by age 25 it makes up only one-eighth. Active growth zones in a child's skeleton appear as orange areas at the ends of the long bones (right). An X ray (opposite) reveals another image of bone growth in the pelvis and upper thigh of a 12-year-old. Marrow fills the spaces between bone cells in this microscopic photo of a growth zone (below). Adolescents shoot up rapidly during puberty, then stop growing taller when the ends of the long bones harden—around age 17 for females and 20 for males.

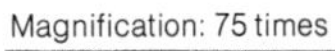

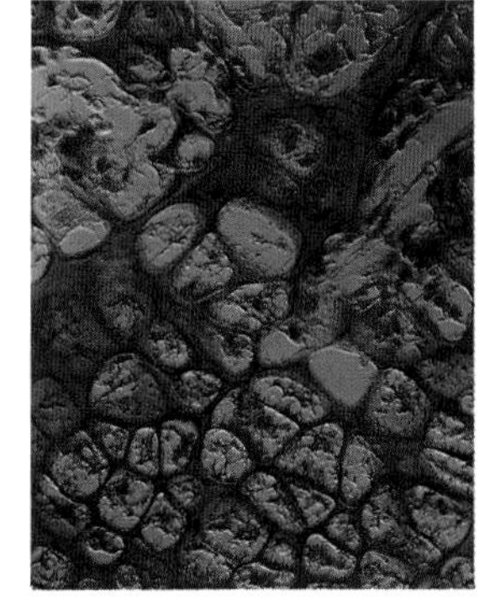

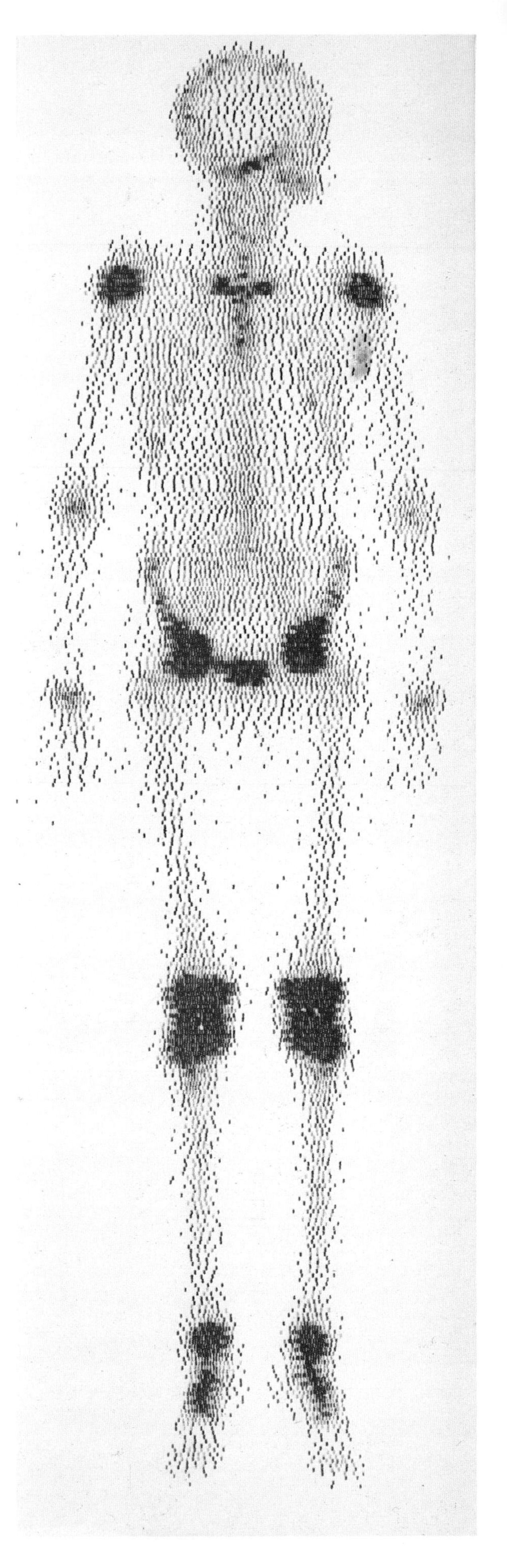

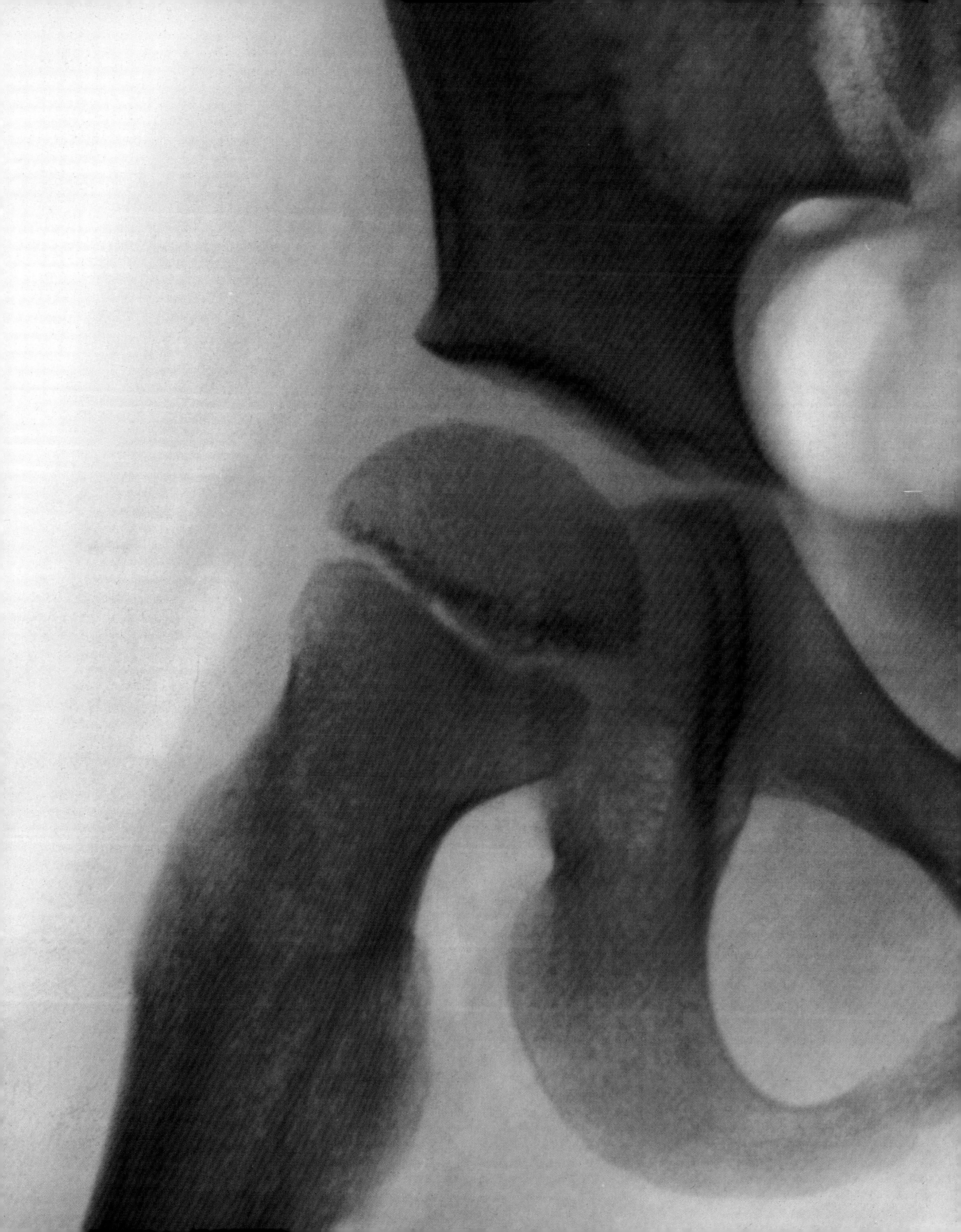

Bones knit together as they mature. The healthy skeleton of a newborn girl (below) shows almost no hips and disconnected leg bones that have a long way to grow before adulthood. In the first year of life, her height will increase by 50 percent. Some bones will continue to grow and harden—a process called ossification—until about age 25.

Two skulls, a newborn's and a 40-year-old man's, reveal how the head changes. The baby's jaws and the cervical vertebrae supporting the head are undeveloped (right, upper). The skull of a healthy, active man (right, lower) has well-defined jaws and cervical vertebrae. Dark areas indicate dental fillings.

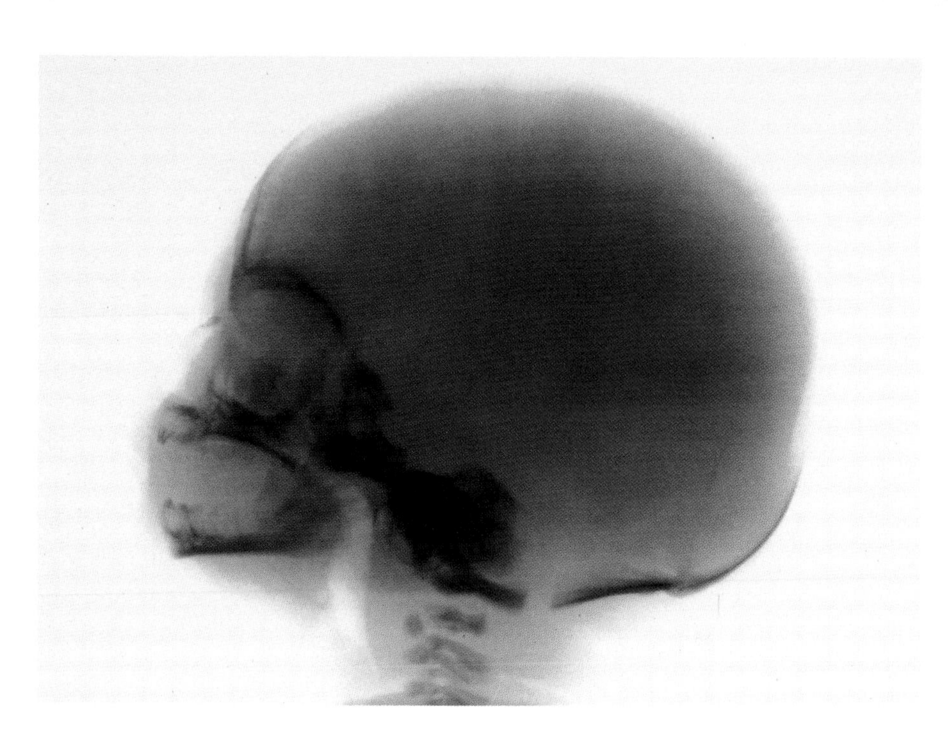

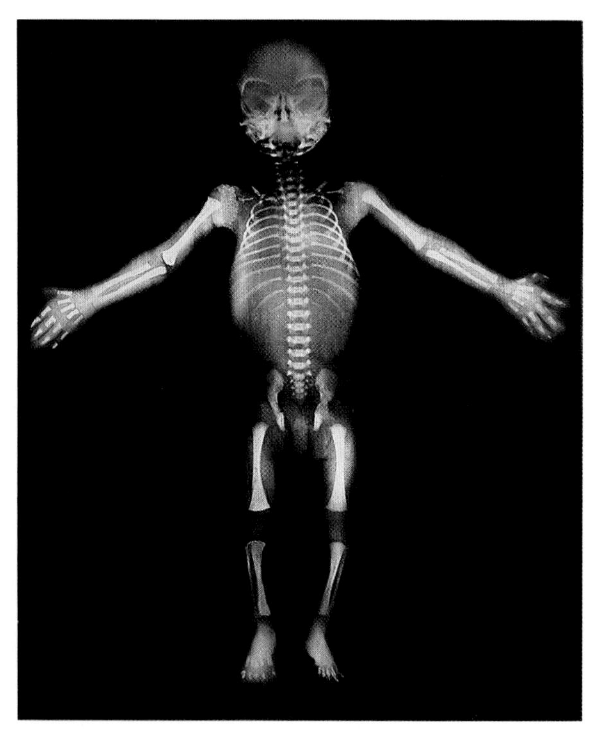

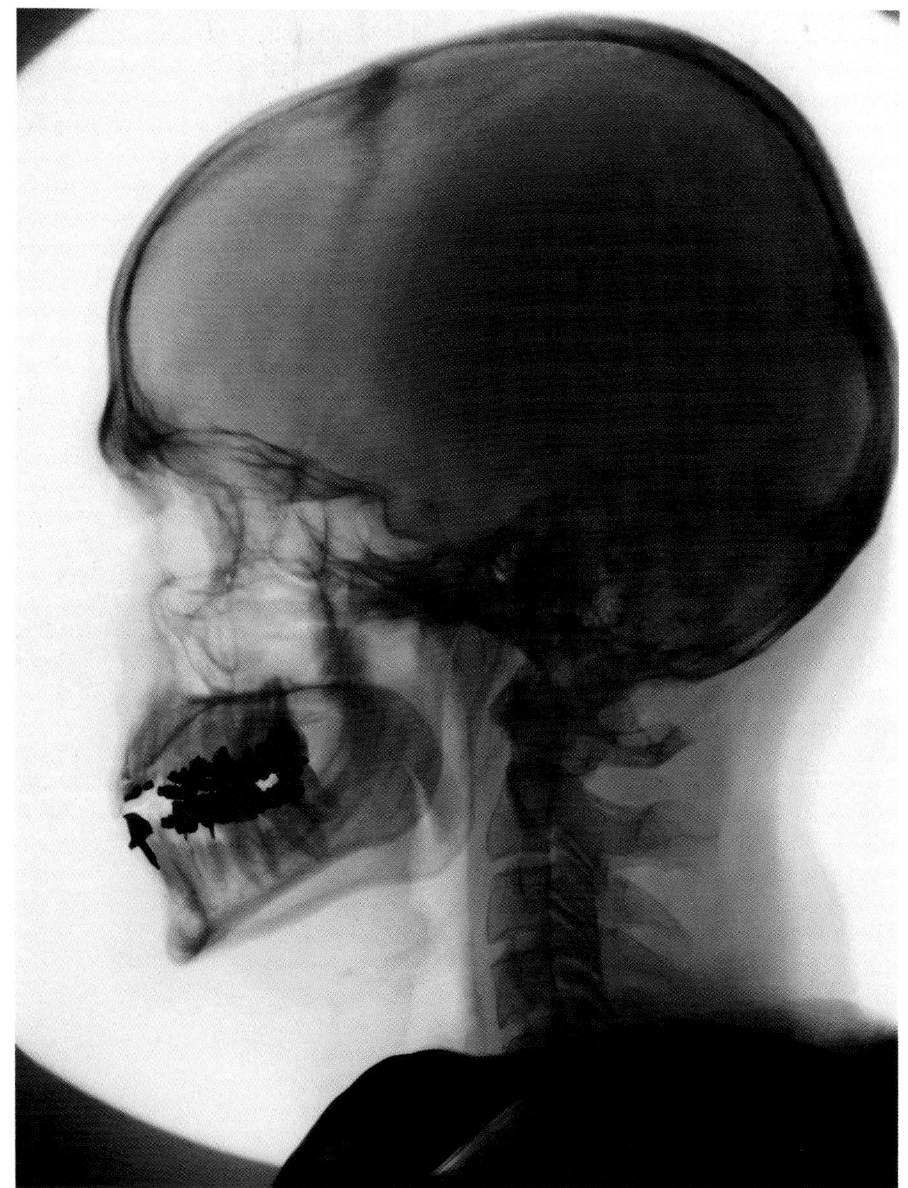

1980 could expect to live about 50 years. A person would be middle aged at 30, old at 45. The various rates of aging depend on physical, psychological, and social factors, as well as the number of years lived.

The Aging Body: Changes Without

We can, however, make some generalizations based on the passage of time. After about age 25 a person's height can go only one way—and that is down. A man or woman might lose an eighth of an inch between ages 25 and 40 as the spongy disks between the vertebrae in the spine shrink, causing the bones to move closer together. The back begins to bend forward after age 40. From age 20 to age 70, a woman may shrink about 2 inches, while a man might lose about an inch.

Why? Women, especially if slim, are more prone to bone loss after menopause. Race is another factor. White and Oriental women are vulnerable to a bone-thinning disease called osteoporosis, which strikes one in four American women over 60. Their bones are brittle and liable to fracture. They may develop a hump in the upper back, often referred to as a dowager's hump. Some lose as much as one-third of their skeletal structure by age 75. This is twice the rate of bone loss in men, who have about 30 percent more bone mass to start with. Osteoporosis afflicts more people than any other bone disease. Extra calcium is needed in the diet, but calorie-conscious American women tend to shun calcium-rich dairy products. Exercise can also help by stimulating the formation of new bone.

Weight seesaws with age. It is likely to go up until middle age—perhaps to age 50—and then start coming down. The rate at which the resting body converts food into energy—basal metabolism—slows down about 3 percent every decade after age 20. When combined with a sedentary life-style and an undiminished appetite, middle-aged spread can result. All the while, we lose muscle tissue at a rate of about .44 percent per year. In the 50 years from age 20 to age 70, a quarter of the original muscle mass can disappear, replaced by fat. In some people the loss of muscle tissue is balanced by the gain in fat. In old age, however, the loss of muscle tissue is more marked. Because muscle weighs more than fat, overall weight declines.

The head and face reveal most about aging. The passing years are etched on the skin, like the bark of a tree. As water disappears from inside the skin and nearby molecules bind together, the skin loses elasticity. The sun's rays inexorably damage the elastic fibers beneath the skin's surface. Young skin is resilient, returning to its original position after being stretched by a frown or a smile. But in an older person, these expressions eventually turn into wrinkles. Crow's-feet

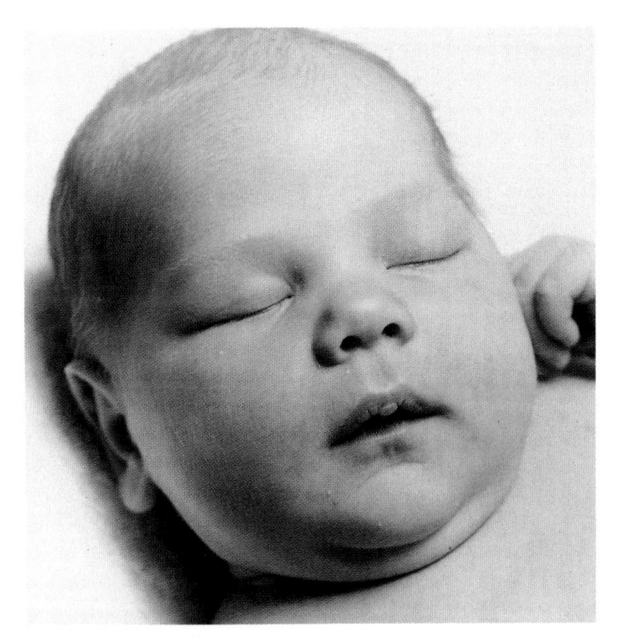

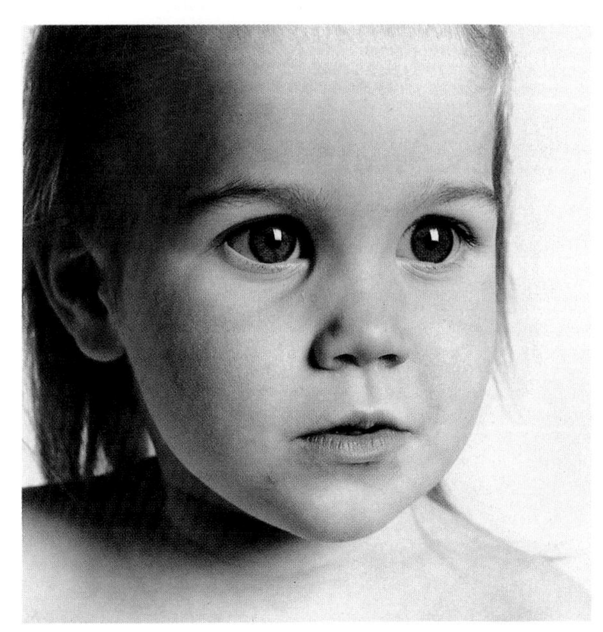

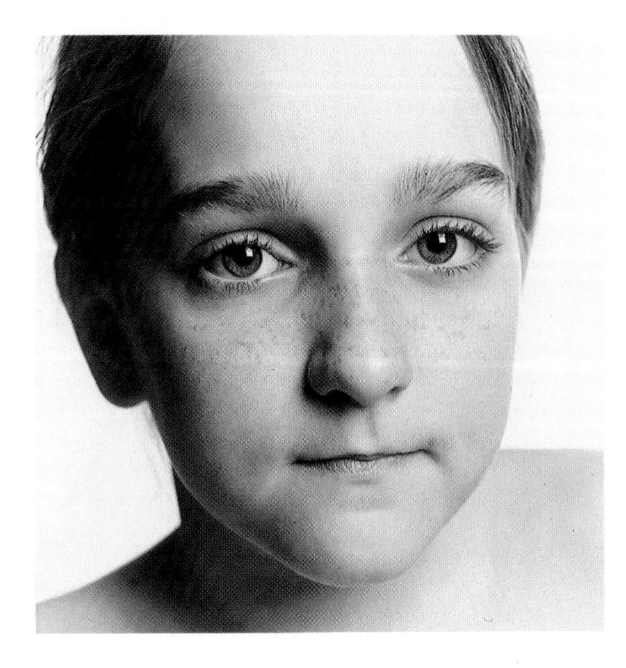

The plump cheeks, button nose, dot of a chin, and rosebud mouth of a nine-day-old baby (upper) typify the immature features of an infant. Her eyes are already two-thirds their adult size.

By the time a child reaches two (center), features become better defined. Cheeks, nose, chin, and mouth take form. By the age of twelve (lower), a child's features lengthen and thin out, giving individuality. Bone structure determines the basic shape of a person's face, but nutrition, hormones, illness, and experiences over a lifetime also etch their imprint.

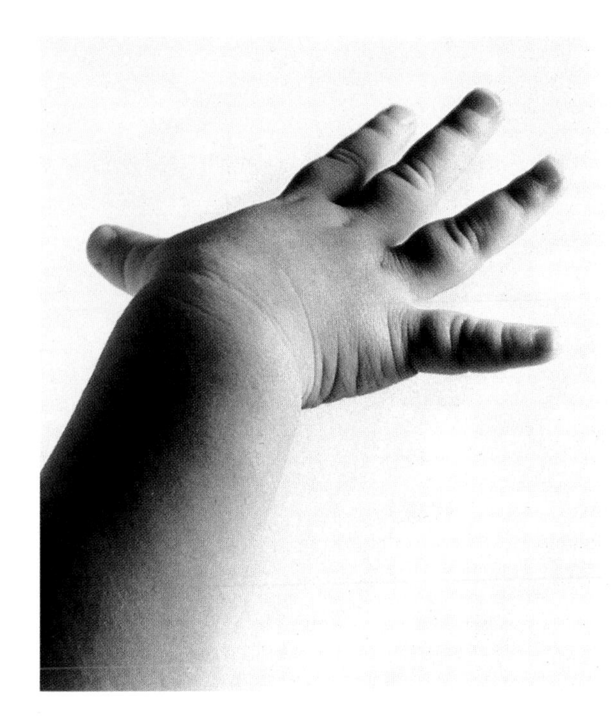

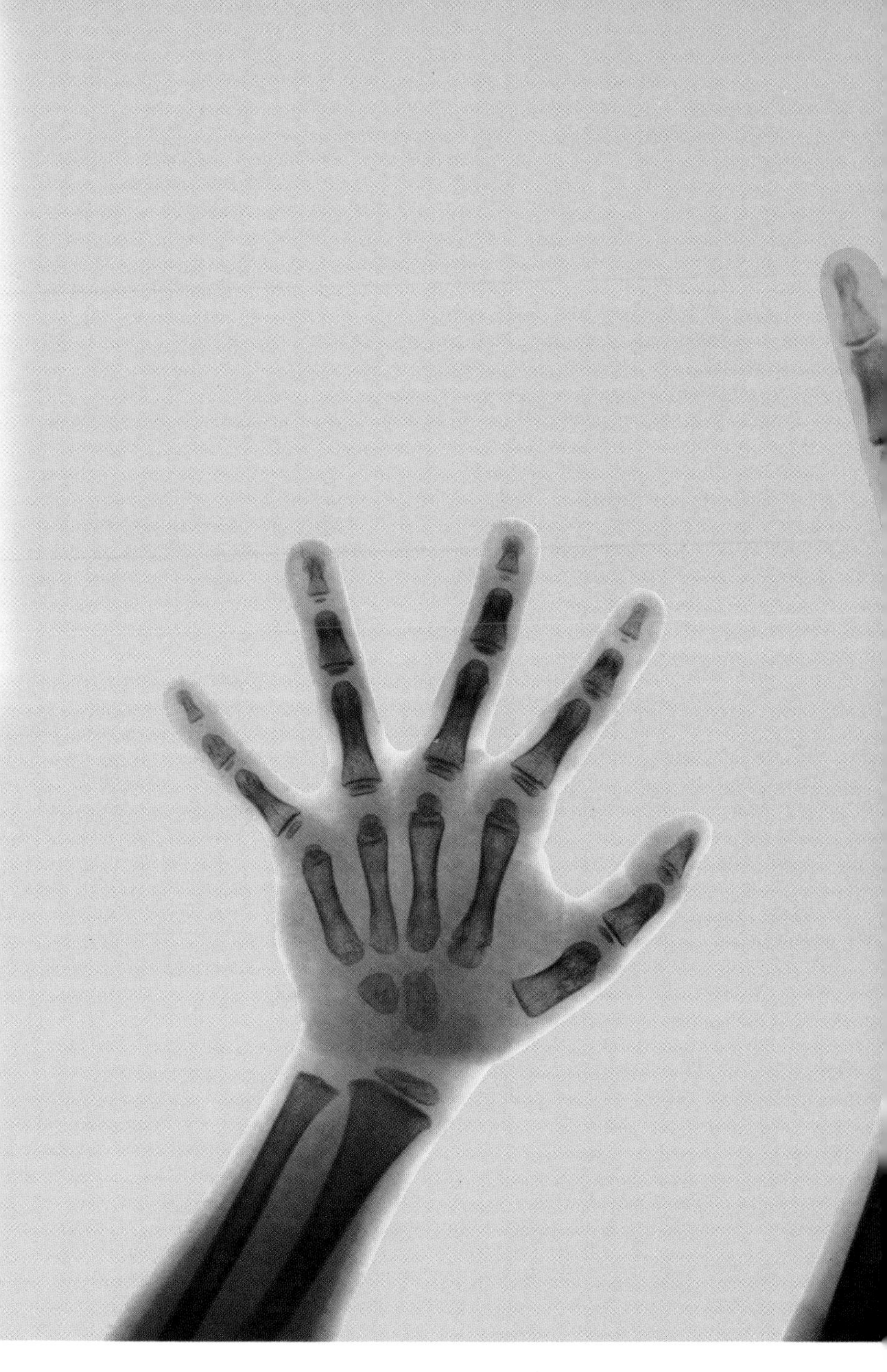

appear near the eyes from squinting. Lines are etched from the nose to the side of the mouth by smiling. The furrows of the brow reveal a worried countenance. The skin also thins and sags, like an overstretched piece of dough. Skin under the chin can turn baggy as the jawbone shrinks. By age 60 bags may appear under the eyes.

The elderly feel the cold more acutely because poor circulation and loss of fat under the skin lower resistance to extremes of air temperature. The skin loses the uniform color of earlier years as the production of pigment decreases. Skin regularly exposed to the sun turns blotchy.

Aging also leaves telltale signs on the nose, ears, and hair. Cartilage is one of the few tissues that continue to grow as we age. Between ages 30 and 70, a nose may lengthen and widen by as much as half an inch. The earlobes droop, and the ears themselves may be a quarter inch longer. Men find hair sprouting from the nose and ears but falling out from the places where it is most wanted. Hairs on the scalp imperceptibly lose diameter: They are thickest in the early 20s, then they begin to shrink; by the time a man reaches 70, his hair is as fine as a baby's. The hairs last 2 to 6 years; baldness occurs when replacement fails to match loss. Male baldness follows a distinct pattern—a receding hairline at the temples that one day may meet up with a thinning spot on the crown. About 75 percent of women also suffer from hair loss, though rarely to the point of baldness. Graying, a gradual process, occurs when cells at the base of the hair follicle stop producing pigment. Ironically, tooth decay declines with age as tooth enamel hardens. But gum disease could ultimately reduce the number of teeth and affect appearance.

Women's breasts change size and shape. Milk ducts and alveoli—milk-producing glandular

An infant's plump hand (left, upper) precedes a lifetime of changes that lead to the gnarled fingers of an old man. Three X rays chart this transformation. The light-colored growth areas, or epiphyses, at the ends of the bones in an 18-month-old baby's hand (above) will gradually turn from gristle-like cartilage to bone. In a 30-year-old male (center), the epiphyses have hardened and fused with the bones. The finger joints of a 75-year-old man (right) are thinner and show signs of arthritis—small outgrowths projecting from some of the bones.

Teeth also change with the years, as documented by wraparound X rays. The permanent teeth of a 7-year-old (right) press on the roots of the temporary baby teeth and gradually dislodge them. By age 13 (far right) most of the permanent teeth are in place. The wisdom teeth at the far corners of the mouth usually erupt between the ages of 18 and 30.

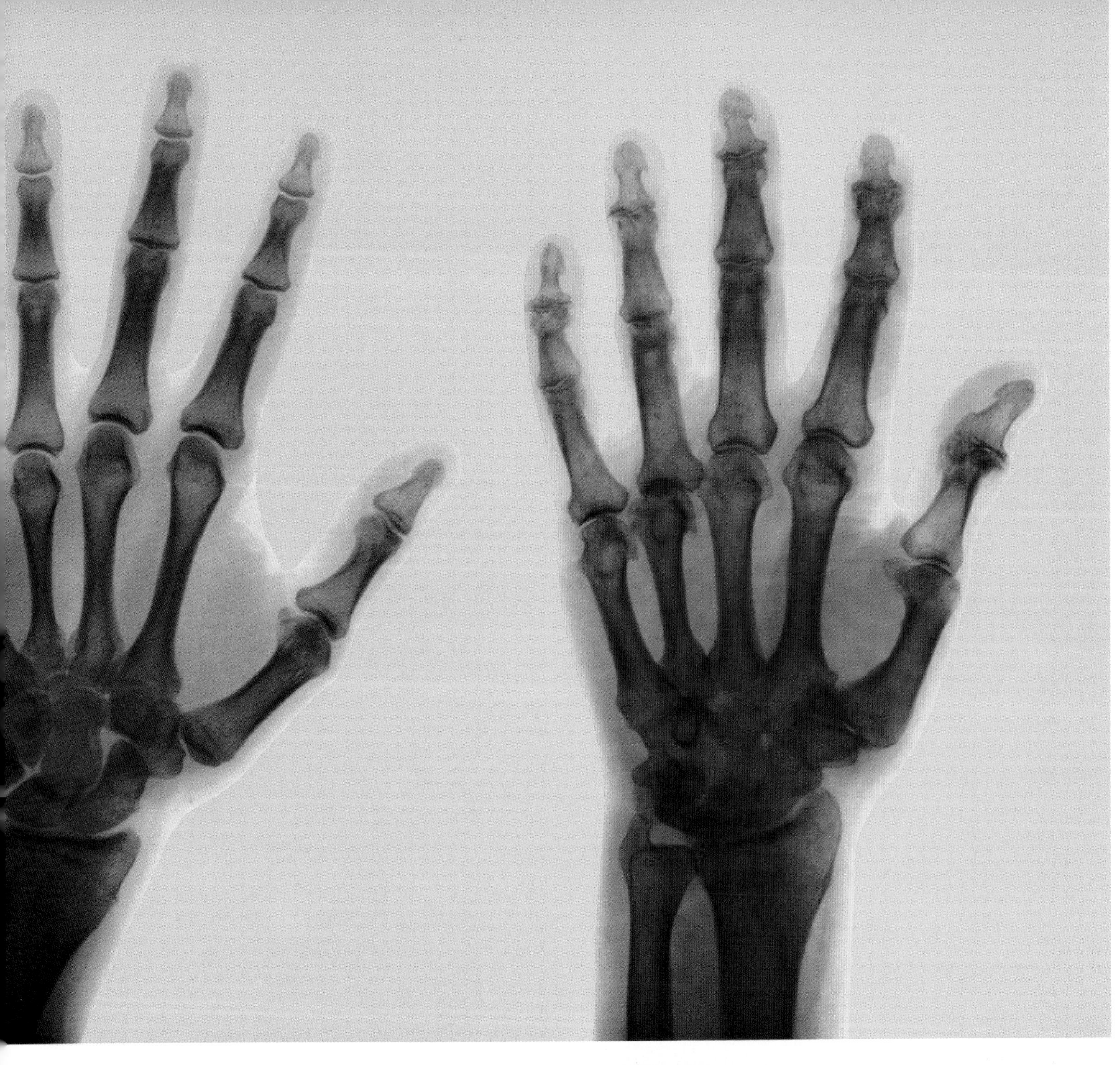

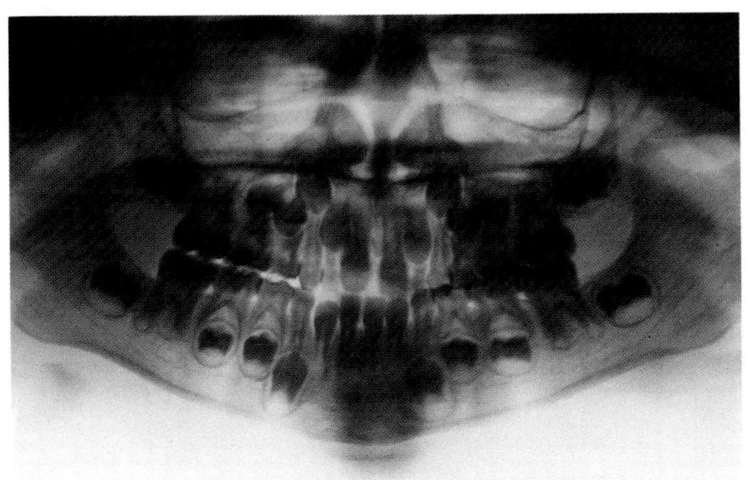

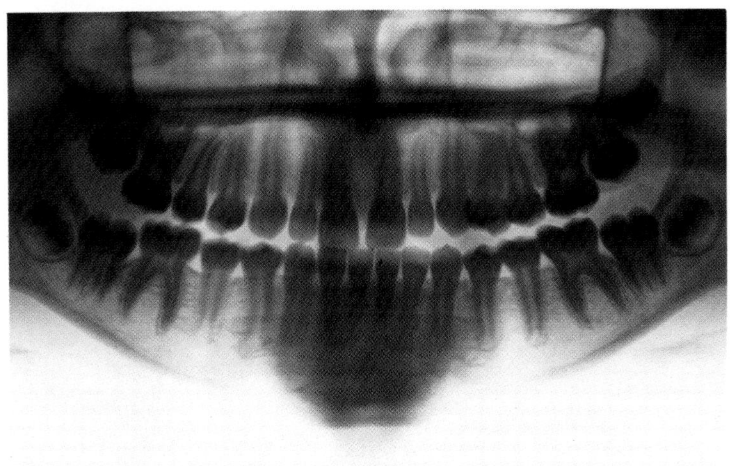

Osteoporosis, a relatively new word in the vocabulary of aging, means "porous bone." Caused by a loss of calcium, osteoporosis results in brittle bones in the elderly. The large, empty spaces in the lattice-like bone tissue of an 80-year-old man (right) show he suffered from osteoporosis. In a healthy 31-year-old man (below), the smaller spaces strengthen the bone.

Researchers believe that fragile bones can be prevented by a calcium-rich diet and by such exercises as bicycling, dancing, and walking, which stimulate the formation of new bone tissue. Estrogen therapy curbs bone loss in women after menopause.

Actual size

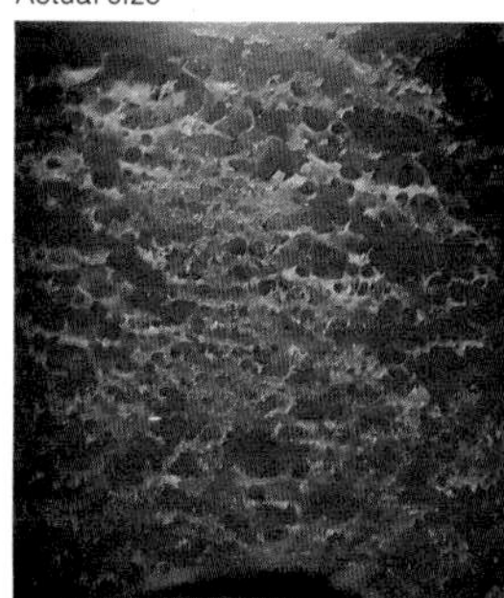

Magnification: 10 times

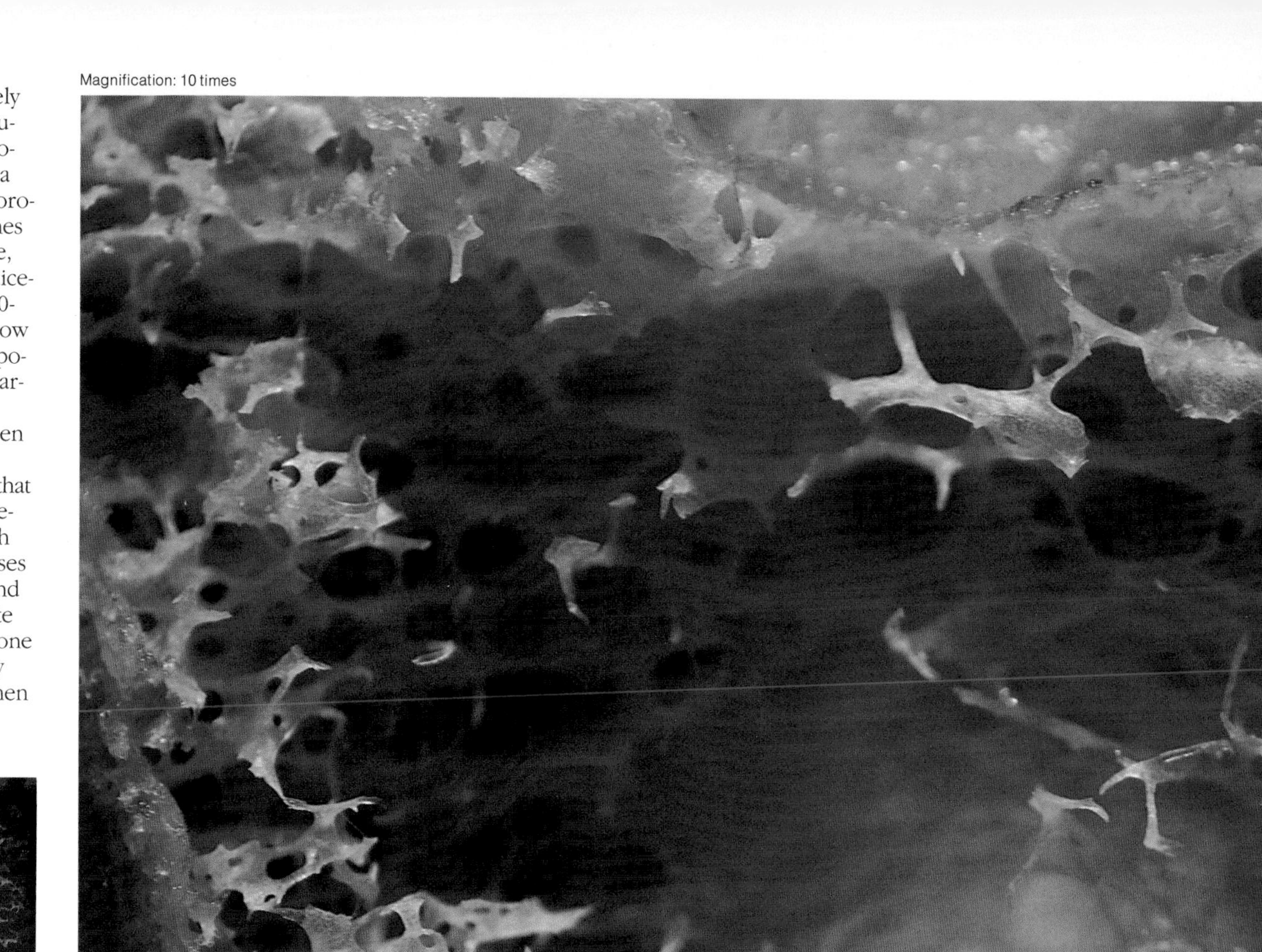

tissues—make young breasts firm and plump, but as hormonal levels drop at menopause, the breasts lose their rounded shape. Ligaments linking breast tissue to the underlying muscle stretch over time, causing the breasts to sag.

The Aging Body: Changes Within

Changes you cannot see are taking place too. Between ages 30 and 40, muscle surrounding the heart starts to thicken. Over the years cholesterol levels rise, fatty deposits build up on arterial walls, and blood pressure increases. The heart of a 70-year-old man at rest will pump a quart less blood per minute than it did 40 years earlier.

The risk of heart disease may increase with age. Cardiovascular problems lead to more than half the deaths of men and women over 65, but aging is not the sole cause. Diet, smoking, and a sedentary life-style all play their part. Smoking by younger women is now blamed for a noticeable increase in heart problems among females. Menopause may add to the risk. The female sex hormone estrogen, which helps protect the arteries, is no longer released in significant quantities after menopause—and is reduced even further by smoking. Why estrogen shields women in this way is an active research topic. Women are also prone to varicose veins, which may stem from increased blood flow during pregnancy.

By age 70 our kidneys will take twice as long to filter blood as they did 30 years earlier. Lungs must also work harder. Our deep breaths become shallower and shallower; a 70-year-old man breathing deeply will inhale half as much air as he did at age 20—about 3 quarts instead of 6. Tissues surrounding the chest harden with age, restricting expansion of the lungs. Less oxygen comes in, and the heart is slower in pumping it around the body. It could take a healthy 70-year-old man about an hour longer to run a marathon than it did when he was 30. The body's built-in vacuum cleaners—white blood cells called phagocytes—gradually lose their ability to clean the lining of the respiratory tract. Pneumonia and other infections can result.

Our senses are not spared either. Hearing peaks at about age ten, but it will be three or four decades before any noticeable change occurs. Higher pitched sounds fade first. Many people in their 70s and beyond, however, retain the ability to hear conversation at normal pitch. In westernized countries men lose their hearing more often than women do, but the reason may not be biological. It could well result from being exposed to more noise in the workplace.

Food gradually tastes more bland as the number of taste buds on the tongue is reduced about two-thirds by age 70. To compensate, the elderly are prone to oversalt their food, which may add to the risk of high blood pressure.

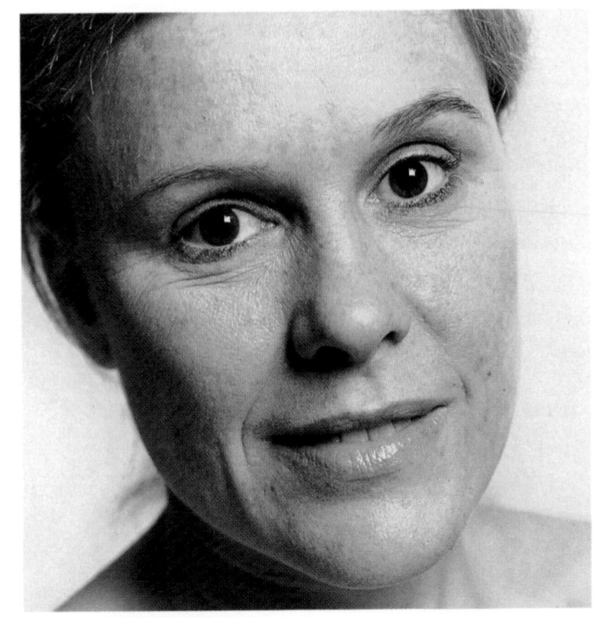

Aging touches every part of the body, sometimes gently, sometimes with a vengeance. The protective wrapper of skin reflects nature's timetable of change as the peaches-and-cream complexion of youth gives way to age spots and wrinkles. Skin elsewhere on the body ages gradually, but on the face it changes in noticeable stages. Smiles, frowns, grimaces, and the ravages of the sun leave telltale signs. Lines on the forehead and around the eyes and mouth deepen, then the skin on the cheeks sags. Bags form under the eyes, and wrinkles eventually appear everywhere.

With age, the skin loses some of its elasticity and becomes coarser and thinner. It stretches as the bones shrink, sometimes fitting like a baggy suit. The production of skin oil remains constant in a man through his 80s but declines dramatically in a woman after menopause, causing her face to age more noticeably. The aging immune system becomes less efficient, so older skin bruises more easily and cuts heal more slowly. Sweat gland activity declines, possibly due in part to hormonal changes.

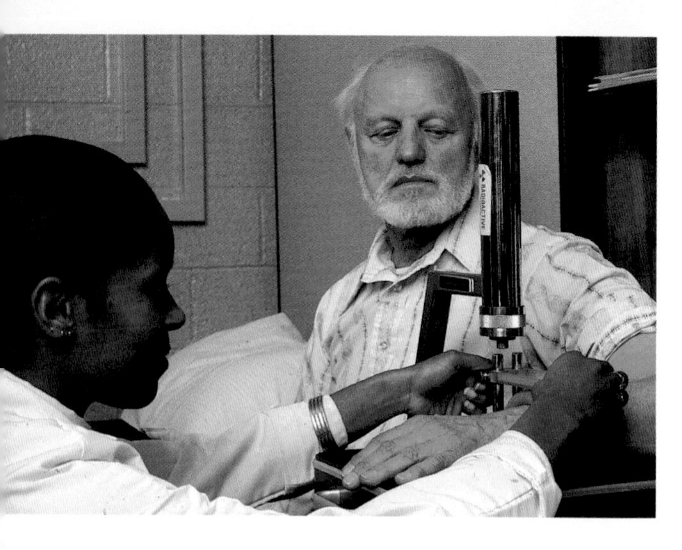

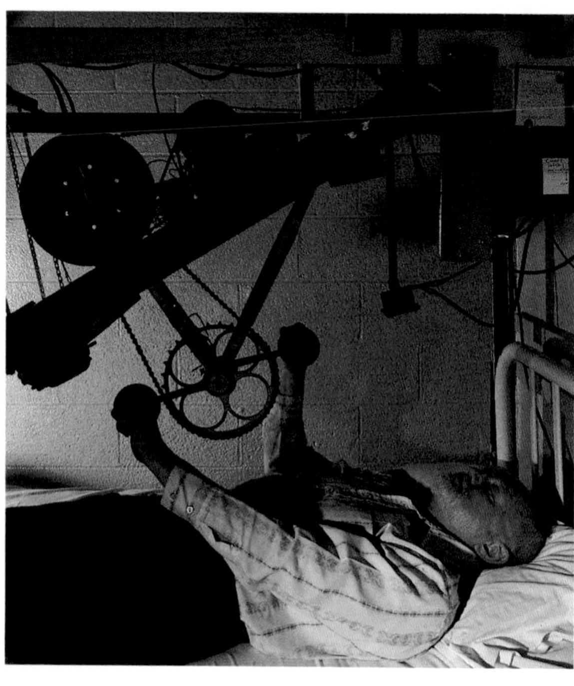

Space age technology measures aging processes at the National Institute on Aging. A volunteer (right) breathes into a unit that gauges his lung capacity. Aided by a computer, a technician (upper) measures bone loss, a serious problem of old age. In another test a volunteer (lower) cranks a device that records the strength of the muscles and joints in his arms and shoulders, as well as the coordinated activity of his heart and lungs under stress. By charting biological and mental changes, this long-term study of healthy adults will help scientists distinguish the changes that occur normally as we age from those associated with disease.

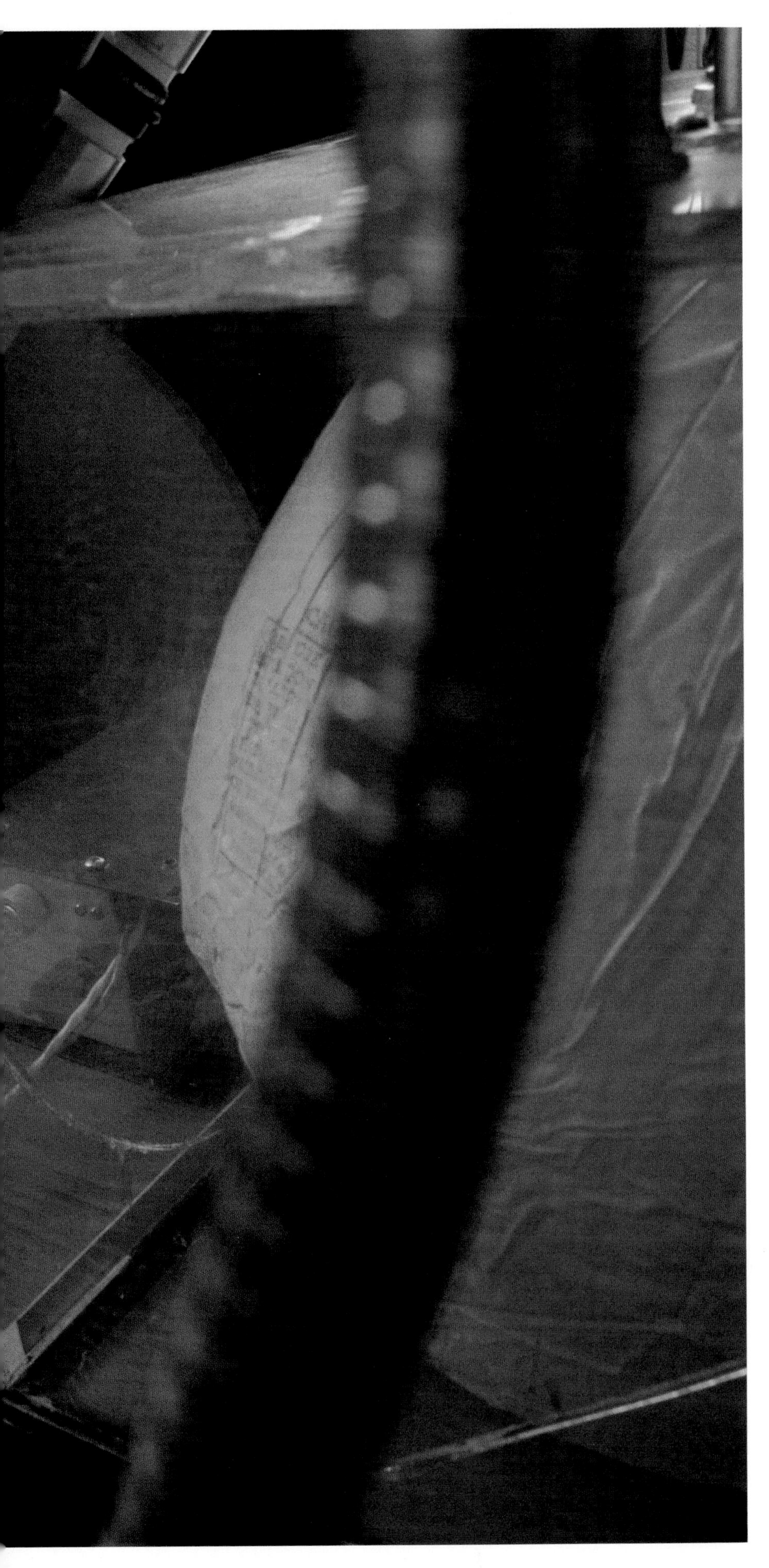

As we age, our eyes cannot adapt as easily to changes of light, and they become very sensitive to bright light. Spots appear in our vision with increasing regularity. The lens of the eye hardens. By age 40 it may be too stiff for the eye muscle to focus on close objects or to read small print. Scientists regard this condition, called presbyopia, as an accurate indicator of chronological age. Visual acuity usually declines: Only one of every six octogenarians has 20/20 vision.

The brain reaches its maximum weight at age 20—about 3 pounds. Over the next 60 years, as billions of nerve cells die within the brain, it loses about 3 ounces. It has long been assumed that the mental capacities of the elderly diminish as the brain shrinks, but recent research challenges this view. Given a liberal response time, a healthy septuagenarian may score as well on tests of intellectual performance as a much younger person. The ability to process information rapidly and to make quick decisions declines with age. So too does the transmission of signals governing coordination. Events from the past are often recalled with more clarity than recent happenings. About 10 percent of those over 65 suffer from Alzheimer's disease and other forms of senile dementia, leading to the progressive loss of mental functions.

Many other physical changes can be cited. A woman's menstrual flow declines in frequency and quantity by about age 45 as a prelude to menopause, when the reproductive organs atrophy. The strength in a man's biceps will be halved between ages 25 and 60. His voice becomes higher by age 50 as his vocal cords stiffen and vibrate more quickly.

Turning Back the Clock

Throughout the ages people have tried various antidotes to aging. All to no avail. It is said that in 1492 Pope Innocent VIII drank the blood of three young donors—nevertheless, he died shortly after. In more modern times, ape testes have been implanted in aging men. Others, including Charlie Chaplin, Winston Churchill, and Christian Dior, have tried injections of fetal lamb cells. Megadoses of vitamins are a more popular, but still unproven, preventive. Exercise, however, may help people feel better and also improve the function of certain muscles and such organs as the heart and lungs. A recent study of middle-aged and elderly Harvard alumni who exercise regularly has shown a correlation with longevity.

Some scientists say the key to increasing longevity, and by implication delaying aging, lies in diet and nutrition. Research with animals supports the view, although there is no evidence that it applies to humans. Scientists have found that by feeding rats a low-calorie, low-protein diet, their immune system takes much longer to mature. It works better later in their lives, and the rats tend to live longer, rather than dying prematurely from illness. But such a stringent, long-term diet may be impractical for humans. Nevertheless, several researchers are imposing the diet on themselves to study its effect.

Certain foods may counteract the devastation caused to the aging body by highly unstable molecules called free radicals, chemicals produced in the cells when oxygen combines with unsaturated fat. Free radicals destroy cells, damage DNA, and trigger destructive reactions. The linking of free radicals with collagen fibers damages artery walls, reduces the elasticity of the lungs, and impairs muscle activity. Free radicals can be blocked, or at least slowed down, by antioxidants such as vitamins C and E, selenium, and the preservative BHT, which are found in vegetables, citrus fruits, cereals, dairy products, meat, fish, and seafood. The cumulative effect of free radicals may be one cause of aging.

The causes of aging remain a mystery. A leading researcher in the field, Dr. Leonard Hayflick, has commented that probably no other area of scientific inquiry abounds with so many untested theories as does the biology of aging. But that has not stopped dedicated scientists from carrying out research.

Theories of Aging

Some researchers suggest that we carry our own death warrants in our cells. Cessation of growth in later years allows the aging process to dominate. A finite life span maintains the balance of nature by providing room for the next generation. The survival of a species depends on its members living long enough to procreate; after that, further life is not necessary. The fact that longevity runs in families implies that genes may somehow control the rate of aging.

Experimental evidence supports the idea that death is programmed in each of us at birth. Dr. Hayflick has shown that human cells cultured in a dish gradually lose the ability to undergo mitosis. Cells from a fetus doubled about fifty times before dying, while those from an adult divided fewer times. Cells from an extremely old person divided only two to ten times before dying. Cells may be programmed to die after a finite life span, but humans probably age and die before the program runs out.

Other researchers say specific aging genes cause decay in our bodies. Genes that code for essential functions in a cell may be switched off, or the proteins they code for may be made incorrectly by these aging genes. One researcher talks of a gene that codes for a "death hormone" secreted by the pituitary. This hormone may interfere with thyroid metabolism and cause the body to deteriorate. *(Continued on page 258)*

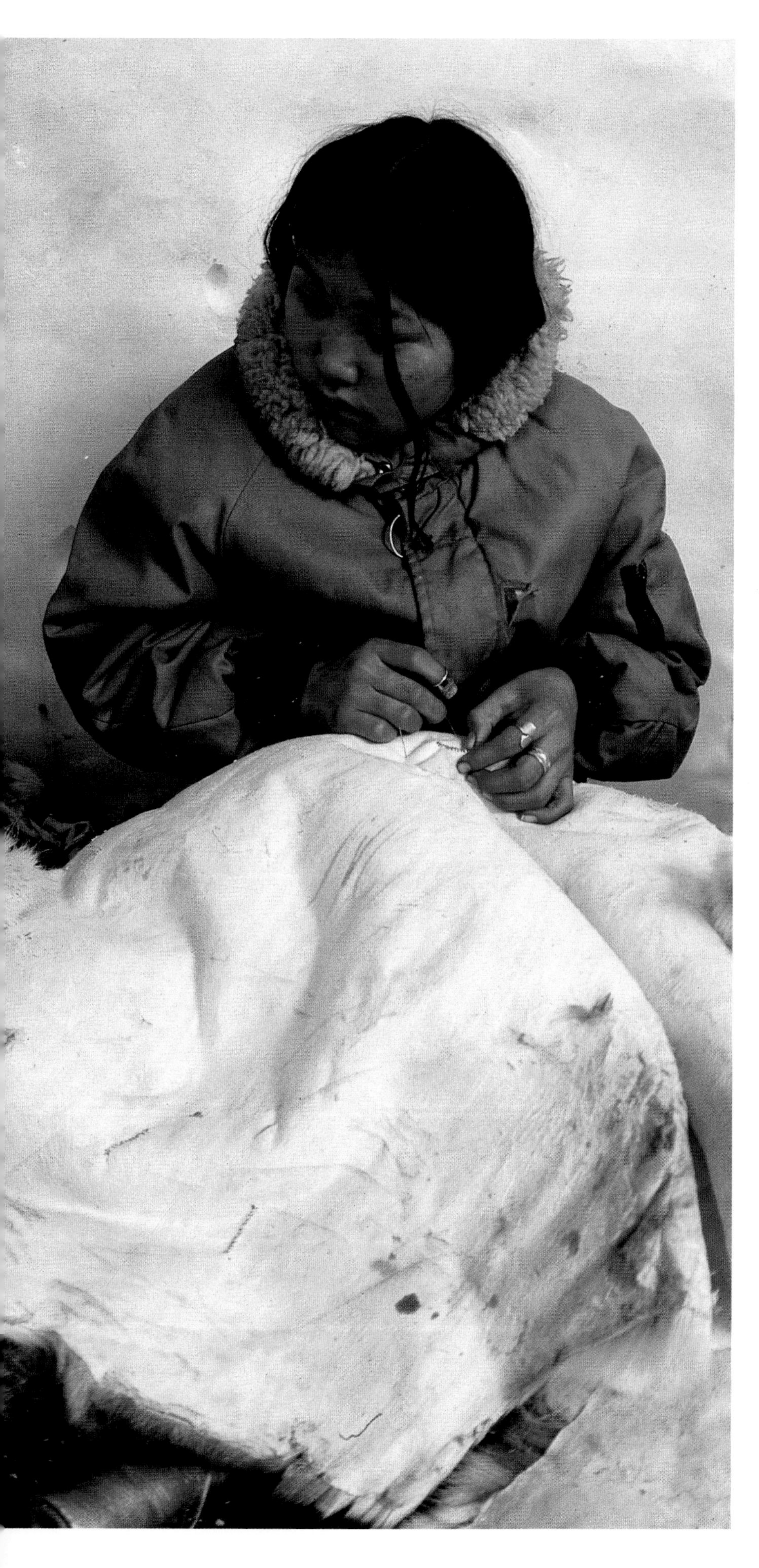

Bundled up inside a snow-house in Canada's Northwest Territories, an Inuit matriarch (left) teaches a young seamstress the ancient skill of stitching caribou skins. A sense of worth and an active interest in life may foster a healthy old age. An elderly trio (above) laughs at a shared joke in the foothills of Russia's Caucasus Mountains, once thought to be the home of some of the world's oldest people.

Recent evidence discredits the theory that a high proportion of people live to be more than a hundred in certain remote areas, including Vilcabamba in Ecuador and Hunza in Pakistan. Earlier studies had relied on faulty or falsified documents, hazy recollections, or calculations based on the birth date of a parent with the same name.

In underdeveloped parts of the world, the elderly are respected for their wisdom and experience. But this honored position is waning as education, urbanization, and industrialization touch even the most remote regions. Tennyson epitomized both the anguish and the aspirations of growing old: "How dull it is to pause, to make an end, To rust unburnish'd, not to shine in use! As tho' to breathe were life!"

Meeting the Challenge of Old Age

Intellectual and physical vitality promote healthy, graceful aging. Violinists with the Dade Senior Citizens Orchestra (below) rehearse in Miami Beach. Calisthenics help keep elderly muscles toned (right); for more than 30 years the city of Miami Beach has sponsored these early morning workouts.

Retirement causes many people to worry about a loss of intellectual ability, since they no longer feel useful. But the National Institute on Aging says healthy people can continue to learn well into their 90s.

Someday we may live to age 150, well beyond the present limit of about 120 years. Some of the disorders of the elderly are diseases that may eventually be treatable. The life expectancy of middle-aged Americans has increased dramatically since the mid-1960s. Factors believed responsible include a decline in cigarette smoking, the detection and treatment of high blood pressure, and changes in diet.

SWIM TO
ISRAEL
NAGE D'ISRAEL

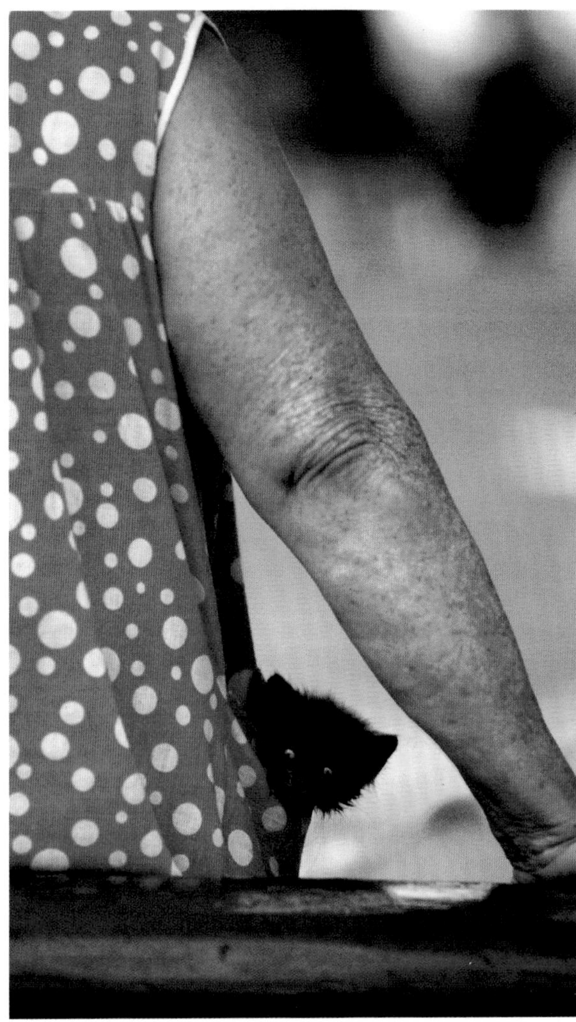

Close family ties, friends, and pets help the elderly combat loneliness, blamed by researchers for some ailments of old age. In South Amana Village, a closely knit religious community in Iowa, a 77-year-old woman teaches her granddaughter how to knit (opposite). Friends meet, mingle, and step out at a community dance in one of Florida's retirement colonies (left). Social programs in some retirement areas would exhaust many younger people. In New York City a curious kitten peeps out from its owner's pocket (above). Pets offer love and affection, adding interest and meaning to the lives of lonely people. Pet owners also receive a soothing side effect: Petting your cat or dog can lower blood pressure.

Wear and tear during life takes its toll on cells. Large numbers of cells die and eventually cannot be replaced because they have reached or neared the limit of 50 divisions. The brain, which stops mitosis early in life, begins to lose cells at a rate of 50,000 per day by the age of 30. Replacement is impossible, and with time the organ may become less effective. But where large parts of the brain have been lost, the rate of aging does not change. Cell loss alone cannot explain aging.

Another theory proposes that an accumulation of mutations or errors in the DNA of cells leads to the production of faulty proteins that do not function as efficiently as the original protein. Exposure to cosmic radiation and chemicals in the environment contribute to damage of DNA and hence to mutations. In older people the body's immune system, while becoming less effective at fighting disease organisms, paradoxically becomes more active in attacking the body's own cells. Some cells, perhaps including those with genetic damage, are regarded as foreign and are attacked by antibodies. Some forms of arthritis are probably autoimmune diseases in which white blood cells mistakenly devour healthy cartilage in the joints. Anemia and multiple sclerosis may also be autoimmune diseases.

It seems likely that a combination of several of these events causes aging. One day medical science may be able to delay aging by manipulating the immune system or by repairing DNA. A maximum life span of 150 has been forecast, with many more people reaching 100. Some question whether the quality of life can be maintained for such periods. Others say that our basic biological processes, such as menopause, would occur later in life. The effects of aging, detrimental or otherwise, would simply be postponed.

Questions of Immortality

Few researchers talk of the inevitability or wisdom of immortality. It would result in stagnation of the species. Aging always leads to the same end—death. The body is a chain of interrelated parts: When a vital link falters as a result of disease, accident, or degeneration, the whole system can break down. It is therefore misleading to talk of people dying of old age; rather they die of cancer, strokes, heart attacks, and other disorders that occur more often with advancing age. Death will take a holiday if the incidence of these biological events, say heart failure, can be postponed. Perhaps one day we will be like the one-hoss shay that "ran a hundred years to a day." Like that carriage immortalized in a poem by Oliver Wendell Holmes, our parts will wear out evenly, then collapse all at once, "just as bubbles do when they burst."

IAN ANDERSON

Mirrored in a portrait by another artist, 89-year-old John de Rosen works in his Washington, D. C., studio. De Rosen continued to paint until the week he died—at age 91. Some scientists believe that retirement to a sedentary life-style initiates or aggravates medical problems, thus shortening life. According to a study of retired people, adults over 65 can learn a creative skill, such as oil painting, as readily as younger students. Grandma Moses did not begin to paint until she was 78—and painted 25 pictures the year after she turned 100. Creativity and mental vitality seem to inhibit the inevitable: the physical aging that begins the moment we are born.

In Touch with the World

Your birthday suit is one outfit you will wear your whole life. Two square yards, six pounds, your skin is your largest organ. Every square inch tingles with a thousand nerve endings, sense receptors by which you probe the world around you and come to know your very self. The skin you're in is both boundary and antenna.

In each fingertip are the sensitive nerve endings which throughout infancy and childhood teach us that we live in a three-dimensional world. By the sense of touch, a baby explores the texture of his environment, his mother's face, his own body. And child or adult, when we look at our body in a mirror, we perceive the contours and curves we know so well, though the image on the glass is as flat as a sheet of paper. Touch has given us an image of reality. What's felt to be in that skin is what is you.

We express ourselves profoundly by touch. In the Middle Ages a symbolic tap on the shoulder conferred knighthood. Rubbing noses or kissing on the lips carries a message of affection or passion. The clasping of hands signifies greeting or ratifies understanding.

The closeness we allow between ourselves and others depends on whether or not we know them and on the culture in which we grow up. Arab businessmen walk arm-in-arm as a gesture of courtesy. Japanese ritually bow in greeting from a distance. Americans, even in crowds, shun body contact with strangers. Yet touching others brings pleasure and comfort. Baby knows the crook of Mother's arm. We welcome a friendly hug. The stimulation of our sense of touch is essential to a normal life.

All the senses—vision, hearing, smell, taste, touch—originate in organs called receptors, specialized to continually instruct the brain about the body's condition and environment. Sense receptors respond to stimuli in the environment by initiating a chain of electrochemical nerve impulses that travel on particular neuronal pathways to regions of the brain that analyze the signals and induce the muscles to produce action. We feel chilly and move toward the fire. We feel the warmth of the fire and smile.

Usually we filter out 99 percent of the sights, sounds, and other sensations around us—they do not seem significant or threatening. If we did not, the sensory overload would drive us crazy. Yet we can call to our consciousness far more sensory data than we ordinarily identify. Any one of us could tabulate a "whole sense" catalog of our surroundings at a hypothetical moment: Tune in on all the sensations around you . . . the hum of a fan . . . the liquid twitter of the mockingbird on the fence outside the window . . . the whir of the neighbor's well-tuned car . . . the perfume of honeysuckle from the yard . . . the roughness of the sandals on your bare feet . . . the blue, red, beige lozenges of the Oriental rug . . . the corner of a page in your book, its crispness remembered by your fingertips . . . the pressure of the chair that supports you . . . soft voices from another room.

Sometimes we accept the existence of a phenomenon only if we can touch it, not trusting our eyes, knowing that our ears can fool us. We use the words "tangible" and "palpable" to emphasize the reality of something immaterial: "The tension in the room was palpable," we say.

Helen Keller was stricken blind and deaf as an infant, isolated from the world and from her fellow human beings. But during childhood the sensitive receptors in her fingertips put her in touch with the world. By feeling, she studied objects, nature, people. And by touching the lips, throat, and cheeks of her teacher, Helen Keller learned to talk and to convey the thoughts and emotions that made her a human being. As a girl of 14 she sat at the side of Samuel Clemens as her fingers "read from his lips" the stories he told.

Touching Tells the Story

The senses of touch, pressure, heat, cold, and pain are called the cutaneous senses, from cutis, the Latin word for "skin." The tongue has a high density of these receptors and a high degree of sensitivity. The center of the back is more sparsely endowed with receptors, and shows a correspondingly lower response. A blind person reads Braille with the fingertips, not the knuckles or the heel of the hand. There are some 640,000 cutaneous sense receptors, distributed unevenly over the body surface.

Everywhere in the skin (and in some other tissues) is one kind of receptor—fibers known as free nerve endings. These have no specialized structure enclosing them. They react to touch and pressure more slowly than other receptors. Another type, Meissner's corpuscles, are nerve endings where the fibers are compartmented in capsules. These exist abundantly in the ridges of the fingertips (9,000 to the square inch), in the lips, the tongue tip, the palm, the sole of the foot, and the genital organs. They respond and adapt quickly, within milliseconds, to even a light brush. Merkel's disks carry continuing signals, such as sustained pressure. They lie along the edges of the tongue and in some hairy parts of the body. The hair end organ responds to the slightest movement of a hair, even before anything touches the skin, by means of nerve fibers that entwine the base of the hair. Ruffini's end organs, deep below the skin surface, contain many-branched fibers, encapsulated nerve endings that respond steadily to heavy, continuous pressure. Other encapsulated receptors, Pacinian corpuscles, lie in the tissue near joints, in the mammary glands, in the genitals, and in some deep tissues like the intestinal walls. Because of their onionlike layers of connective tissue, they

Brunch abounds with sensory messages. You sip your wine and gaze at your companion, overhearing a conversation next to you. Warm sunlight bathes your hair, you catch the scent of a certain flower and suddenly remember a friend not seen in years.

You might have learned in school that human beings perceive the world by means of what they taste, see, hear, feel, and smell. Five senses—and you know that the wine is mellow, the coffee strong.

But *how* do you know? Mechanical, chemical, and electrical processes transform the wine on your tongue (and its bouquet in your nose) into electrical impulses and speed them along nerve fibers. Jumping microscopic gaps, these messages become chemical, then turn electrical again. They plug in at your brain and, somehow, you have sensations.

Scientists now describe as many as 20 sensory systems. One system detects pheromones—chemical signals that convey fear, identity, sexual receptivity. Another responds to light and synchronizes internal body rhythms to the rhythms of the sun.

We respond easily, yet the scheme is stunningly complex. With our senses we know the world.

react to vibrations and pressure changes within a fraction of a second. Finding the receptors for cold and warmth has not been easy. Once, good candidates were the Krause end bulb and the Ruffini end organ—but no longer. Cold receptors, examined under the microscope, just look like free nerve endings.

Each type of sensation may seem to have its own type of receptor, but actually our bodies possess a continuum of touch receptors that respond to a spectrum of stimuli and sensations, where there may be a fine line between a tickle and a twinge, between pleasure and pain.

When some stimuli are present over a period of time, we adapt to them. We put on clothes every morning and, at first, various receptors send messages to the brain that make us conscious of their weight, texture, and pressure. But before long the messages dwindle and disappear, switched off because continuing stimuli of constant intensity will stop activating the receptors. You can accept an affectionate but heavy cat curled in your lap, not because the cat gets lighter but because for a time you become oblivious to the pressure. A change must occur to reactivate the receptors. The wristwatch we are so used to that we forget its presence will suddenly attract our attention if the clasp breaks and it threatens to fall off. At the end of the day, receptors will signal the pleasure of removing ties, jackets, and tight shoes.

The Alarm of Pain

Pain, however, is a sensation we seldom get used to. Pain is an alarm that warns of tissue injury. The several million free nerve endings are our pain receptors, and the more that get hit, the more it hurts. Some pains prick, some burn, some ache. A pricking sensation travels to the brain fastest, up to ninety-eight feet a second, and locates its source precisely in the skin's outer layer. A signal of burning pain or of an ache travels more slowly, no more than about six and a half feet a second; originates deeper in the skin; or seems to come from a more diffused, generalized site, such as an arm or the abdomen. Thus we feel a sharp pricking pain first (from a wasp sting, for instance), then the slow burn. The simplest pain response is a reflex that travels only to the spinal cord—an even speedier means of protection when it is necessary to snatch back one's hand from a hot frying pan.

Our sensory apparatus, with the autonomic nervous system, monitors functions inside the body. Digestion proceeds, blood circulates, lungs expand and contract. We are seldom conscious of these messages.

But if something is amiss, receptors will alert us. Hunger pangs and thirst we know how to remedy. On a very hot, humid day, the torment

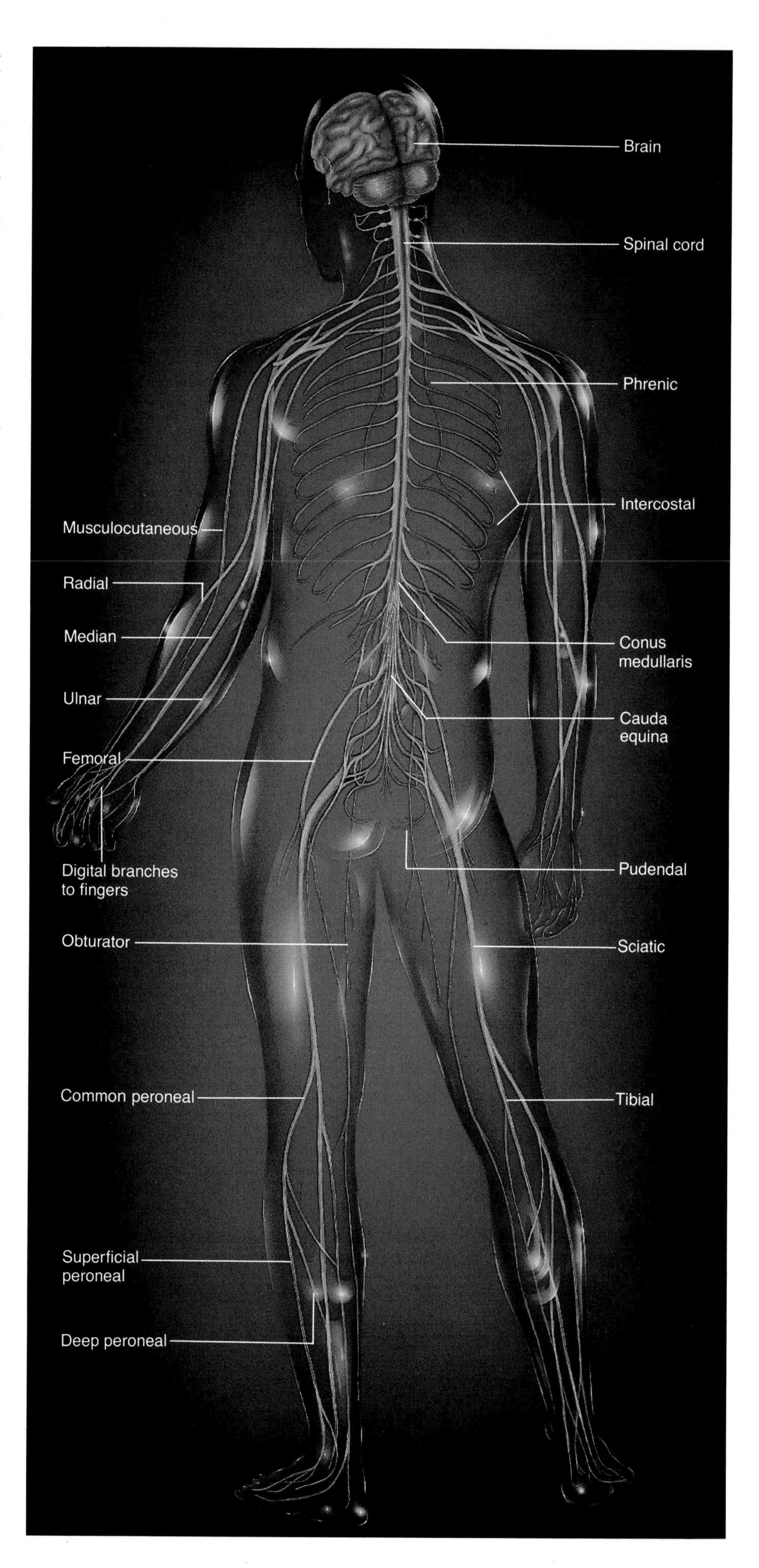

Magnification: 100 times

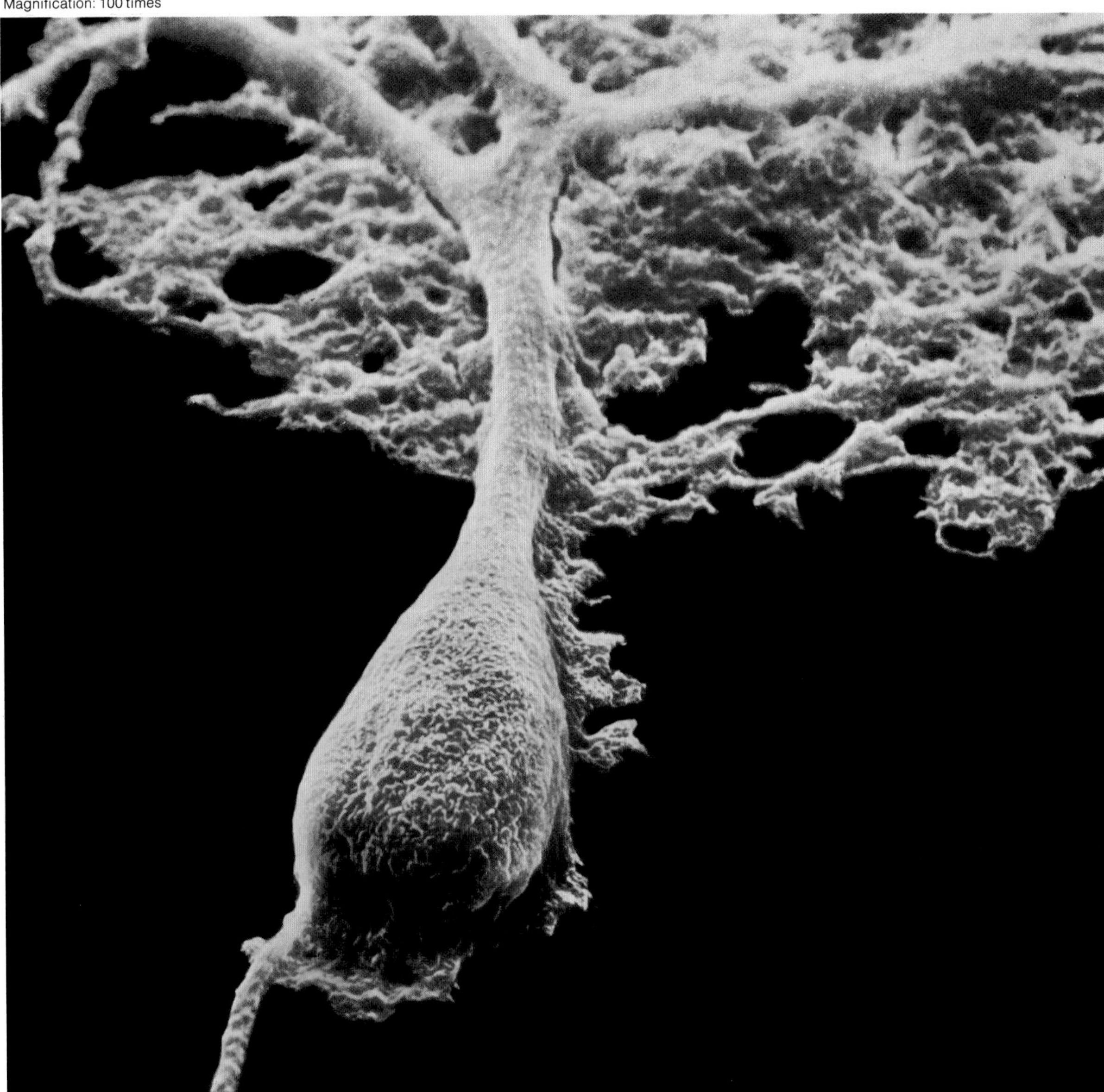

Every few seconds in every day of a lifetime, tens of billions of sensory messages travel as electrochemical impulses along the slender branches of the human nervous system (opposite). They make their way to communication headquarters in the central nervous system—the brain and spinal cord. Here, 50 to 100 billion nerve cells, or neurons (right, upper), act as information specialists. Each receives messages on branching arms called dendrites and sends signals via a single nerve fiber, or axon (ropelike structure at bottom of photograph). Axons outside the brain and spinal cord often form cables (right, lower) that bring the brain information from sensory receptors or carry commands to muscles, glands, and viscera.

Most nerve fibers are

Magnification: 10,000 times

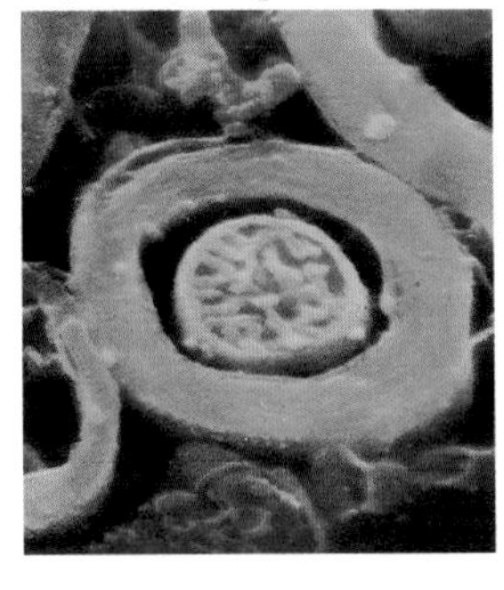

sheathed in myelin, which forms a thick outer covering (cross section, above). Myelin acts as an insulator and allows the nerve impulses to move faster. Along large nerve fibers—such as the 3-foot-long branches of the sciatic nerves in the legs—impulses travel at up to 290 miles per hour.

Magnification: 10 times

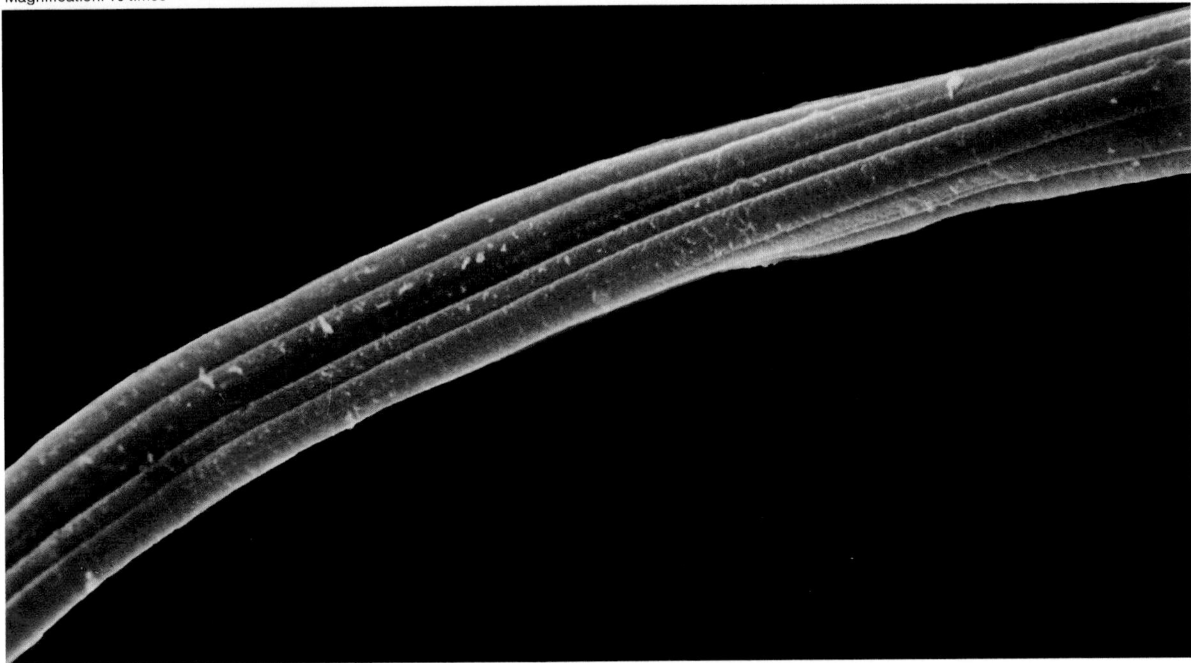

Magnification: 5 times (below); 3 times (right); 7 times (opposite)

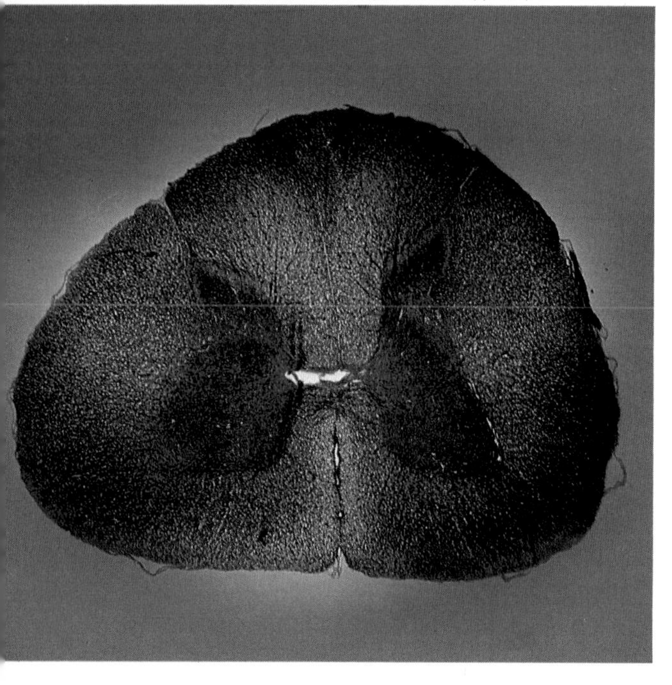

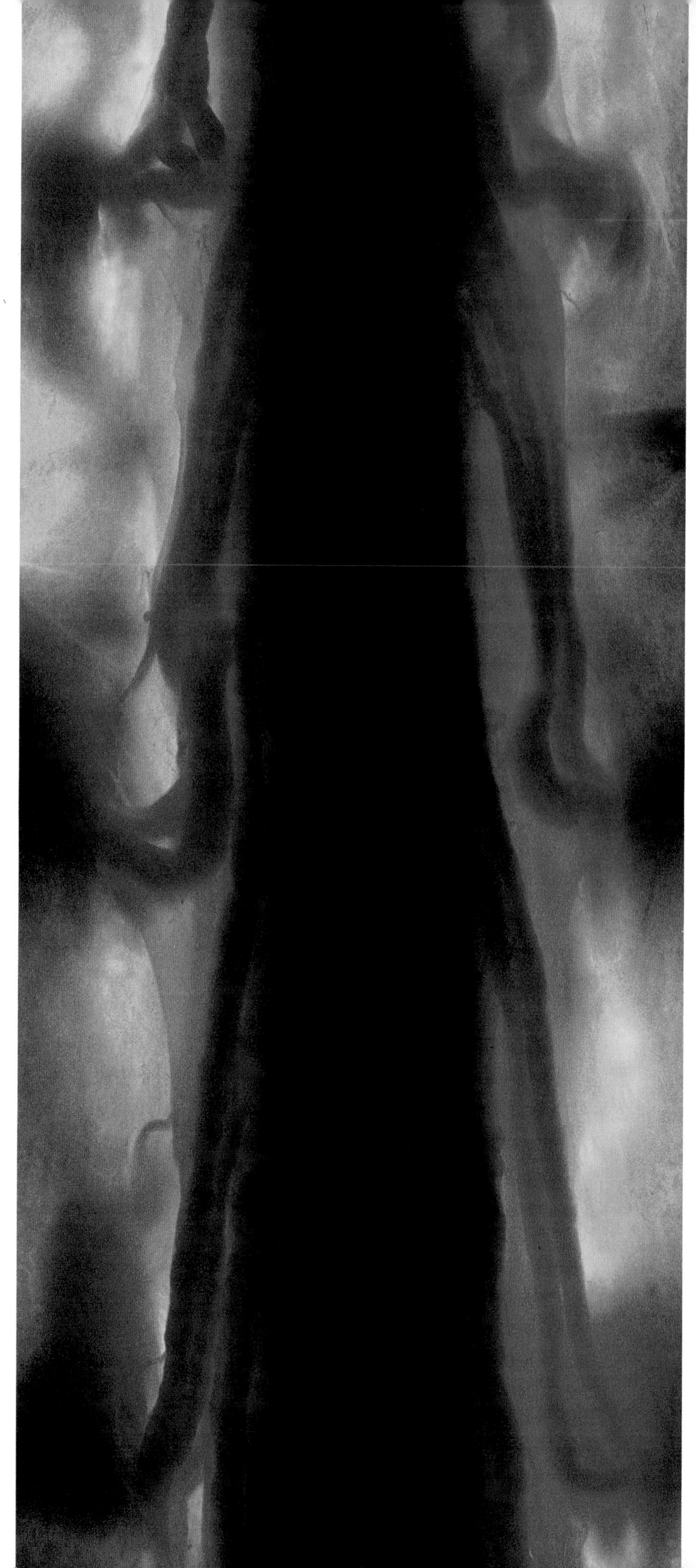

About seventeen inches long, up to three-quarters of an inch thick, and as flexible as a rubber hose, the spinal cord (right) provides the main link between the brain and the rest of the body. Thirty-one pairs of spinal nerves containing hundreds of thousands of individual nerve fibers emerge through holes in the spinal cord's bony protector, the spinal column. Thousands more fibers project from the bottom of the spinal cord in a cluster called the cauda equina, or horse's tail (opposite), before they, too, exit through the spinal column.

Inside the spinal cord, millions of nerve cell bodies in the gray matter (tinted red in cross section, above) process sensory and motor input and handle automatic reflex actions. Touch a hot stove, and your hand jerks back instantly because of a spinal cord command. Conscious reactions occur when the spinal cord relays messages to and from the brain via nerve fibers in the white matter (tinted brown in cross section). You feel pain and know you have burned fingers because a report was sent to your brain by way of the spinal cord as the spinal cord mediated a reflex action in your arm muscles.

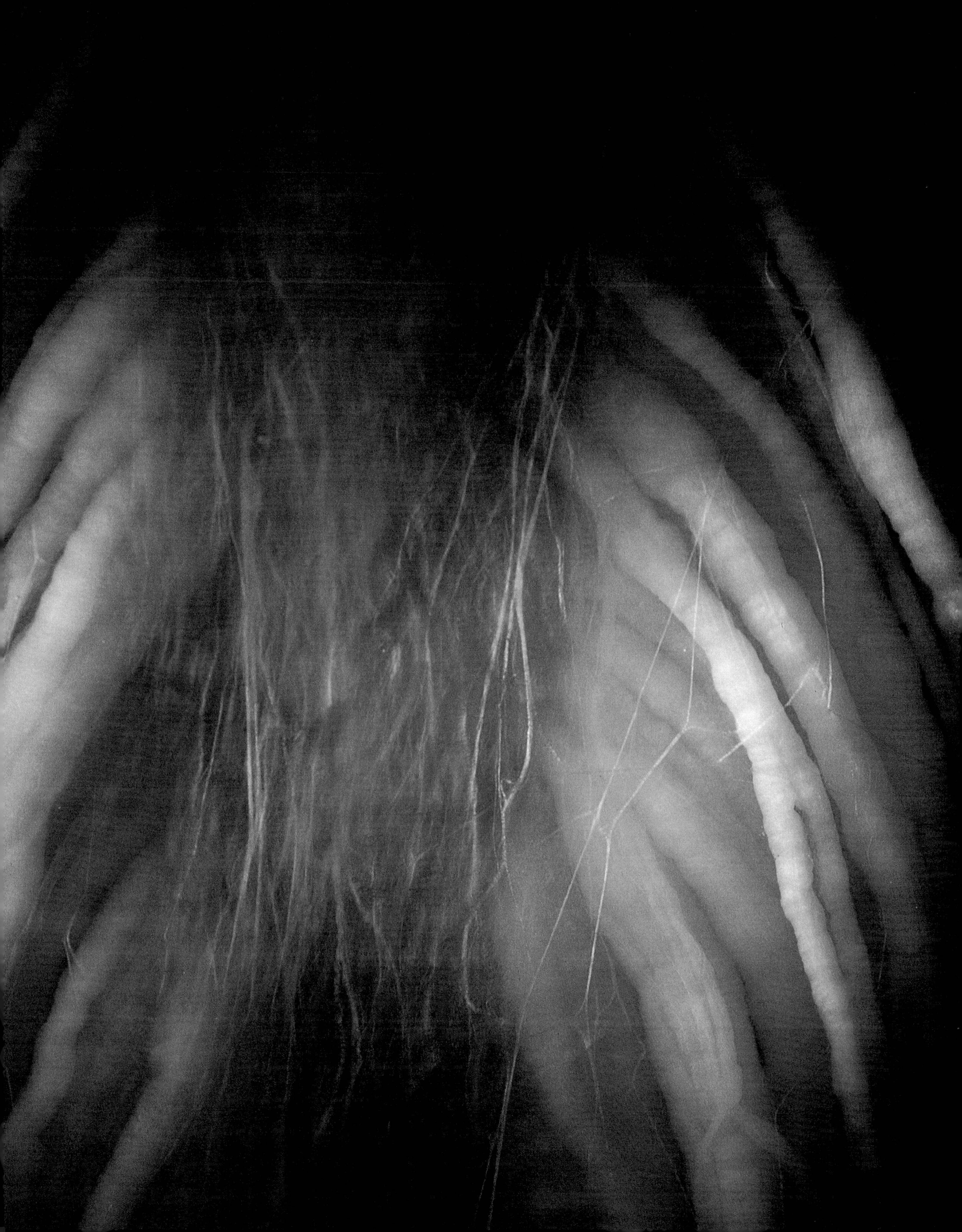

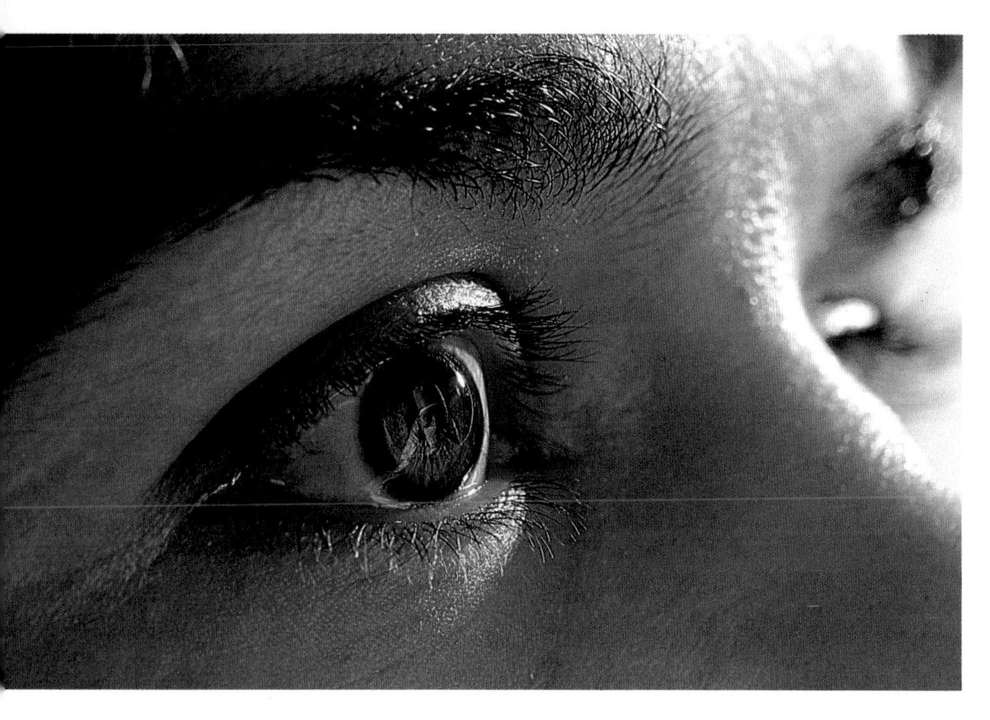

of exhaustion accompanies depletion of essential body fluids. Blinding headaches of altitude sickness are an extreme example of oxygen deprivation in a person whose lungs are not used to the rarefied atmosphere of high regions. But most of us know the painful, heaving gasps for air that our muscles enforce following unusual exertion at any altitude.

As we all know, pain goes beyond a purely physical experience. It can be an emotional or cultural one as well. Women's perception of the intensity of pain from menstrual cramps or childbirth has differed throughout history and in various parts of the world. To the stereotypical Victorian lady, the monthly period was an element of her identity as the frail sex. An affirmation of that identity by an "acceptance" of pain is not surprising. Fear can be a part of pain—apprehension of a visit to the dentist only makes the experience more difficult. Yet powerful social or religious rituals can make the body endure what would otherwise be crushing pain, as when North Dakota's Mandan braves suspended themselves from the lodge rafters by means of iron hooks through their flesh.

The human eye observes its surroundings as tiny, upside-down images. But we perceive a world of right-side-up objects because vision is largely the work of the mind. We analyze available data and base our interpretations on what we have learned to expect. In a distorted room (right) the vertical back wall recedes to the left, but the windows and floor are designed to make the room appear rectangular. Because we are so accustomed to rectangular rooms, we are tricked. We expect to see the figures on the same plane, so they appear to be the wrong sizes. Only familiar figures such as spouses are not distorted by the room.

A Sense of Your Self

Can you touch your fingertips to each other behind your back? Can you close your eyes and find your feet? Can you walk in a straight line? Probably you can. The reason is that our bodies pos-

Mask of a human face? Yes, but it's actually concave, as the lower view shows. In the process of seeing, the brain interprets ambiguous information by referring to a storehouse of learned hypotheses and models of external objects and events. A concave face is so improbable that the brain chooses to see it the way a normal face would appear—convex.

The brain memorizes configurations too, like the pattern of human facial features. Such an arrangement exists in the vegetable bowl (opposite). Your brain remembers a face right side up, so it will not recognize the expected configuration until you turn the page upside down.

sess a sense sometimes called kinesthetic (from the Greek for "perception of motion"), and served by its own proprioceptors. (*Proprio* comes from the Latin for "one's own.") The information that the proprioceptors send to the brain creates some of the most fundamental components of our sense of self. Few of us go through the day giving close attention to what each part of our body is doing and where it is. Yet we are aware, subliminally, and continually process information from proprioceptors, consciously using it to direct action. Look around at people you know. One may be a piano player, another a roller skater, another a champion runner. Proprioceptors are at work here, as they are in the performance of daily chores like washing dishes and driving the carpool.

Pacinian corpuscles and other receptors in the joints, ligaments, muscles, and tendons respond to the stimulation that occurs when we move a joint. Some monitor the rate of movement and the tension of muscles. Others signal our position in space. Still others measure pressure changes—when you turn a steering wheel, then relax your grip, or when a quarterback reaches for the ball, then takes it—and continually tell the brain what is going on. These messages travel very fast and produce varying degrees of coordination. They help make Olympic champions of some of us and Walter Mittys of others.

A Lifeline to the World

If you stop and think of the skin you're in, and try to analyze the sensations, you will appreciate something beyond even proprioception. Imagine being deprived of the pleasure of touching things, and of physical contact with other people. Unless one has seen such deprivation, it may be hard to realize that the sense of touch is a lifeline, a need as basic as the need for food and water. To grow normally, infants must be handled and caressed. If they are not, respiration, blood flow, the development of vision and hearing, growth, and mental health all suffer. Stimulation of all the sensory systems is necessary to make them work right. People blind from birth who are given corrective surgery usually cannot see well even though their eyes move normally. The Skylab astronauts complained of the sensory sterility of their orbiting environment. They missed the colors, textures, and smells of Earth.

Most of us have learned that human beings have five senses: sight, hearing, taste, smell, and touch. Aristotle usually gets the credit for starting this idea. But today we don't just count to five. We look at a continuum. The complex connection of senses that make up "touch" are only the beginning. We see interrelationships among all the organs that respond to the environment, some in ways not dreamed of only a few years

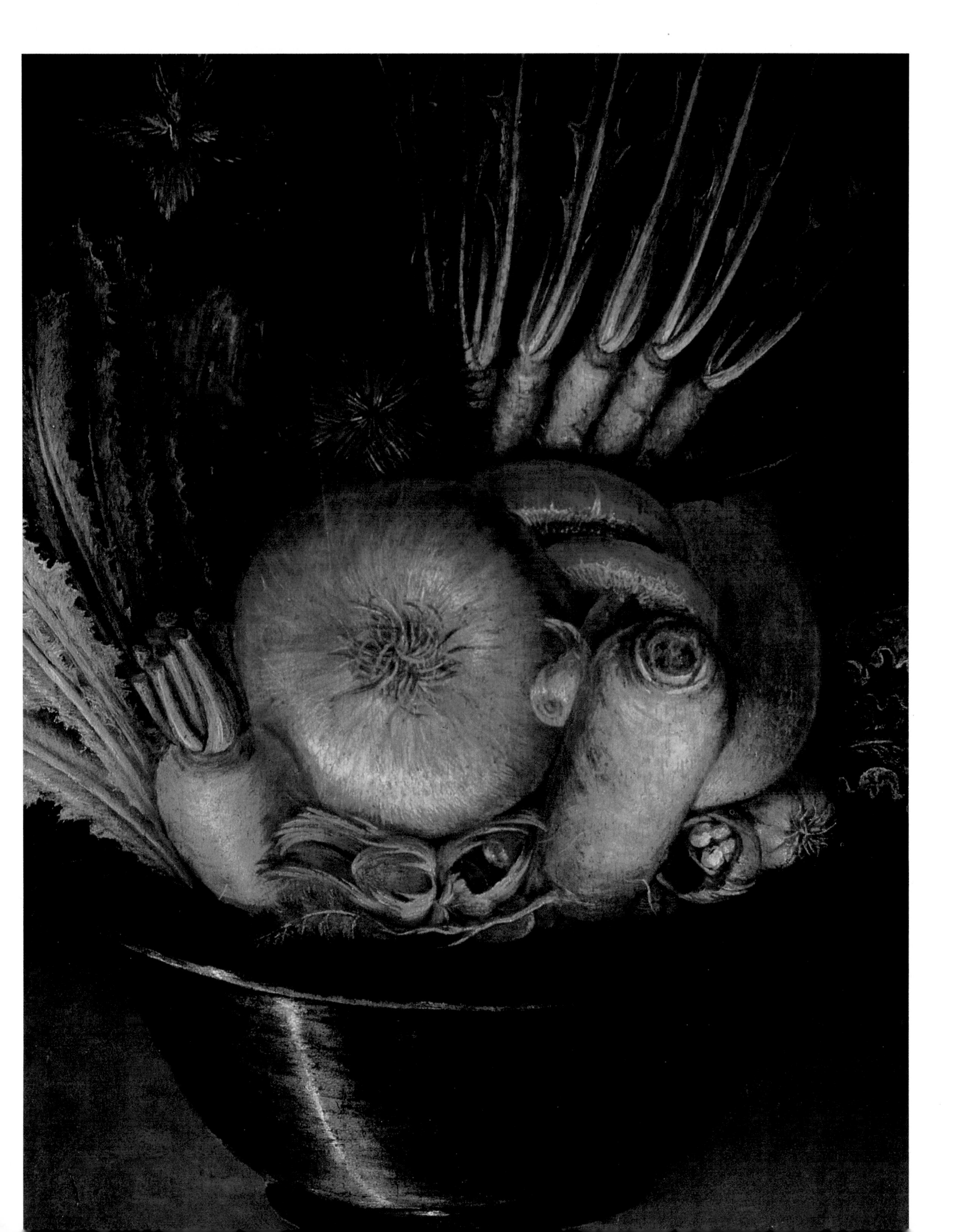

ago, and some in ways no one yet pretends to understand. Can human beings orient themselves by Earth's magnetic field as birds do? Some researchers think so. Do we use our sense of smell to recognize each other or to choose a mate? It seems that we may.

Pleasant sensations mean so much to us that we write poems, compose songs, and paint pictures to immortalize or enhance them. The feel of the world is the theme of continual communication between the sense receptors and the central nervous system. Suddenly you realize that you've been sitting in one position too long . . . a twinge induces you to shift a leg . . . nostrils twitch as they receive the fragile molecules drifting on the air, molecules that translate into a summons to run to the kitchen to taste a pot of chicken soup.

The first thing you might do is to sip, then decide to add salt and pepper. If you inhale some of the pepper you will, upon sneezing, meet further evidence that none of our senses is a simple matter. Pepper and other pungent "smelling" or "tasting" things like ammonia and hot chilies do not really present us with a smell or taste. The receptors that respond to their stimuli are pain receptors and the response is a protective reflex. The sneeze stops inhalation and expels a possibly dangerous substance. People who have peeled chilies, then accidentally rubbed an eye, know to their sorrow that the fiery result is not one of taste. Eyes and nose water to wash away

the irritation. It's no joke. A sign hangs in the Varanasi, India, airport: "No guns. No knives. No red chili powder."

The senses of taste and smell—gustation and olfaction—are known as chemical senses because they respond to chemical energy in the environment. (Sound waves, light waves, temperature, and pressure are forms of mechanical energy.) For a substance to be smelled, it must be volatile, that is, either as a gas or as particles suspended in the air it must move to the nose and make contact. Also, it must be water soluble, to dissolve in the mucous lining of the nostrils and thus reach the nerve fibers, which lie in a dime-size patch of yellow epithelial membrane in the upper part of each nostril. These receptors

Why is the sky blue? For that matter, why are there rainbows, why are trees green, why are painted figures on a white wall red, yellow, blue, green, black? Objects in nature contain molecules called pigments that absorb certain wavelengths of sunlight. Paints, too, contain light-absorbing pigments suspended in a transparent medium, such as linseed oil. Wavelengths not absorbed by these pigments are reflected in varying combinations; we perceive them as colors when they stimulate receptors in the eyes and the receptors send messages to the brain. Other color producers contain no pigments at all. Water droplets act as tiny prisms, refracting the spectrum of visible wavelengths in a beam of sunlight—slowing them down and bending them at different angles—to create a rainbow.

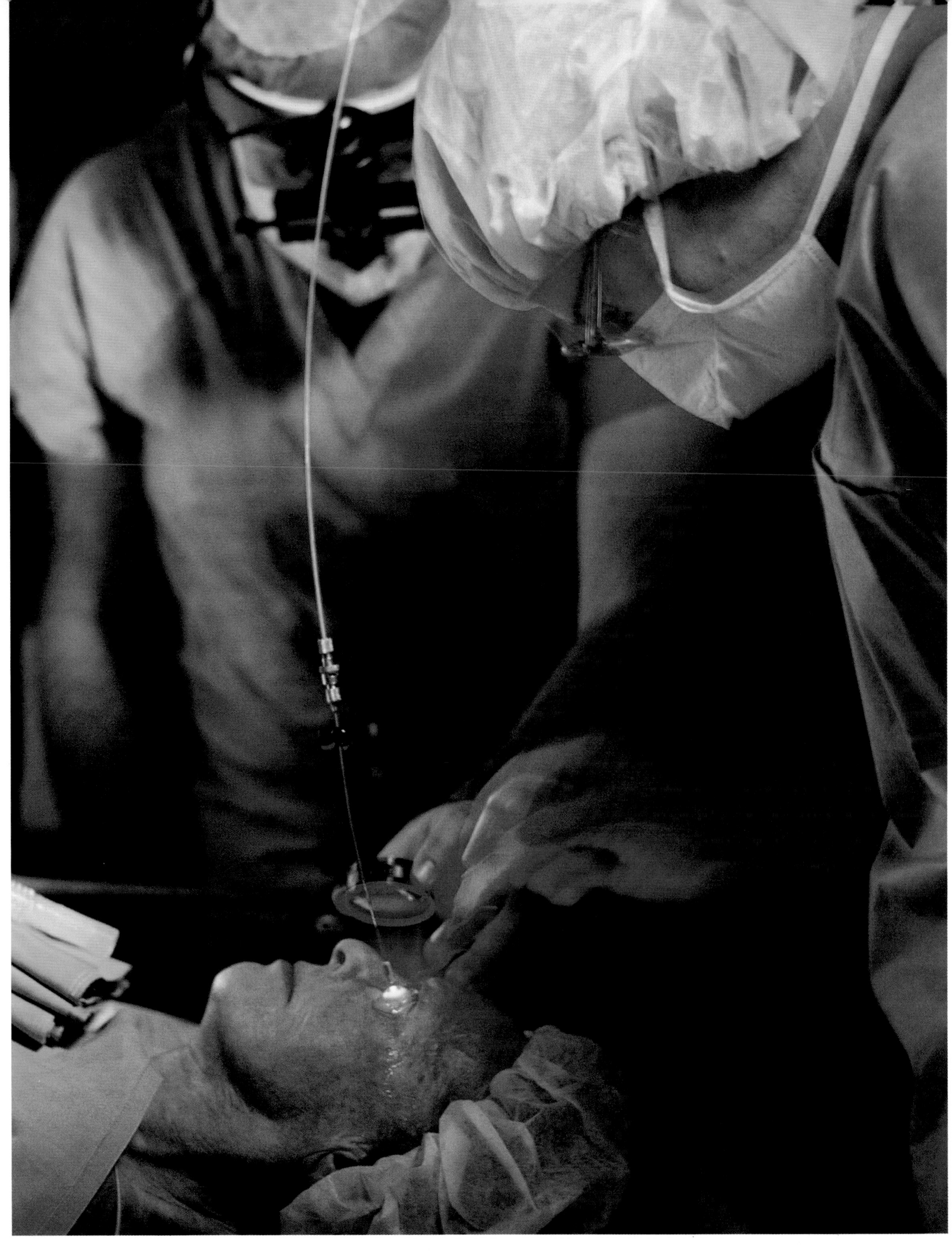

The white light of an incandescent bulb is composed of many wavelengths radiating in all directions. But the laser beam, usually of one wavelength and moving in one direction, is so pure and intense that, focused on areas no larger than a pinpoint, it can make delicate incisions, seal torn tissues, or destroy alien cells.

Within a laser, stimulated atoms and molecules of crystals, gases, or dyes release energy as photons that bounce back and forth between two mirrors—one a full mirror, the other partially silvered—gathering intensity until they burst through the partially silvered mirror in a powerful beam of parallel light rays. The beam may be carried to its destination by a flexible optical fiber.

Eye surgeons (opposite) attack a malignant tumor with a red light generated from a laser. Three days earlier, the patient was injected with the dye hematoporphyrin derivative (HPD), which is retained longer by tumors than by healthy tissue. Irradiating the cancer cells containing HPD, the laser initiates a photochemical reaction that almost immediately begins to destroy the blood supply to the tumor.

The blue-green light of an argon gas laser (right) penetrates an eye in simulation of a surgical technique. Argon laser light is particularly useful for eye surgery because it travels through clear portions of the eye without being absorbed. Striking the retina full force, the beam releases heat to weld vision-impairing breaks or leaks in blood vessels.

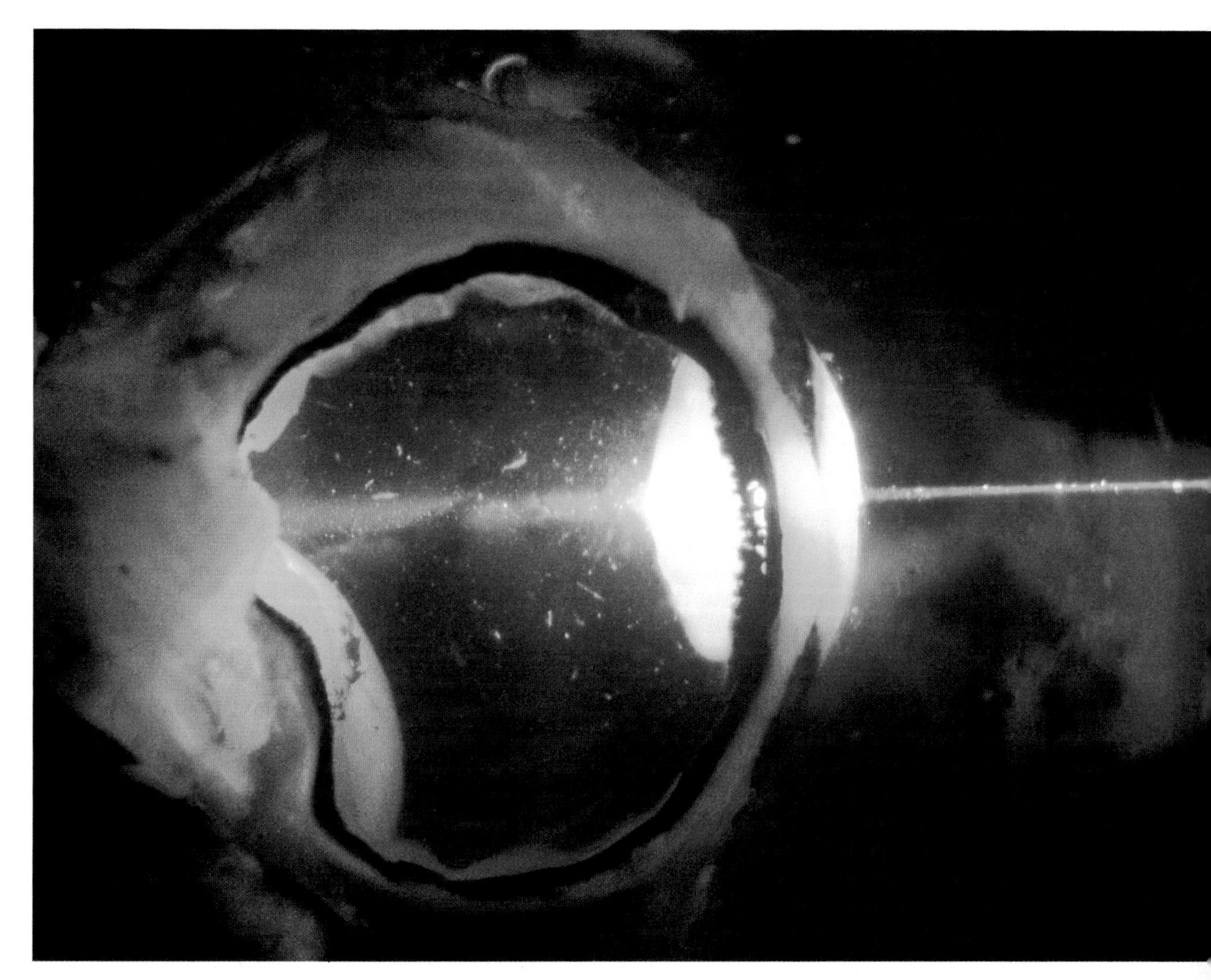

are just a little higher than the pathway of air flowing through the nose to the lungs during breathing. Normal inhalation brings in a moderate amount of nearby odor molecules.

When we exhale, three small, bony plates called turbinates partly block the air passage, pushing upward the molecules of whatever we are smelling. A greater amount of the odor thus makes contact with the epithelium when we exhale than when we inhale. When we're especially interested in some scent we've picked up, we sniff, bringing in even more air and more volatile molecules. Does this mean, in theory, that every satisfyingly deep inhalation of the odor of your new car depletes its supply of molecules and makes the car age faster?

In each nostril are ten million olfactory receptors. These are neurons. The forward end of each lies in the epithelium, the other end reaches inward as an axon to the brain. Projecting into the epithelial mucous membrane are the receptor sites, hairlike fibers called cilia.

Smell is our most mysterious sense. How does a brain identify and remember up to ten thousand odors? One theory is that odor molecules vibrate and are picked up like sound waves. Another is often called the lock-and-key theory. Every substance has its own particularly shaped molecule. It is possible that each molecule finds a receptor site of a comparable shape and fits into it, binding with a receptor protein, as a key fits a lock. Then, the message from the neuron to the brain somehow identifies the molecular configuration, and the nose knows: chicken soup! Just like Mom used to make.

Remembering with Your Nose

Aromas from the past can evoke nostalgia in anyone. A displaced New Englander remembers, can even create anew in his mind, the smell of hemlock. A Russian grown up along the Volga will be homesick at the odor of crushed wormwood underfoot. One whiff of a wax crayon transports us back to a grade school classroom.

There is a physiological reason for these precious moments. The olfactory neurons reach into a portion of the brain known as the limbic system, a seat of emotions and memory. That is why you can lean over a pot of soup, breathe deeply, and sigh, "It's just right." Even smells we can't specifically label remind us of places, people, events. We sometimes find the smells hard to name; this is because the olfactory system's link with the brain's *(Continued on page 282)*

Pathways to the Brain

Our eyes and ears are transformers. They sense the light and sounds around us and turn them into electrical impulses that the brain can interpret. Each organ is designed to handle its own medium: The eye admits light waves, bends them at the cornea and the lens, then focuses them on the retina; in the ear, sensory cells within fluid-filled canals are acutely sensitive to sound vibration. The pathways that light and sound take through the eyes and ears are outlined on the following pages.

Once the eye and ear have done their work—once light and sound are electricity—how do their messages reach the brain? Nerve fibers from the retina at the back of each eye bundle together to form optic nerves, which meet at the optic chiasm (lower right inset). Here, the optic nerves join, and before splitting into two optic tracts some fibers cross so that part of the input from the right visual field goes to the left side of the brain and vice versa. The brain fuses all the visual messages into an integrated whole, but it is the partial crossing of nerve fibers at the optic chiasm, along with the slightly different angles of view provided by the two eyes together, that produce three-dimensional, stereoscopic vision.

Before the visual pathways arc along nerve fibers toward their destination in the brain's visual cortex, they pass through relay stations, called lateral geniculate bodies, buried deep in the brain. Here, some scientists believe, visual information is coordinated with impulses sent by other sensory organs. These bodies may also help to turn off visual messages to the brain when you are concentrating on some other kind of input.

Sound transformed into electricity leaves the cochlea in the inner ear and travels along nerve fibers in the auditory nerve (upper left inset). At processing centers, the fibers synapse with neurons that carry the messages to the auditory cortex on each side of the brain. Auditory signals from each ear travel to both sides of the brain, so that a dysfunction in one auditory pathway will not affect hearing significantly in either ear.

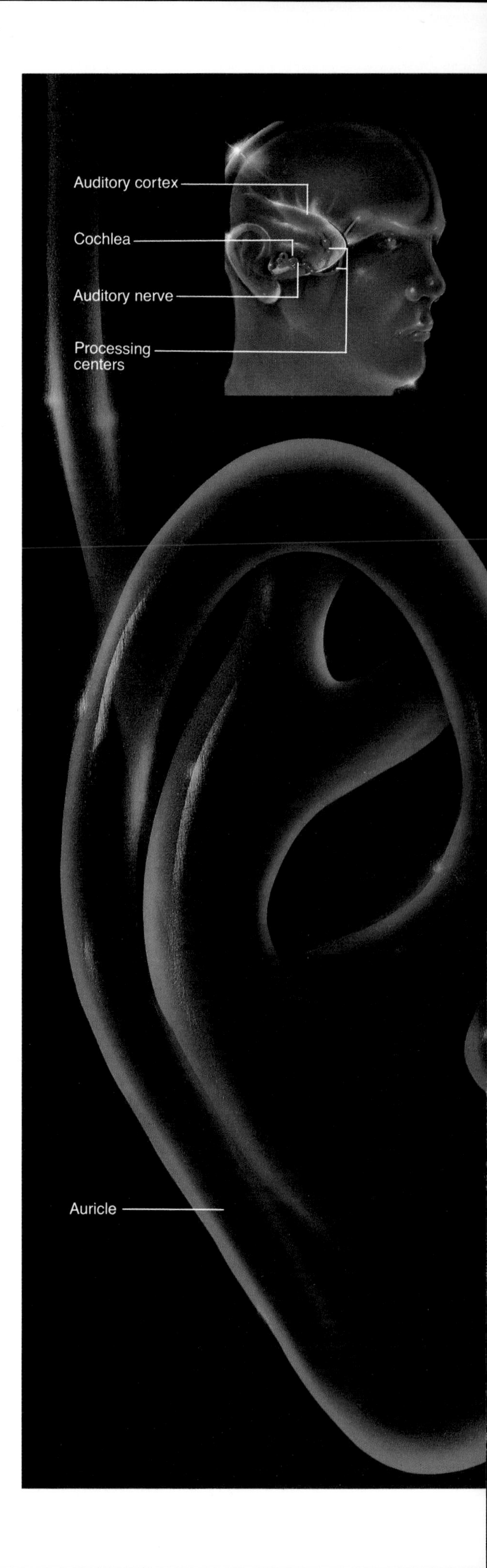

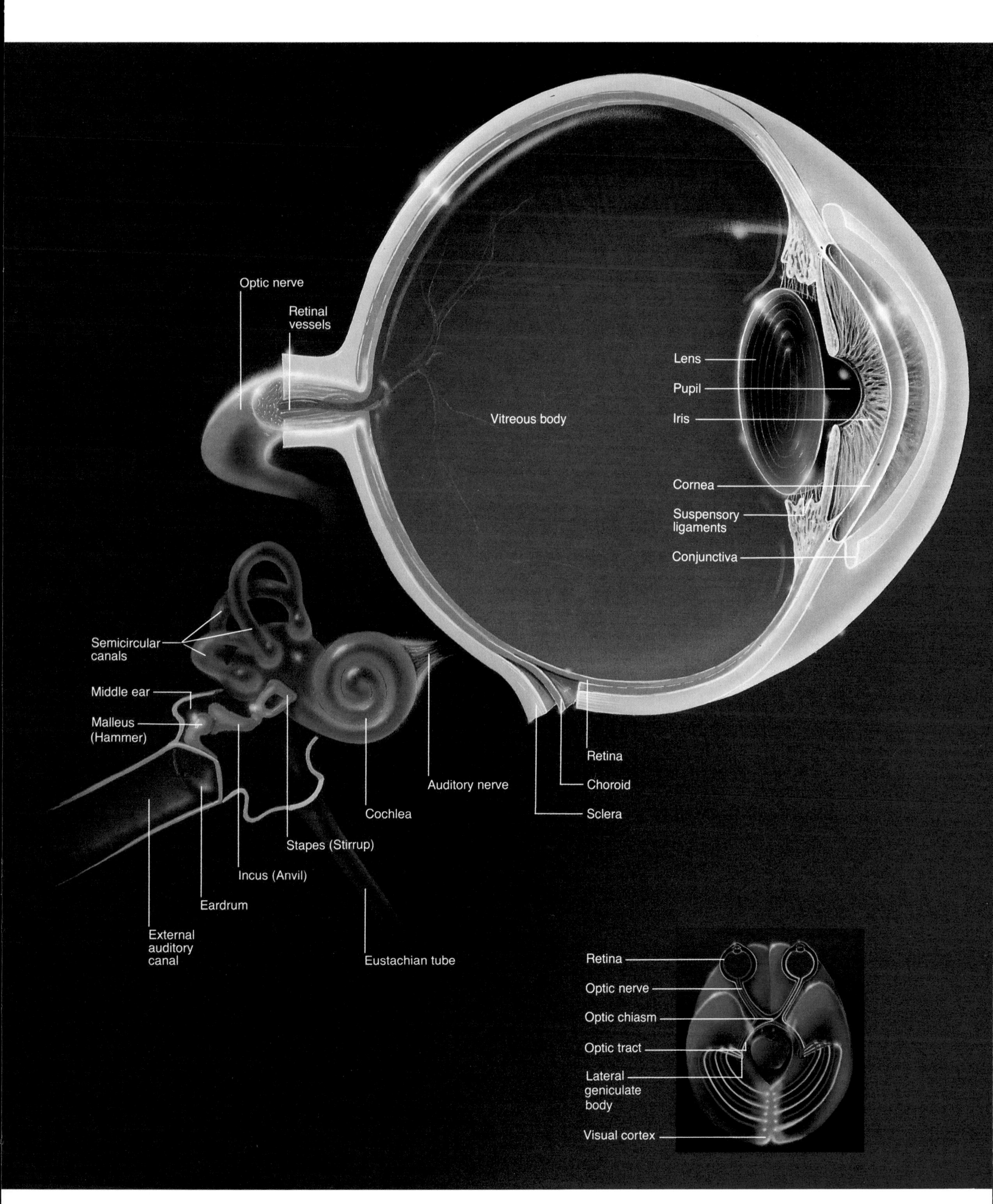
Optic nerve
Retinal vessels
Lens
Pupil
Iris
Vitreous body
Cornea
Suspensory ligaments
Conjunctiva
Semicircular canals
Middle ear
Malleus (Hammer)
Retina
Choroid
Sclera
Auditory nerve
Cochlea
Stapes (Stirrup)
Incus (Anvil)
Eardrum
External auditory canal
Eustachian tube
Retina
Optic nerve
Optic chiasm
Optic tract
Lateral geniculate body
Visual cortex

Magnification: 7 times

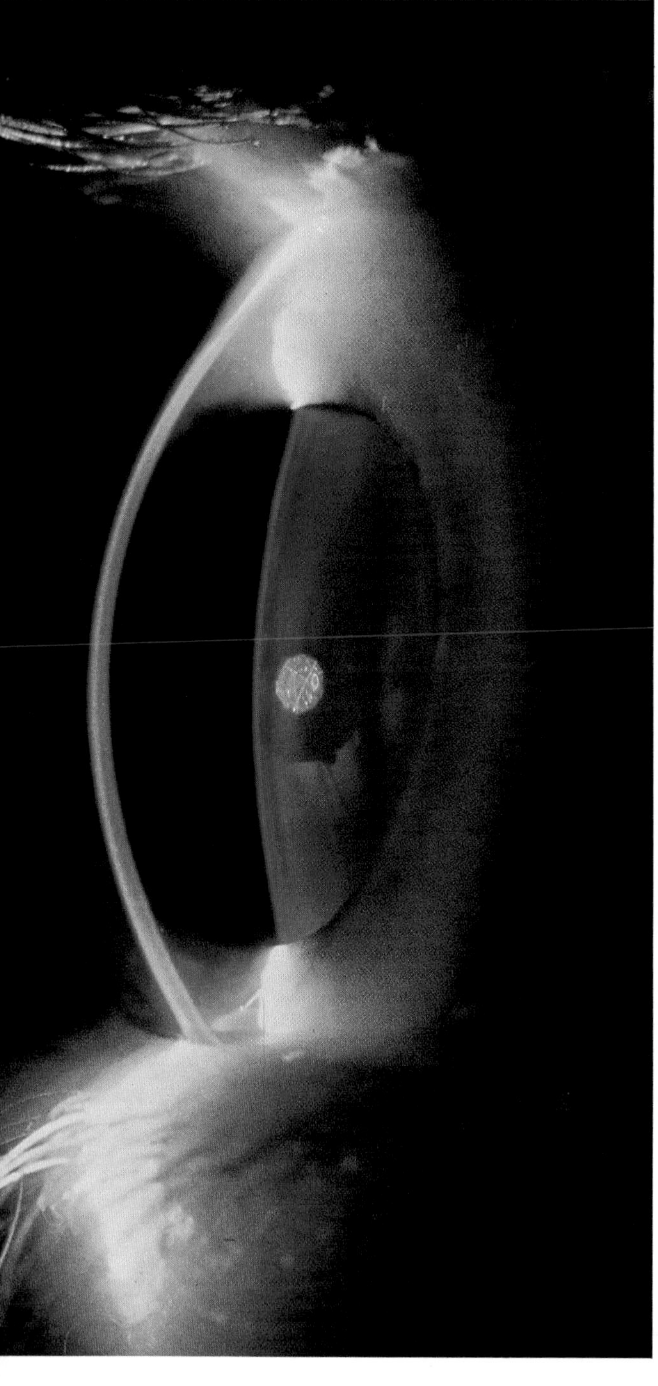

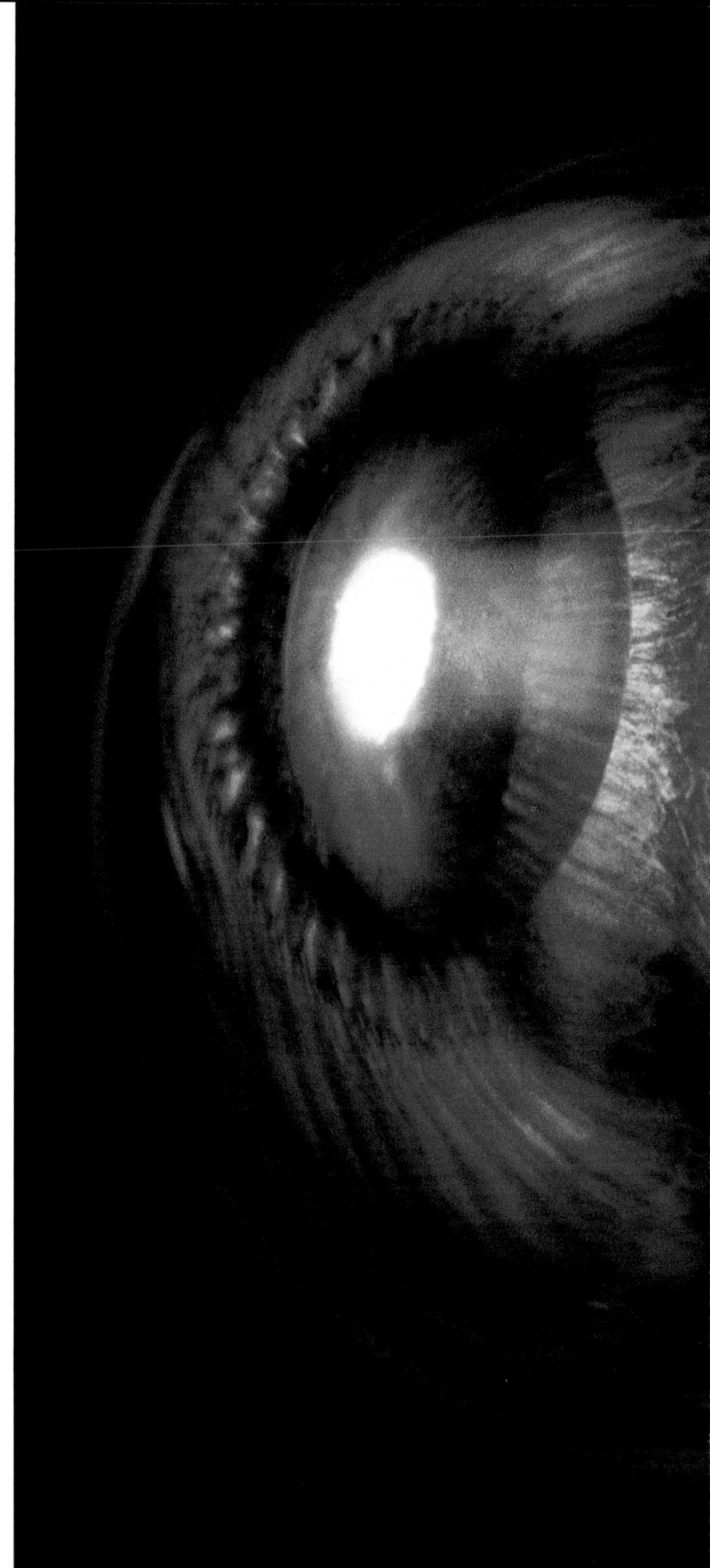

The mind has a thousand eyes,
And the heart but one,
wrote a turn-of-the-century poet named Francis William Bourdillon. Perhaps. But the daily business of seeing is performed by a single pair of gelatinous orbs that gather, guide, and filter light, then translate it into electrical impulses—the language of the brain.

Light first penetrates the cornea—a thin, transparent section of the eye's outer coat (above). Behind the

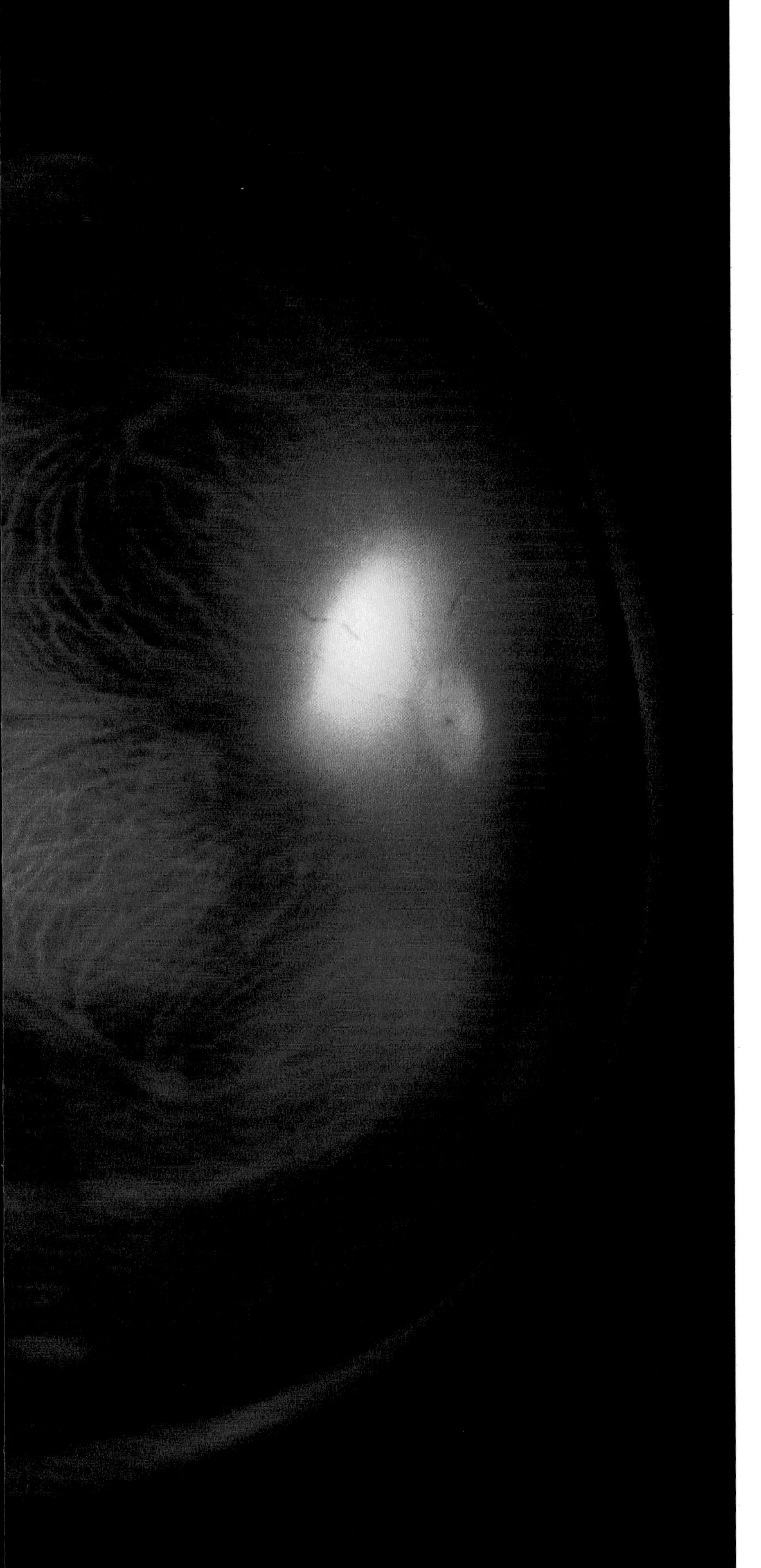

Magnification: 8 times (left); 10,000 times (below)

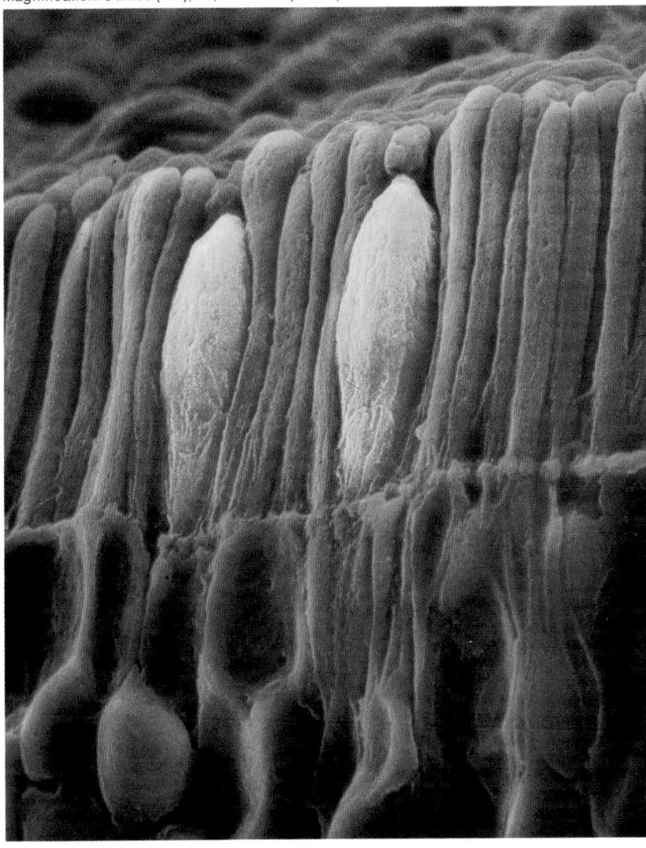

cornea a chamber of clear fluid, the aqueous humor, presses the cornea into a bulge so it can bend incoming light and direct it through the pupil (the bright white light) onto the lens (the grayish oval behind it). The iris—a circle of muscle around the pupil—expands and contracts to keep light passing through the pupil at an ideal intensity. At the lens, light bends into final focus; then it travels through the vitreous humor, a body of clear jelly that forms the largest part of the eye, and arrives on the retina as an inverted image. Here, light penetrates a layer of photoreceptor cells called rods and cones (above) where it is encoded for color perception in the brain. Long rods translate dim light into shades of gray so we can see shape and movement but not color; shorter cones absorb spectral light, enabling us to see myriad hues.

Minute changes in air pressure, caused by vibrations, create waves that travel through the three sections of each ear and become electrical impulses that alert the brain to a world of sound. The auricle—the visible flap of skin on each side of the head—and the auditory canal just beyond, funnel sound waves inward to the eardrum. Tiny hairs and droplets of wax (right, upper) snag insects or dust particles that stray into the auditory canal.

Beyond the canal, sound waves arrive at the middle ear (right, lower), striking the eardrum (tinted yellow). The eardrum vibrates at the same frequency as the air waves that strike it, causing the vibrations to continue through the three small bones of the middle ear—first the malleus, or hammer, attached to the eardrum; then the incus, or anvil; and then the stapes, or stirrup—all named for objects whose shapes they resemble. Designed to transmit even very faint sounds, these bones diminish the amplitude of the vibrations but increase their force as they enter the inner ear cavity and the cochlea, a coiled tube with three canals partially filled with fluid. In one canal, the scala tympani (far right), sound vibrations become liquid waves that move at the same frequency as the air that entered the outer ear. Deep within the cochlea, in the organ of Corti, the fluid waves move a membrane which bends delicate hairs; these hairs stimulate nerves that carry messages to the brain.

Magnification: 20 times; 25 times (right)

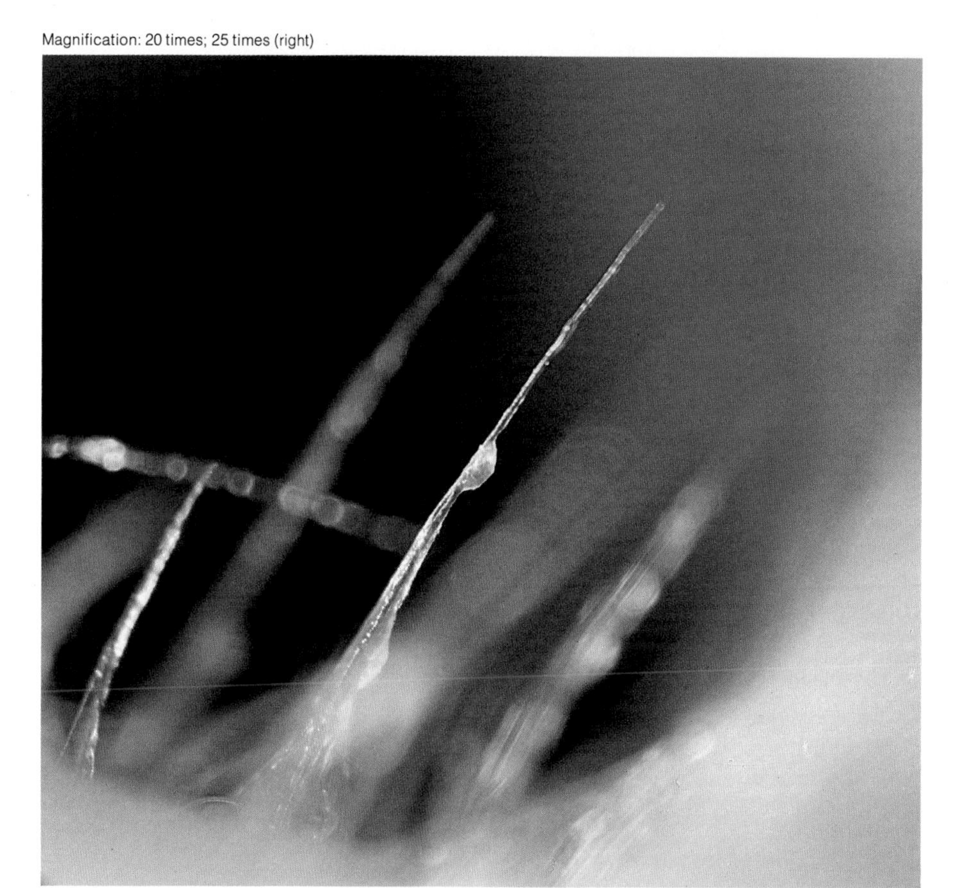

Magnification: 10 times

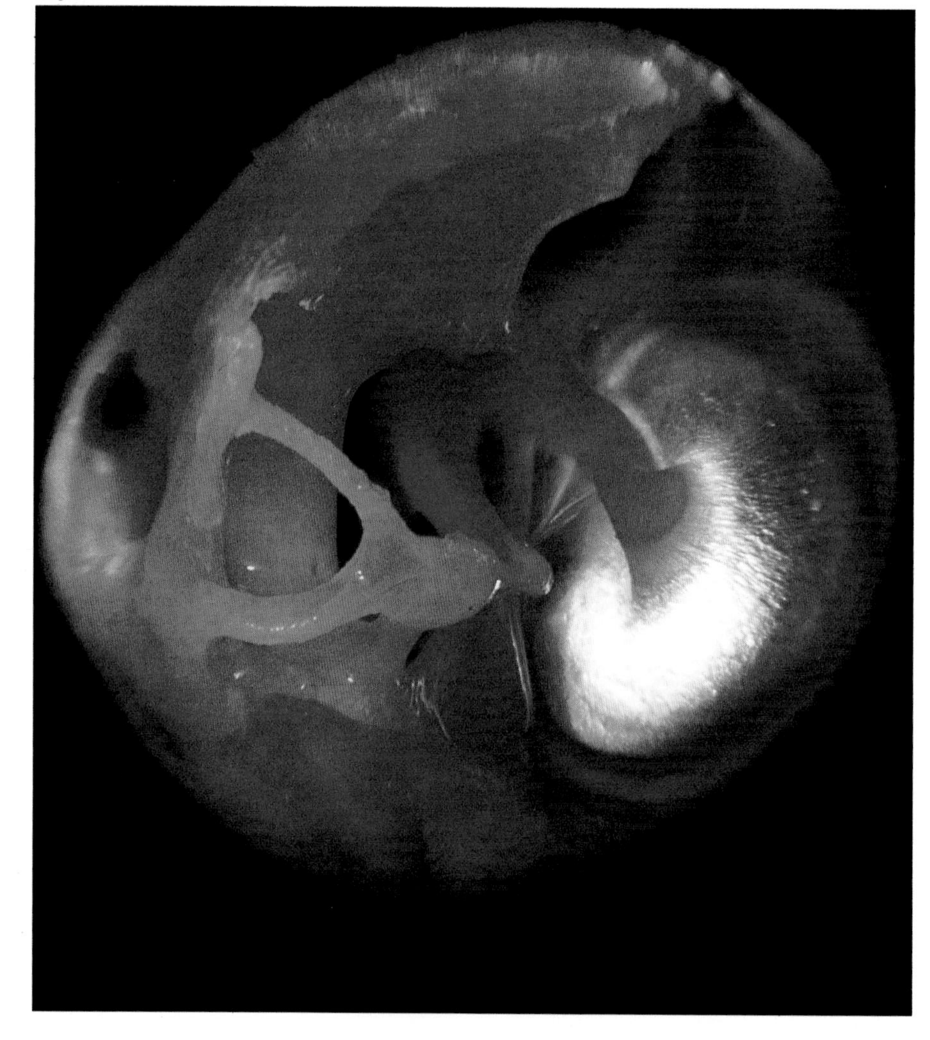

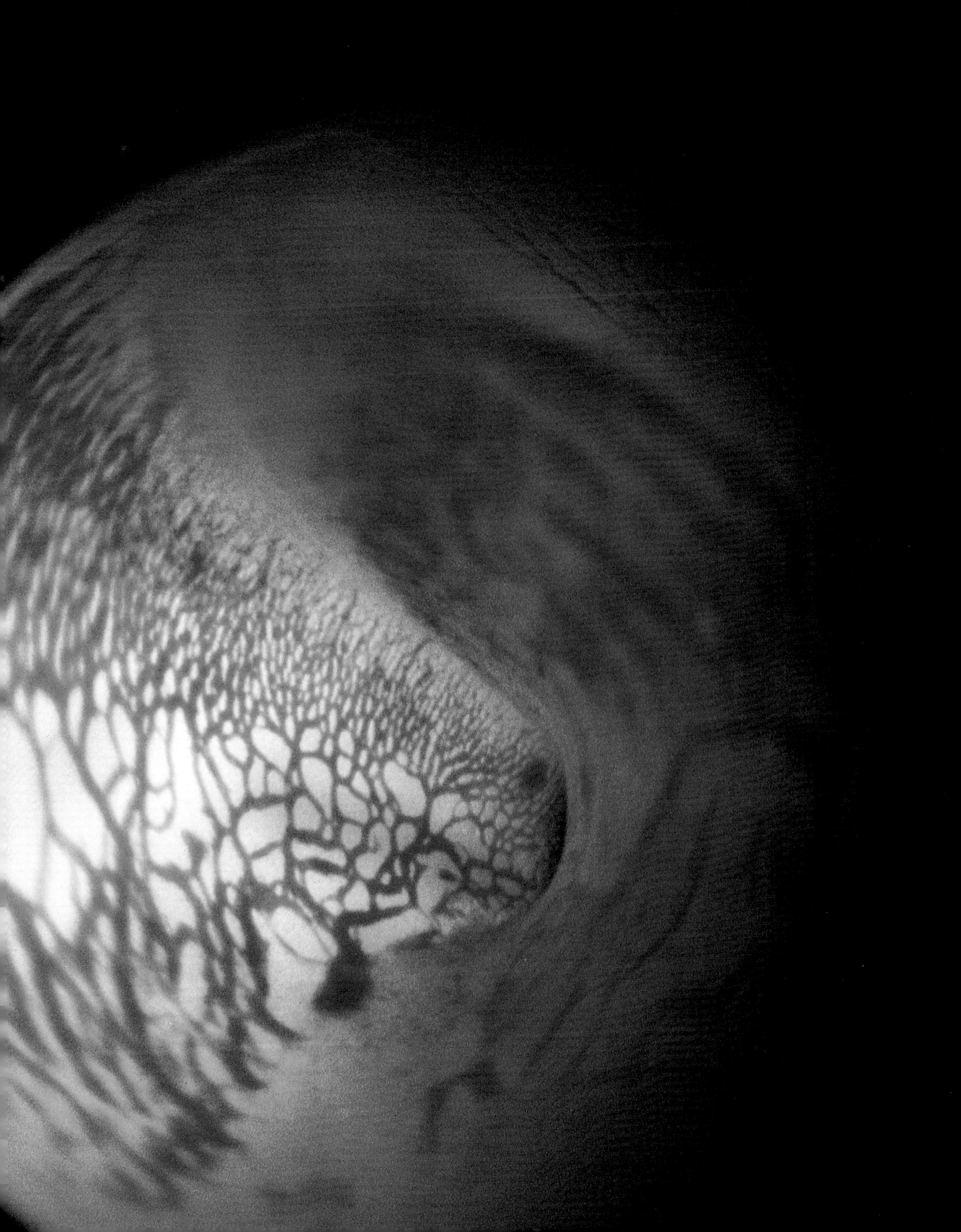

Music to your ears, noise to someone else's—it all depends on what you find pleasing. And what you find pleasing depends, in large part, upon what you perceive as familiar. The plaintive twang of the *komiz* may not start American toes tapping, but the Kirgiz people of China's far west dance to it. Likewise, the syncopated rhythms and blues scales of jazz may fail to fire the Chinese soul.

Musical perception is a complex human activity, and no one knows exactly why arrangements of tones and rhythms, sensed by the ears and translated into messages for the brain, become so indelible, or why music has the power it does to arouse such a range of emotions. Voltaire said it simply: "It is impossible to translate poetry. Can you translate music?"

language centers seems to be a weak one.

Ever since Aristotle, people have tried to classify the odors that human beings recognize. Linnaeus, the 18th-century taxonomist, had an obviously sliding scale of seven types: aromatic, fragrant, musky, garlicky, goaty, repulsive, and nauseous. Later other kinds were added: ethereal, camphoraceous, acid, spicy—and caryophyllaceous and santalaceous. And these are just a few. An objective analysis of odors is impossible. We don't know enough about how the olfactory system works, so all lists are impressionistic. The sense of taste is a different matter.

A simplified map of the tongue locates messages it delivers to the brain: sweet and salt from the tip; sour from the sides; bitter from the back.

Whatever we taste, too, will be water soluble. The tongue's surface is a mucous membrane, covered with tissue projections called papillae. The papillae contain the gustatory receptors—taste buds. There are also taste buds on the palate and the epiglottis. Adults have some 10,000 taste buds, infants and children many more. Taste buds die as we age.

Each taste bud contains about fifty taste cells. These cells don't live long—from a week to ten days—and new cells continually replace the old. So if you burn your tongue, don't worry. It'll be better soon. The taste receptor cells cluster around a taste pore, one to a taste bud. Microscopic hairs emerge from the tips of the cells to the taste bud and into the mouth. When you eat something, the food—as dissolved chemicals—enters the taste bud through the pore. The receptor cells make a link with a neuron at the base of the taste bud. The neuron informs the brain about the taste of what you have eaten.

Our taste sense is far less sensitive than the sense of smell. The olfactory receptors are themselves neurons, with a direct link to the brain, while the taste receptors must take an extra step to the neurons that serve them. Recall the experience of smelling soup from another room in the house. Only a few molecules did that. To taste the soup, we need much more—maybe a drop. It can take as much as 25,000 times more soup to taste it than to smell it.

Like taste cells, olfactory neurons replace

themselves. These are the only neurons in the body that can do this. Because the taste and smell receptors have this way of maintaining good performance, unless we're injured we can always find food and enjoy it.

When the taste-identifying impulses reach the brain, they end up in the cerebral cortex, the same area that receives information from all the senses, including data from the mouth about temperature and touch. Also, the mouth and nose share a common air passage, so that we smell odors from food in the mouth as well as from food on the plate. If the sense of smell is out of commission, as it often is when we have a cold, the sense of taste hardly functions at all. Thus, what most of us refer to when we say something "tastes good," is an assembly of sensations that might more aptly be called flavor.

The generally accepted four basic tastes—sweet, salt, sour, and bitter—combine with each other and with other sensations to make all the other flavors we know and may ever know. Even sound plays a part in gustatory appreciation. Steaks sizzle, stew bubbles, salads crunch. When you consider the acuteness of human hearing—only a little more sensitivity would enable us to hear the collisions of molecules in the air—you have to admit that we judge dishes not only by flavor but by their snap, crackle, and pop.

The Energy of Sound

Our sense of hearing depends first of all on the mechanical energy of sound waves in the air. A snap of your fingers creates energy in the form of pressure against the air molecules around your fingers. This force pushes the molecules out a little, into space already occupied by other air molecules, forming a densely packed area called a shell of compression. This shell pushes against nearby air molecules, which in turn form another shell, and so on. The air around your fingertips is temporarily rarefied—less dense—because it has lost some molecules, but the molecules rush back to fill this partial vacuum—and create another in their wake. Behind them, the molecules of the second shell return to fill that space. These alternating shells of compressed and rarefied air are the sound waves—the energy created by the snap of your fingers. When you move a one-ounce pencil, you use 140 million times more energy than the energy of the sound waves that reach your ear from a soft whisper. Sound waves move about 1,100 feet a second in air, and even faster through liquids and solids (the Indian scout with his ear to the ground knew this).

For the human ear, sound has three important physical attributes: frequency, intensity, and quality. Frequency, or pitch, refers to the number of vibrations, or shells of compression, that

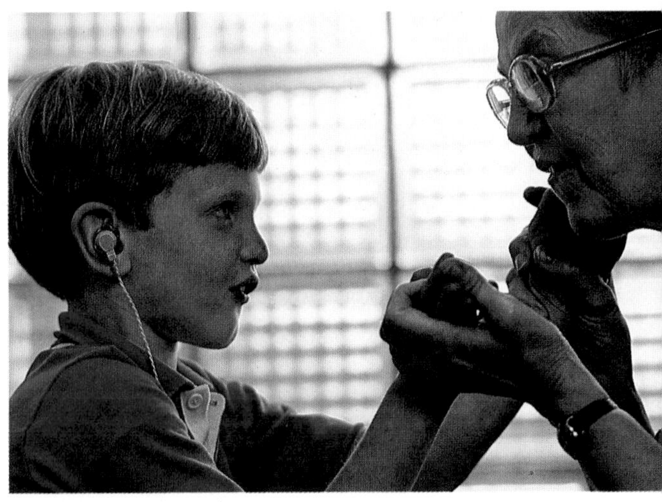

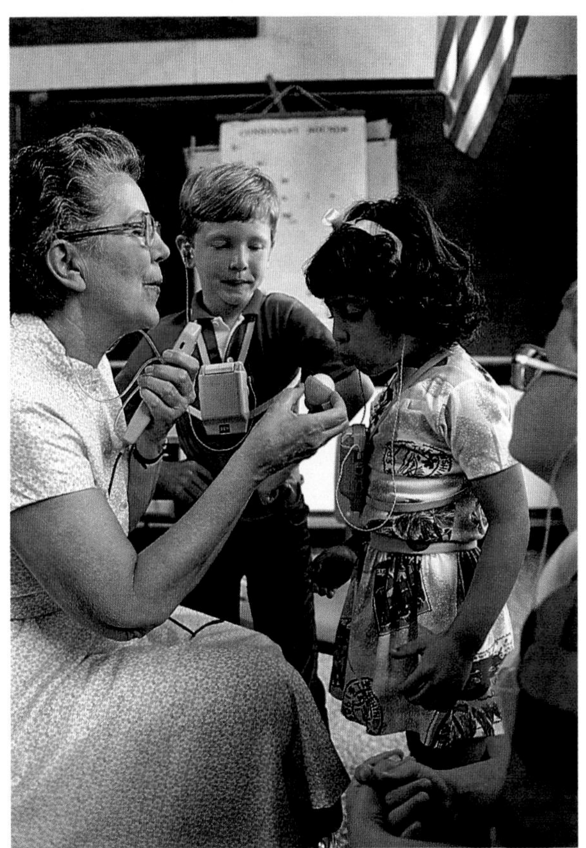

Deaf children at the Clarke School in Northampton, Massachusetts, use sight and touch to learn to talk. Mimicking their teacher's facial movements while watching their own, two children mouth sounds (left); touching the teacher's face (upper) reinforces the visual information. Children also practice breath control by watching objects as they blow on them (lower).

Human speech ranges in frequency from 400 to 3,000 vibrations per second, but the entire range of sounds audible to humans is greater—from 20 to 20,000 vibrations per second. So the deaf or hearing impaired may hear some sounds, even if they cannot hear speech. Amplifiers on shoulder straps make more audible any sounds the children are capable of hearing.

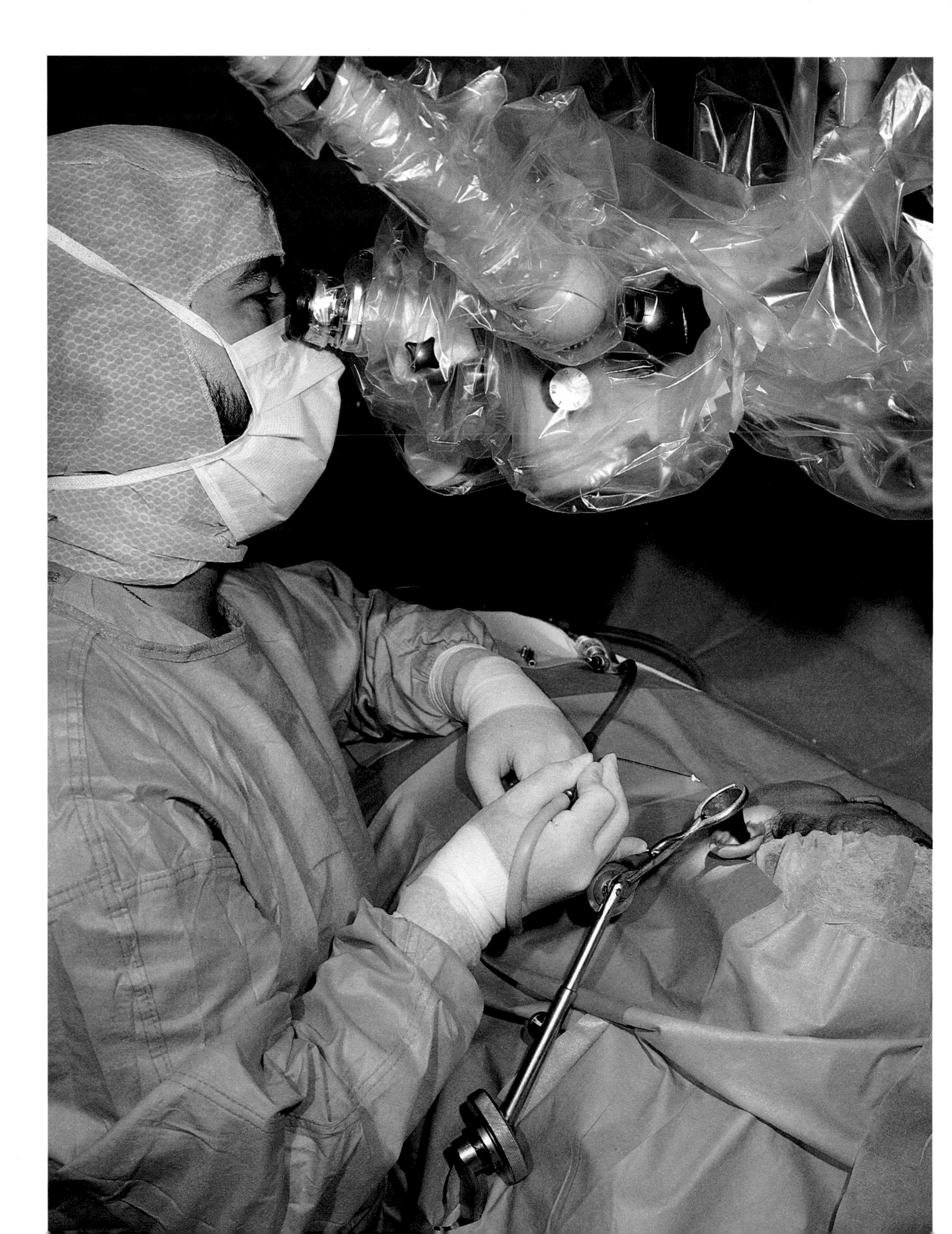

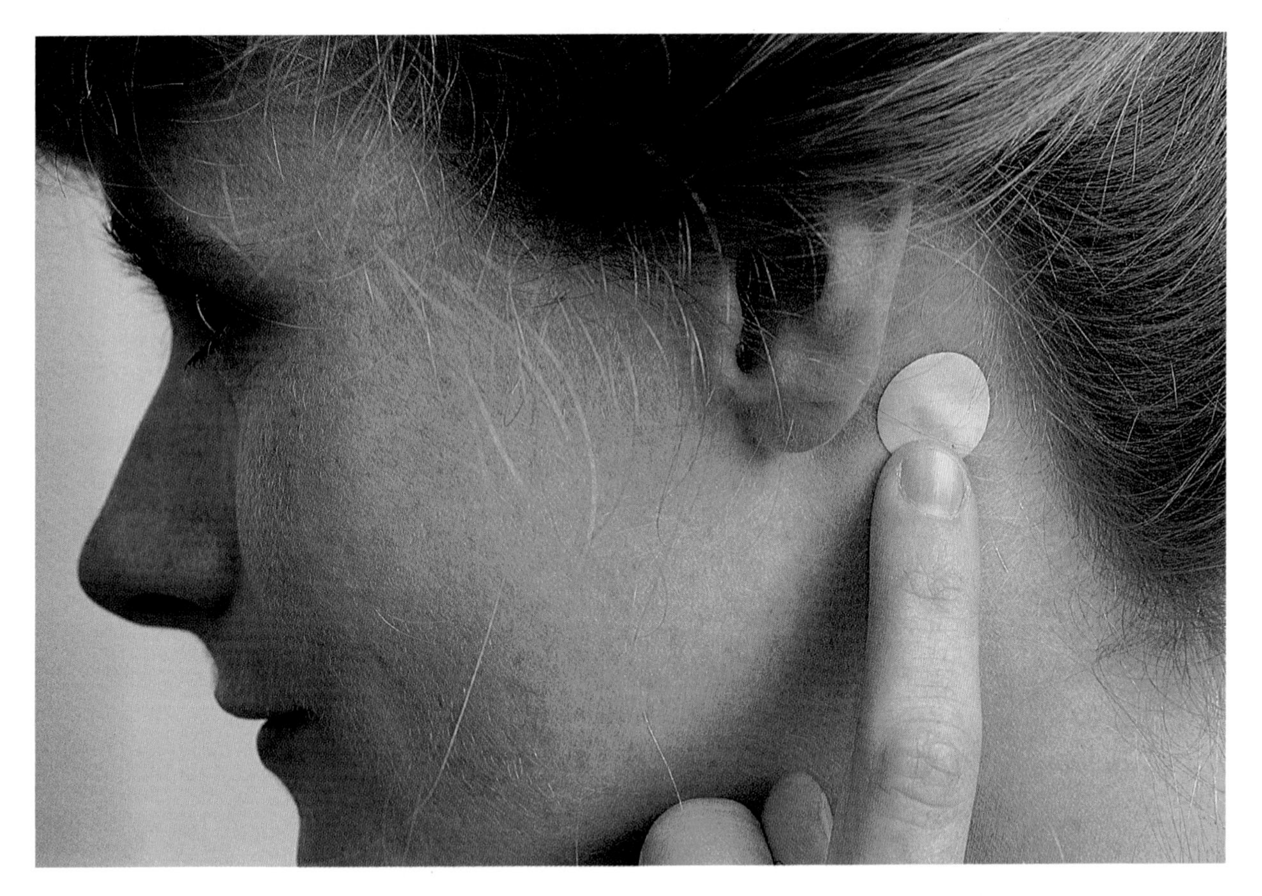

In the minute passages of the middle and inner ear lie the delicate mechanisms for hearing and balance. When things go awry—from a minor case of motion sickness to a profound hearing loss—modern devices and surgical techniques often lead to total recovery.

Conductive hearing loss results from interference with the mechanical transmission of sound to the inner ear. This may happen when the three small bones of the middle ear become disconnected, or damaged by disease. Surgeons (opposite) can replace damaged middle-ear bones with implants made of a custom-fitted glass called Bioglass. A Bioglass prosthesis makes a better replacement than other artificial implants because it resists rejection by the body and bonds successfully with natural tissue. And it is easier to shape than natural bone.

Motion sickness, manifested in nausea and vomiting, occurs when several senses send the brain conflicting signals about body movements and positions. A transdermal patch (left) allows a controlled dose of a preventive drug to enter the bloodstream directly through the skin, thus eliminating many of the side effects that develop when the drug is taken orally.

pass a given point in a second. Frequency is measured in units called hertz. The range of the human ear enables us to pick up sounds that are pitched higher than a flute—up to 20,000 hertz—and lower than a bass fiddle—down to 20 hertz. You can create a visual example of sound frequency by taking a whip and lazily waving it in the air. It moves slowly and generates sound waves below the frequency that we can hear. But move it fast—crack the whip—and the frequency rises to an easily audible level. Be grateful that the human ear is not sensitive to sounds of very low frequency. If we could pick them up, we would be forced to listen to our muscles creaking, dinner digesting, and bones vibrating with the shock of every step.

Sound waves have size, or amplitude. The greater the amplitude, the louder or more intense the noise. Sound intensity is measured on a logarithmic scale of units called decibels. A whisper in a quiet library (30 decibels) is ten times more intense than the ticking of a watch (20 decibels). The hum of a refrigerator (50 decibels) is a hundred times greater than the whisper. Zero on the scale measures the lowest sound that the average human ear can pick up. Near the top of the scale, where pain and hearing damage may result, is the blast of a jet plane taking off—140 decibels, a hundred trillion times the zero threshold and one trillion times the sound of breathing.

The quality of a sound mixes the basic frequency with secondary vibrations. When a piano, a trumpet, and a flute all produce the same note, the frequency is the same but, to the listening ear, overtones give each note a different quality.

That cartilaginous, skin-covered funnel on the side of the head is not the "ear." It is just the beginning. This trumpet of an outer ear is an enormous device (the better to catch the sound waves that surround us) compared with the exquisitely miniature parts of most of our acoustic apparatus. The ear flap, or auricle, opens to a downward curving tube about an inch long, the auditory canal. The auricle and the canal are fairly simple things, too, compared with the structures of the middle and inner ear. The canal, cartilage at the outer end and bone at the inner, ends at the eardrum. Despite its simplicity, the auditory canal serves several functions. The middle frequencies of sound resonate there. It is a buffer between the environment and the more elaborate parts farther in—near the opening, hairs and glands that produce earwax help keep foreign materials out of the ear. The outer ear is really just a hole in the head—a fold and a tube of skin that lead to the eardrum, a membrane that has an outer surface of skin and an inner surface of mucous membrane.

Shaped like a flat cone, the eardrum closes off the outer ear. It measures a third of an inch in diameter and is less than 1/50 of an inch thick, as thin as paper, and as tense as the head of a tambourine. When the sound waves generated by a dinner table conversation travel down the auditory canal and strike the eardrum, the displace-

Magnification: 25 times (opposite); 4,000 times (below)

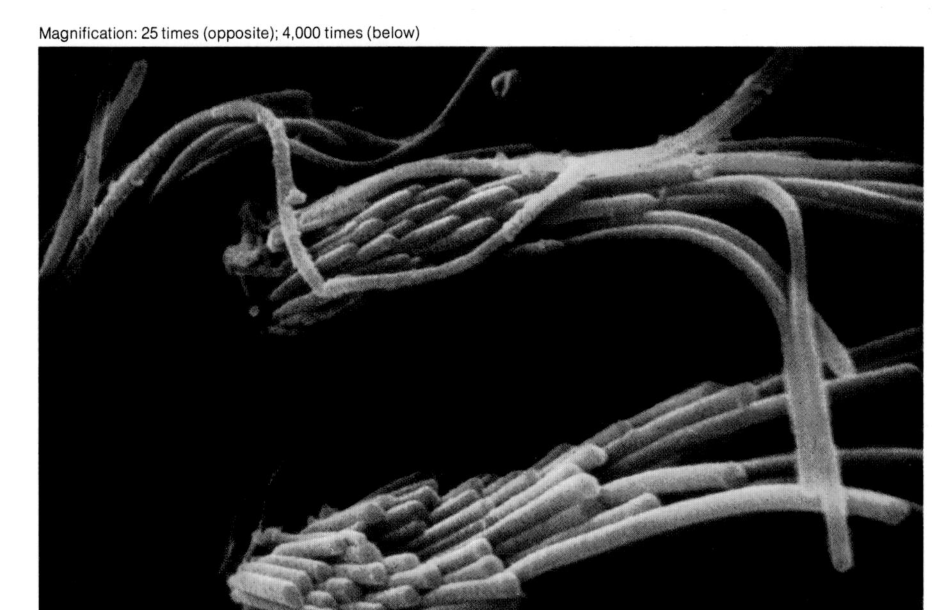

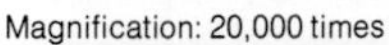

Magnification: 20,000 times

How do you remain standing despite the perpetual pull of gravity? Why can you whirl around suddenly without falling down? The vestibular organs in the inner ear help maintain equilibrium by sending the brain information about the motion and position of the head.

Three semicircular canals (opposite) contain fluid and hair receptor cells (left, upper). Head motion causes the fluid in the canals to move and to bend the hairs; the bending hairs trigger messages that travel to the brain. The canals are set at right angles to one another in three planes of space and so react separately and in combination to different types of movement. One is most sensitive to up-and-down motion, one to side-to-side motion, and one to tilting motion.

Two other vestibular organs, the saccule and utricle, hold hair receptors whose tips are covered with a jellylike membrane containing tiny crystals called otoliths (left, lower). When you accelerate straight forward or when your head changes position relative to the ground, otoliths bend the hairs which send nerve impulses to the brain.

The vestibular organs work together with receptors in the eyes, neck, muscles, and ligaments; none of these alone would be enough to keep you balanced. When you look out the window of a stationary train and see another train pull away, your eyes might suggest that you are moving until other receptors in your body inform you that you are not.

A well-balanced welder wields his torch while perched confidently 27 stories above downtown Pittsburgh, Pennsylvania. His poise testifies to the reliability of the human vestibular system.

Sometimes, though, the system does not function. In outer space, for example, receptors in the inner ears that normally inform the brain about the head's position relative to gravitational pull have nothing to report because there is no gravity. But the eyes still perceive position changes, and the receptors that report motion and acceleration keep working. The brain receives conflicting messages and is confused; the result can be space adaptation syndrome, with symptoms such as headache, nausea, and cold sweating. Some of the same discomforts accompany seasickness and car sickness, brought on when the inner ear registers movement that the eyes might not see.

At least half of NASA's space shuttle astronauts have experienced space sickness during the first few days in orbit; afterward the problem generally disappears, perhaps because the brain learns to disregard confusing signals from the inner ear.

In experiments like the one simulated at left, specialists who staff the shuttle's Spacelab place a subject in a special chair to test the effects of weightlessness on vestibular function. They are seeking ways to prevent the initial discomfort and disorientation of space flight.

ment of its vibration is about equal to the diameter of the hydrogen atom. (In a drop of water—H_2O—there are 100 billion billion atoms.)

Beyond the inner surface of the eardrum are the three smallest bones in the body, operated by tiny muscles. The bones, named for their shapes, are the malleus (hammer), the incus (anvil), and the stapes (stirrup). The malleus is about eight millimeters long, the incus seven millimeters, and the stapes four millimeters—shorter than a grain of rice. These three bones, the ossicles, amount to a chain of levers that transmits energy across the air-filled chamber of the middle ear to the inner ear, a distance less than the width of a paper clip. The "handle" of the malleus is attached to the eardrum. Its head, secured by ligaments, fits into a tiny socket in the incus, which loosely joins the stapes.

The middle ear is an energy transformer. It absorbs sound and lowers the amplitude. The bones vibrate at the same frequency as the eardrum but, because the arena of action is smaller, the force is proportionately intensified by the time the sound waves reach the footplate of the stapes and its attachment to the oval window. This is one of two small openings in the bony boundary between the middle and the inner ear. The other is round in shape and compensates for vibrations of the oval window. When the stapes thrusts inward, the membrane of the round window moves outward, and vice versa.

In order for the eardrum to vibrate freely, and thus accurately transmit sound waves, air pressure must be just about equal on both sides. Unequal pressure makes it bulge tautly inward or outward. The one-and-a-half-inch long eustachian tube links the pharynx and the middle ear. When we swallow or yawn, the tube opens and air enters the ear cavity. We all feel the change in air pressure as we ride up or down in a fast elevator, the stuffiness in the ears, and the consequent feeling of slight deafness. A similar obstruction often results from a head cold, but in that case a swallow will not solve the problem.

Labyrinths in the Ear

Where the footplate of the stirrup rocks in the oval window, sound waves cross into the inner ear, an irregularly shaped cavity in the hard temporal bone of the skull, deep behind the eye socket—a secure site for the delicate machinery at work there. In the cavity are serpentine networks of bone and membrane—channels that fit one inside the other, with the bony labyrinth

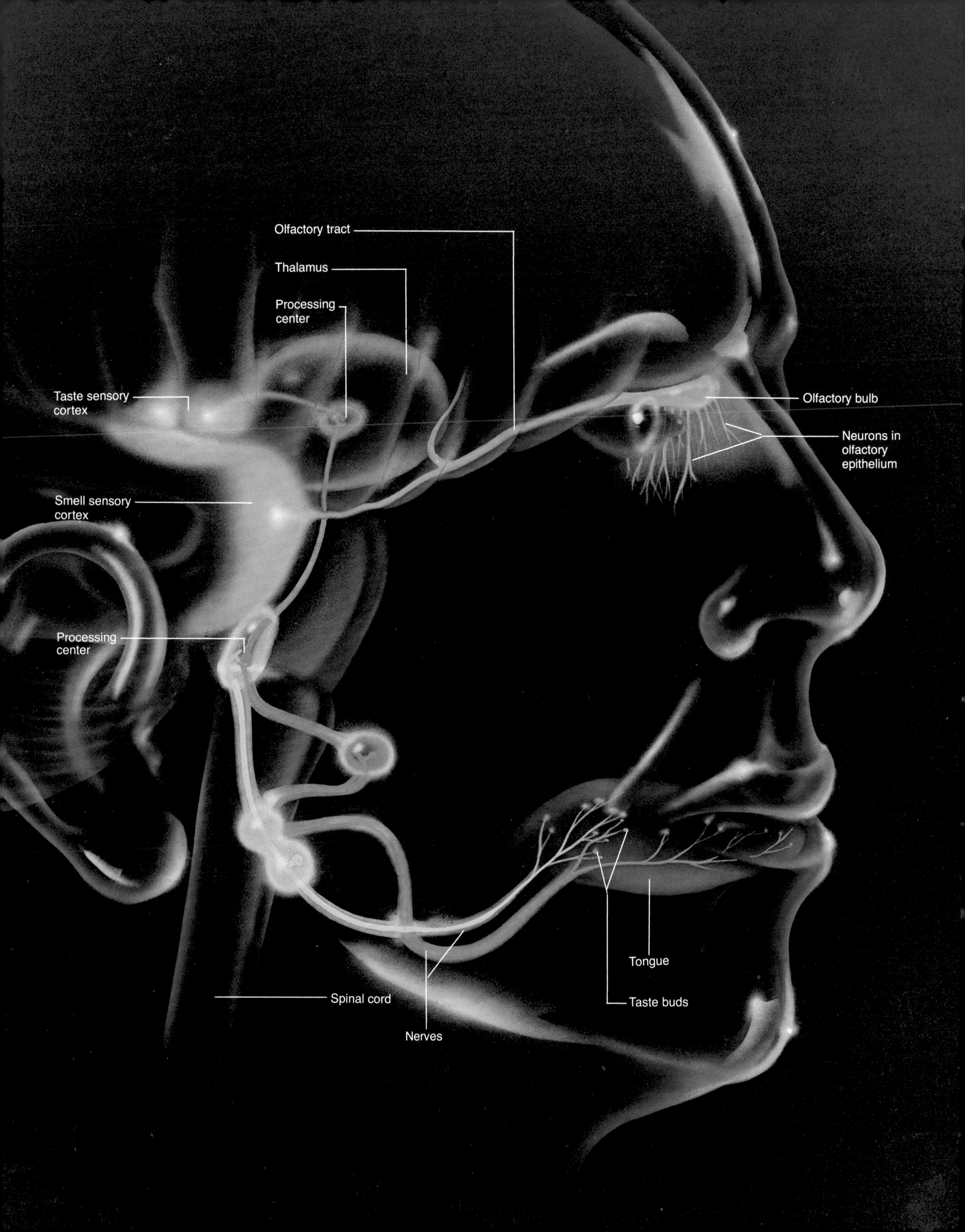

Olfactory tract
Thalamus
Processing center
Taste sensory cortex
Olfactory bulb
Neurons in olfactory epithelium
Smell sensory cortex
Processing center
Tongue
Spinal cord
Taste buds
Nerves

wrapping the membranous labyrinth. Both are fluid filled. Sound now travels as pressure waves in the liquid. Just inside the oval window is the vestibule of the bony labyrinth and, at the vestibule, the cochlea, where hearing gets down to some pretty tricky business.

The cochlea, a pea-size, snail-shell spiral, contains a duct that winds from the oval window around and back to the round window, and a piece of tissue extending the length of the spiral along the base. On this tissue, the basilar membrane, are some 25,000 auditory receptors—rows of hair cells known as the organ of Corti. Waves in the fluid of the cochlea ripple the basilar membrane and bend the hairs with a pluck about the width of an atom. The tips of the hair cells reach into a gelatinous membrane. When the hairs bend—so slightly—electrical impulses relay the message along the 30,000 nerve fibers of the auditory nerve and thence to the brain.

The cochlea discriminates. Hair cells near the oval window are more effective at sensing high sounds of high frequency. Cells farthest from the oval window react best to low-frequency sounds. The brain knows the notes of this tiny keyboard, reads them as they come rushing in as fast as 100,000 times a second, and identifies, perhaps, the opening chords of the B minor sonata. When several frequencies vibrate harmoniously—in periodic oscillation—that's rhythm. The frequencies of everyday din and clatter lack this harmony. That's noise.

Magnification: 8 times

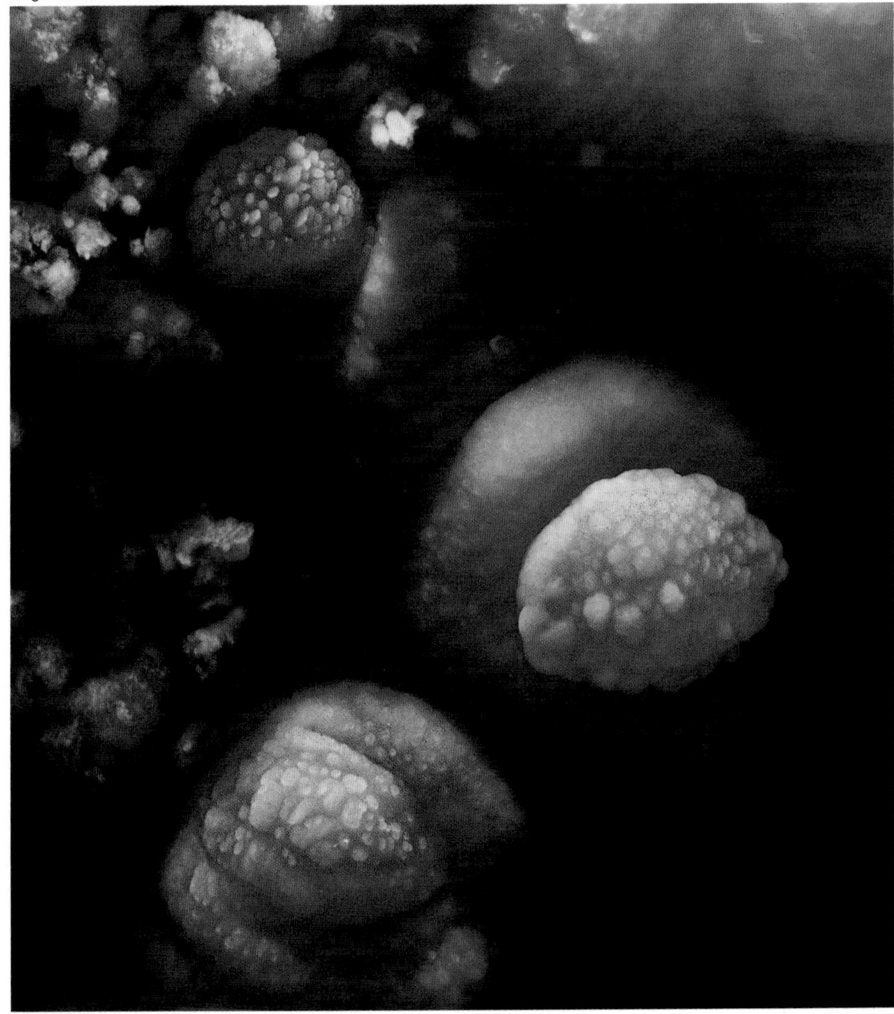

Although they work together when we eat, taste and smell are separate sensory systems. Some 10,000 taste buds, clustered in tiny bumps called papillae (above) and spread throughout the mouth, contain receptor cells that sense sweet, sour, salty, and bitter tastes in varying degrees and combinations. Stimulated receptors send electrical impulses along nerve fibers to the taste cortex in the brain.

The papillae shown here contain taste buds especially sensitive to bitterness. Bitter tastes register 10,000 times more strongly than sweet ones. For good reason: Many poisonous substances taste bitter.

For you to smell something, air containing scent molecules must reach two small patches of membrane called olfactory epithelia above and behind the bridge of your nose. A sniff or two sends scent-laden eddies of air up to both patches. Each olfactory epithelium, about half a square inch in area, contains some 10 million receptor neurons that send messages to two olfactory bulbs at the base of the brain, then on to the smell cortex.

Assaults on Our Ears

Very loud noise (and some illnesses) can damage hair cells, possibly destroying their ability to bend according to the frequencies that stimulate them. Broken hair cells are dead, gone forever, and with them the joy of acute hearing. Constant exposure to the clamor of a busy restaurant or the din of traffic—both rated at about 70 decibels—is generally safe. But a day behind a loud lawn mower or a couple of hours over a chain saw—90 to 100 decibels—can jeopardize hair cells. Stereo players worn with headphones can attack the ear with 100 decibels. And the acoustic suicide spot, 120 decibels, in front of the speakers at a rock concert, makes permanent hearing loss inevitable. These assaults are known to hearing researchers as "leisure noise."

Deterioration of hearing is not a necessary plague of middle and old age. Certain people, such as the Mabaan, who live in Sudan, have sensitive hearing even when they are old. Their world is a quiet one. But most of us, swamped by the cacophony of modern life, can look forward to losing some hair cells and nerve fibers.

"Shriveled and destitute of neurina." In these words the autopsy pathologist described the auditory nerves of Ludwig van Beethoven after the

The nose knows what the tongue cannot, as any wine taster will agree. The sense of smell is thousands of times more sensitive than taste, so that to detect the flavor of alcohol without your nose, you need 25,000 times the concentration. No wonder the wine expert's first impression comes from a sniff, and the final judgment from the scent that rises behind the palate into the nasal cavity.

Wine tasters can also rely on their noses to help distinguish among wines because the sense of smell is well attuned to subtlety. An odor may be made up of a single chemical or hundreds of different substances. Of the countless possible combinations, most of us can recognize a few thousand smells, and a few trained noses distinguish up to 10,000. Naming them, though, is a different matter altogether: The link between odor perception and language is weak.

Why are smells so evocative? Because the nerves that carry olfactory messages travel to those parts of the brain involved with cognition, emotion, and sexuality. Odors associated with events and people can trigger intense feelings and vivid memories for a lifetime—one whiff of a particular perfume, even if you have not smelled it for years, can generate sadness over a lost love. Perhaps the woman test-sniffing toilet-bowl cleaners (right) remembers her mother's rose garden in one chamber, a childhood walk through a pine forest in another.

composer's death in 1827 at the age of 57. Almost 30 years earlier, possibly after an illness, Beethoven had begun to notice "an infirmity in the *one sense,*" he lamented, ". . . which I once possessed in the highest perfection." The first signs of damage had appeared by 1801, when he realized that he could not hear the high frequencies even in performances of his own music. Gradually Beethoven descended into profound deafness. In later years he explained how he had been able to compose his opera *Fidelio,* his *Eroica* Symphony, and other masterpieces even as his hearing deteriorated. The music lived in his head: "It rises, grows upward, and I hear and see the picture as a whole take shape and stand forth." Ideas, said the deaf genius, "are roused by moods . . . transmuted . . . into tones, that sound, roar, and storm."

Even we ordinary mortals hear not in our ears but in our brains. We can summon up remembered melodies, even if we cannot compose them. People are fond *(Continued on page 302)*

The Neuron Pathway: on the Trail of a Touch

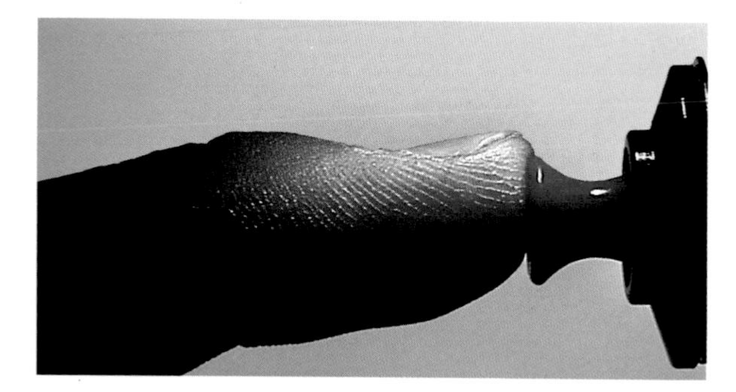

Push the button. Feel it? Of course. Your brain told you. But how did your brain find out? Keeping in touch with the world around you, however tenuously, is a complicated business involving millions of nerve cells, a mass of nerve-cell circuitry, and uncountable rapid-fire signals in the brain.

The illustrations on these pages trace the pathway of a single nerve impulse as it moves from fingertip to brain. Actually, pressing a button activates thousands of touch receptors thickly clustered in the fingertip.

Meissner's corpuscles (1), which encapsulate nerve endings near the skin surface, are among the first of several kinds of touch receptors to react. Pressure distorts them, triggering an electrical discharge in the corpuscle. The discharge pulses into a nerve fiber called the axon (2), and travels toward the spinal cord at up to 425 feet a second. A jellyroll-like layer called the myelin sheath (3), insulates sections of the axon. Each length of sheathing is built by a specialized cell, the Schwann cell. The unsheathed interstices between the myelinated segments, known as the nodes of Ranvier, serve as foci for the nerve impulse—it speeds through myelinated portions of the axon, jumping from one node to the next.

Once past the neuron cell body in the spinal ganglion, the impulse enters the spinal cord (4) between two cervical vertebrae. It then crosses to the opposite side at the brain stem (5) and, after a relay in the thalamus deep in the brain (6), races to its destination in the sensory cortex (7)—all in less than the blink of an eye.

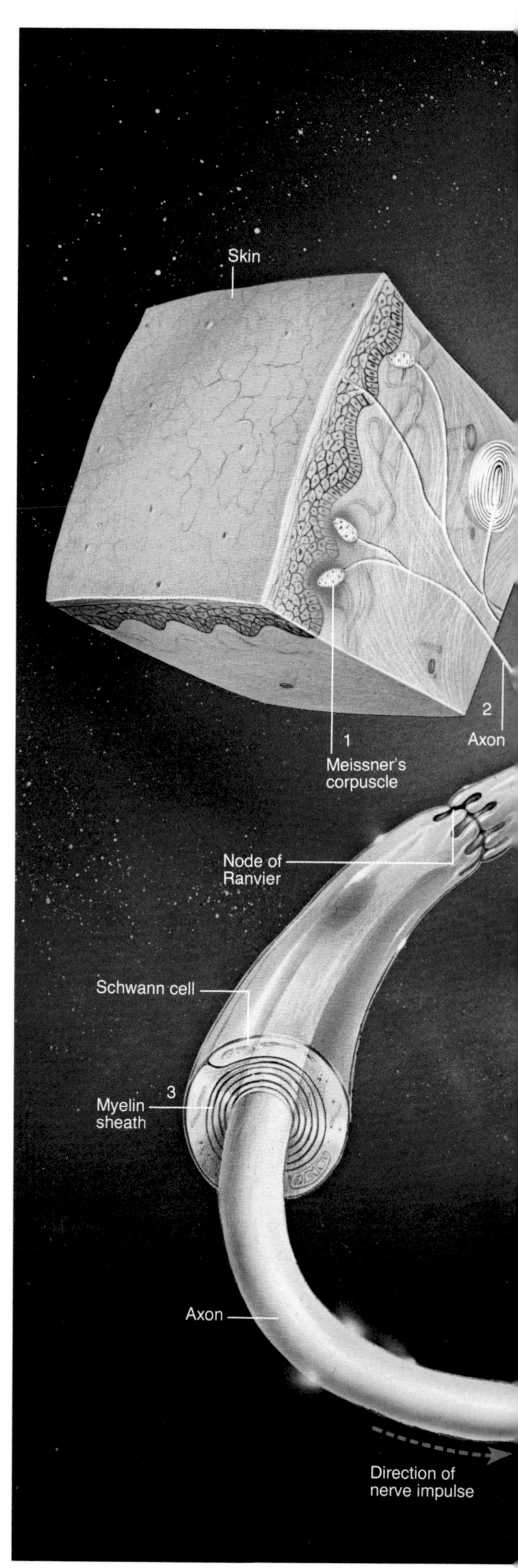

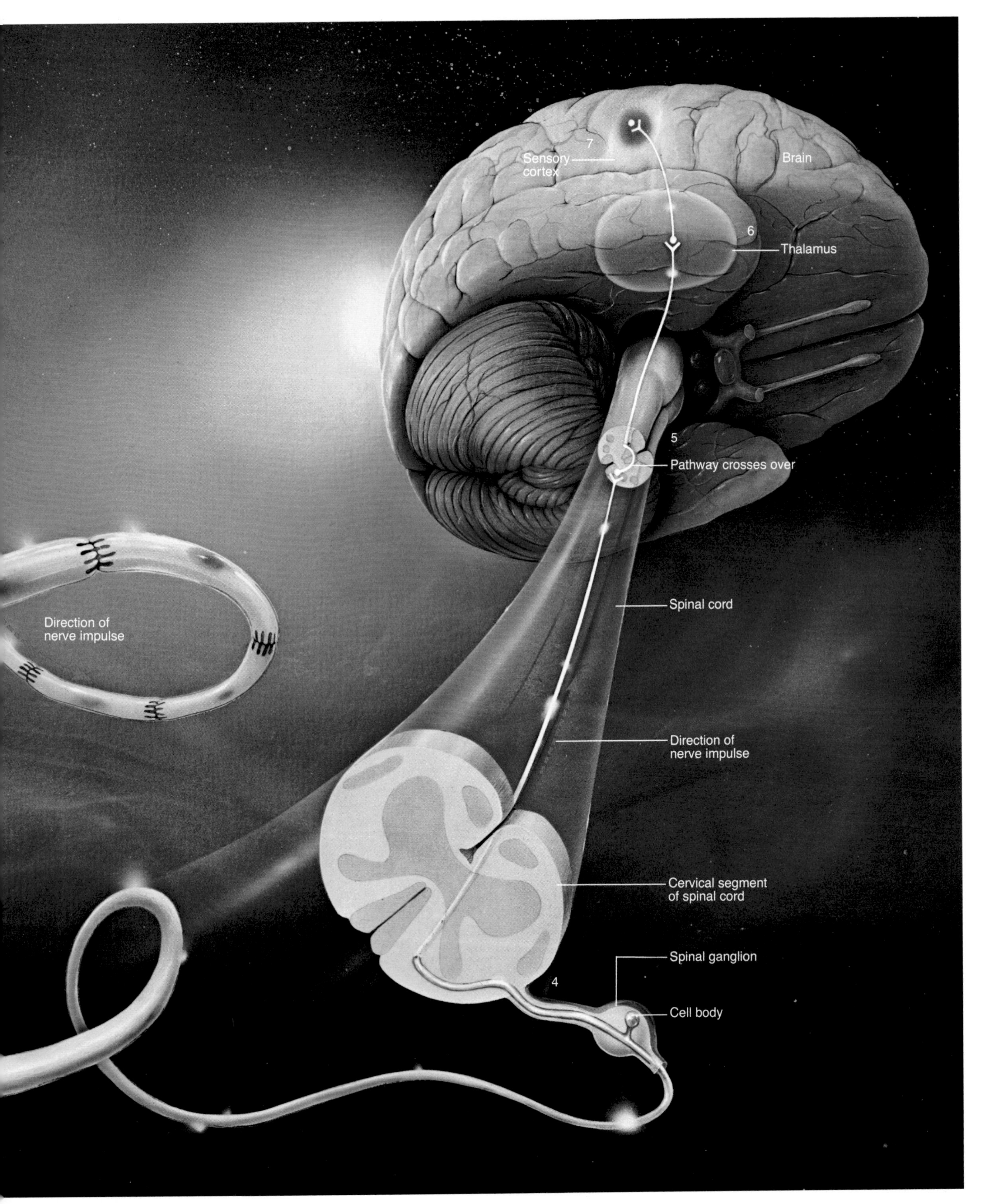
7
Sensory cortex
Brain
6
Thalamus
5
Pathway crosses over
Direction of nerve impulse
Spinal cord
Direction of nerve impulse
Cervical segment of spinal cord
Spinal ganglion
4
Cell body

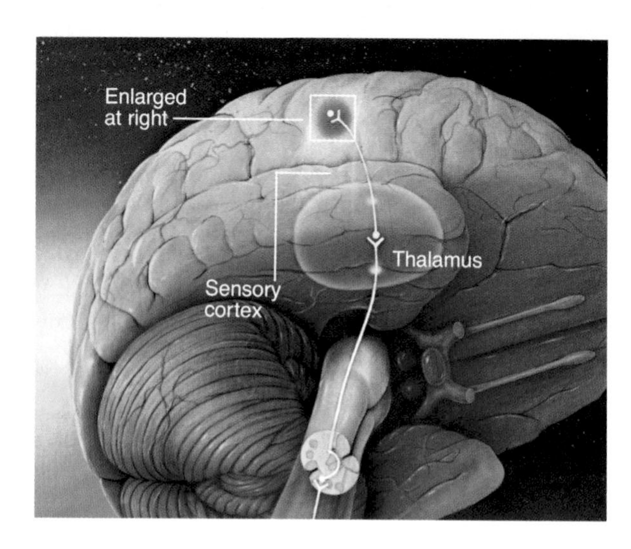

Once past the thalamus and into the sensory cortex (above and right), the electrical impulse diffuses through the area devoted to receiving and processing touch signals from the left forefinger—a small patch of cerebral cortex high in the brain's right hemisphere. For an instant this part of the brain's electrochemical circuitry becomes, as one scientist puts it, "a seething mass of patterns going off and on."

The pyramidal, stellate, and fusiform nerve cells, which number in the millions, receive incoming signals. In ways not fully understood, they process and sort the neuronal messages. Other cells, such as axoaxonic and basket cells, act as inhibitors to prevent the spread of impulses that might overload the sensory circuits. Still other cells, including Martinotti and horizontal cells, bring additional information to the brain's surface.

The constant interplay of impulses goes on until the touch of the button is recognized for what it is, and not, say, the painful prick of a pointed object.

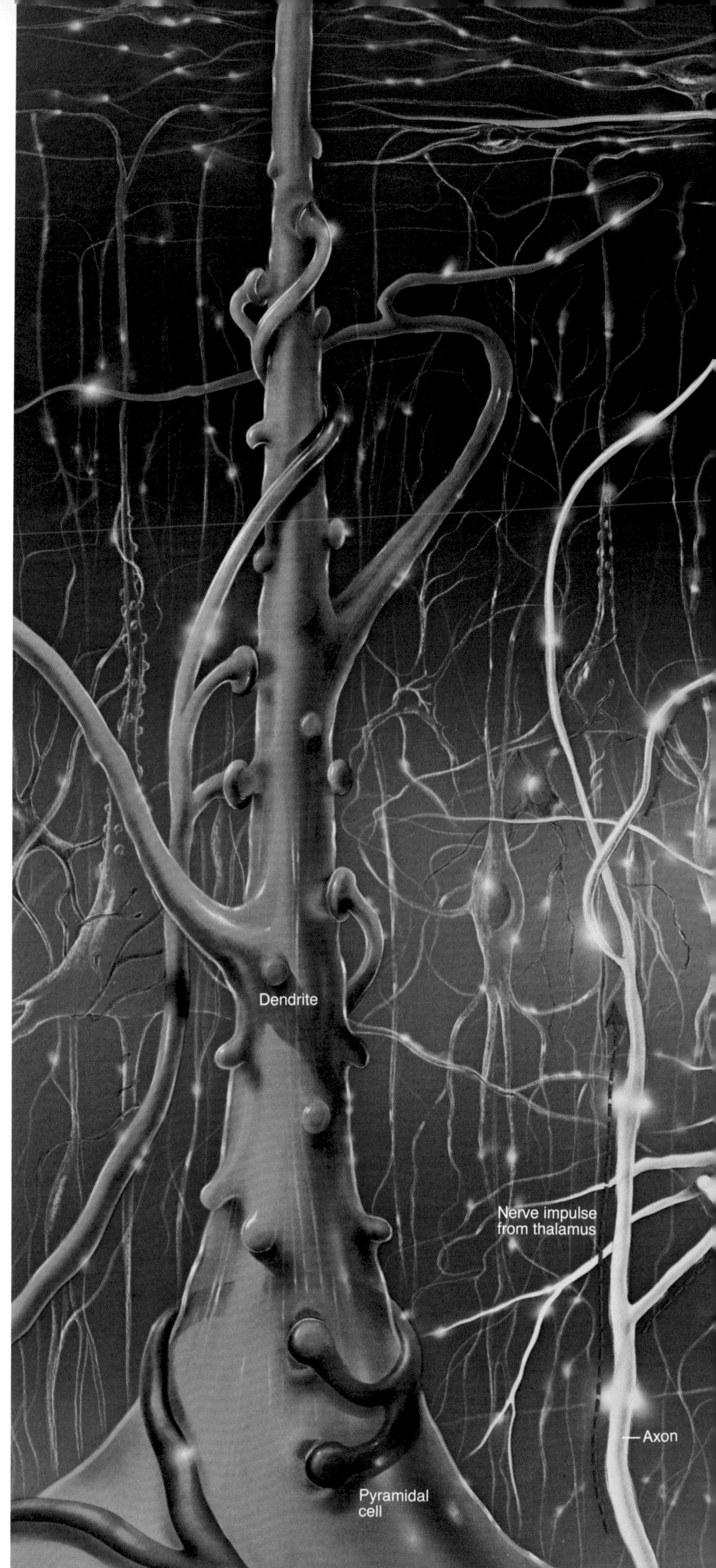

Horizontal
cell
Dendrite
Stellate
cell
For detail,
see page 301.
Axoaxonic
cell
Basket
cell
Pyramidal
cell
Fusiform cell
Pyramidal
cell
Dendrites
Axon of
pyramidal cell
Martinotti
cell

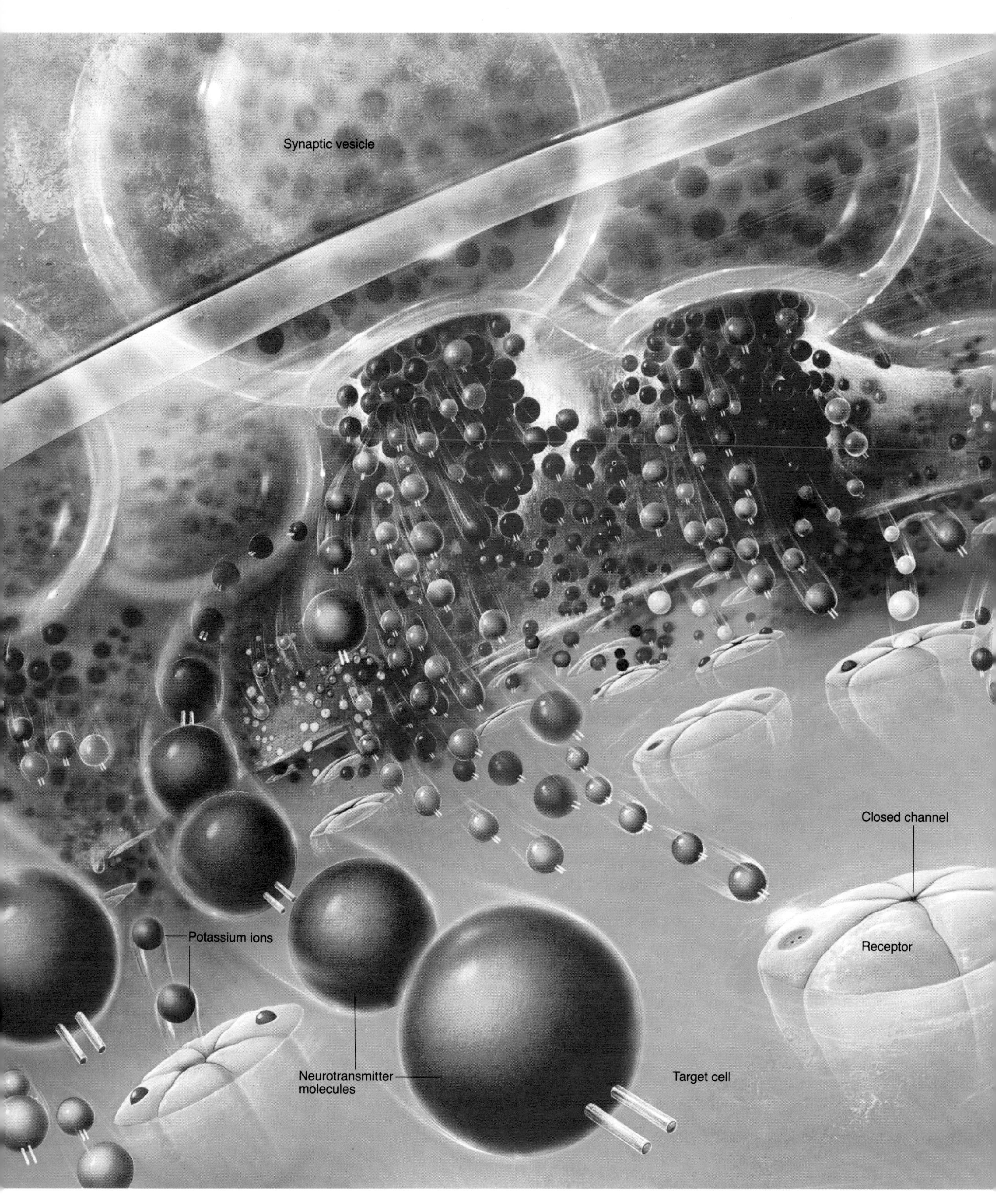
Synaptic vesicle
Closed channel
Potassium ions
Receptor
Neurotransmitter molecules
Target cell

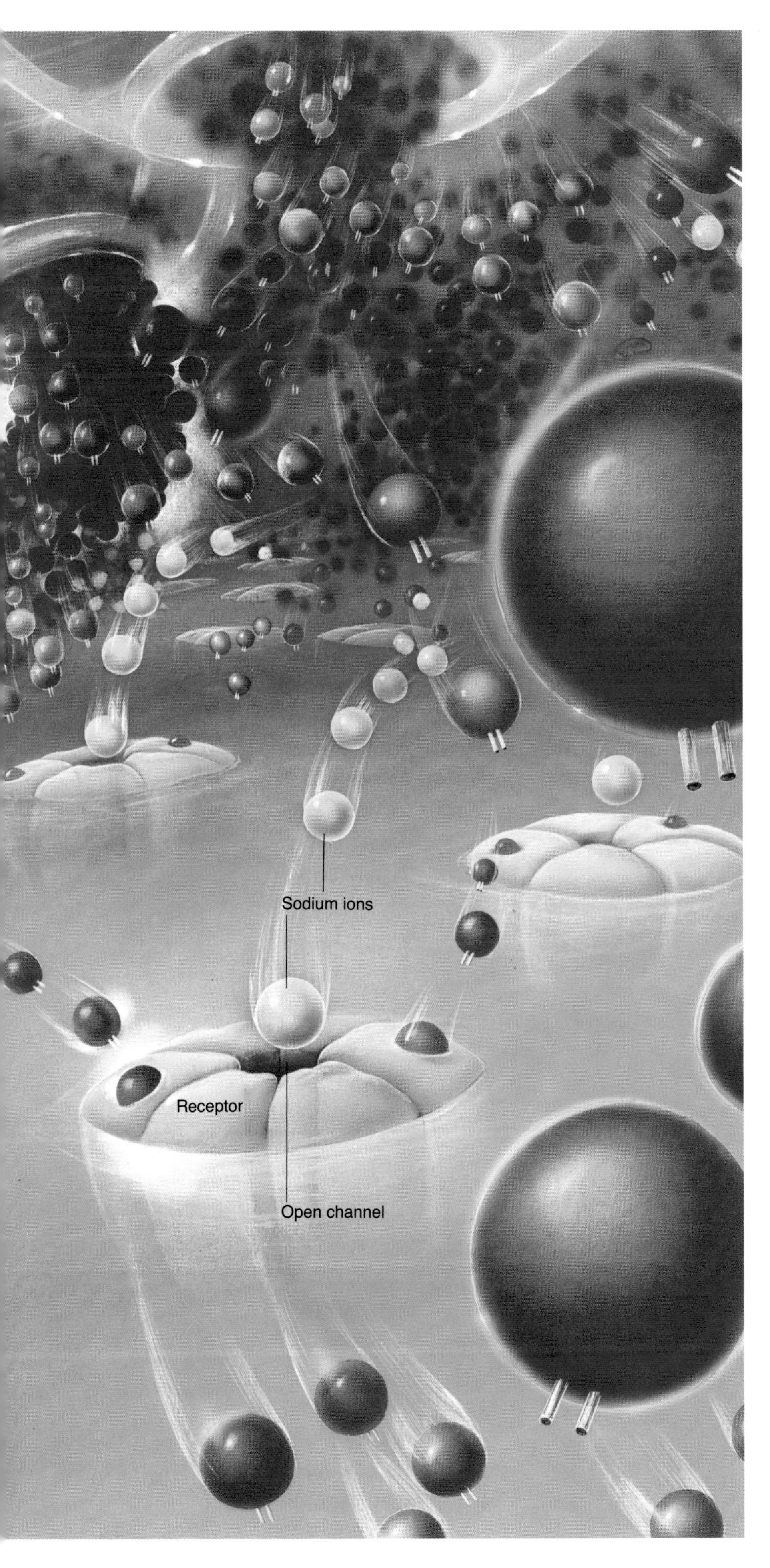

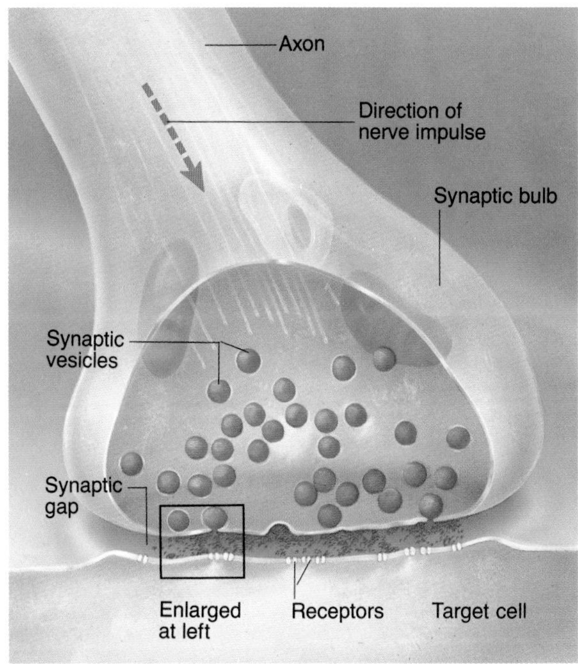

The synapse serves as a biological switch to transmit electrochemical messages. Each synapse is made up of two parts (above)—a knoblike synaptic bulb at the end of the transmitting neuron, and a receptor region on the membrane of the target cell. A synaptic gap, about a billionth of an inch wide, separates the membranes of the two cells.

The synaptic bulb contains tiny globules called synaptic vesicles, each holding thousands of neurotransmitter molecules.

When the nerve impulse reaches the synaptic bulb, the vesicles fuse with the membrane, spilling their contents into the synaptic gap (left). The neurotransmitters bind with receptors on the target cell; this opens certain receptor channels that allow sodium ions to rush into the target cell, and potassium ions to leave. The flow of ions excites an area of the target-cell membrane and generates an electrical impulse in the target cell.

of asking: If a tree falls in a forest, and no one is nearby, is there any sound? Waves there are. But to *hear* them, a brain must identify the mixture of frequencies, amplitude, and quality as the phenomenon that our combined senses have taught us is the crash of a tree falling . . . over there.

A pair of ears, one on each side of the head, means that our hearing apparatus is stereo. We locate the source of a sound by the minute time difference between the arrival of sound waves at one ear and the other. If they hit the right ear first, we may react by turning in that direction so as to improve or verify the stimulus. If the sound aims directly front or back, again we will turn, seeking its origin with our ear.

The brain—specifically the reticular activating system—also filters sounds. Imagine that you are walking through a crowded shopping mall. Your eyes indiscriminately pick up almost everything in their field of vision. But your ear chooses: music from a doorway; snatches of the interesting conversation just behind you; then a friend calls your name, and your auditory receptors instantly fix on the direction from which the call came and the identity of the voice. (This ability has been labeled, after arduous investigation by dedicated researchers, "the cocktail party phenomenon.") When we sleep, the lower brain screens out noises that are part of our normal environment—street traffic, for instance—but the reticular activating system will rouse us if the auditory nerves pick up and relay new sounds that may spell danger or emergency.

Some sounds travel inside the head by vibrations of the skull and not because sound waves have entered the outer ear. When you crunch on toast, you hear it through bone conduction. If you stop up your ears with your fingers, you can hear your jaw muscles expand and contract; click your tongue—the tick-tick-tick reaches your hair cells by the internal route. Imagine the tom-tom that can beat in the head, masking sounds of the world, if something goes wrong with the vessels pumping blood by the organ of Corti. Who of us does not recall the dismay at hearing our own voice on a tape recording? (You thought it was grander than that.) What we didn't realize is that we hear ourselves both through the air and through the resonator of the skull.

The Cosmic Sense

Our sense of which way is up reminds us that we are cosmic creatures. In the inner ear are the organs of balance that tie us to the same force that holds the planets in orbit. Gravity pulls us toward the center of Earth. We resist by trying to hold ourselves erect and in a state of equilibrium. The vestibular system does the work.

It is a proprioceptive system consisting of parts that keep us oriented when we are moving

A man may see how this
world goes,
with no eyes.
Look with thine ears.

As Shakespeare suggests, we can identify many things by the sounds they make: We can recognize people; we can tell the distance, size, structure, or the direction of movement of objects. We usually verify what we hear with our eyes, but the blind or visually impaired can learn to rely on hearing, in fact on all of their available senses, to gather environmental data that keep them oriented and mobile.

Walking down the street is Johnny Allen, at 14 rendered almost completely sightless by a disease called retinitis pigmentosa. Johnny receives a wealth of information from his other senses. The sweep of his cane tells his arm and his brain that the next step is clear. The sound of his cane and his shoes, and the pressure on the soles of his feet tell him whether he walks on concrete or grass. A drop in temperature—and the smell of garbage cans—may mean he is passing an alley.

In school Johnny reads Braille with his fingertips. After school he practices judo, where he uses all of his senses together, including highly developed kinesthesia—the same body awareness that allows him to know where his foot is or how far to duck after touching the car roof with his hand. Johnny's kinesthetic awareness provides such excellent balance and coordination that he won a state championship . . . in competition with sighted opponents his own age.

A gust of wind, a swift temperature change on the top of your head, the sight of your hat somersaulting skyward, and your hot pursuit—all occur in a split second. Remarkably fast, considering that your scalp and your eyes must transmit sensory messages to your brain, which translates them and routes them through several information-processing areas before you know what has happened and how to act.

The parts of your brain that feel your head suddenly bare, note your hat's location, and send you springing after it with outstretched arm are located in two parallel arches across the cerebral cortex, the outer layer of the brain. The sensory cortex receives signals from the skin, bones, joints, and muscles everywhere in the body; the motor cortex activates the muscles.

The illustrations below indicate how much of each arch is devoted to each area of the body—the size of a section corresponds not to the size of the body part but to the precision with which it must be controlled for various tasks. The lips do a great deal of moving, and less sensing, so they take up more space in the motor cortex than in the sensory cortex.

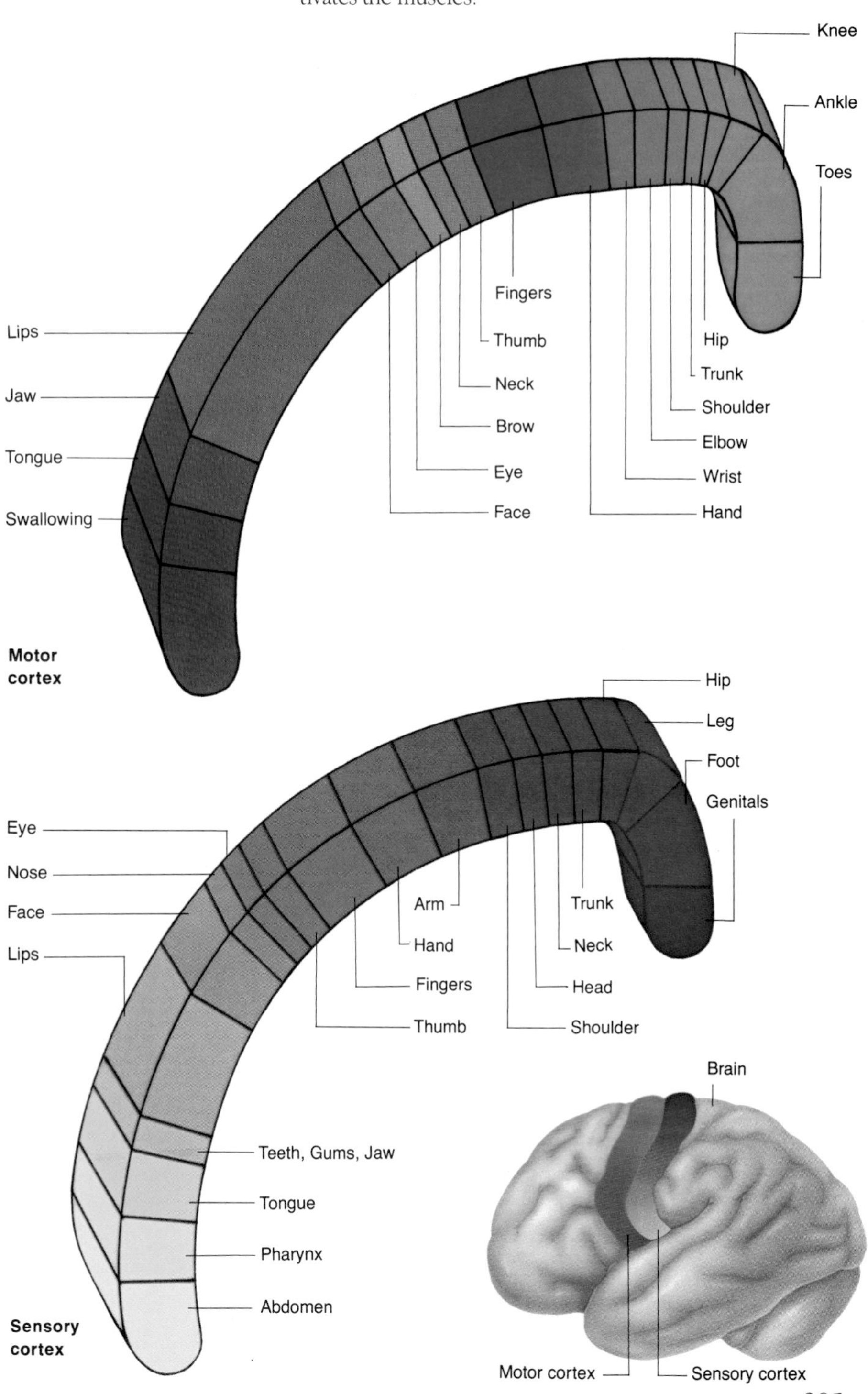

The skeletal muscles that keep us moving—and that make minute adjustments to maintain our posture even as we sit perfectly still—could not do their work without the nervous system. Like a maestro, the brain directs the muscles in a symphony of contractions and relaxations, in myriad sequences and combinations, so that we glance at an attractive stranger, turn a cartwheel, write a letter.

For every movement, the central nervous system sends a barrage of orders as electrical impulses along nerve fibers (tinted white; right, upper) to the appropriate muscle fibers (tinted brown) that, bunched together, form muscles. Each nerve fiber branches out and terminates in a motor end plate (right, lower), where the electrical message is transferred to the muscle fiber by release of a chemical. This triggers a contraction. When the impulses stop, the muscle fibers relax.

and other parts that keep us oriented while we are at rest. Because of these receptors, you can walk through a pitch-dark room and, with a little help from pressure cells in the soles of your feet, make a decently steady job of it.

Part of the cochlear duct are three fluid-filled, bony loops called the semicircular canals. The canals are perpendicular to each other and parallel, respectively, to the side of the head, the plane of the face, and the horizon. In the ampulla, the bulbous end of each canal, are hair cells. Moving the head sets the fluid in motion, and pressure waves bend the hair cells. If you shake your head "no," parallel to the horizon, fluid in the lateral canal will begin to swirl. If you nod or bend your head sideways, the other canals will respond. The head movement sets the fluid in motion, the way a suddenly accelerating car will jerk passengers against the seat. So there is a tiny lag between bending over to tie your shoe and your perception of bending over to tie your shoe. When the same lag occurs after you've been twirling around, you say when you stop that you're dizzy. That's because the fluid in the semicircular canals continues to swirl and fools the brain into the belief that you're still spinning.

Between the cochlea and the semicircular canals are two other organs of balance—sacs known as the saccule and the utricle. They also contain fluid and hair cells and a unique gelatinous layer enclosing the tips of the hair cells. On this membrane are minute calcium carbonate crystals—otoliths or "ear stones"—that lean like little magnets to the pull of gravity. They keep your head on straight. When you stand up straight, the nerve fibers of the hair cells stand erect too, monitoring, quietly sending impulses to the brain that all is stable. When you bend over or stumble, the otoliths—drawn to the hair cells closest to the ground—deflect the nerve fibers. By the angle, the brain knows that the head has been lowered (indeed that the brain itself may be in danger). Impulses from all the organs of balance cooperate with proprioceptors in the muscles and tendons to quickly help you right yourself.

If you've ever watched someone else whirling, you know that the eyes are doing funny things, snapping to the right, then to the left. As the twirler goes around, the eyes try to fix on a stable spot. The body and head whirl but the eyes seem to want to stay behind or jump ahead. (Stop the world. . . ?) What keeps a pirouetting ballerina from collapsing like a spent top? She whips her head faster than her body. She has taught herself to suppress the messages of the whirling fluids in the semicircular canals by keeping control over the gaze of her eyes.

Every day your eyes take a fifty-mile hike—or the equivalent—in their muscular workout. Eyes move continually. The muscles around the

Magnification: 1,700 times

Magnification: 1,000 times

Lennart Nilsson, © Boehringer Ingelheim International GmbH.

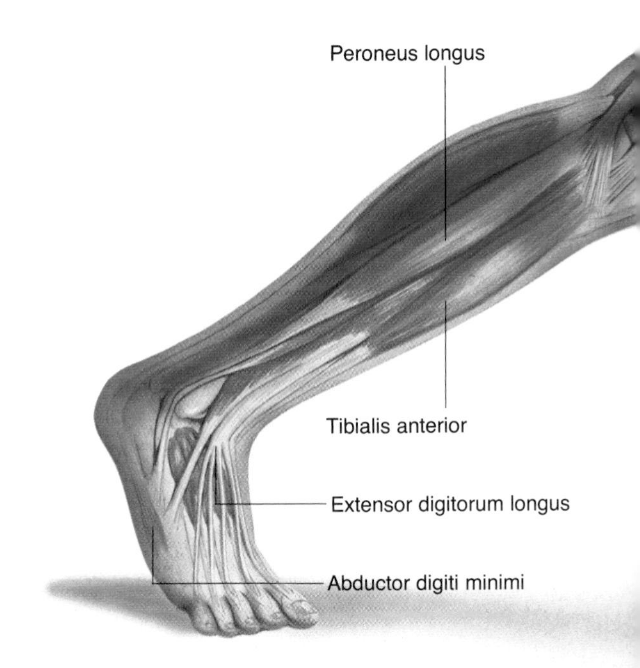

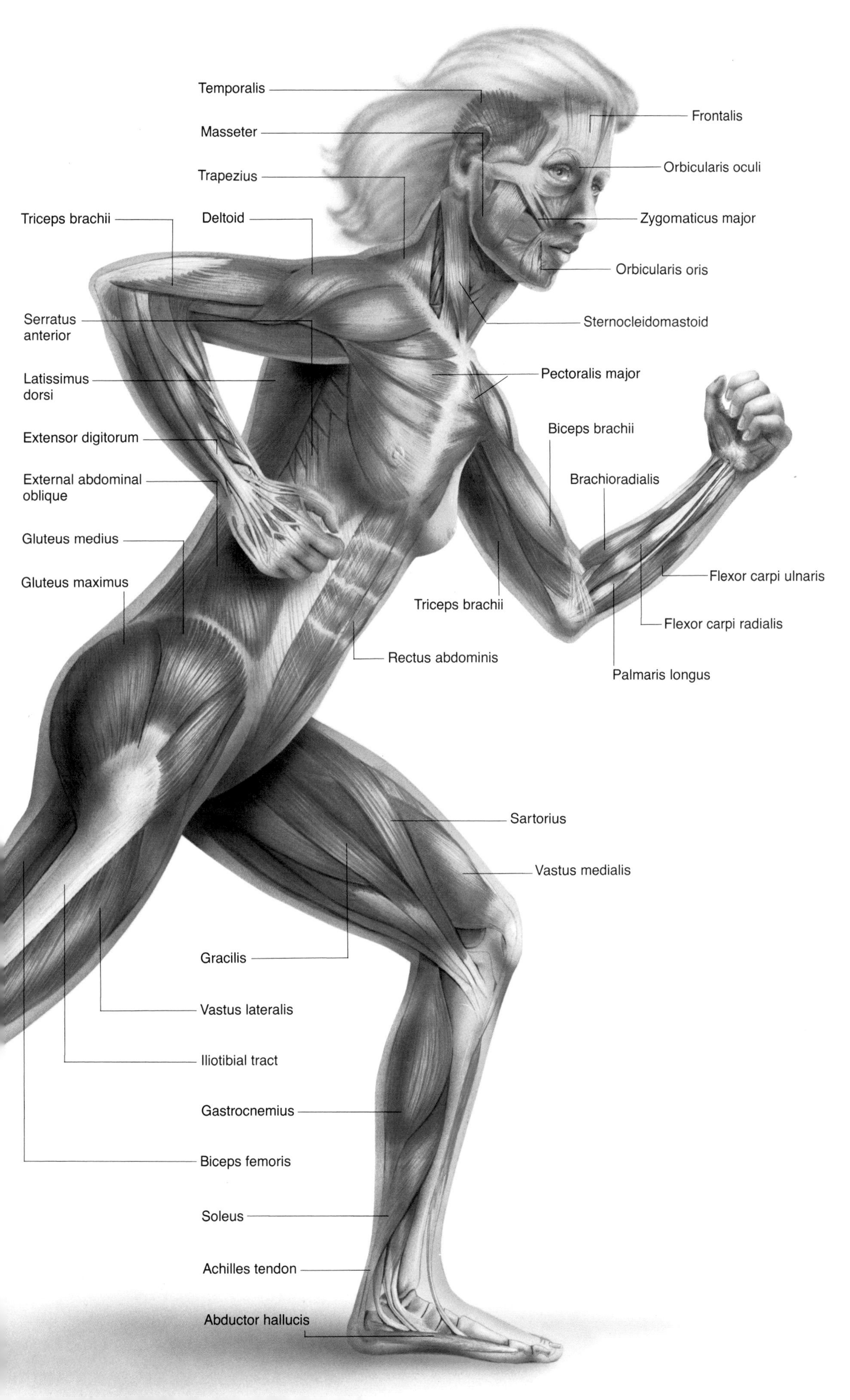

Some 650 muscles sheath the skeleton, creating our contours and, by their contractions, coordinating our every motion. Fastened to the bones by means of tough tissue called tendon, they pull our bones into thousands of poses.

Nearly always, skeletal muscles work in groups in response to instructions from the brain. It takes about 200 muscles to achieve one walking step. Forty or so lift your leg and move it forward; muscles in your back immediately compensate, pulling on your shoulders and chest to keep you from losing your balance with the forward momentum. This accomplished, abdominal muscles contract to keep you from falling backward. All of these forces operate each time you walk; you can see them exaggerated in anyone striving to remain upright on an icy sidewalk.

lenses contract or expand a hundred thousand times a day as they focus on objects as close as the end of your nose and as far away as the stars. You blink every two to ten seconds. As you focus on each of these words, your eyes swing back and forth a hundred times a second, and, every second, the retina performs ten billion computer-like calculations.

As you watch a butterfly dance, children play, traffic going by in the street, there are no little movies playing in the brain. If there were, we would need little eyes in there to see them. And perhaps little chairs to sit on and little bags of popcorn. Such a scene would be no stranger than the things that do go on to make us see. There *are* little images in your head, projected in code on a postage-stamp screen on the retina. They are upside down, reversed left to right, and last half a second. As quickly as the images hit the "screen," millions of light-sensitive receptors convert them to electric impulses in the nerve fibers. As often as a thousand times a second, sodium ions and potassium ions exchange places to inform the brain about size, distance, patterns, color. Without light, none of this would happen.

Images from Light

Everything we see is light—packets of energy called quanta that move in waves as a small part of the band of electromagnetic radiation that reaches Earth from the stars and galaxies. As few as ten quanta, roughly the glow of a candle flame at ten miles, will stimulate the human eye. When you go out at night, if there is a bright moon, you may be able to read the largest newspaper headlines, but it will be difficult because there may not be enough quanta striking the retina to create a clear image of the words in the tenth of a second that your eye needs for the task. Should you care to read the paper in the moonlight after a fresh snowfall, it will be easier of course.

The eyeball surveys the world from a bony socket in the skull. Fat cushions it. Six muscles hold it in a sling that rotates in whatever direction we wish to look. The outer layer, the white of the eye, is the sclera, a tough, opaque film of connective tissue. At the front, transparent tissue forms the cornea, which covers the iris, the colored part of the eye. (Anyone who is vain about their big baby blues might consider that the blueness means only that there is less pigment than in darker eyes.) The iris has circular and radial muscles by which it expands and contracts the pupil, the dark hole at its center. Bright light and near vision constrict the pupil; for far vision and in dim light the pupil expands, letting more quanta in. Emotions affect the pupil. Sometimes an unpleasant scene can make it contract—to shut out the pain? The pupils can dilate by 30 percent at the sight of a sexually interesting

Bundles within bundles of fibers form our skeletal muscles. Within the biceps brachii, severed here from a tendon by the artist, are muscle fibers grouped into cylindrical packages, or fasciculi, of varying sizes and configurations. Electrical impulses from the brain and spinal cord travel to the muscle fibers via motor nerves, then transfer to the fibers via motor end plates. From the end plate on the surface of a muscle fiber, the message is conducted inward to a network of sacs called the sarcoplasmic reticulum, and then to the myofibrils. These are composed of many threadlike fibers called myofilaments, which are made mainly of two proteins—myosin and actin. Within repeating sections, or sarcomeres, of the myofibrils, the work of the muscle takes place. The myosin has connecting structures called cross bridges which attach to the actin filaments and then move in unison like oars in a racing shell. This cross-bridge movement pulls the thin filaments forward, leading to a shortening of the sarcomere. As this process is repeated many times, the entire muscle shortens.

One motor nerve can stimulate simultaneous contractions of numerous muscle fibers within a fasciculus, but fasciculi within a muscle can work independently. Generally, the fewer muscle fibers innervated by a single motor neuron, the subtler and more finely graded are the movements they can produce.

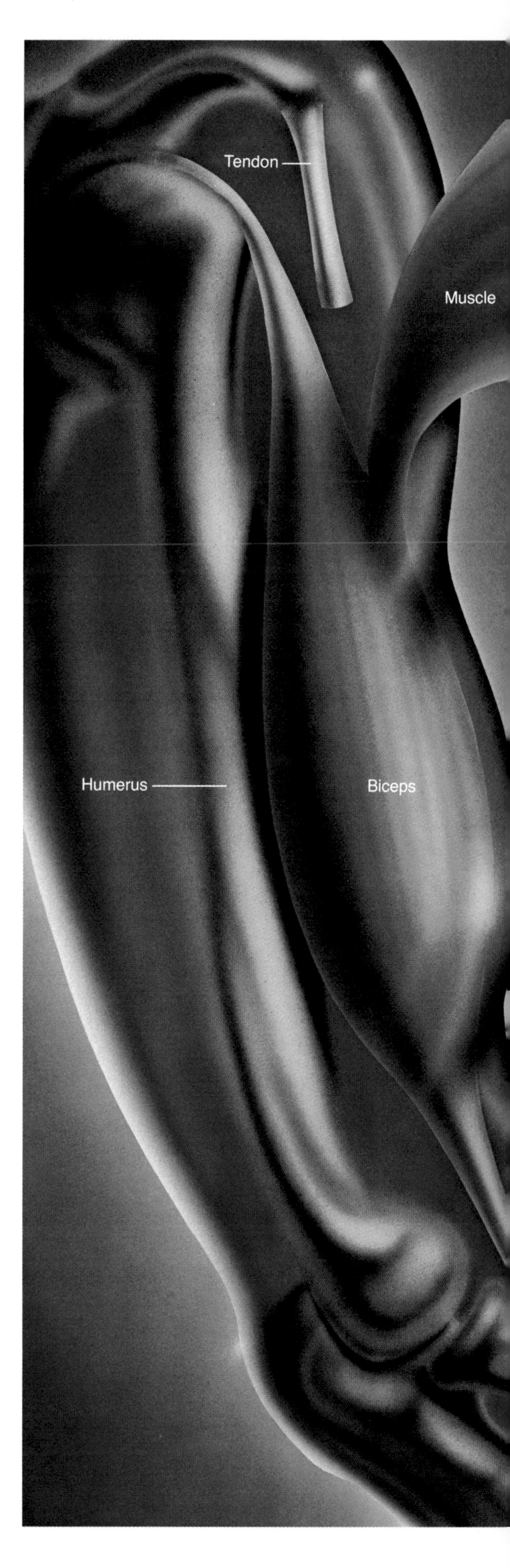

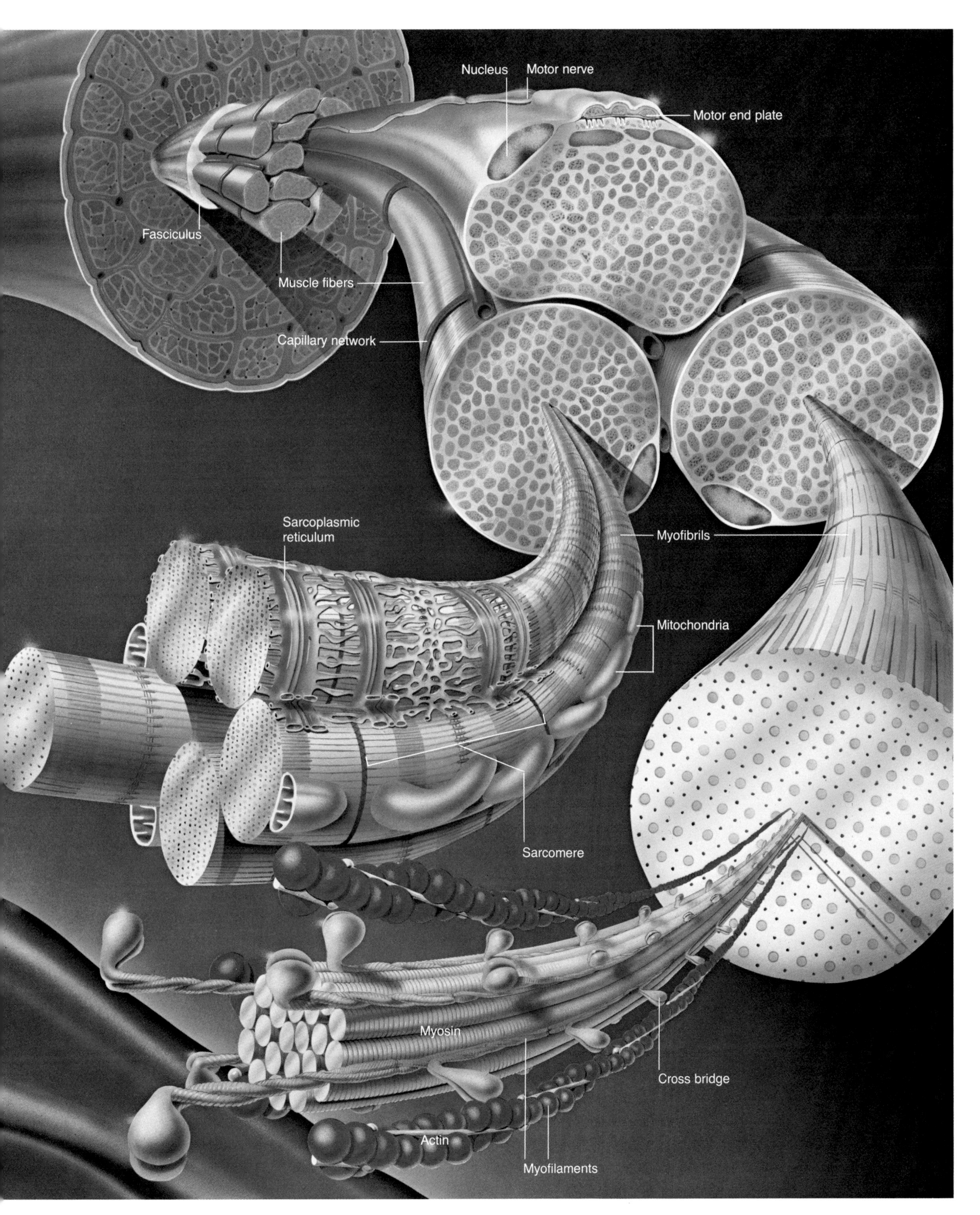
Nucleus
Motor nerve
Motor end plate
Fasciculus
Muscle fibers
Capillary network
Sarcoplasmic reticulum
Myofibrils
Mitochondria
Sarcomere
Myosin
Cross bridge
Actin
Myofilaments

HAWAII
83
'84 SUMMER
Los Angeles

member of the opposite sex, an occasion upon which, it's safe to say, the chemistry is right. The pupil also widens when we sense danger. Possibly there's some connection there.

When light enters the eye, it passes through the parts that control it to achieve sharp images: the cornea; the anterior chamber filled with a liquid quaintly known as the aqueous humor; the lens (which lies behind the iris); a second humor, called vitreous, or glassy, which fills the eyeball; to the retina, the back of the eyeball. The retina converts light to nerve impulses that travel, by the optic nerve at the back of the eye, to the brain. Both the cornea and the lens receive nourishment from the aqueous humor. Like the organ of Corti, they could not do their job if blood vessels got in the way.

Bending Light into Images

Light travels in straight lines at 186,000 miles a second until it hits something. The rays may then be reflected from whatever they hit—the house next door, a face, an airplane at 20,000 feet. Or they may penetrate the new medium—into water or glass or cornea. Then they are slowed and bent. This refracts them. It is the slowing and bending that makes images.

The convex cornea and the biconvex lens bend light rays to a focus. Light rays reflected from close objects, such as the words on this page, diverge, and the cornea needs more help from the lens to get a sharp image. Unlike the lens of a camera, which moves back and forth to focus, the lens of the eye changes shape, flattening to focus on distant sources, thickening and rounding to focus on nearer ones. It is the close work that becomes difficult as we age, since the lens, like too many other parts of the body, loses moisture and elasticity.

Like all organisms, human beings are sensitive to light. This is not a passive thing. It is part of the cycle of nature. In order to maintain our light sensitivity we absorb light. That's why you need to eat your carrots. Foods containing vitamin A enable the eye pigment to absorb light by replenishing the supply of rhodopsin and other light-sensitive chemicals in the retina.

Sunlight is white light, a combination of all the colors of the visible spectrum. When sunlight flows through a prism it breaks into all of its parts—red, orange, yellow, green, blue, violet. Each part has a different wavelength. A thing has color because it absorbs some wavelengths of light and reflects others. Things that look white are reflecting all wavelengths; things that appear black are absorbing them all. This is the same as saying that an apple is not really red, it just looks as if it is. You can experiment with this puzzle by placing an apple under a variety of natural and artificial lights. *(Continued on page 316)*

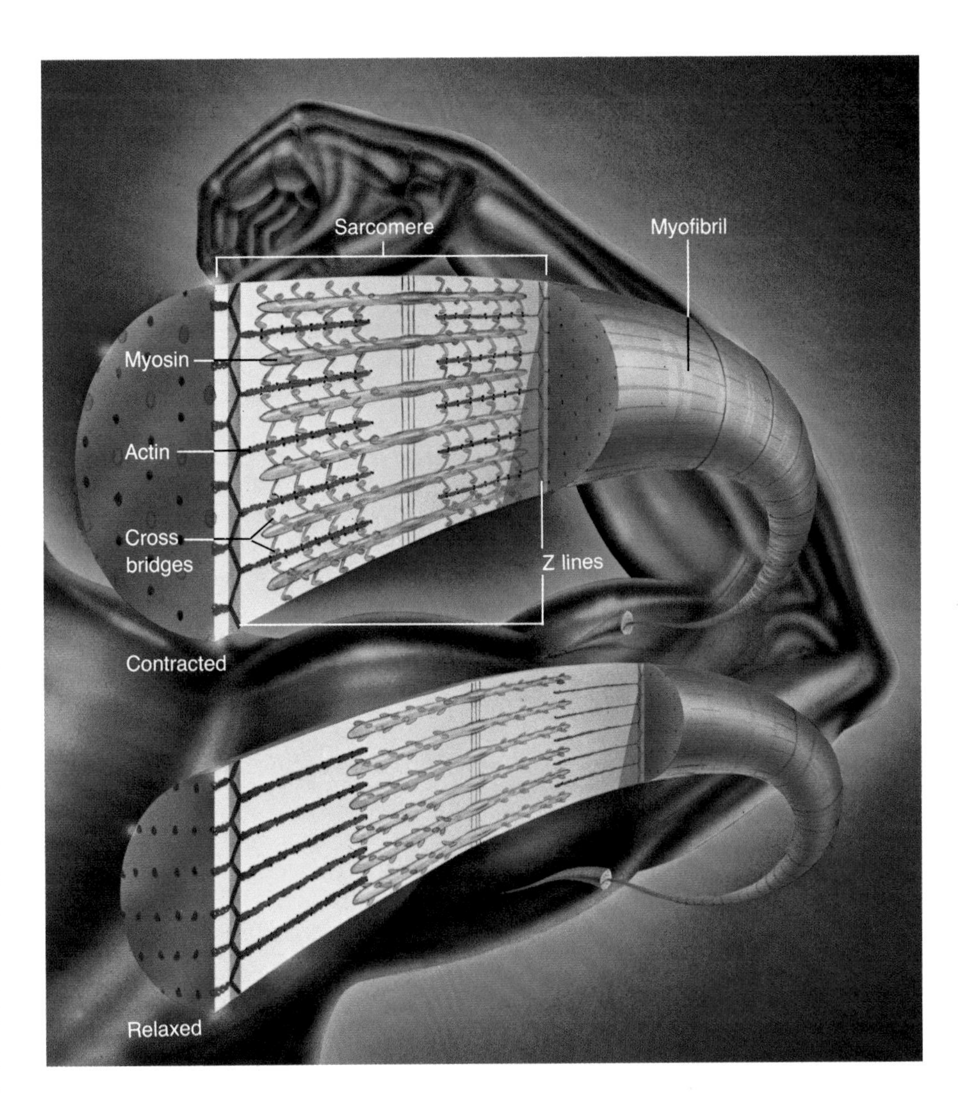

Actin and myosin—strands of protein within the muscle's myofibrils—are the mechanisms for motion. They overlap within segments called sarcomeres, which are divided by boundaries called Z lines.

When the brain signals "contraction," cross bridges on the myosin strands attach to the actin strands and pull them toward the center of a sarcomere. Actin strands are attached to Z lines, so the pulling draws the Z lines together to shorten the sarcomeres and contract the muscle. Lacking an impulse, the cross bridges let go (as they do in the triceps brachii when the biceps brachii contracts), the sarcomere lengthens, and the muscle relaxes.

Exercise stimulates the production of actin and myosin and also adds undifferentiated satellite cells to existing muscle fibers. This causes the muscle fibers to thicken, and produces the enlarged muscles of a weight lifter (opposite).

Monitoring the Body's Every Move

Most of us are coordinated enough to get through the day, but athletes elevate coordinated movement to an art form by relentless practice and with machines that provide feedback designed to help them move more expertly and efficiently. Computer images help a rower modify his technique (right); analysis of exhaled gases leads to more efficient aerobic efforts (below).

The rest of us have no machines to help us, yet we can walk, jump over puddles, put food into our mouths without spilling too much. This is remarkable, considering what is involved in every movement we make. Only after a deluge of sensory information has been received and analyzed by the central nervous system can even the simplest movement occur.

To take just one step, we first need to know the positions and spatial relationships of the parts of our body. Data are provided by the eyes, and receptors

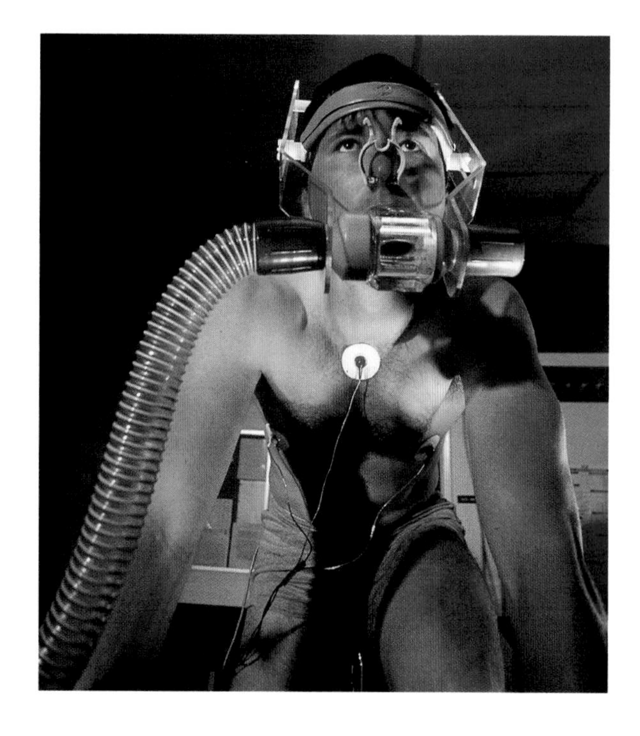

NIKE

in the skin, muscles, joints, and vestibular organs—the balance mechanisms in the inner ears. The central nervous system processes these messages and sends orders for movement, all the while monitoring new feedback. All of this information travels continuously, and unimaginably fast, as we plod along.

Despite this complex system, many of our movements seem automatic. Scientists theorize that the cerebral cortex develops mental models of learned patterns of motion. The cerebellum compares these models to what the muscles and joints are actually doing, and corrects discrepancies. Practice makes this process a habit, so that the intense concentration it takes at first for a gymnast to perfect her handspring, a high jumper his approach and takeoff, a swimmer his start—or a toddler his walk—becomes unnecessary.

Try violet light if you like purple apples.

If you say that something is green—a palm leaf, for instance—a member of the Jalé tribe of New Guinea will disagree, even though his eyes are the same as yours. He will describe the leaf only in degrees of light or dark—whiteness or blackness. His language has no specific word for the color green. On the other hand, the Maori of New Zealand have scores of words for red, and use color words to describe stages of plant growth and cloud formations.

The failure to label colors as we've been taught to do in our linguistic shorthand has nothing to do with so-called color blindness, which means simply that the retina is deficient in certain color receptors. It's not that the "colorblind" person can't *see* the color, because there isn't any color to see. There is no red to the rose, no yellow to the bumblebee, no green to the bean. It's all in your head.

Cells that See Color and Shadow

Vision depends on two kinds of light-sensitive cells in the retina, named for their shapes, along with their specialized support cells. There are about 125 million rod cells that give vision in dim light—they produce shades of gray—and about 7 million cone cells that give vision in bright light. The cones produce the sensitivity to the various wavelengths of white light that we call color vision.

Suppose you've gone camping overnight. You open your eyes as darkness gives way to dawn. Everything looks gray as the first light strikes your rod pigment, rhodopsin. The pigment bleaches as it undergoes the chemical changes that signal your brain of the looming outlines of trees and tents. Simultaneously, the bleached molecules are regenerating. If you get up and wander around, the dim-light sensitivity of the rods will enable you to see where you're going. When the morning brightens, the rods retire, and support cells alert the cones to take over.

There are three types of cones, each receptive to wavelengths in the red, green, or blue portions of the spectrum. The three work in combination to create the ever changing moods and hues of nature. Cones, too, have light-sensitive pigments that also undergo a bleaching and regeneration process, but they do it a little faster. That's why, when you enter a dark theater from the bright outdoors, your dark vision comes slowly; and when you go back on the street, your eyes adapt to the sunlight a little more quickly.

A map of the retina would show rods mostly around the sides, with some cones. This is the area that gives you peripheral vision, a primitive capability by which, in days gone by, you might have spied the fleeing deer that would be dinner for the family back in the cave. Near the center of

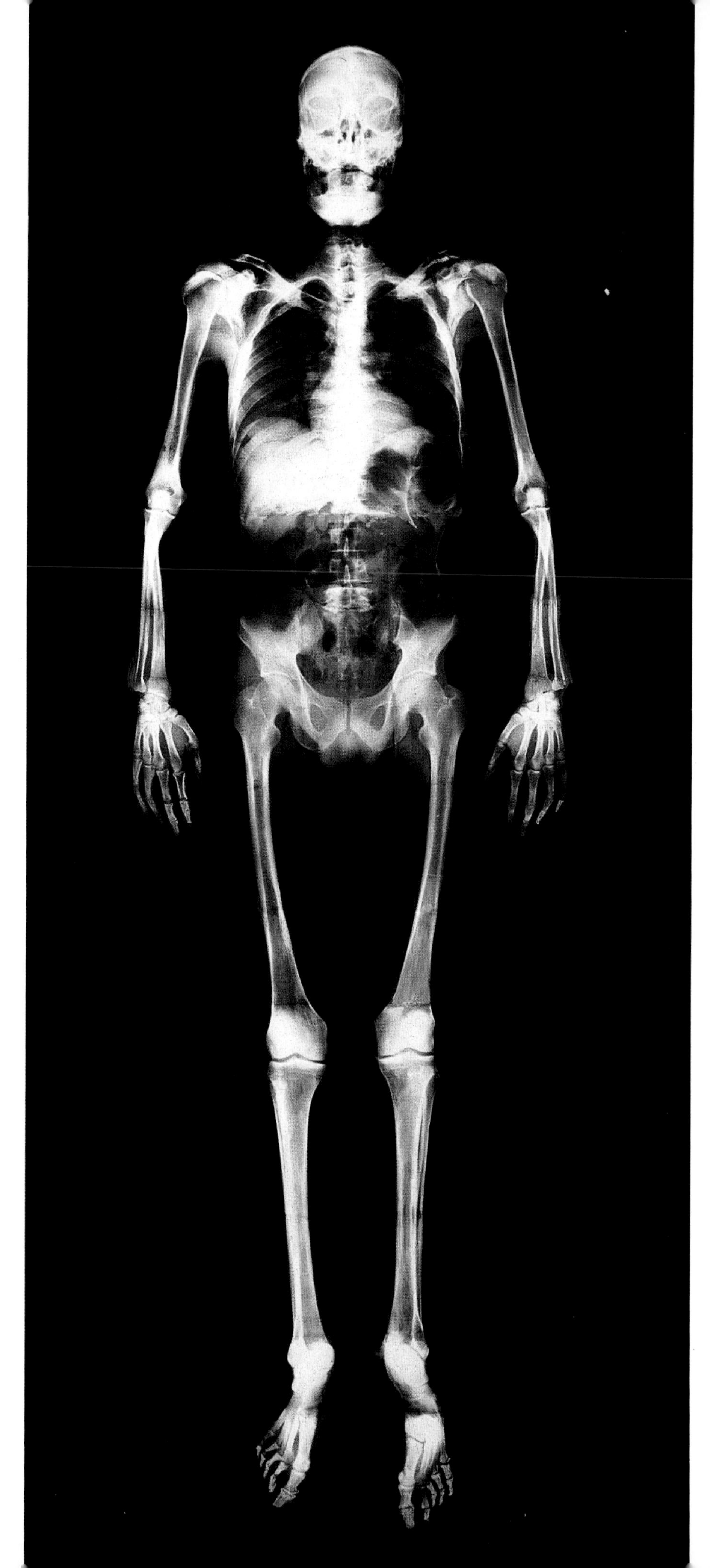

Magnification: 2.5 times

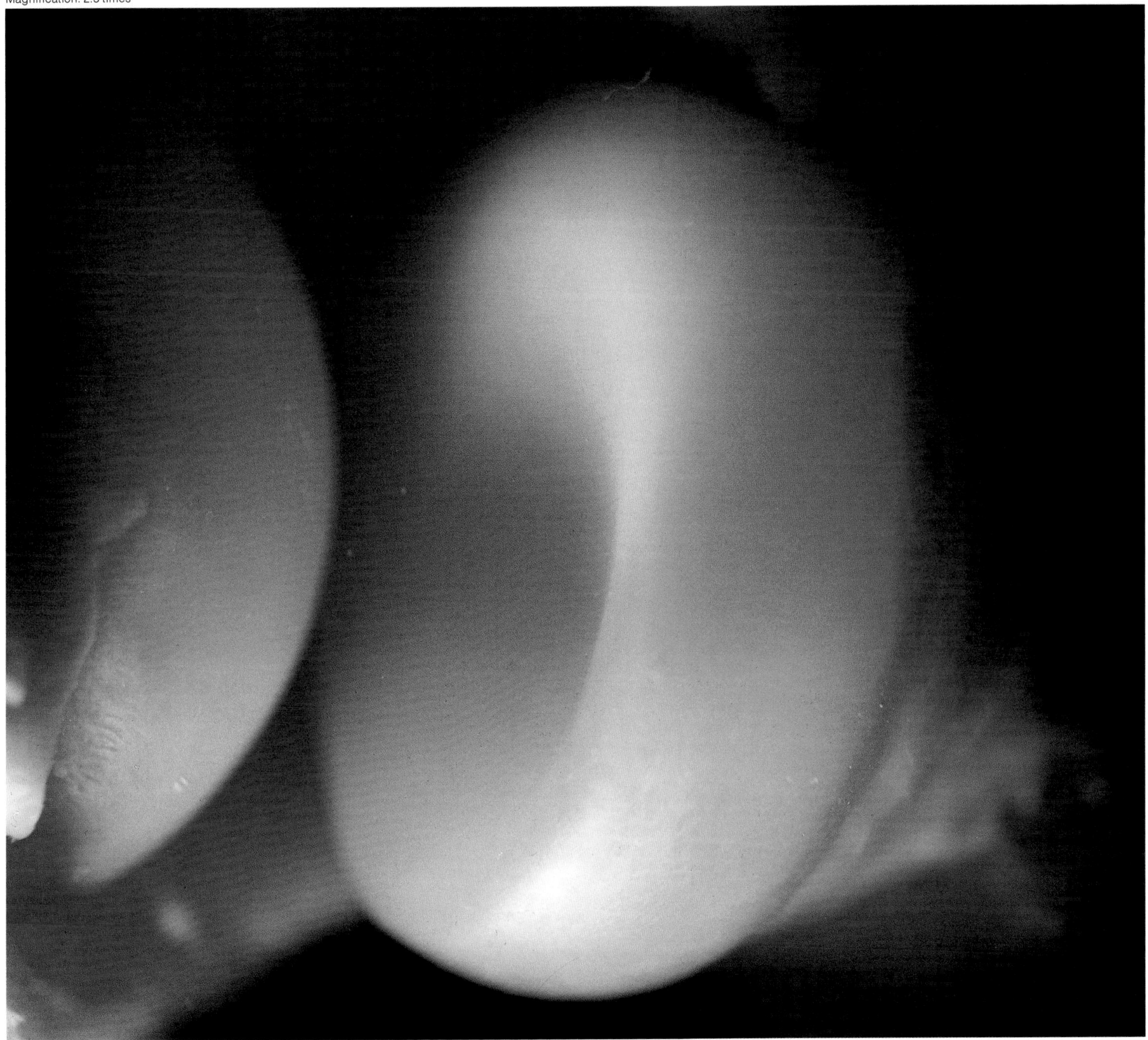

The intricate and harmonious workings of the senses, the nervous system, and the muscles would all be for naught without the approximately 206 bones of the human skeleton. For these make the movable framework that gives shape to motion.

Rigid bones are capable of fluid movement because of joints—68 of them—which permit various types and degrees of action. Pivot joints, hinge joints, and ball-and-socket joints allow the widest range of motion; the ball-and-socket joint at the shoulder (above) is among the most maneuverable.

Normally the ends of bones at joints are coated with a tough tissue called cartilage, which provides a wear-resistant surface for a lubricating fluid that surrounds the joint. Infection, injury, disease, or wear and tear can cause cartilage to ulcerate or deteriorate resulting in arthritis. Patches of scar tissue (right) are typical of an advanced case of arthritis.

Magnification: 150 times

left: Lennart Nilsson, © Boehringer Ingelheim International GmbH.

Magnification: 2,000 times

the retina is the macula lutea, with most of the cones, and in a pinhole-size cup at its center, the fovea, the area of keenest vision. The fovea has no rods, only some 4,000 cones as small as a hundredth of a millimeter thick. In the fovea, almost every cone uses a separate nerve fiber, making an interference-free path to the brain. On the rest of the retina each nerve fiber serves many receptor cells, and vision is not so sharp. When you really want to focus on something of interest, the fovea goes to work.

To understand the fovea's importance, remember that your eye is never still. Whether you are reading, looking at a picture, or watching a scene on the street, your eye flicks from one image to another several times a second. If you try to fix on a single spot, you will see what a small focal area is available to your artificially steadied eye. Whatever lies in the peripheral field is hazy. Only in the fovea are the receptor cells dense enough to clarify what you need to see. So you turn your eyes, and when necessary your head, to keep bringing new, sharp images to the retina.

One area of the retina has neither rods nor cones. In a way it is the climax of the whole operation, yet it is blind. This is where the dense nerve fibers of the entire retina gather one million strong into the optic nerve. You can easily find your own blind spot: On a piece of white paper, make a quarter-inch-high black X. Two and a half inches to the right of the X, make a quarter-inch solid black dot. Then cover your left eye and

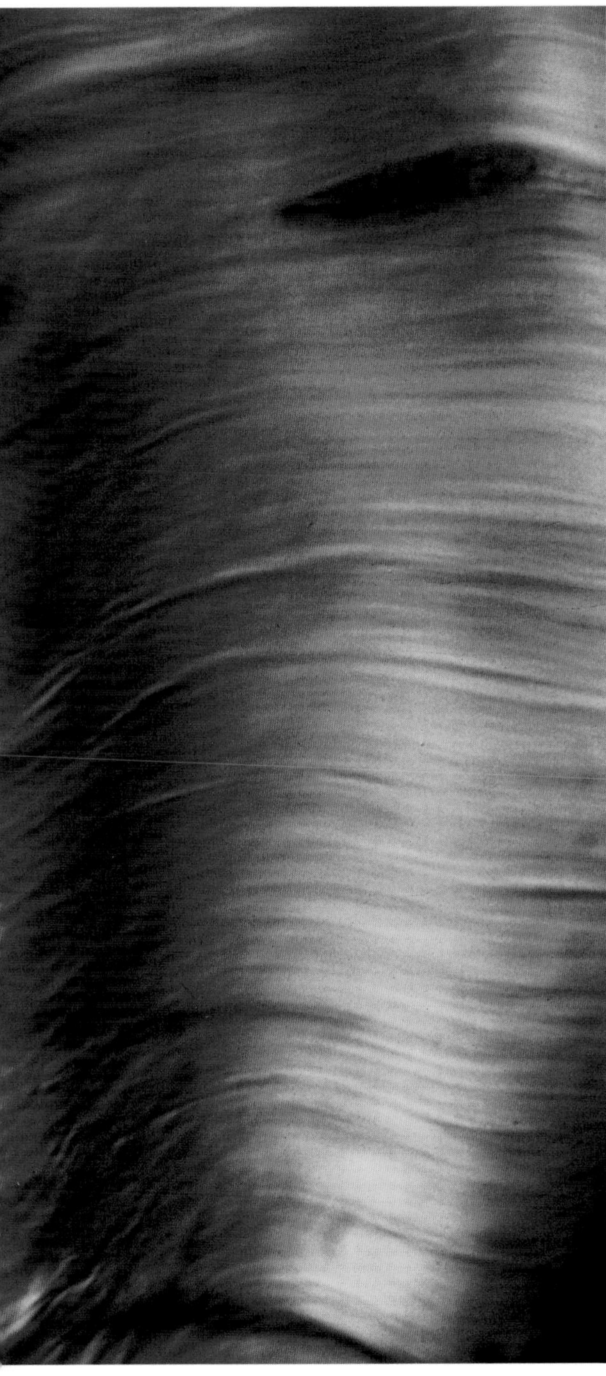

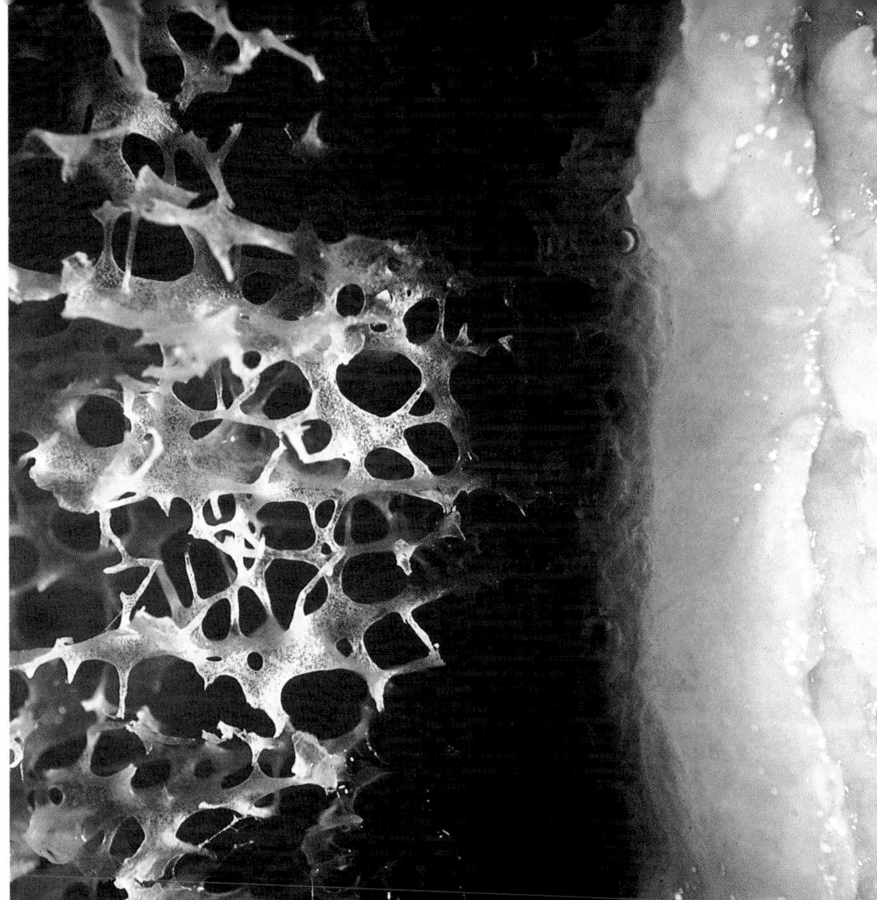

Magnification: 5 times (above); 300 times (below)

Tendons are the shiny, white bundles of fibers (tinted above) that anchor muscles to bones. This flexible connective tissue displays about half the tensile strength of steel.

Bone tissue (right, lower) gets its strength from a structure that resembles reinforced concrete—threads of a fibrous protein called collagen reinforce the hard cement of calcium and phosphorus compounds. In concentric formations called haversian systems, rings of bone fibers surround canals containing nerve fibers and blood vessels. Small cavities contain osteocytes, cells that maintain bone structure.

Layers of bone and cartilage alternate in the human spine (right, upper). The lattice-like structure of cancellous bone in a vertebra absorbs external pressure. Compressible cartilage disks between the vertebrae absorb shock and keep the vertebrae from grinding together when the spine bends.

The "thumb alone would convince me of God's existence," said Isaac Newton of the digit that makes us dexterous. Of the thousand or so different functions we perform daily with the 19 bones in each hand, 2 are demonstrated in an X ray (right); neither would be possible without the thumb. In a precision grip, a flexed finger and opposing thumb grasp an object in a posture that assures accuracy and fine control; in a power grip, the object is held between flexed fingers and palm while the thumb exerts counterpressure. Each hand can perform both grips at once—with a small object grasped in your palm, you can pick up another with thumb and forefinger.

The 26 bones of the foot are devoted less to precise movement than to weight bearing and locomotion. The bones of the toes correspond to the bones of the fingers, but the big toe does not work in opposition like the thumb. Instead, you use its strength to push off each time you take a step, and to balance when you stand on tiptoe (opposite, lower left).

In contrast to flexible feet and hands is the skull; two views represent a consolidation of 70 CAT scan images (opposite, lower right). Of the 28 bones fused to form the cranium, or brain case, the face, and the ear bones, only the lower jaw can move freely.

Put everything together with a computer and you have the human skeleton, jumping and striding through space like a parade of ghostly marionettes.

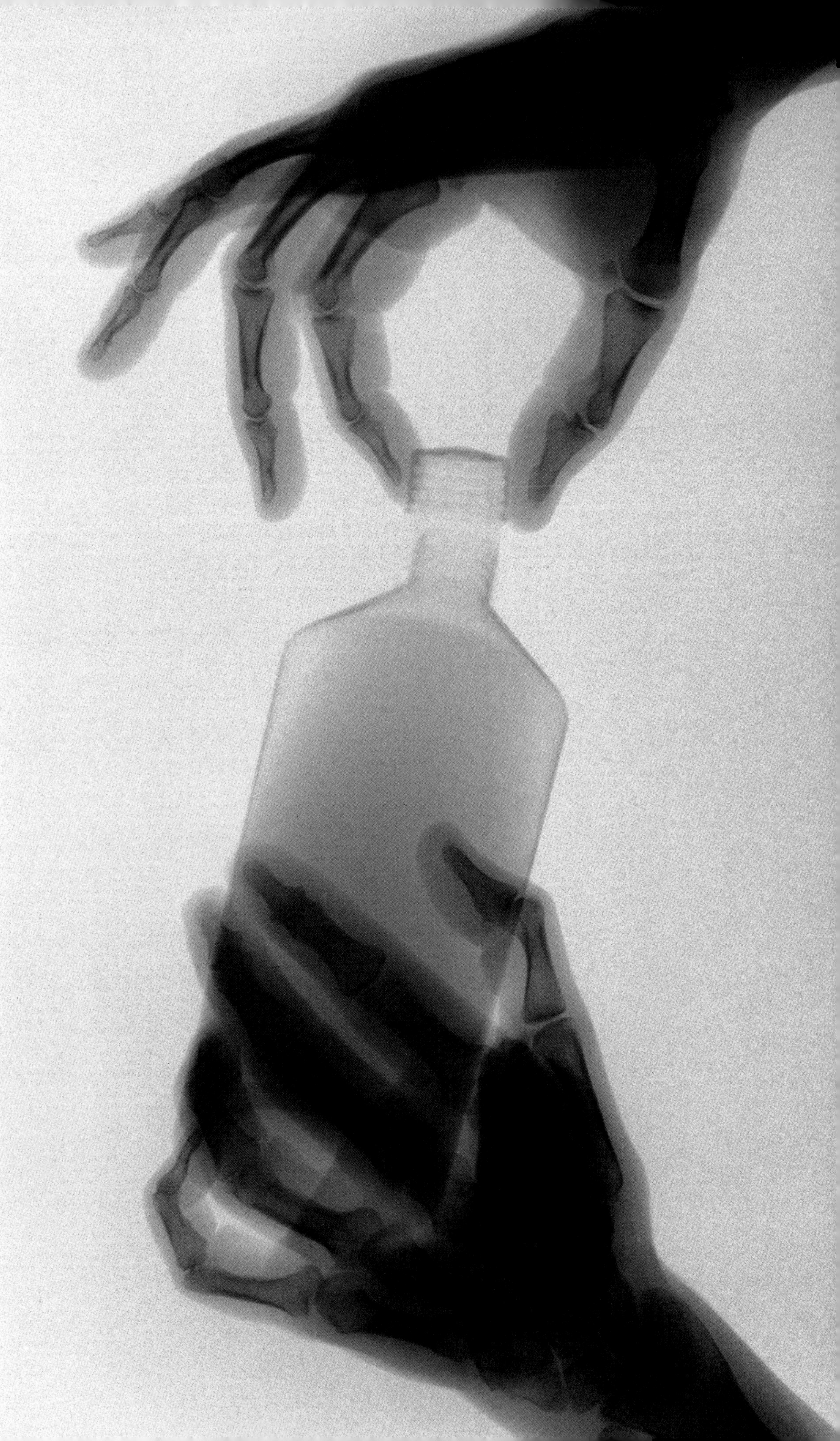

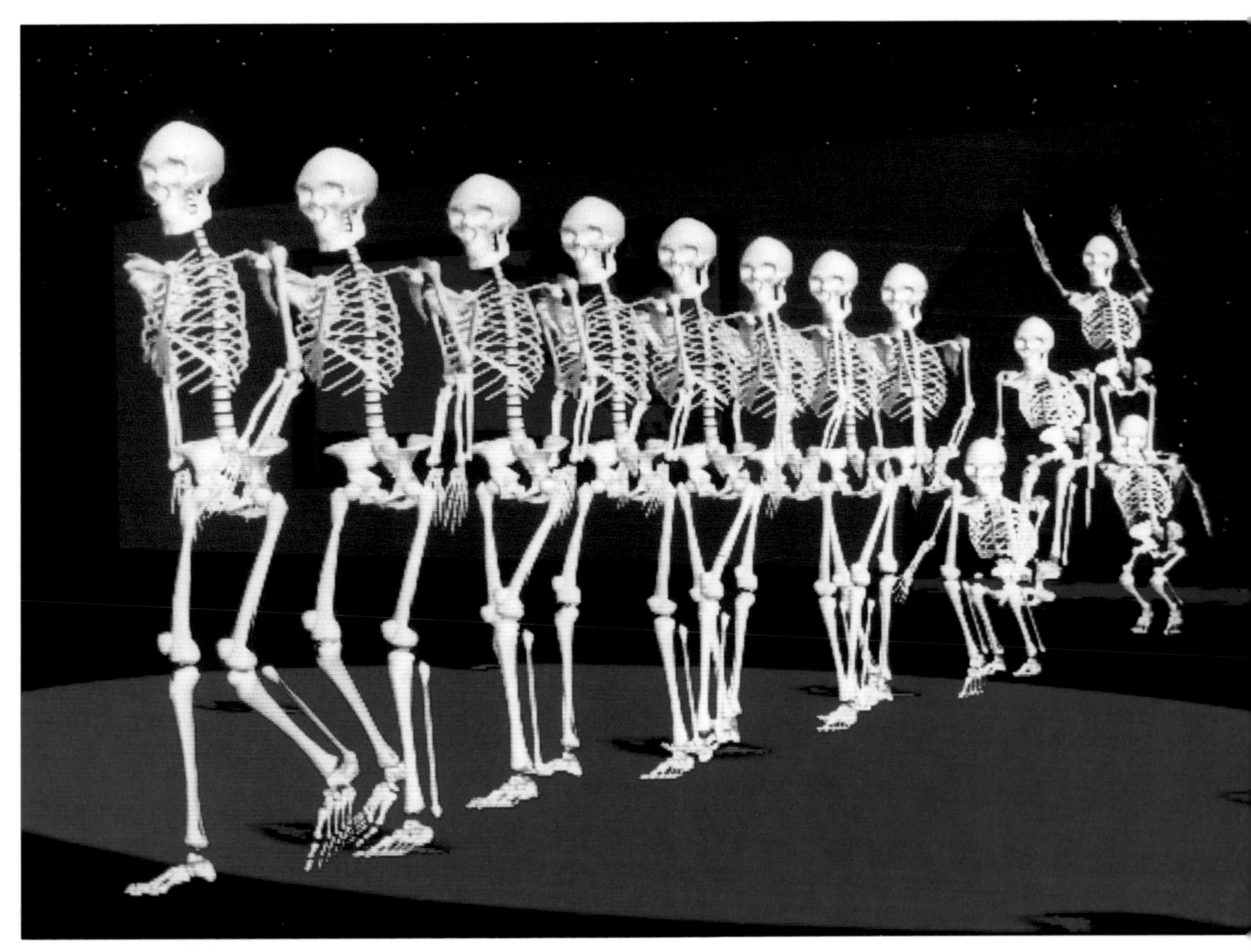

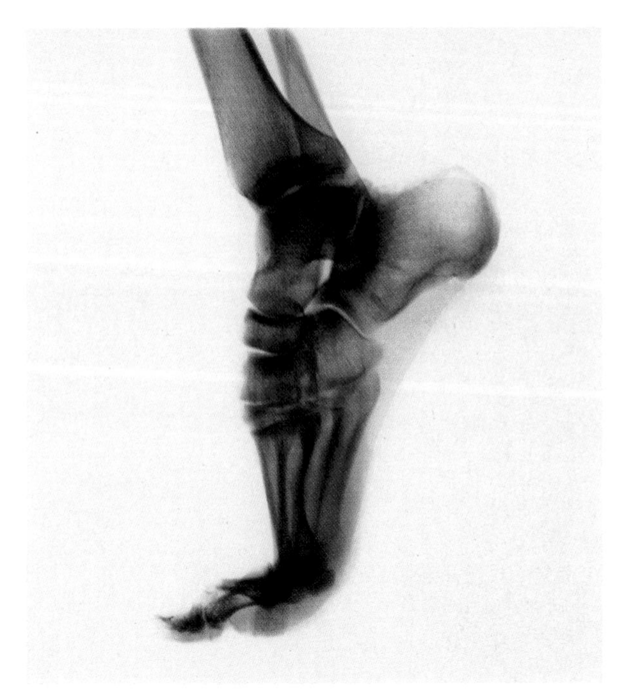

focus on the cross with your right eye, holding the paper at arm's length. Then slowly move it closer until the dot vanishes. Of course, when you aren't playing games, the blind spot never causes blindness—because of the constant motion of your eyes and because the seeing area of one eye overlaps the blind spot in the other.

Trying to see with one eye emphasizes what we gain from stereoscopic vision. Each eye has a slightly different field of vision: If you stare straight ahead, stretch your arms backward, and then move them forward just to the point where you first see them, then close one eye, then the other, you will see the divergence of the angles of your two eyes. Normally the two images overlap because the brain looks for stimulation from matching points on the two retinas. The brain measures the angle of convergence to determine whether you are focusing just past your nose on a printed page, or across the street to hail a bus.

From the back of each eyeball, the optic nerves travel to a meeting place just in front of

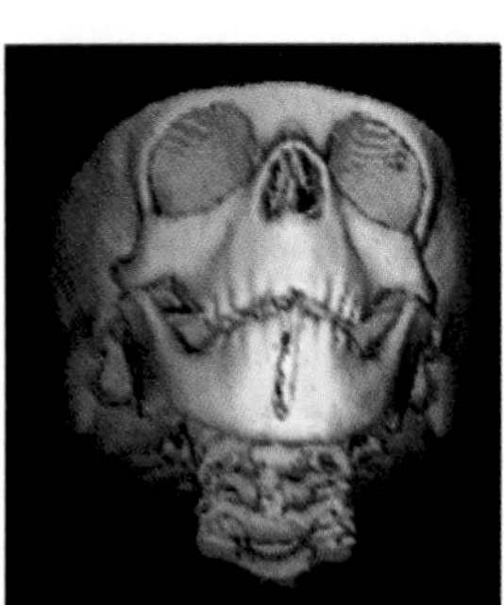

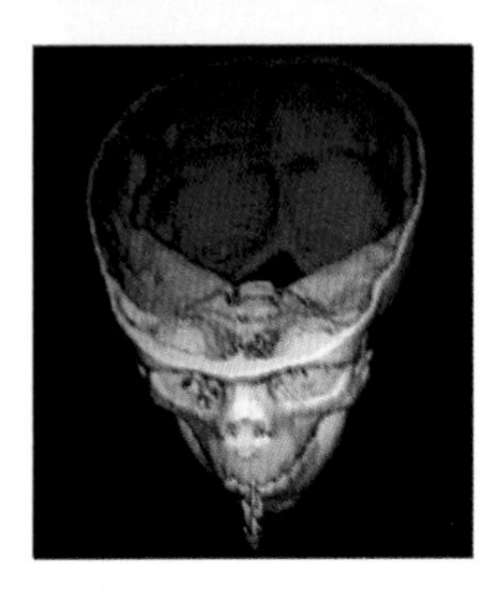

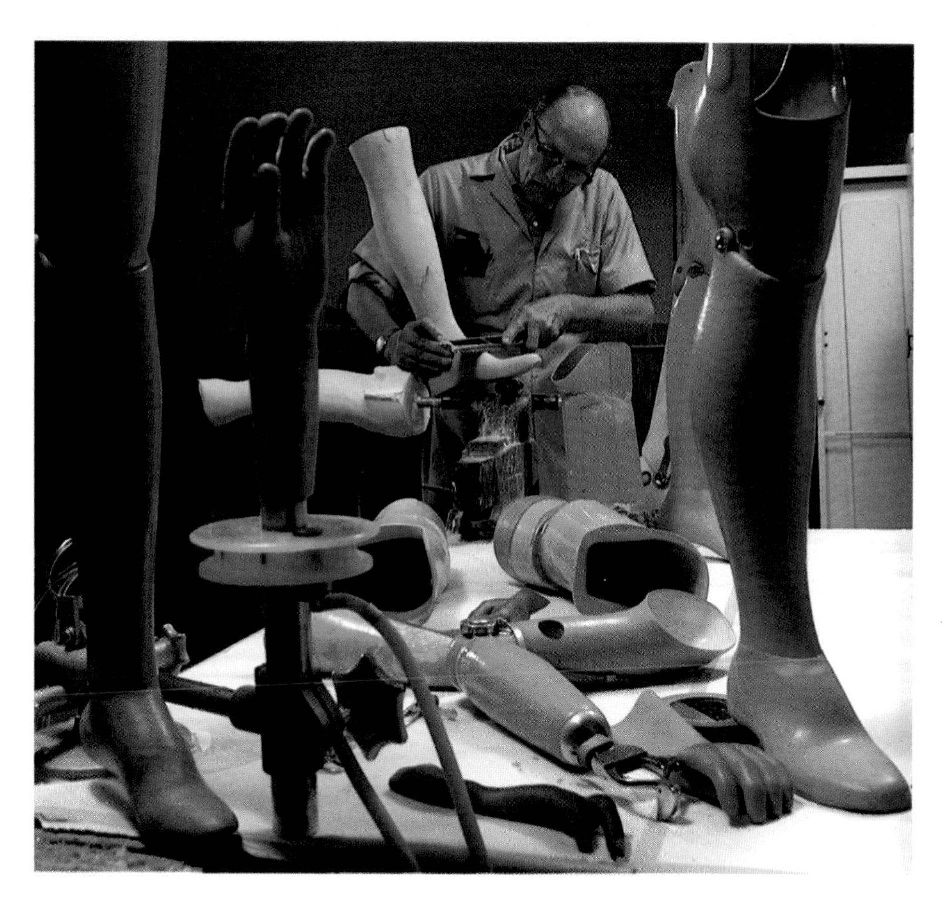

the pituitary gland, the optic chiasm. Here some of them cross over to the opposite side of the brain. Impulses from the right side of each eye go to the brain's right hemisphere, impulses from the left side of each eye to the left hemisphere. The two images fuse into one with depth and solidity.

The art world has a phrase: trompe l'oeil. Fool the eye. Camouflage, illusion, optical tricks, fooling the eye are everyday things for all of us. Yet we humans use our senses to copy and elaborate on the environment. Why we do is a question for philosophers. Biologists tell us that eyes do not exist to make replicas of the world, but to help us live in it. The retina is an exposed part of the brain, a rotating antenna that senses, processes, and encodes the motion, pattern, and color of our surroundings for a brain that invests those signals with meaning and fashions its own images. The entire spectrum, all our fragile, tough, marvelous, ordinary sense organs, are exposed parts of the brain as they cooperatively process data. We feel in the soles of our feet the percussive rhythms of a symphony orchestra, even as our ears sort its harmonies. On the front of our faces is a lump of bone and cartilage that has the power to revive memories of a childhood we thought lost. With our hands we envision sculptures, pyramids, cathedrals. Are the sense organs only receivers and transmitters? How can we know the messenger from the message?

MARGARET SEDEEN

Replace the lost leg of an amputee with a state-of-the-art artificial limb, and he can walk, drive, even ride a bicycle. But he cannot feel warm sand between his toes. Replace an assembly-line worker with a robot, and you free that worker from tedious or dangerous tasks. But if two parts do not fit together, the robot stops operating until a human solves the problem.

Imitation of human sensory systems presents a formidable obstacle to scientists. Prosthetics today can react to electrical impulses from the brain; perhaps tomorrow a robot facing an obstacle will think about it and try a different solution.

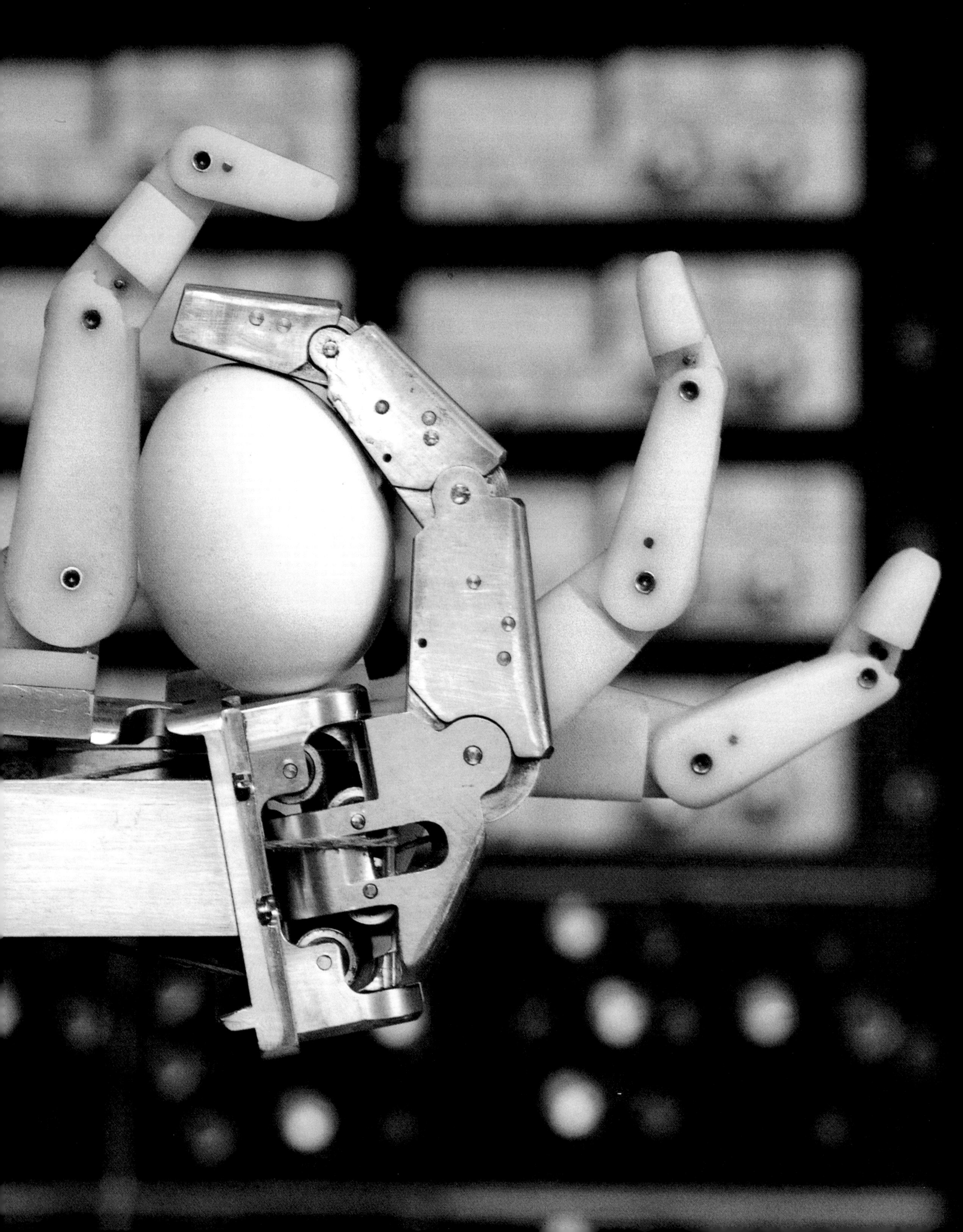

Landscapes of the Mind

Suppose it were possible to transplant human brains the way doctors now transplant kidneys or hearts. After this operation, would the recipient of the new brain still be his original self? Or would he have become the person whose brain was implanted in him?

While such transplants are unlikely ever to occur, these questions point up the difference between the brain and all other organs in the body. We don't just use or depend on our brains, we *are* our brains. Everything we feel, say, think, or do comes from our brains. We cannot take a breath or wake up without directions from these three pounds of soft, wet, pinkish gray tissue—an unprepossessing mass which somehow manages to hold all our memories, our understanding of the world, our dreams.

There are some who say that the whole purpose of our bodies is to supply food and oxygen to sustain our brains. Our brains are voracious consumers of energy, which they get from blood glucose, and of oxygen, which is also borne by the blood. Although the human brain makes up only 2 percent of the body's weight, it uses up 20 percent of the body's oxygen. It needs this oxygen day and night—sometimes even more at night, because dreaming is a particularly active phase of mental life. These requirements make the brain exceptionally vulnerable. If the supply of oxygenated blood to the brain stops for as little as 10 seconds, we lose consciousness; a longer delay may produce permanent brain damage.

How do 10 or 12 watts of electricity and some chemicals pumped in by the blood produce our plans and intentions, our mental life? Though much has been learned about the brain in recent decades, we still know only the barest outlines of its operation. Some philosophers doubt that we shall ever be able to fathom how mind and body interact in this convoluted organ. It is biology's ultimate challenge.

Encased in a bony armor—the skull—and guarded by a series of chemical barriers, the brain has long defied analysis. For centuries it was ignored or deemed unimportant. Although Hippocrates correctly surmised some 2,400 years ago that "Not only our pleasure, our joy and our laughter but also our sorrow, pain, grief and tears arise from the brain, and the brain alone," most of the ancients believed that thoughts and feelings came from the heart, rather than the brain. Aristotle decreed that the heart was the seat of the soul while the brain—which he assumed was composed largely of water—just cooled it. Aristotle's notion held sway for over a thousand years.

The famous Greek physician Galen performed hundreds of experiments on animals in the second century A.D., and showed that the spinal cord was essential to many sensations and movements. But dissection of the human body was forbidden by the church and no real

Magnification: 2 times

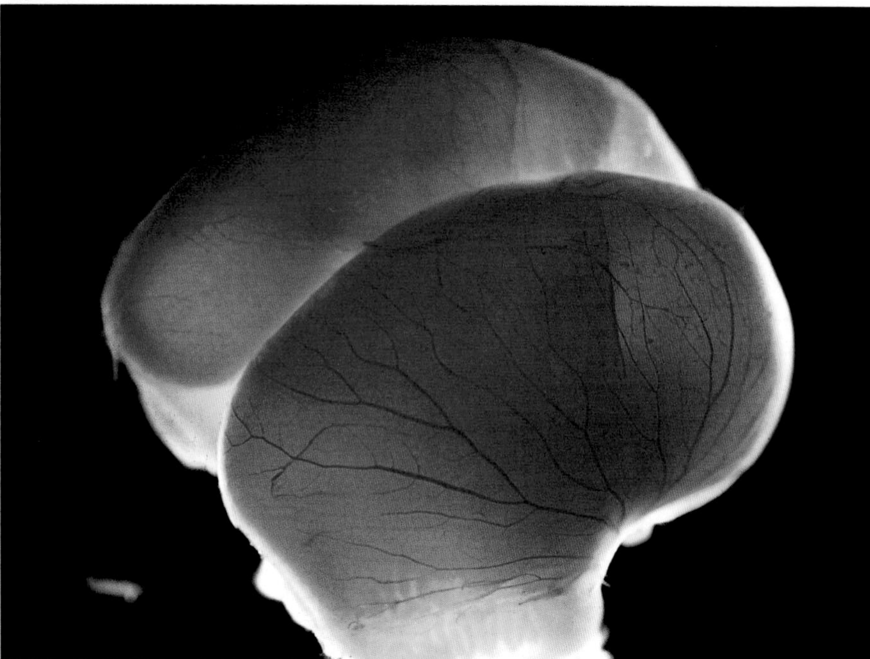

Two-thirds actual size

The human brain distinguishes Homo sapiens, "man the wise," from all other life—and makes each person unique. The deeply folded cerebral cortex, seen here from the front (above) and in a cutaway (opposite), makes up about 40 percent of the brain's total weight. The cortex enables us to perceive, communicate, remember, understand, and appreciate a melody.

During fetal development, the brain generates about 360 million new nerve cells a day, a process that probably halts at birth. At 4 months the cerebral cortex of a fetal brain (upper) is ready to develop its ridges and fissures.

studies of the human brain were done until the Western world emerged from the early Middle Ages. When the Flemish anatomist Vesalius dissected "the heads of executed criminals . . . still warm" in 1543, he firmly established, at last, the overarching importance of the human brain and spinal cord.

Through the centuries it has been natural for people who believed in the power of the brain to explain it in terms of the most advanced technology of their time. Aqueducts and sewer systems based on fluid mechanics were major achievements in Galen's day, so Galen assumed that the most important part of the brain was not its tissue but the fluid-filled cavities within it. As we now know, these cavities contain the cerebrospinal fluid which bathes and cushions the brain. Around the start of the industrial revolution, people assumed that the brain worked much like the simple machines then being developed. Vision, for example, they explained in the same fashion as some optical instruments, as if light traveled directly to the back of the brain through simple lenses, rather than exerting its effect through the complex activity of nerve cells.

When telephone switchboards became a mark of progress, the brain was compared to a busy telephone switchboard with an operator who made decisions. And now that computers reign, many researchers think of the brain as a sort of computer. This comparison is still inadequate. Even the most sophisticated computers that we can envision are crude compared to the almost infinite complexity and flexibility of the human brain—qualities made possible by its intricate, calibrated system of electrochemical signals.

The Command Center

The millions of signals flashing through your brain at any moment carry an extraordinary load of information. They bring news about your body's inner and outer environments: about a cramp in your toe, or the aroma of coffee, or a friend's funny comment. As other signals process and analyze information, they produce certain emotions, memories, thoughts, or plans which lead to a decision. Almost immediately, signals from your brain tell other parts of your body what to do: wiggle your toe, drink the coffee, laugh, or perhaps make a witty reply.

Meanwhile your brain is also monitoring your breathing, blood chemistry, temperature, and other essential processes outside your awareness. It sends out commands that keep your body on an even keel despite constant changes in your environment. It also prepares for future demands. Signals from the brain direct the testicles to prepare sperm, or the ovaries to ripen eggs, at just the right time without our having an inkling of it. *(Continued on page 334)*

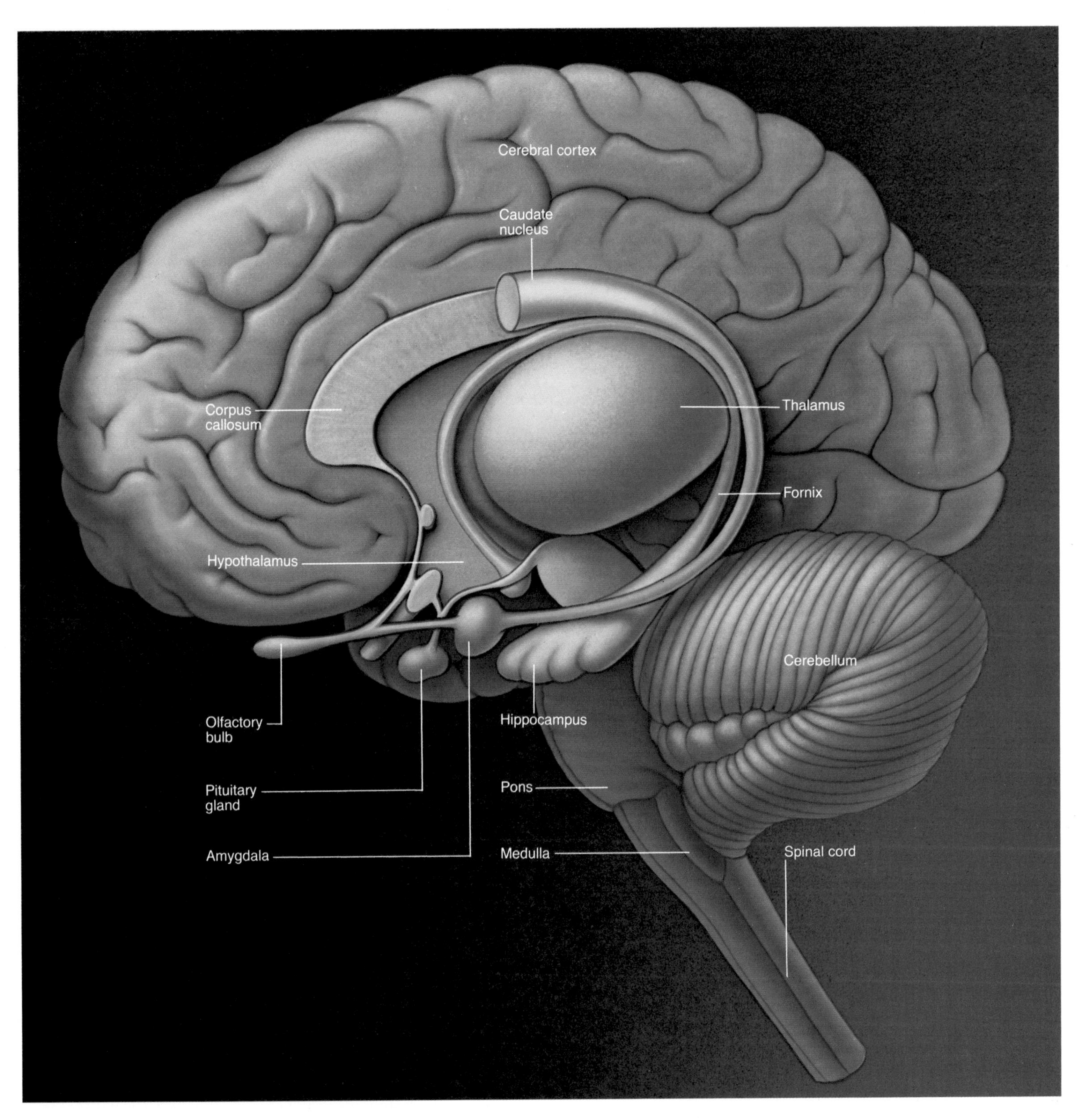

The convoluted cerebral cortex crowns the brain, which, at maturity—around age 18—weighs 3 pounds and numbers 50 to 100 billion nerve cells.

A bundle of nerve fibers called the corpus callosum connects the two cerebral hemispheres, enabling the individual to function as a coordinated whole.

Beneath the cortex lie the brain's more primitive structures, which govern many basic bodily functions. The medulla helps to maintain life-support systems such as breathing, heart rate, and blood pressure. The various parts of the limbic system—including the hypothalamus, hippocampus, amygdala, fornix, and olfactory bulb—control emotions and rule the basic drives for food, sex, and survival. They also help maintain the body's internal environment and may be involved in consolidating memories. The thalamus acts as a relay station for information related to sensation and movement. It processes the data and passes them to the cerebral cortex. The cerebellum, at the back of the head, and the caudate nucleus, part of the basal ganglia, help coordinate body movements and maintain balance. The pons acts as a communication bridge between the cerebellum and the cerebral cortex.

Within the Skull: Care and Keeping of the Brain

Twenty-eight bones make up the skull, eight of them thin, domed, interlocking plates (right) that form the bastion of the brain—the cranium. Curved like an eggshell, the cranial bones provide maximum strength with minimum weight. Their tightly knit joints, called sutures, grip like pieces of a jigsaw puzzle, forming a bond that can withstand a heavy blow. The skull's 20 other bones help shape the face, jaws, and inner ear.

At birth the bones of a baby's cranium are not yet fully formed or joined: During birth the cranial plates may squeeze together or even overlap so that the head can fit through a narrow birth canal. Six membrane-covered gaps in the cranium, called fontanels, will come together and fuse by the time the infant reaches about eighteen months of age.

The cranium and the facial bones grow at different rates and over different periods of time. The braincase itself continues to expand until the child is about seven years old. Facial bones keep growing until adulthood to make room for teeth, muscles, and sinuses.

Openings in the floor of the skull (below) give passage to nerves and blood vessels. The largest opening, the foramen magnum, accommodates the medulla, the structure that joins brain stem to spinal cord.

One-fourth actual size (below); magnification: 30 times (right)

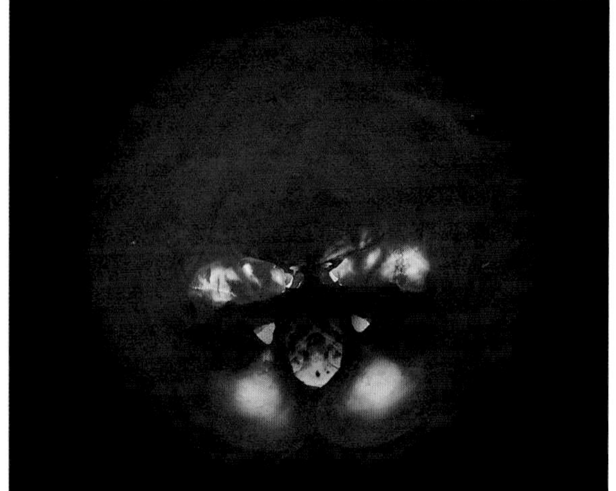

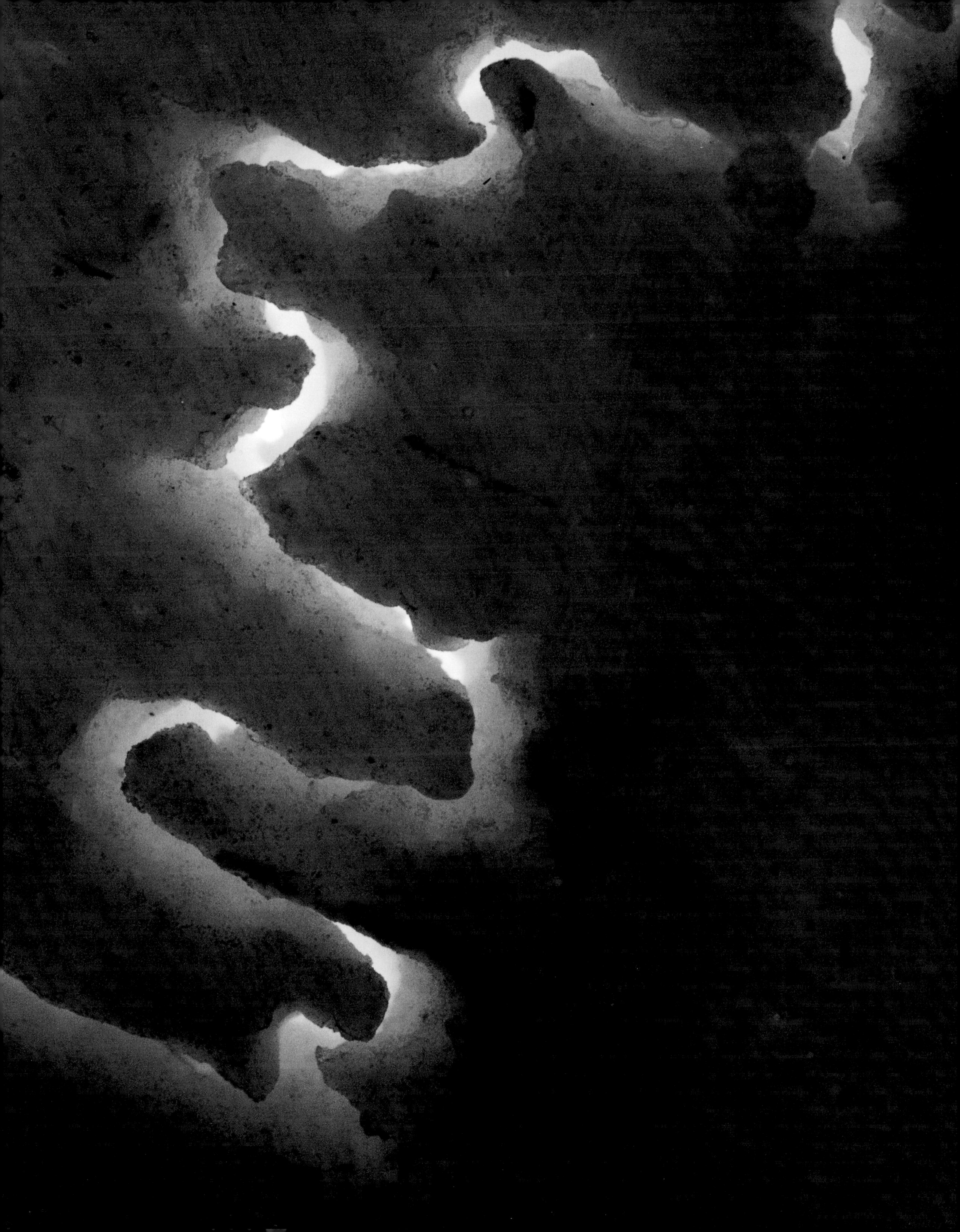

Magnification: 10,000 times (below); 5 times (right)

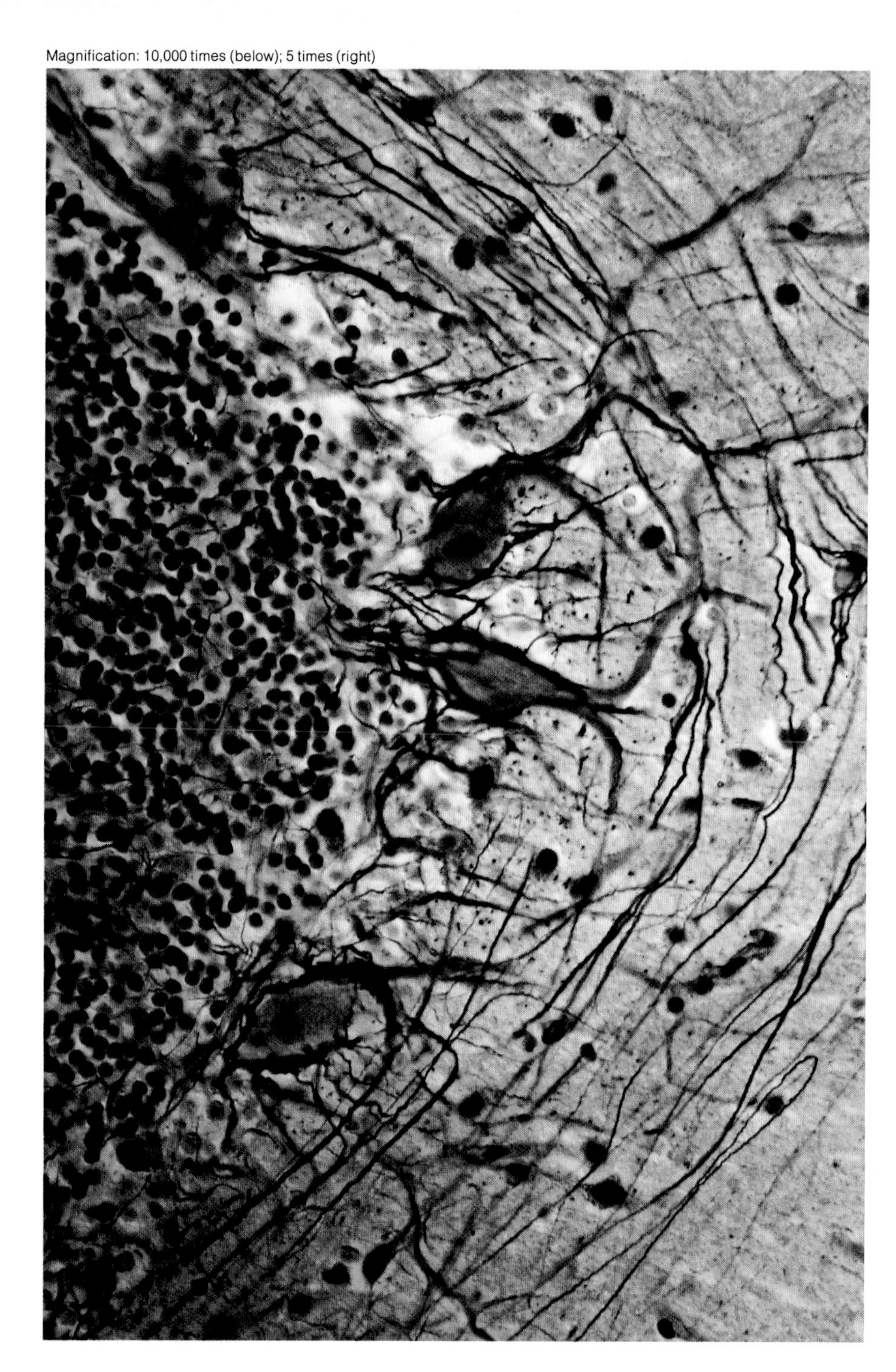

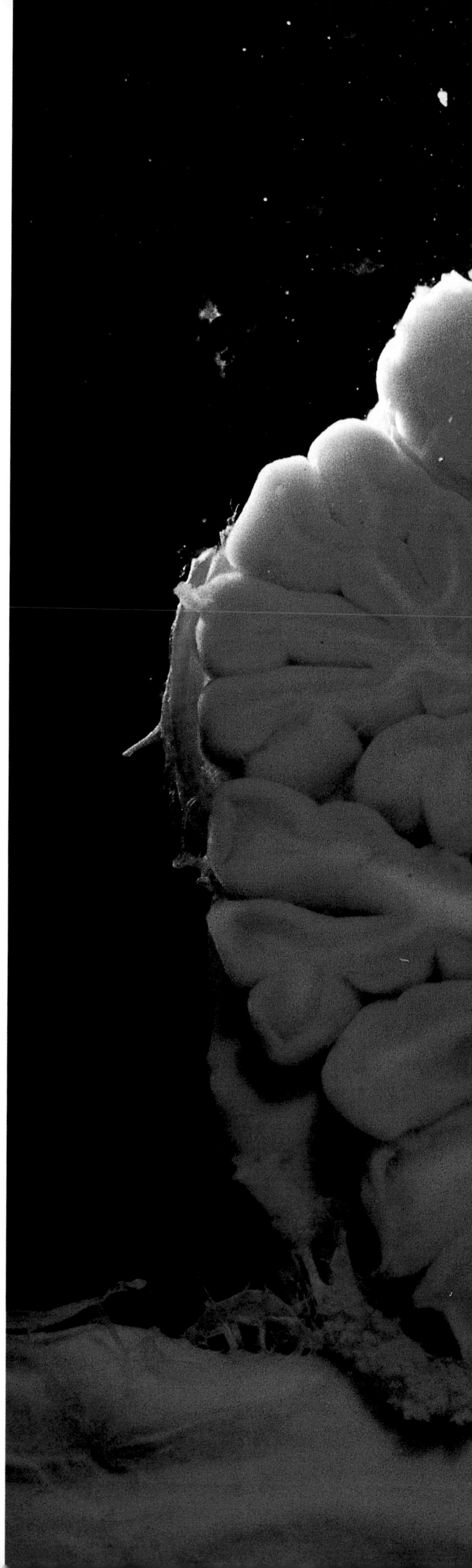

Resembling the stony fronds of a coral reef in this cross-sectional view (right), the cerebellum, or "small brain," acts as a control and coordination center for movement. Without this bun-size lump, it would be almost impossible to dance, play the piccolo, or write a letter. The cerebellum does not by itself initiate motion. Instead, it governs and regulates the action of those muscles involved in the body's conscious and automatic activities.

The cerebellum carries within it "programs" of previously learned movements which are activated by the cerebral cortex. The cerebellum also tracks the position of head and neck, limbs and trunk via messages received from the muscles, joints, and tendons. The messages play back and forth along the tendrils of the arbor vitae, the "tree of life" visible here as ghostly branches.

Large neurons in the cerebellum called Purkinje cells (tinted orange, above) integrate this information, constantly monitoring a "map" of body position and location. Each cell receives input from up to 100,000 other neurons, enabling the cerebellum to tell the body where it is and where it is going.

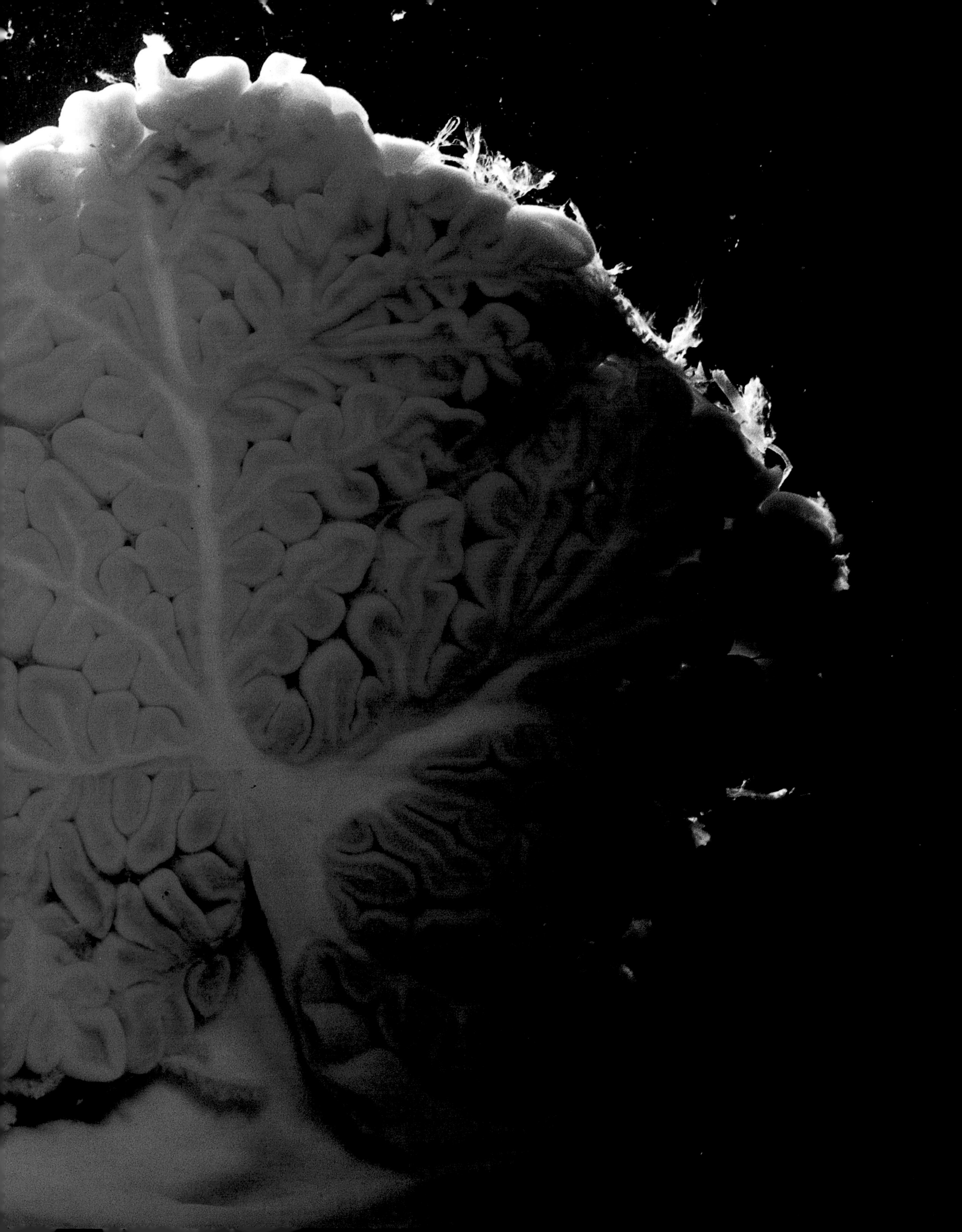

Magnification: 7 times (below); 6 times (right)

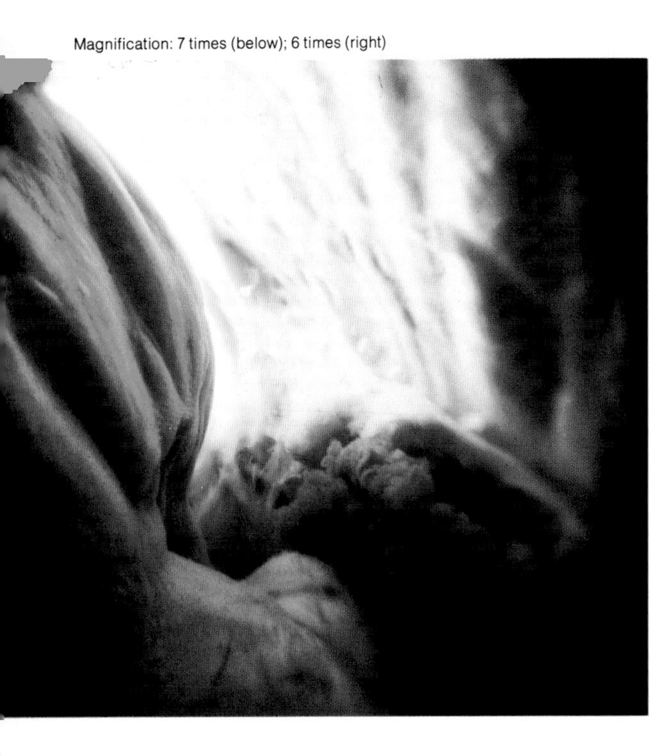

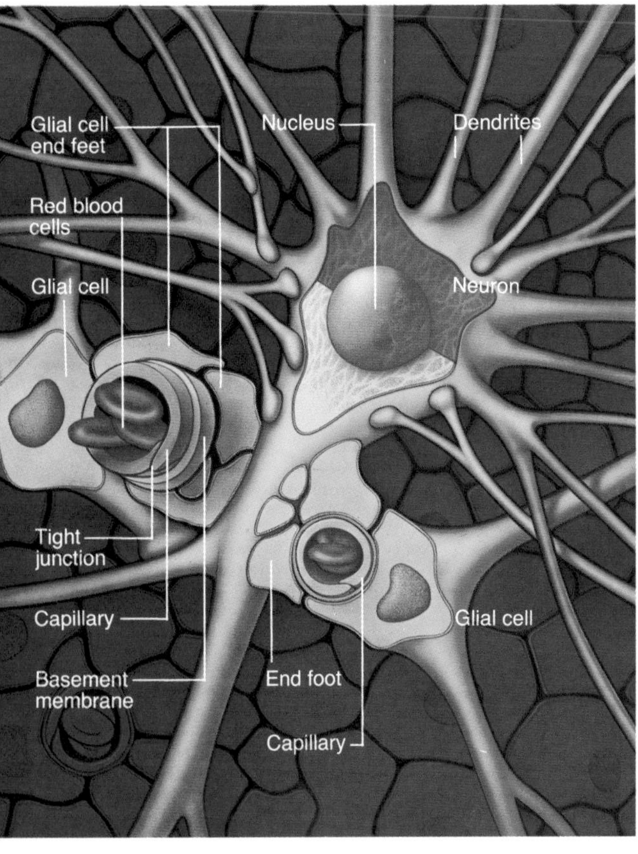

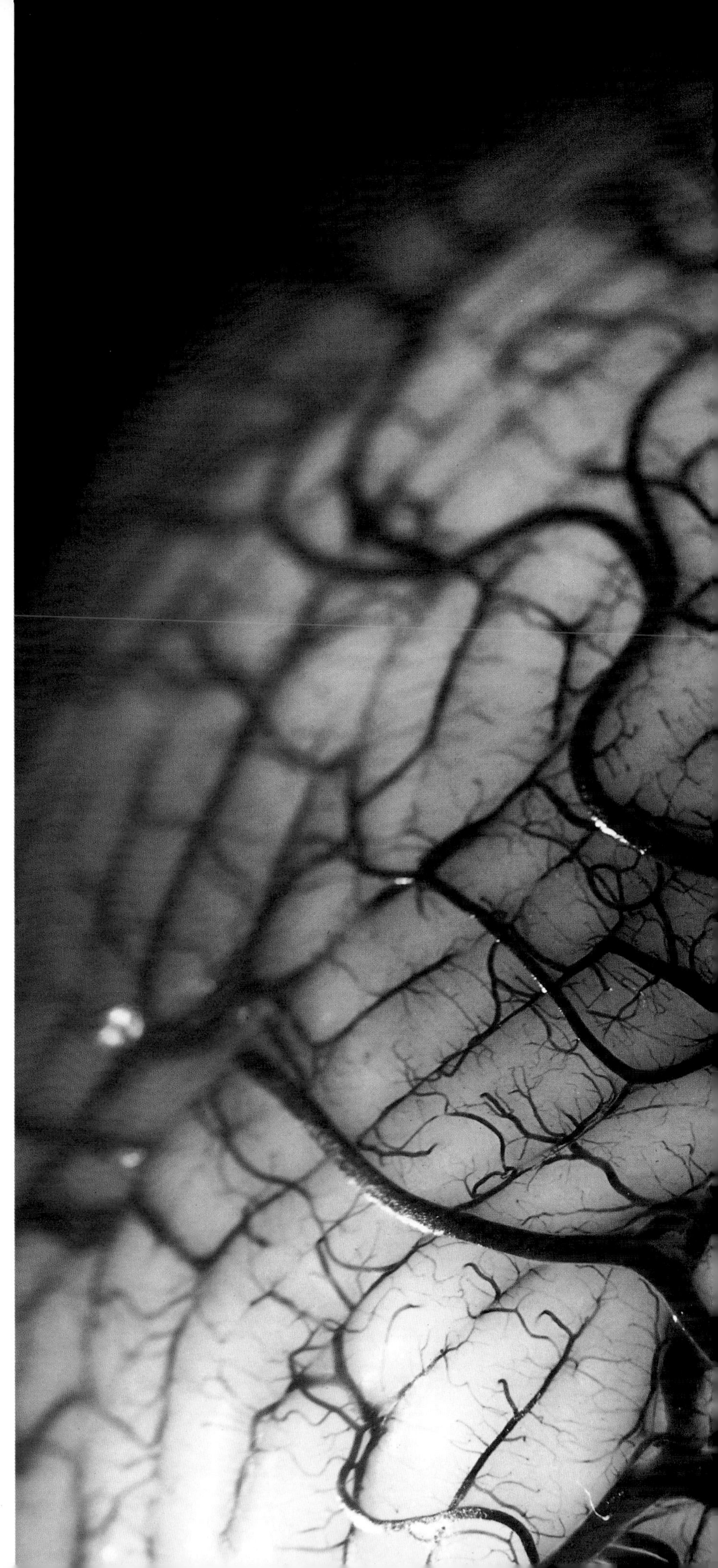

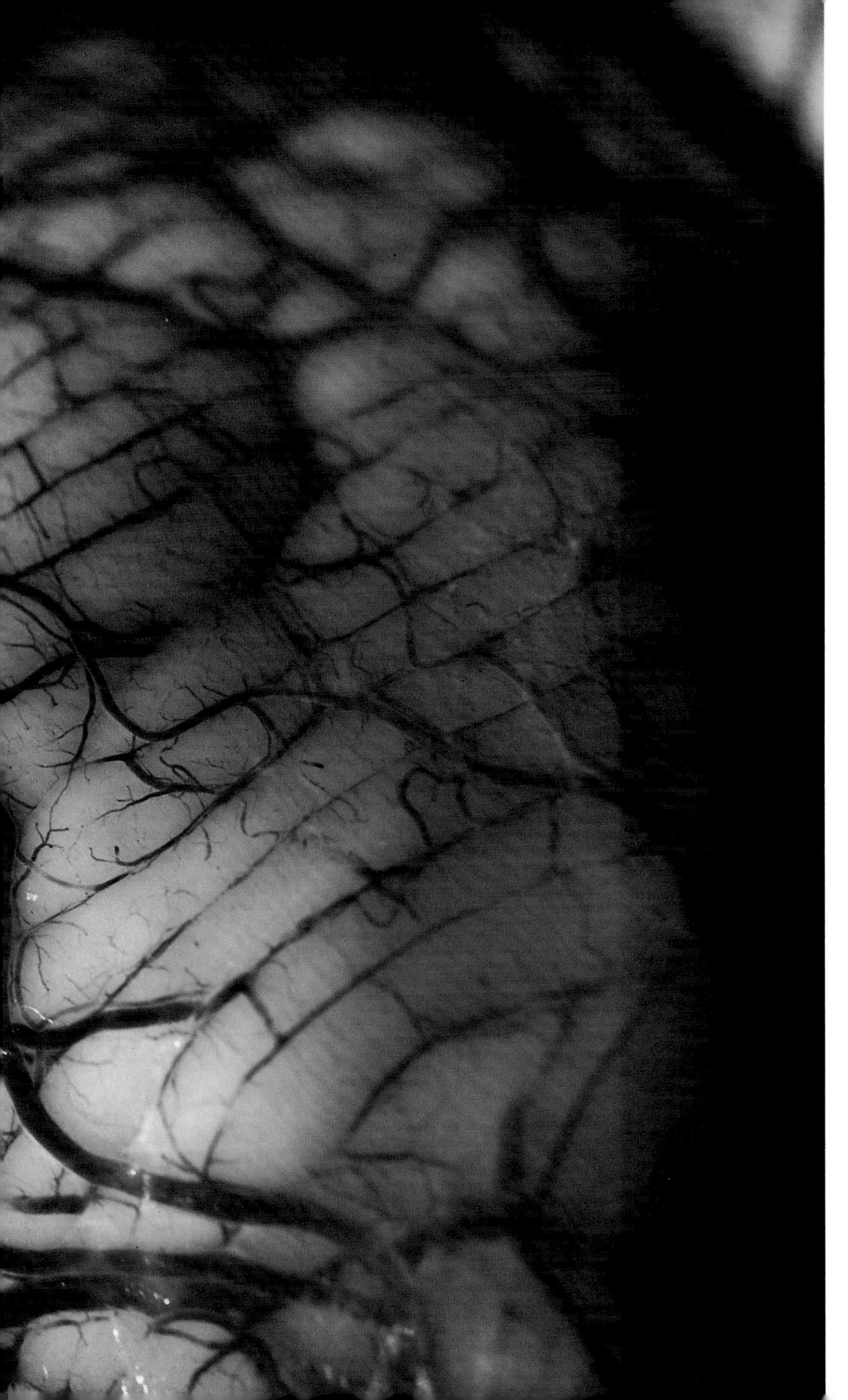

The brain abides in a world of liquid that both cushions and nourishes it. Nearly a fifth of the blood pumped by the heart surges through the brain's intricate network of blood vessels (left) to meet its unflagging demand for oxygen and glucose.

Specialized blood vessels within the brain called choroid plexuses (tinted orange; opposite, upper) produce protective cerebrospinal fluid. Each of the brain's four cavities, the ventricles, contains a choroid plexus. The cerebrospinal fluid continuously washes over the brain and spinal cord, suspending these organs in a liquid cushion that protects them from injury.

The blood-brain barrier, another protective feature, consists of a network of uniquely structured blood vessels (opposite, lower). These capillaries are nearly impermeable to many harmful chemicals carried by the blood, but do allow oxygen, water, and glucose to enter the brain. The cells of these capillaries are more tightly joined than cells of other blood vessels. The vessels themselves are wrapped twice—first by a layer called a basement membrane, then by the fatty extensions, or end feet, of special glial cells.

All these tasks depend on a flow of signals through variously branched circuits of nerve cells that are activated in constantly shifting patterns. Each task requires the cooperation of several distinct regions of the brain.

Deep in the center of the brain lies its oldest and most primitive region: the brain stem, a bulging extension of the spinal cord which controls such basic functions as breathing, blood pressure, swallowing, heart rate, and arousal.

A relatively small but powerful structure, the hypothalamus, sits right above the brain stem. It oversees many brain stem functions and commands such basic drives as hunger and sex, as well as our responses to threat. The hypothalamus also controls the pituitary, once believed to be the body's master gland. The pituitary produces hormones that circulate throughout the body. The hypothalamus releases chemicals that turn on these hormones in the pituitary, providing a major link between the brain's nerve cells and the biochemistry of the body. Directly above the hypothalamus, the thalamus acts as the major relay station and processing center for messages between lower brain regions and the cerebral hemispheres above it, especially for information about sensations and movement.

These central structures—the brain stem, hypothalamus, and thalamus—developed early in evolution. They produce the rigidly programmed, instinctive behavior necessary for our survival, and they exist in the less complicated brains of birds and reptiles, as well as in mammals. On top of these structures, mammals grew a sort of "thinking cap" or primitive cortex, the limbic system, which plays a key role in emotions and in the formation of memories. One reason so few people keep reptiles as pets is that these creatures can show neither affection nor pleasure. Dogs and humans can—because of their limbic system.

Hemispheres of the Brain

The limbic system forms the inner border of the two bulging cerebral hemispheres—almost mirror images of one another—which are the most prominent features of the brains of higher mammals. As one goes up the evolutionary ladder, one finds that the size of these twin hemispheres expands dramatically. Barely visible in the brains of frogs, they are quite clear in rats and appear swollen in the brains of cats. In human beings they dominate the brain, taking up more than five-sixths of the total brain mass.

The most magical element in the brain, the cerebral cortex, makes up the top layer of these ballooning hemispheres. The cortex is a sheet of deeply fissured and convoluted gray matter which produces our thoughts, language, and plans, controls our sensations and voluntary

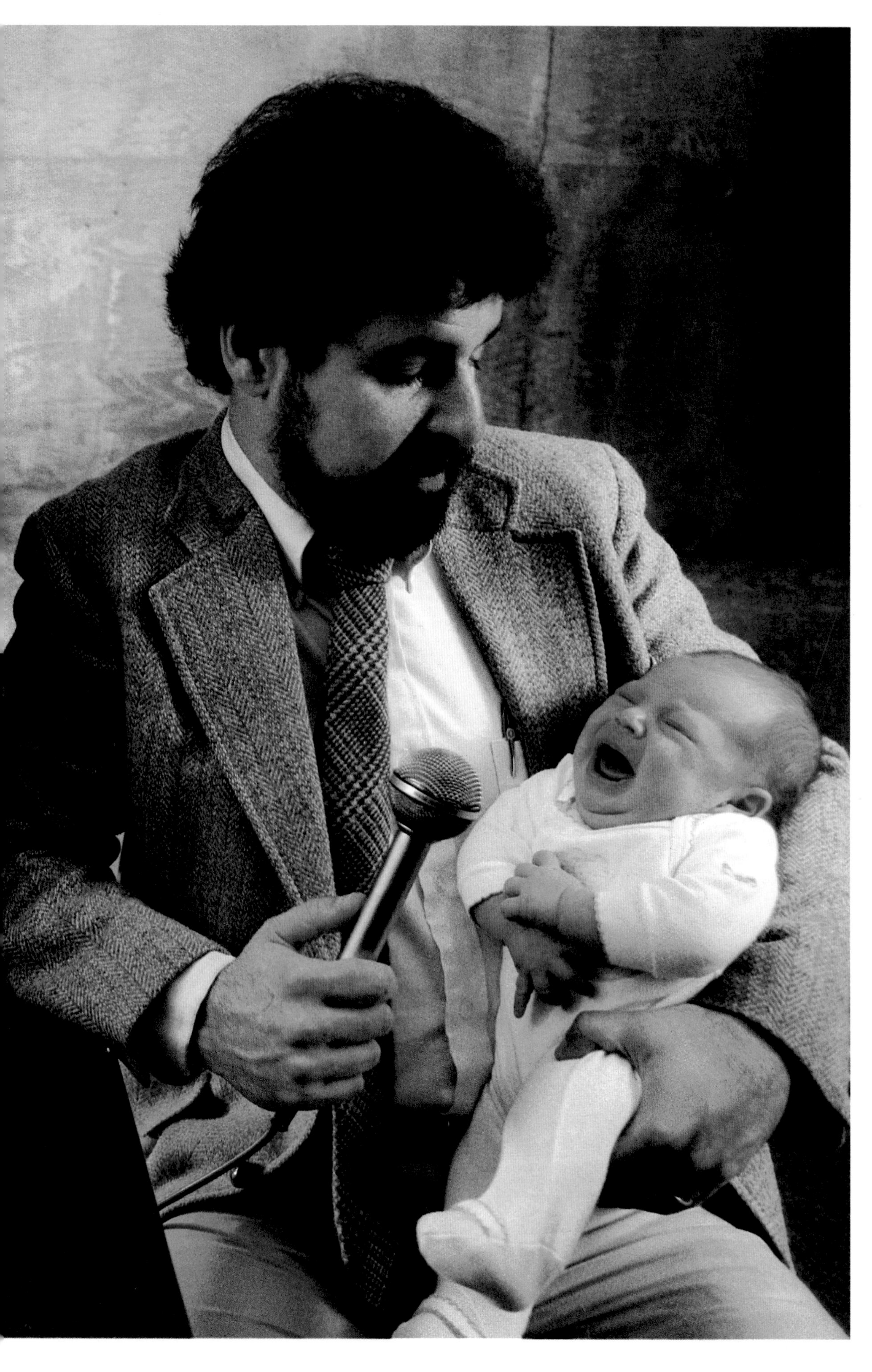

Hearty dissent provides grist for a researcher's studies of crying and brain development. Using a computer programmed to analyze sound, Dr. Barry M. Lester of Brown University monitors pitch, frequency, loudness, timbre, and other acoustical qualities of a baby's wail to detect possible abnormalities in its brain and nervous system. A high-pitched cry, for example, may signal brain damage. Overly monotonous or variable cries may predict a delay in nervous system development—and potential retardation. A healthy infant's lusty yells signal that all is well—brainwise.

His face a televised blur that looms above a monitoring technician (right), an astronaut takes the first step in learning to control the space sickness that often accompanies weightlessness. Strapped into spinning chairs and hooked to sensors (left), queasy astronaut trainees are analyzed for physiological responses related to motion sickness—heart rate, depth and rate of respiration, flow of blood to the hands. Then biofeedback equipment (below)—devices such as electrocardiographs that translate biological data into audio or visual displays—helps them monitor and learn to control their symptoms.

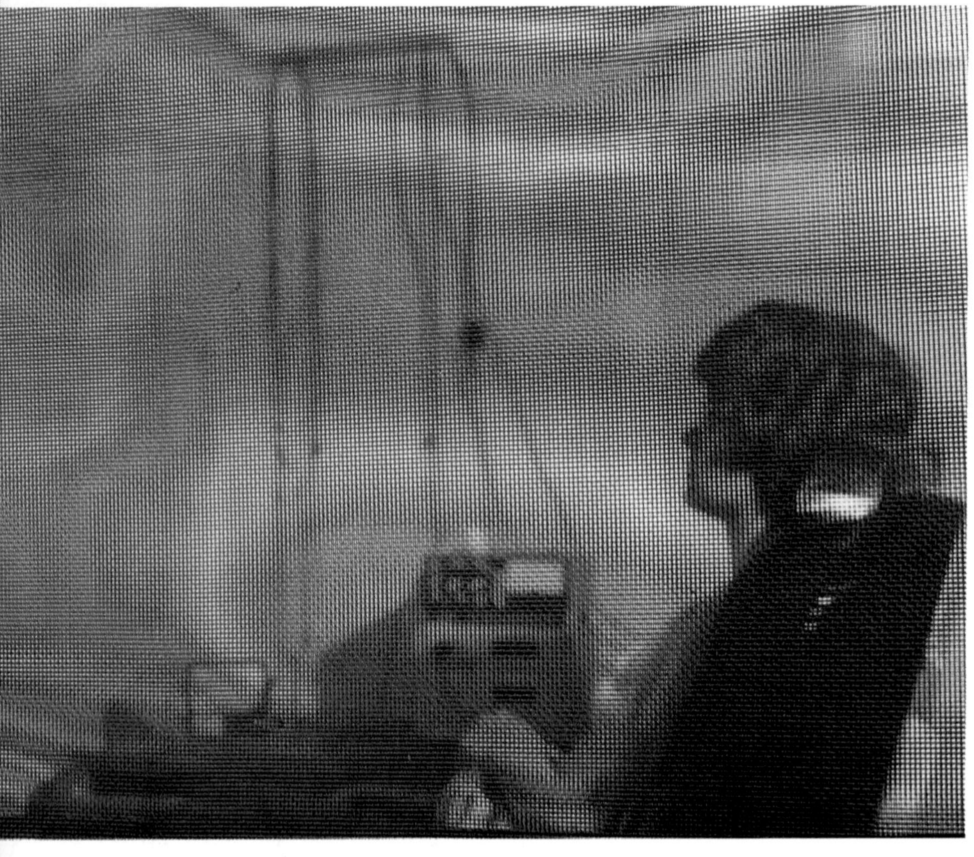

Down but not out, football players at the University of Wisconsin reduce stress before a practice game with relaxation techniques learned through biofeedback training.

Biofeedback equipment that monitors internal functions such as blood pressure and circulation enables some people to develop techniques that help regulate those functions. With practice, some individuals can learn to lower blood pressure or ease a muscular tension headache by imagining themselves in a relaxed state. Or they can warm cold hands by imagining themselves sprawled on a sunny beach.

movements, stores our memories, evokes our flights of fancy—in short, makes us human. Its many folds and convolutions allow the cortex to be packed very densely into the restricted volume of the skull; they triple its surface area. Although the human cortex is only about an eighth of an inch thick, it contains some 75 percent of the brain's 100 billion neurons, or nerve cells—a figure that approximates the number of stars in our galaxy. These neurons are surpassed at least 10 to 1 by another kind of brain cell, the glial cells. These housekeeping cells surround, support, and nourish the neurons.

The Neuron—a Tree of Life

The neurons hold the key to the brain's subtle and efficient system of communication. Typically a neuron has a gray cell body topped by a sort of tree with numerous delicate branches, or dendrites, through which the neuron can receive signals from other cells. Each neuron has a long, white-sheathed tail called an axon which transmits nerve impulses. A single axon may branch out into as many as 10,000 terminals, and each terminal can connect with a separate neuron. Since each neuron can also receive signals from more than a thousand other neurons, a single neuron could conceivably carry on several million separate conversations at the same time—a mind-boggling feat, considering the astronomical number of neurons in the human brain.

During these conversations neurons never actually touch, but they are so close to each other at the linking sites—synapses—that nerve impulses can travel from one neuron to the next as rapidly as if there were electrical continuity between them. (This and some other aspects of brain anatomy and operation are illustrated in Chapter 7.) When an impulse reaches an axon terminal, it releases a packet of message-bearing chemicals, or neurotransmitters, which jump the tiny synaptic gap and fit into protein receptors on the dendrites of the receiving cells. These neurotransmitters may serve as either excitatory or inhibitory signals: They either impel the neuron to produce an electrical impulse or prevent it from firing. Certain neurotransmitters can also modulate the sensitivity of neurons to the signals of other neurotransmitters.

Each neuron keeps a running account of all the signals it receives, both excitatory and inhibitory. While in a resting state, the inside of a neuron maintains a net negative charge in relation to the fluid that surrounds it. This salty fluid—which may be a carryover from the primordial oceans in which life began—contains a high level of positively charged sodium ions. Incoming excitatory signals change the properties of the neuron's membrane, opening tiny channels that allow the sodium ions to rush into the neuron

from the extracellular fluid. This influx of positively charged ions brings the neuron closer to initiating an impulse. Inhibitory signals have the reverse effect, suppressing the neuron's tendency to fire. If the signals received by the neuron add up to sufficient excitation and the neuron reaches a certain excitatory threshold, it fires. The impulse then flashes down the axon, releasing packets of a specific neurotransmitter at its tip. These chemicals zip across the synaptic gap to another cell and start the process again.

In this fashion, electrical impulses can travel clear across the brain in milliseconds. Some signals follow express routes. When your pupil constricts in response to light shining on your retina, for instance, this reflex occurs almost instantaneously: The command to constrict the pupil involves only four or five synapses of cells between the eye and the brain stem neurons controlling the muscles that constrict the iris.

Apart from such reflex actions, most activities—even something as simple as drumming your fingers—involve thousands of synapses in linked but different parts of the brain. First you must *plan* these movements. Studies of cerebral blood flow in different parts of the brain recently showed the difference between actually moving your fingers and just planning to do so. The researchers measured how much oxygen was consumed in these different brain regions, a sign of neuronal activity. When a man moved his fingers, the researchers recorded increased activity both in the supplementary motor area and the motor area of his frontal lobes. When the man merely *thought* about moving his fingers, however, his motor area was quiet and only his supplementary motor area remained active—a graphic demonstration of a purely mental process at work.

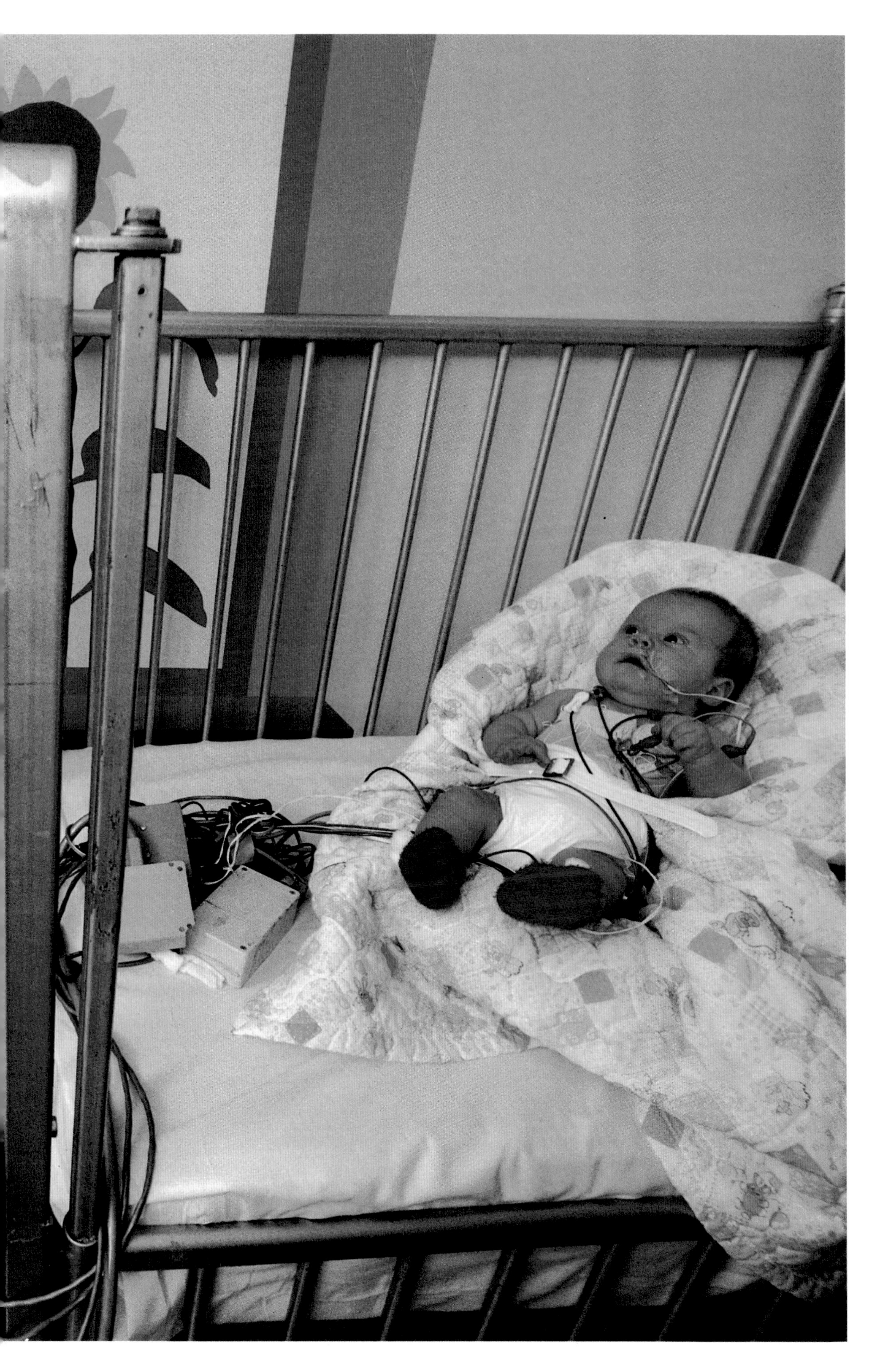

Sleep, the blissful balm of a child in Ōsaka, Japan (opposite), presents a potentially deadly threat to a baby at a California research center (left).

Each year more than 7,000 American infants die in their sleep—victims of sudden infant death syndrome, commonly known as crib death. No one has pinpointed the cause of this malady. Some researchers believe it may result from a delay in the development of the nervous system, or some as yet undiscovered brain abnormality. The sleeping baby simply stops breathing—and dies if not immediately awakened.

This youngster at Stanford University's Sleep Disorders Center, suspected of being at risk for crib death, has its heart rate, breathing, and blood-oxygen levels continuously monitored by a polysomnograph. Home monitors and respiratory stimulants may help such children survive the period of risk—usually the first 6 months of life.

Scientists know little about the biological basis for sleep. Shakespeare called it "nature's soft nurse," and for many years sleep was thought to result from a lack of sensory stimulation.

Modern science recognizes that sleep is by no means a passive process. A sleeper's brain often is as electrically active as that of someone awake and alert.

While scientists probe the mysteries of sleep, one fact remains clear: Most people spend about a third of their lives asleep.

After you have made the decision to move the fingers of your right hand, specific cells in the motor cortex of the left hemisphere are activated. The motor area is a strip of neurons at the back of the frontal lobes, parallel to the sensory area. All along this strip, vertical columns of cells control groups of muscles on the opposite side of the body in a very precise array. When activated, many of these cells send impulses all the way down to the brain stem, where they cross over to the opposite side of the spinal cord and excite special motor neurons. These motor neurons then send impulses to the appropriate muscles of the right hand, which may be as far as two feet away. At the junction of nerve and muscle—the motor end plate—they release acetylcholine, a neurotransmitter which makes the muscles of the fingers contract.

Two other systems of brain cells are also needed to regulate the fingers' movement—to start and stop them correctly, coordinate them with posture, block unwanted movements and monitor muscle tone. The basal ganglia—a collection of structures next to the limbic system in each hemisphere—and the cerebellum, a curiously shaped mass that sits just below and outside the cerebral hemispheres, take over these tasks. At the same time, several intricate feedback systems continually bring the brain information about how much the fingers have moved and how much farther they need to go. Deeper centers of the brain may also get involved if any emotions are aroused by this movement.

The uncanny ability of neurons to make just the right connections with one another may have been gained at the expense of their ability to reproduce. Nearly all other cells in the body can divide; when some of them die, they are replaced by new ones. But if the brain's neurons divided, their connections with other cells might be hopelessly disrupted. We are born with nearly our full complement of brain neurons, and those that die—some 18 million a year between the ages of 20 and 70—are lost forever.

In a cavern world devoid of external time cues, French geologist Michel Siffre in 1972 took part in a study of the natural body rhythms that govern human life. For 6 months he lived by himself in a Texas cave (above), isolated from sunlight, temperature changes, clocks, radio, and television.

Here (left) with only books for company, he reads before drifting off into an electronically monitored sleep. Siffre found that his body continued its "daily" cycle of sleeping and waking—except that his "day" averaged about 28 hours. This sleep-wake cycle, known as a circadian rhythm—from the Latin *circa,* about, and *dies,* day—reflects a biological adaptation to the cycles of light and dark created by the Earth's daily rotation. Other circadian rhythms include cycles of body temperature, blood pressure, hormone levels, and about a hundred other rhythms—all probably controlled by small clusters of cells located in the hypothalamus.

Because many of these rhythms relate to the sleep-wake cycle, disrupting that cycle often creates a disharmony in the body. When Siffre's rhythms fell out of synchrony, he suffered a clouded memory and impaired coordination. His worst experience? Loneliness. "Beyond all bearing," he wrote.

A meditating Indian holy man embarks on an inward journey into the realms of consciousness. Researchers have found that meditation can also produce physiological effects. By engaging in mental and physical exercises, especially breath control, a meditator can reach an altered state of consciousness measurable by changes in the brain's electrical activity.

The practice can also lower blood pressure and decrease heart rate and oxygen consumption, inducing relaxation.

Skeptics charge that simply resting can achieve similar effects, but do allow that meditation reduces stress—reason enough, perhaps, for this ancient Eastern practice to find favor in the West.

In Granada Hills, California (right), fire walkers stride across a bed of hot coals suffering only sooty feet and occasional blisters. Such feats derive from ancient devotional rites—but defy full scientific explanation. Perhaps the answer lies with the power of the brain and nerves to alter the body's normal response to heat; or perhaps it involves a quirk of physics. Whatever the reason, the practice testifies to our unbounded fascination with the powers of the human mind.

While they live, our irreplaceable neurons go on making some new connections or rearranging old circuits in our brains. The neurons that do the most work seem to grow richer. Their cell bodies become larger, and their dendrites develop new branches on which to accept additional connections with other cells. Judging from experiments with animals, new experiences which activate specific parts of the brain will make these brain regions grow heavier and thicker. Certain skills and certain kinds of intelligence may develop simultaneously with enriched networks of brain cells.

Every human being has a pattern of brain activity as distinctive as a fingerprint. Everything you have ever done or experienced has left its mark in your brain cells, influencing their growth and connections. Your genetic endowment is engraved in every brain cell. So are the effects of your culture. As a result, you see the world a little differently from anyone else. Your brain is unique among brains.

The Cycles of Sleep

Your brain never rests, not even when you sit quietly or when you sleep. Throughout the night it continues to regulate your breathing, digestion, and heart rate and to some extent, monitor your environment. You don't react to routine noises, but you may jump up with alarm at some things that go bump in the night. Apart from these maintenance duties, your brain goes through five different cycles of activity as you sleep. These cycles involve changes in the patterns of electrical activity of large groups of brain neurons—that is, changes in brain waves.

When a German psychiatrist, Hans Berger, first tried to eavesdrop on the brain, in 1924, he pasted two small pieces of silver foil on his 15-year-old son's scalp, attached wires to them, and connected the wires to a galvanometer, an instrument that measures electrical current. He saw at once that some strange electrical signals—recorded as spikes on a roll of paper—emanated from his son's brain. The signals followed a regular pattern, which he dubbed the alpha rhythm. Berger's contraption was the forerunner of today's electroencephalograph. He tested others, who also produced alpha waves, but the rhythm of their brain waves changed when they concentrated on arithmetic or other problems.

Alpha waves show that the brain is idling; the alpha state is calm, serene, relaxed. On an electroencephalogram these waves occur 7 to 12 times per second, and we go through this stage as a precursor to sleep. As we fall into a light sleep we enter a slower stage: theta waves, at only 3 to 7 cycles per second, with a low amplitude. These slow waves are interrupted by frequent bursts of activity called sleep spindles, at 12 to 15 cycles per second. Regular, high-amplitude delta waves follow at only 1 to 2 cycles per second, with a few sleep spindles. Then sleep deepens and the brain produces wave patterns

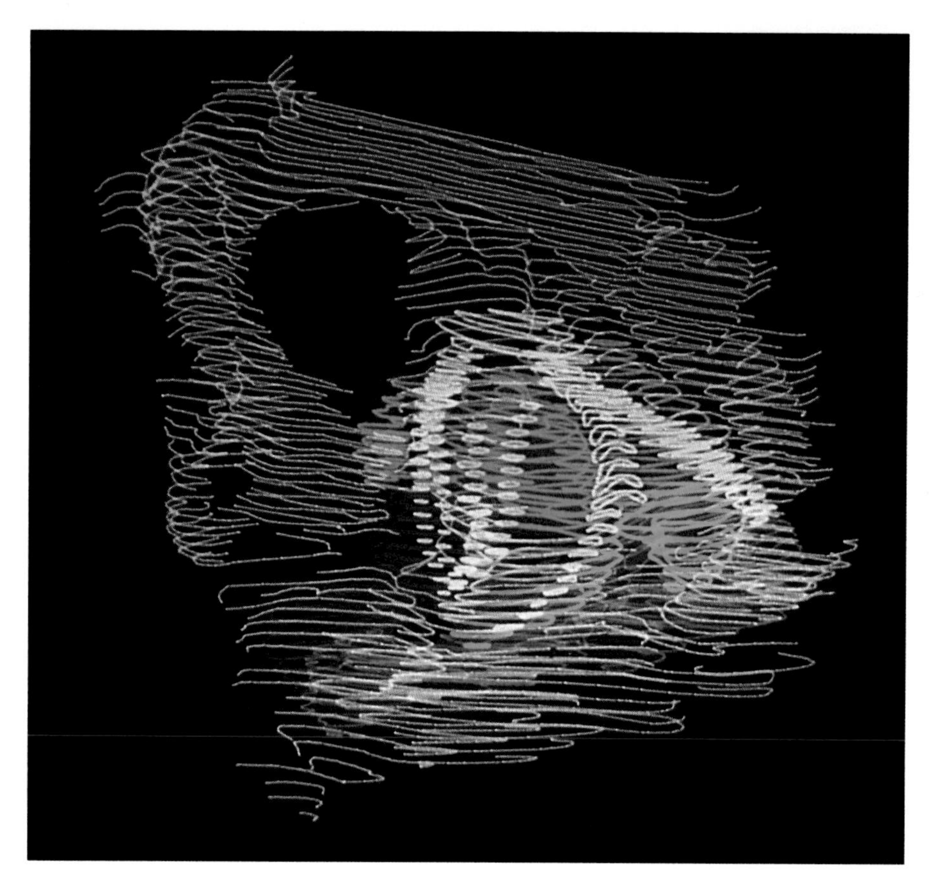

The sight of a foreigner—the first one she has seen in her 100-year life—taps wellsprings of emotion in this woman of Kashi, China. Such spontaneous responses to a chance encounter stem largely from the brain's limbic system, whose color-coded parts appear in a computer-generated image (above). The orange, handlelike structure at the top represents the limbic region of the cortex. The yellow fornix seems to reach across the green septum to pluck a cherry hypothalamus. Blue hues represent the hippocampus; pink, the amygdala. Together, this alliance of cerebral structures interacts with other parts of the brain and nervous system to mediate deep-seated emotional responses such as fear, rage, aggression, and pleasure.

similar to those occurring in a coma, at 1 to 3 cycles per second.

At this point we go into an extremely active stage with rapid, low-amplitude waves, during which our eyes move under closed eyelids as if we were looking at something—and in fact, people who are awakened during this rapid eye movement (REM) sleep almost invariably report that they were dreaming. It is a paradoxical state in which the brain is as active as when we are awake. The heart rate, breathing rate, and blood pressure go up, the penis becomes erect, and yet most of the body is paralyzed.

REM sleep recurs in regular cycles every ninety minutes throughout the night and lasts from about ten minutes to half an hour. Babies spend about half of their sleeping time in the REM stage, but as we grow older the time devoted to it decreases. Adults average about one or two hours of REM sleep every night.

Why we dream—and why we sleep at all—is not clear. Several scientists have suggested that REM sleep gives the brain the opportunity to analyze and file the day's events, update its programs, or discard useless information. During this stage, perhaps, certain connections among neurons are strengthened while others are weakened or lost.

The electrical changes in the sleeping brain are accompanied by changes in the activity of certain neurotransmitters. The neurons containing norepinephrine, for instance, fire less frequently during the early stages of slow-wave sleep and do not fire at all during the REM stage. During waking—and especially during emotional arousal—they show increased activity. These norepinephrine-containing neurons make up less than one percent of the cells in the brain, and their cell bodies are concentrated in the brain stem; nevertheless they have far-reaching effects through their connections with many different parts of the brain.

When levels of norepinephrine and a related transmitter, serotonin, are too low, people tend to become depressed. Reserpine, a drug used in the 1950s to reduce blood pressure, made some people extremely depressed and even suicidal; it was later found to reduce concentrations of norepinephrine and serotonin in the brain. The fact that this chemical could produce depression came as a revelation to psychiatrists and helped to establish that behavior problems can have a biochemical basis. Most of the drugs now used to fight depression increase the availability of norepinephrine and serotonin in the brain.

Another neurotransmitter that can affect our emotions and even our sanity is dopamine. Dopamine-containing neurons in the brain stem connect with the limbic system and the frontal lobes, the site of thought. The symptoms of schizophrenia have been linked to an oversupply of dopamine, particularly in the frontal lobes, and nearly all the drugs that control these symptoms fit lock-and-key fashion into dopamine receptors on brain cells; by blocking these receptors in the brain, the drugs reduce the effect of dopamine.

There is a price to pay for this reduction. Through a different set of connections, dopamine-containing neurons reach the basal ganglia, which control complex movements. When dopamine levels are reduced there as a result of damage to some neurons, or when dopamine receptors are blocked by antipsychotic drugs, some people develop motor problems resembling those of Parkinson's disease, a condition caused by a lack of dopamine.

The Brain's Barrier

Ideally, the way to treat the genuine Parkinson's disease would be to give patients more dopamine. But dopamine, like many other neurotransmitters, cannot get into the brain from the bloodstream because of the curious structure of the blood vessels in the brain—the so-called blood-brain barrier. To protect the brain against toxic chemicals that might circulate in the blood, only a few selected substances are allowed to seep into the brain through the walls of these blood vessels.

Besides oxygen and carbon dioxide, only some very small nutrient molecules such as the essential amino acids *(Continued on page 355)*

Short-circuiting the Body's Pathways of Pain

Preparing for brain surgery, a patient at Hotel Dieu Hospital in New Orleans has his head aligned within the grip of a stereotaxic ring. Surgeons seeking to relieve the agony of a chronic back problem will remove part of the patient's skull (opposite), permanently implanting a battery-powered electrode deep inside his brain. The energized electrode will stimulate specialized brain cells to produce natural painkilling substances. Such a procedure is used only as a last resort to treat chronic, long-term pain.

In a much rarer operation, used almost exclusively to eradicate the pain of advanced cancer, electrodes temporarily implanted in the frontal lobes (below) obliterate the brain cells that register pain. In both operations the patient remains conscious, receiving only a local anesthetic. Strangely enough, the brain itself, seat of all sensation, remains almost completely insensitive to pain.

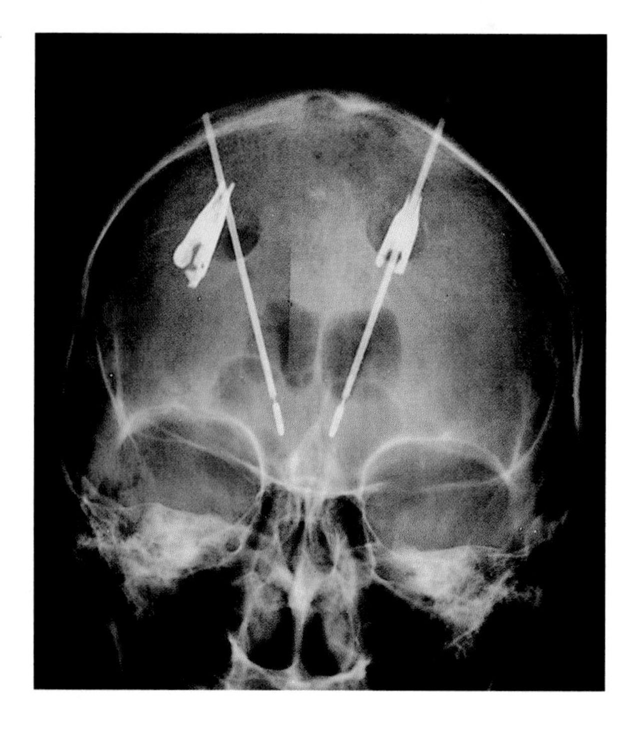

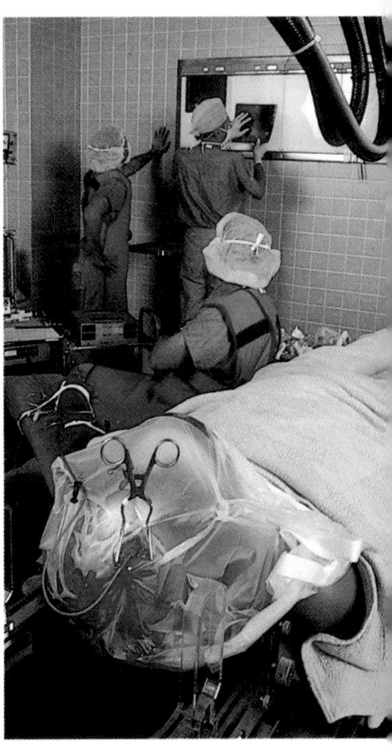

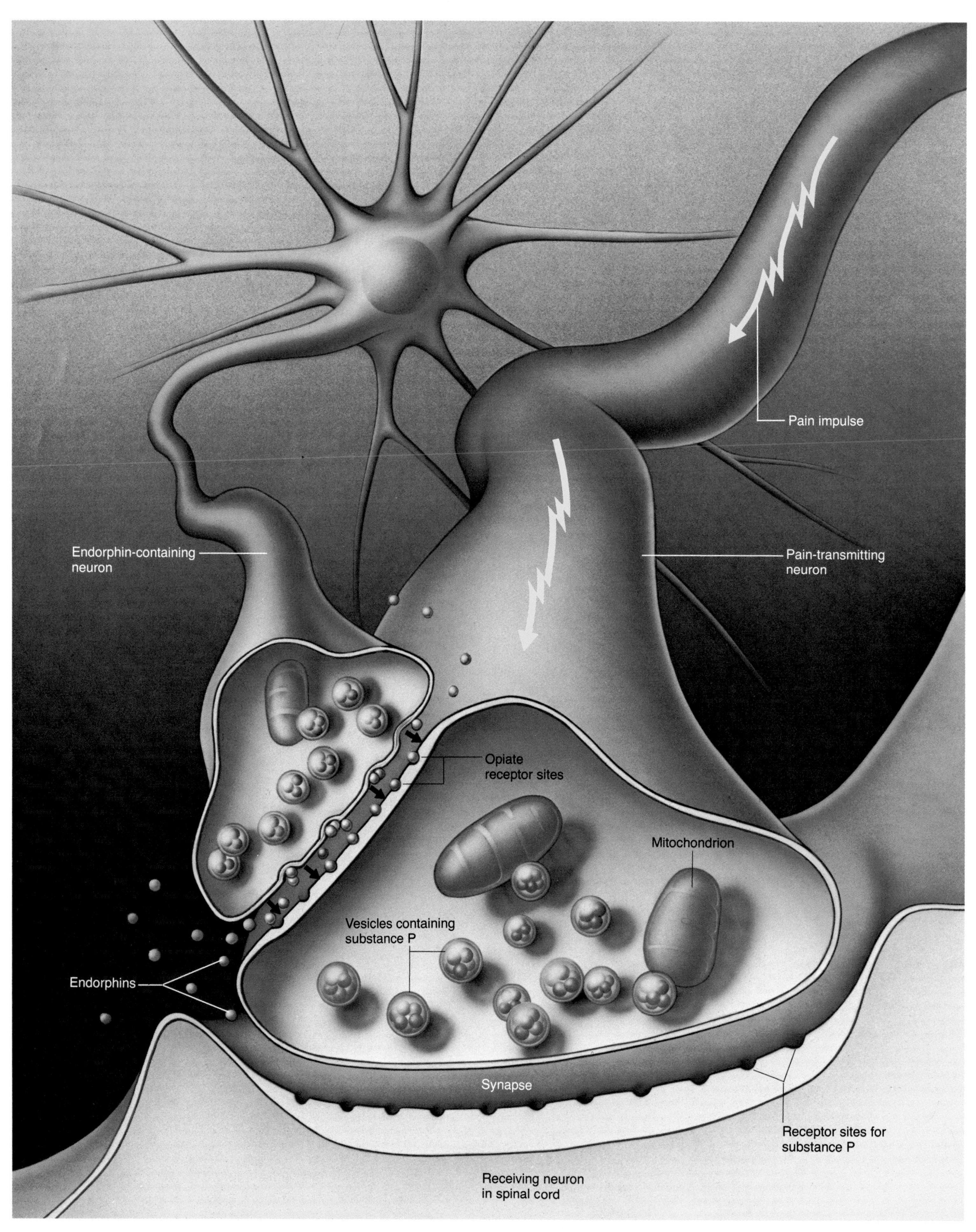
Pain impulse
Endorphin-containing neuron
Pain-transmitting neuron
Opiate receptor sites
Mitochondrion
Vesicles containing substance P
Endorphins
Synapse
Receptor sites for substance P
Receiving neuron in spinal cord

The body's own painkillers, a family of natural opiates called endorphins, ease pain by inhibiting its transmission (opposite). Stub a toe or pull a muscle and the pain impulse travels along special nerve fibers from the trauma site through the spinal cord to the brain. A chemical called substance P transmits the bad news.

In response to the message, special neurons in the brain and spinal cord may release endorphins. These substances latch onto opiate receptors on the pain-transmitting neurons, reducing the output of substance P and quelling the pain.

To relieve a football player's aching back (right, upper), a trainer applies brief electrical charges to nerve endings under the skin, a process known as transcutaneous electrical nerve stimulation. The treatment stimulates the flow of endorphins and other natural painkillers.

The computer-generated colors of a thermogram (right, lower) reveal areas of body heat and coolness. Asymmetries in such a temperature map can point to physiological abnormalities sometimes associated with pain.

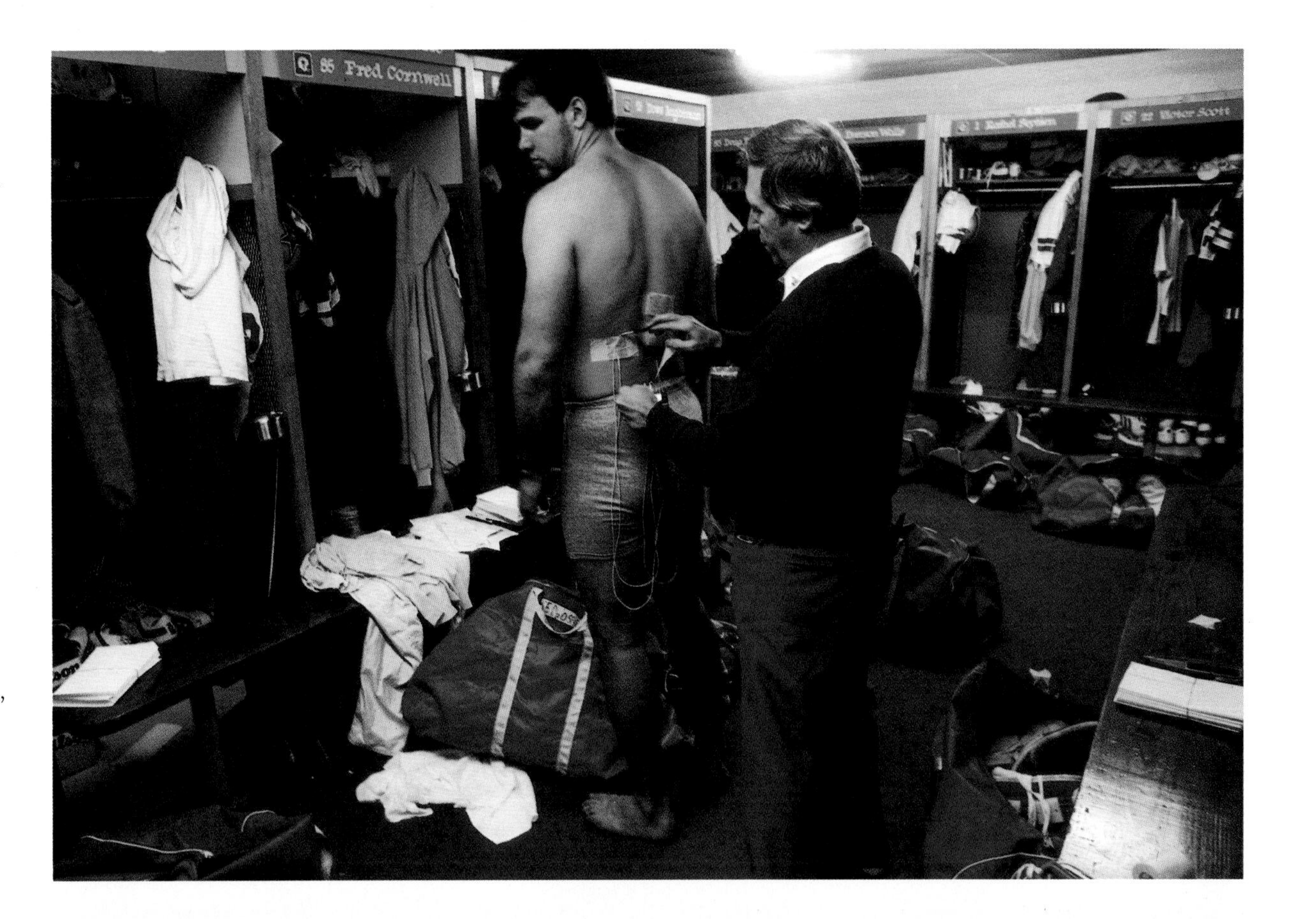

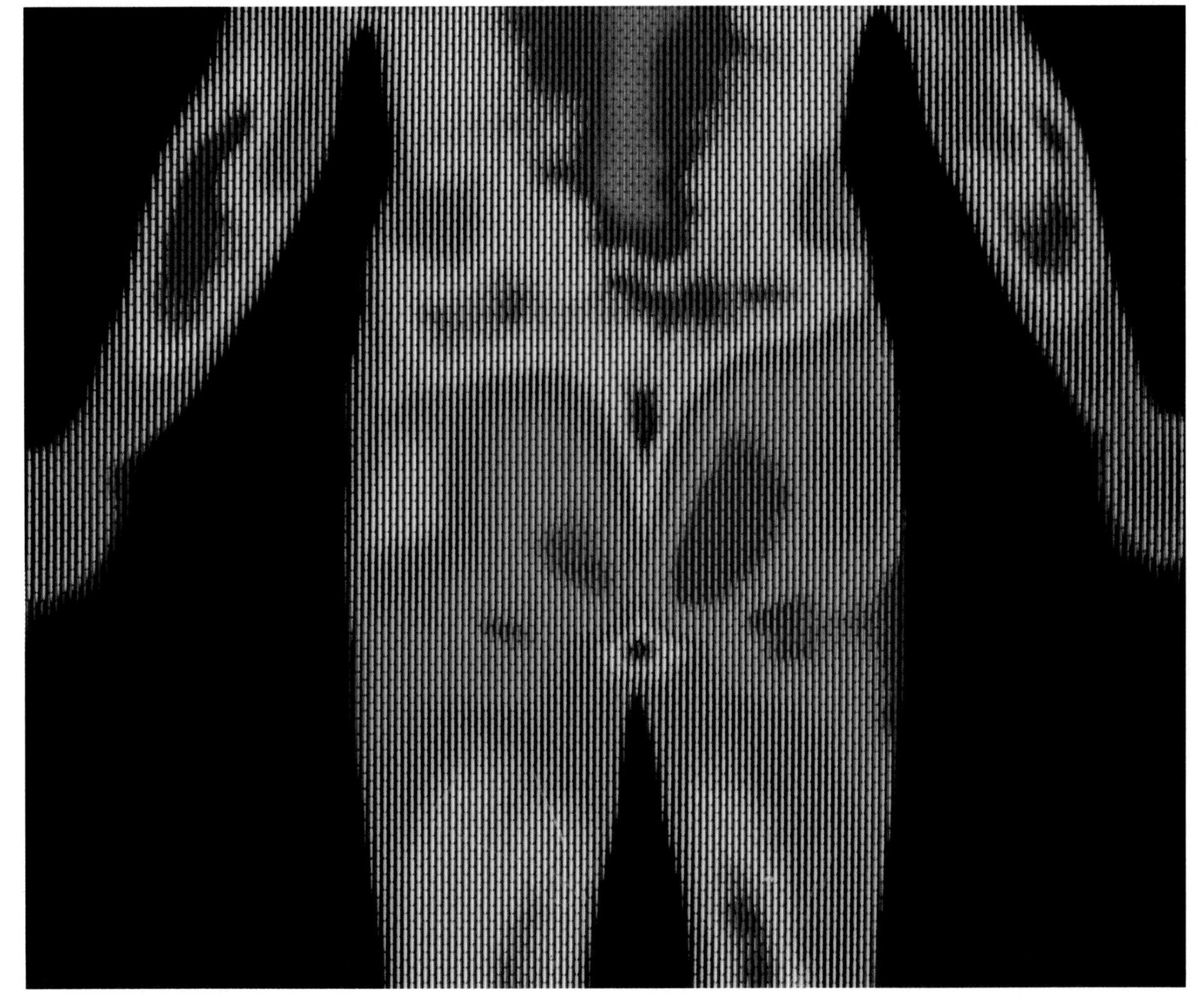

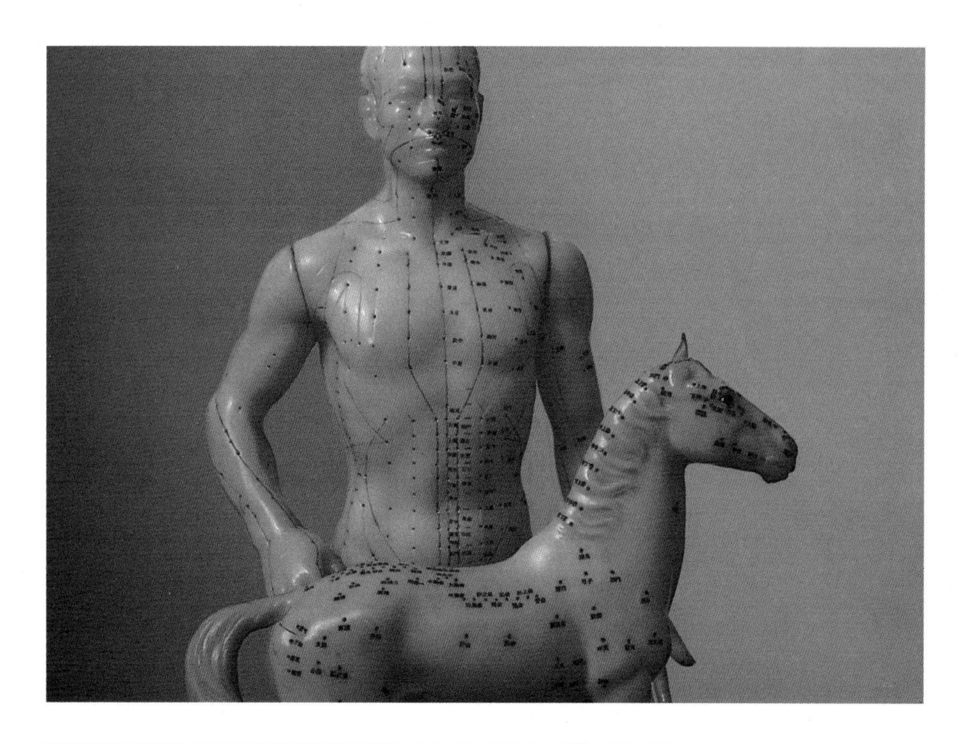

Acupuncture, a healing art practiced in China for thousands of years, may have the power to stimulate the brain's pain control mechanisms, according to medical researchers. Chinese belief holds that inserting fine needles under the skin at specified points (left, upper) helps rebalance life forces, ending pain and restoring health.

In Shanghai (left, lower), a child appears unfazed during a prickly treatment for headaches.

Scientists now believe that stimulation of the puncture points by twisting or agitating the needles triggers a release of endorphins and other painkilling neurotransmitters. Some 40 conditions—ranging from hiccups to diarrhea—respond to acupuncture, but most American doctors use it to relieve pain from osteoarthritis, migraines, and toothaches.

In Venice, California, a woman undergoes acupressure treatment (right), a variation of acupuncture technique.

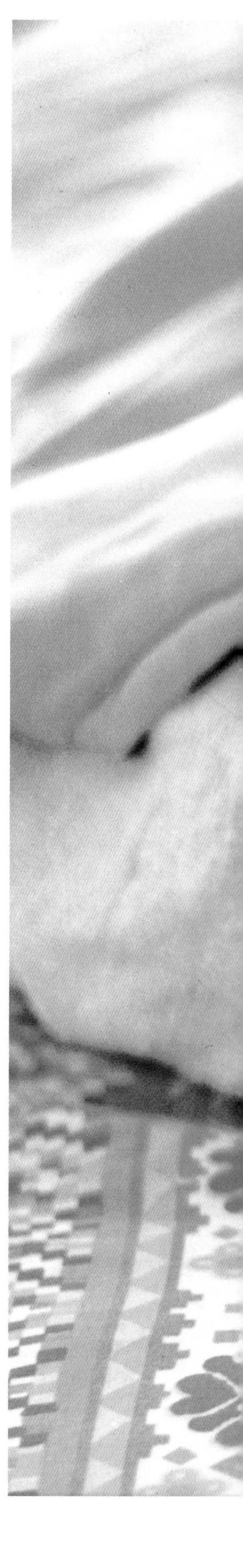

Engaged in animated conversation at a café in Jerusalem, this couple calls into play a uniquely human ability—the power to communicate by talking.

In talking, air expelled from the lungs rushes up the throat to the larynx. There the air creates sound by vibrating a pair of vocal cords, more precisely called vocal folds. Muscles tighten the folds (below, upper) to produce high-pitched tones. Relaxed folds (below, lower) create deeper tones. The lips, tongue, and jaw, all working together, further modify the sounds into intelligible words.

The listener's inner ear converts words and sounds to nerve impulses which travel to the brain. Here, the auditory cortex processes the information and sends it to a language area—usually in the brain's left hemisphere.

Actual size

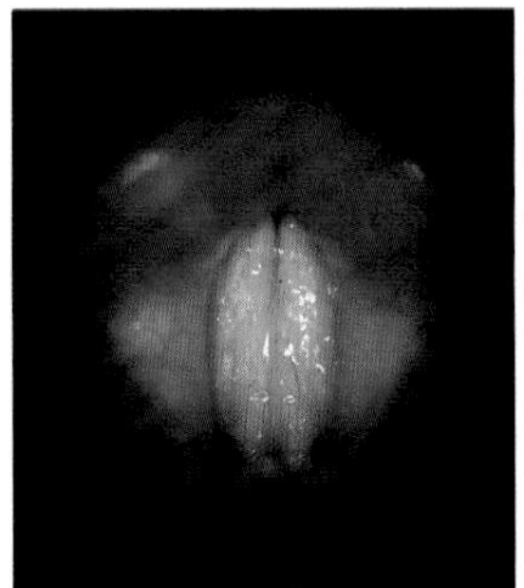

Actual size

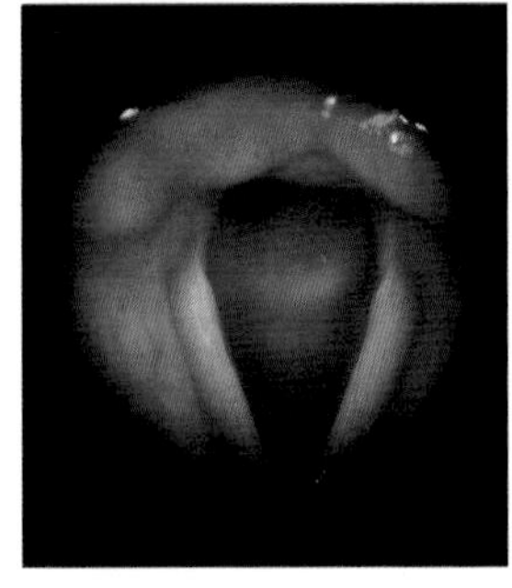

The computerized scans (right) demonstrate changes in the brain's electrical activity when jolted by an unexpected sound—in this case a high-pitched audio tone. A startling word or phrase might trigger a similar reaction.

The two specialized regions of the brain that enable people to talk to one another are named for the 19th-century doctors who discovered them. Wernicke's area, located in the left temporal lobe next to the auditory cortex, allows a listener to recognize and understand spoken words. It also helps to shape a response, if necessary, conveying the appropriate information to a part of the brain known as Broca's area. This egg-shaped patch, in turn, devises a vocalization program which it passes on to the motor cortex area that controls the muscles of speech.

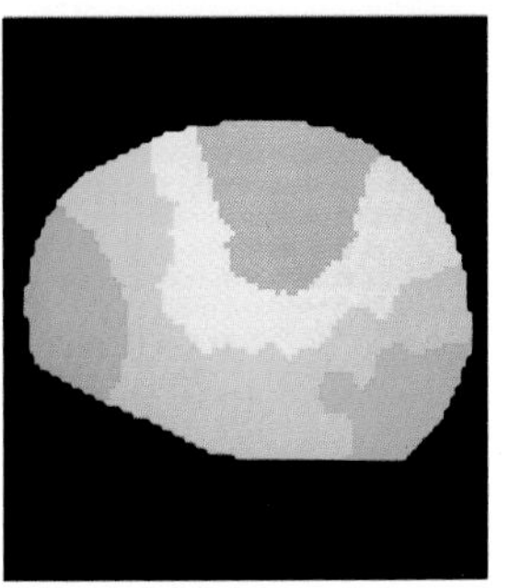

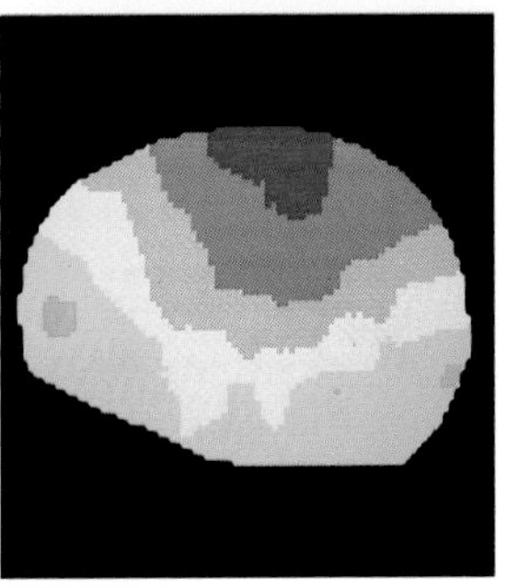

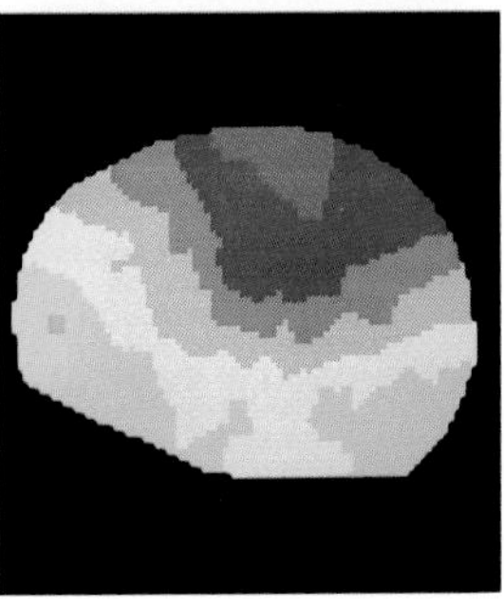

(from food), glucose, and a few other substances of a particular structure, electrical charge, and solubility can get past this barrier. They include such mood-changing substances as caffeine, nicotine, and alcohol. Certain foods, too, can influence moods because their amino acids pass through the blood-brain barrier. Eating carbohydrates can make you sleepy because this allows more of the amino acid tryptophan to enter the brain, where tryptophan is converted to serotonin, a neurotransmitter that is involved in sleep. Similarly, the chemical L-dopa passes through the blood-brain barrier and is converted to dopamine in the brain, so it is now the standard treatment for Parkinson's disease.

Many of the brain's chemical circuits have been traced through the forbidding jumble of neuronal connections. With the help of fluorescent dyes, certain neurotransmitters can be made to glow in different colors in slices of brain tissue, revealing their pathways through the brain. Thus, norepinephrine glows bright green, while serotonin glows yellow. Radioactive chemicals can be injected into the brain, to be taken up by certain cell bodies and transported along their axons, where they accumulate at the terminals and show up on film. Other chemicals used by researchers travel in the opposite direction: Taken up by axon terminals, they reveal the cell bodies from which the axons originate.

A new map of the brain, based on chemistry, can now be superimposed on older models of anatomical connections. But this map must be revised almost daily as scientists continue to discover new brain chemicals. A decade ago, science knew only a handful of neurotransmitters. Today about 50 have been identified, and researchers expect to find hundreds more.

Emotions and Chemistry

Most of the new ones are neuropeptides—short chains of amino acids linked together—which play a prominent role not just in the brain but throughout the body. In the brain they are neurotransmitters of the emotions. Elsewhere in the body they act as hormones and influence the immune system. They are extremely potent even at low concentrations.

One recently identified neuropeptide causes anxiety, the unpleasant feeling that impels an animal to watch out for itself in threatening circumstances. Valium and similar tranquilizers block the cell receptors for this neuropeptide, preventing it from having any effect.

Another neuropeptide, enkephalin—one of the endorphins, a coined term meaning "morphine within"—has nearly miraculous powers. It relieves pain and also produces euphoria, mimicking the action of opiates. For thousands of years, human beings have combated pain with

The mastery of written language eludes a dyslexic child, who struggles to copy words from the board. Dyslexia renders reading a lifelong challenge to otherwise able individuals. Its cause has escaped scientists so far.

Apparently adept at language, the gorilla Koko signs "teeth" as she views a picture of a grinning chimp. The subject of one study aimed at teaching language to primates, Koko knows more than 500 words in American Sign Language.

Whether Koko and other "talking" apes actually use language—or simply show an elaborate form of conditioned behavior, as critics claim—stirs debate. Nevertheless, studying the way apes learn the elements of language has helped determine how such skills are acquired by children, and how human communication evolved.

substances taken from the opium poppy, the original source of such addictive drugs as heroin and morphine. In 1973 researchers at Johns Hopkins University found and measured the cell receptors to which these drugs attach themselves in the brain, and shortly afterward other scientists isolated enkephalin, a natural brain substance which also fits into these receptors.

It is good to know that we produce our own painkiller. Apparently we always have a small supply of it, but it may be released in large doses at times of extreme stress, as in battle or in childbirth and, it seems, during acupuncture.

Pain and pleasure exist entirely in the brain. Strange as it may seem, even orgasms occur in the brain. Though we may think that pain comes from a wound or illness, in fact the reason that any damage to the body produces pain is that a neuropeptide which transmits pain-related signals, substance P, is released from peripheral nerve fibers and stimulates certain neurons in the spinal cord. The neurons then forward these pain signals to the brain. The transmission of substance P can be stopped, however. Other neuropeptides—either enkephalin or one of the opiates—can attach themselves to receptors on neurons that contain substance P and prevent them from firing, so that fewer pain impulses reach the brain.

The sense of well-being or euphoria created by the opiates and by enkephalin seems to involve opiate receptors in a different pathway of the brain from that of substance P. In addition, another pleasure pathway or "river of reward" may be activated by the neurotransmitters dopamine and norepinephrine. Both pleasure systems probably evolved to reinforce activities essential to survival, such as eating and mating.

When runners in a marathon experience a high, it might reflect a rush of enkephalin; and some runners appear to become addicted to the experience. Addiction to opiates may result from a feedback system of the cells that have receptors for natural enkephalin in the brain. Narcotics such as heroin and morphine occupy many of the receptors that would normally be filled by enkephalin. A kind of thermostat then signals the brain that no more enkephalin is needed, and the brain produces less of it. After a while, many enkephalin receptors become unoccupied, creating a craving for more narcotics to compensate for this loss. If these drugs are then stopped, withdrawal symptoms arise because of a lack of natural enkephalin.

The more scientists learn about the brain's myriad chemical circuits and feedback loops, the more complicated the system appears. It is totally unlike the way other organs operate. In the brain, many different types of neurons collaborate with one another, and their function depends entirely on their intricate anatomical and chemical connections. Not only are there far more neurotransmitter systems than was originally believed; not only do some neurotransmitters modify the effects of others on particular cells; but in addition, many neurotransmitter

ABCDE
abcde
bad
dog
went

Bach becomes child's play at the hands of young violinists on the grounds of Matsumoto Castle in Matsumoto, Japan (opposite). The youngsters, ranging from three to five years old, study under Shinichi Suzuki, a violin teacher who years ago observed that "all children in the world can speak their native language." Applying his observation to musical instruction, Suzuki has taught children to "speak" music as easily as they learn their own tongue—and at a rate associated with genius-level intelligence.

Less common is a phenomenon known as eidetic imagery—sometimes called photographic memory. Studies indicate that the ability may be present in many young children, but that it fades away rapidly around the onset of puberty. People with eidetic imagery can scan the hillside photograph (right, upper) for thirty seconds—and remember the entire scene in detail.

Scientists believe that humans have at least two memory systems which may reside in different parts of the brain. Procedural memory, a memory for skills, probably develops earlier in life than declarative memory, the ability to recall facts. Fact memory appears to center in the hippocampus, amygdala, and part of the thalamus. Procedural memory, believed more widely dispersed, is therefore less subject to impairment by illness or injury. Hence a victim of amnesia can solve a complex puzzle (right, lower), a task involving procedural memory, with the same efficiency as a normal subject—though unable to remember having seen the puzzle before.

circuits actually interconnect. Furthermore, a given neuron may produce more than one type of neurotransmitter. And for each neurotransmitter, there may be several types of receptors. So far we know of at least four different kinds of dopamine receptors. As a further refinement, cells often try to compensate for an oversupply or a lack of a given neurotransmitter by changing the number of receptors on their surface.

The brain thus has an almost infinite number of options to choose from in reaching a particular goal. How these are juggled as the brain performs its tasks is not yet clear. The most baffling operations involve such complex, higher functions as language, memory, and learning.

Language depends largely on two regions of the left hemisphere that are unique to human beings. A French surgeon, Paul Broca, identified one of these areas in 1861 when he did an autopsy on a man who had lost the ability to speak. Now known as Broca's area, this egg-shaped region of the left frontal cortex produces a detailed program for vocalization (involving muscles of the face, tongue, jaw, and throat) which it sends to the motor cortex. The motor cortex then transmits these commands for muscular activity to the motor neurons that control the muscles involved in speech. If cells in Broca's area are damaged, normal speech is impossible. However, the victim can still understand what is said and can communicate through facial expressions, hand gestures, or writing.

When Language Goes Wrong

A different part of the brain, Wernicke's area of the left temporal lobe, is essential for the understanding of language. When people are injured there, the results are catastrophic. They can still speak, but only gibberish. In a strange parody of normal speech, they go on conversing fluently with correct rhythm, intonation, and even grammatical form, but their words are meaningless and they cannot grasp what is said to them. Their reading and writing are also impaired.

Then there are people who can read, write, and speak normally, but cannot understand spoken language—they are "word deaf." Their problem is that the connections between Wernicke's area and the part of the cortex that receives messages from the ear have been damaged; the signals that the brain receives about the sound of other people's speech can no longer be interpreted as language.

For some reason, language is the specialty of the left hemisphere, at least in 95 percent of right-handed people and 70 percent of left-handers. Besides Broca's and Wernicke's areas, several other parts of the left hemisphere may be involved in language as well. The right hemisphere does make a contribution, however: It

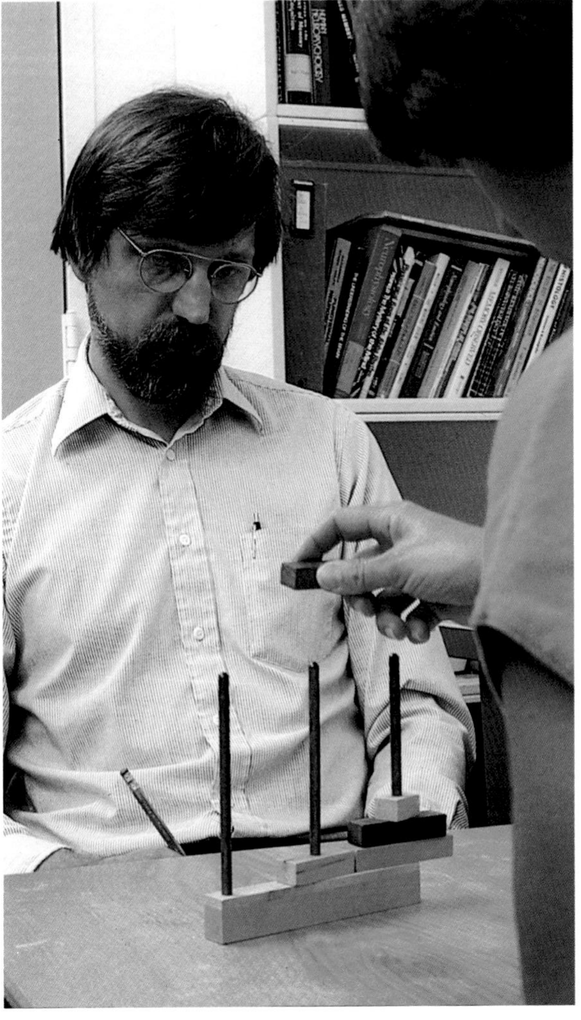

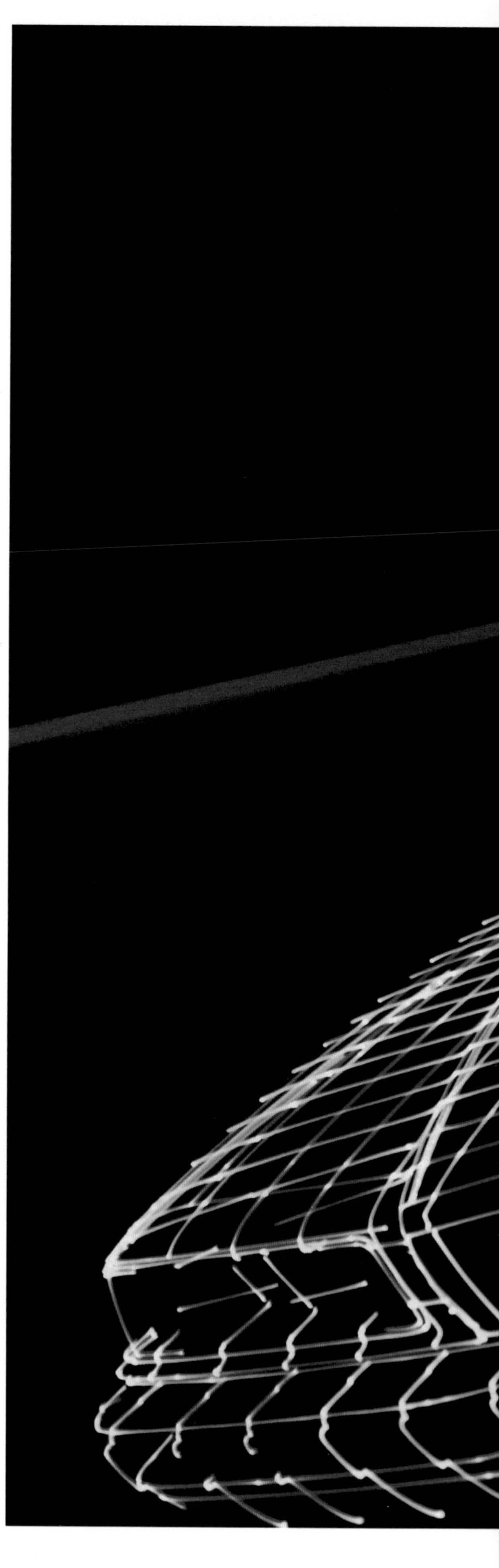

Sublime inspiration or plain hard work? Science seeks to define the creative process—whether expressed in the studied simplicity of Zen art or the curvilinear lines of a new car model. Some researchers view creativity as a matter of unconscious drive or sudden, blind inspiration. Others suggest creativity is just a different way of thinking, an outgrowth of normal perception, memory, and understanding.

Studies show that creative people tend to have certain traits in common. Many were drawn to their fields at an early age, inspired by an experience that moved them deeply. They also share a capacity for hard work, giving credence to Thomas Edison's view of the matter. "Genius is one percent inspiration," wrote the inventor, "and ninety-nine percent perspiration."

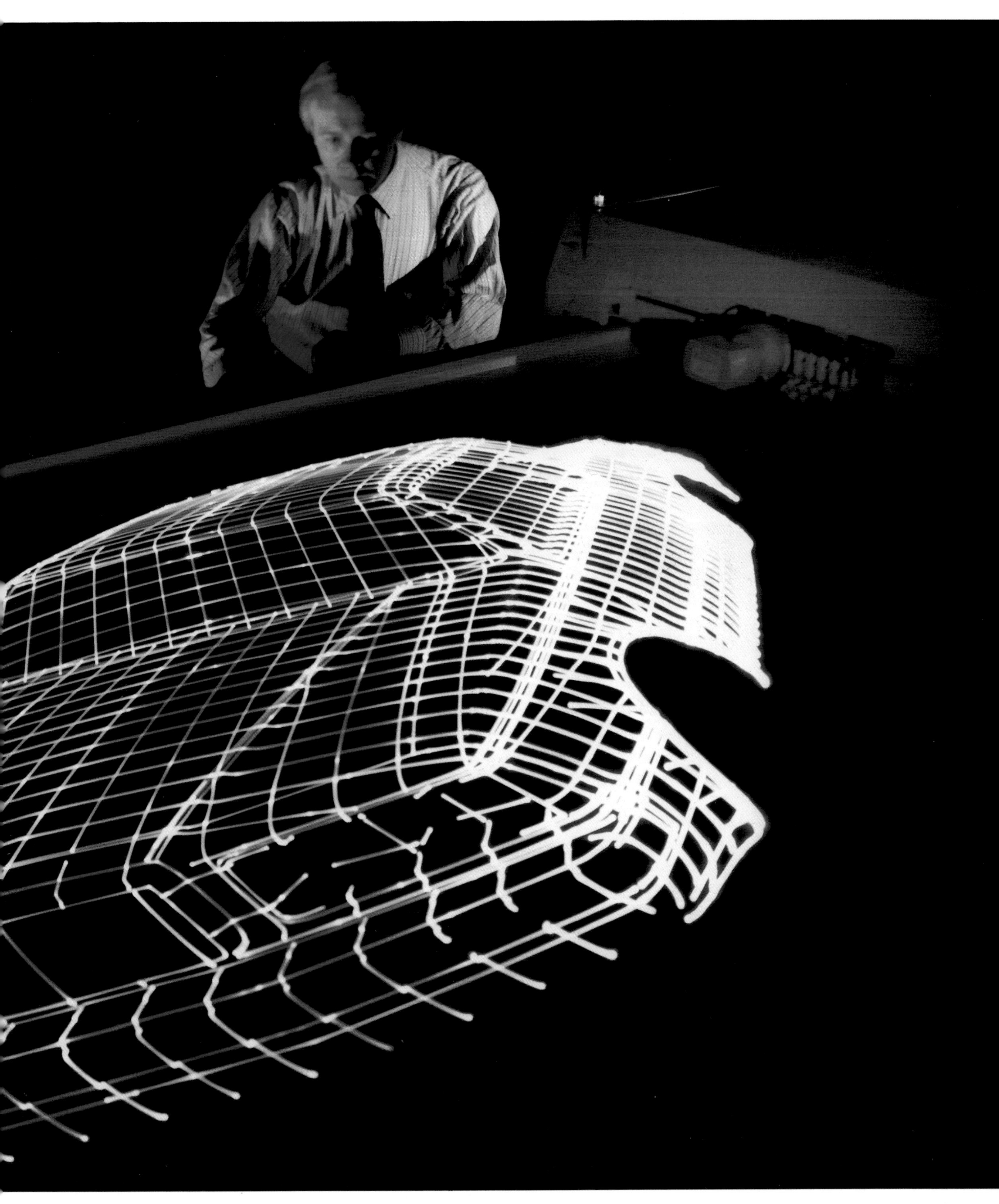

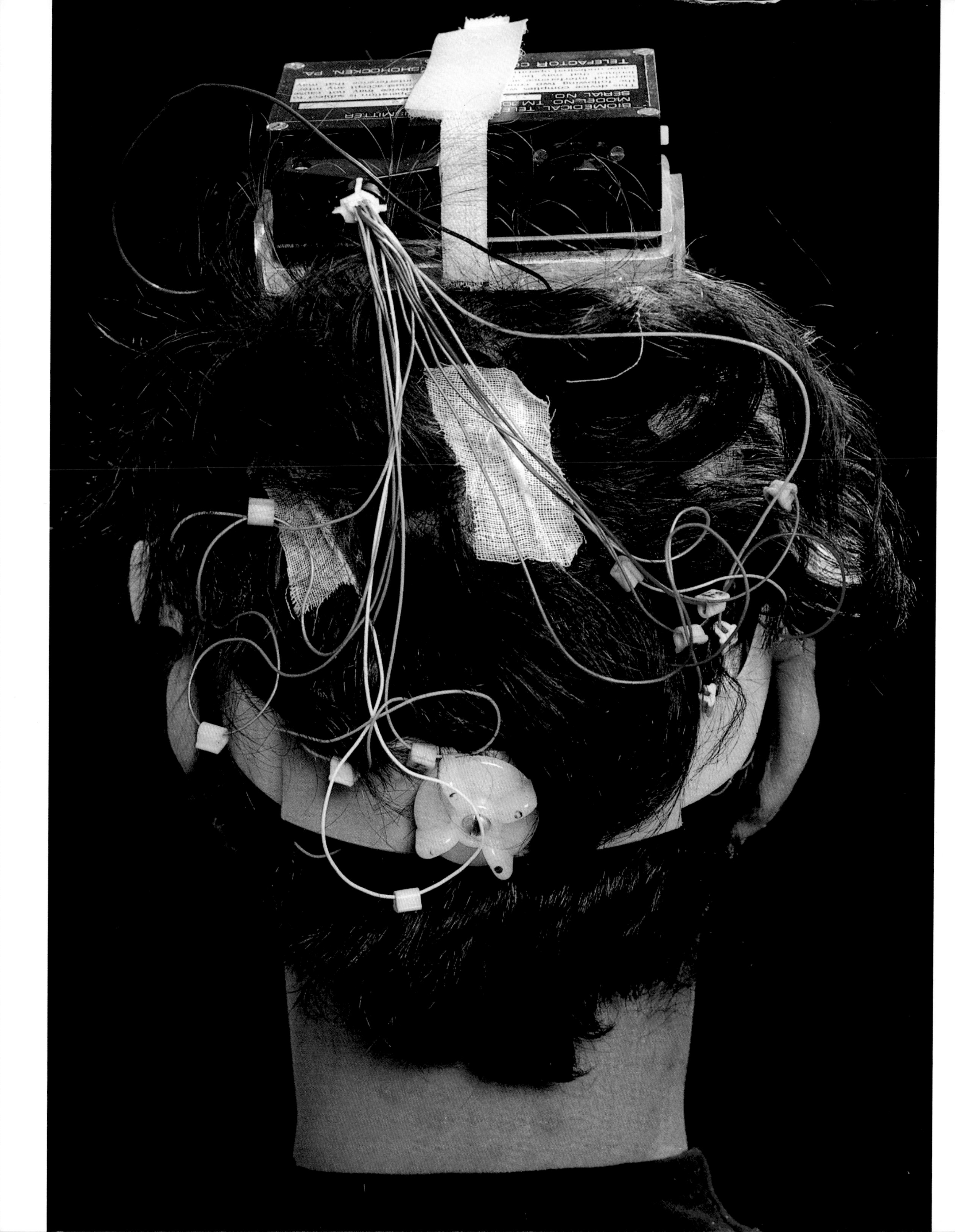

Epilepsy, the equivalent of an electrical storm in the brain, affects one of every hundred Americans—more than two million people in the United States. This young patient wears headgear that allows him to move about freely while doctors monitor his brain waves and behavior over an extended period.

Epilepsy has no single cause. Some victims inherit a propensity for it. Others develop symptoms as a result of infections, head injury, tumors, or other conditions that can affect the brain.

Normally, brain cells produce small bursts of electrical impulses in a constantly changing pattern. But in epilepsy, large clusters of nerve cells fire simultaneously. This disrupts normal brain function and sometimes causes a seizure.

The seizures themselves vary widely in form and intensity. The so-called simple partial seizure, centered in a particular area of the brain, often results in motor or sensory disturbances—the jerking of limbs or numbness. The grand mal, a generalized seizure which involves the entire brain, can cause violent convulsions and loss of consciousness.

Once over, the seizure often leaves the victim dazed, exhausted, and with no memory of the event.

Doctors today tailor treatment to the specific type of seizure suffered, usually with antiepileptic drugs. If drugs fail, surgeons may remove areas of diseased tissue. In extreme cases, they may sever the connection between the brain's two halves in an attempt to confine the electrical rampages to a single hemisphere.

deals with intonation, emotional tone, and other nonverbal aspects of language. A small area of the right frontal cortex, the mirror image of Broca's area, controls vocal nuances and gesturing. After an injury there, people sound completely flat; they have difficulty expressing anger, surprise, or joy. Injuries to another part of the right hemisphere, the mirror image of Wernicke's area, lead to difficulty in understanding what it means when someone's voice inflection, word stress, or gestures change.

The left and right hemispheres act much like joined twins, physically linked but with personalities of their own. Their link is the corpus callosum, a large bundle made of over 200 million nerve fibers, arching for four inches along the midline of the brain. Along this nerve bundle runs an endless stream of messages that unify our thinking by letting one side of the brain know what the other is doing. The corpus callosum coordinates the two sides of our bodies and two different ways of dealing with the world.

The left hemisphere tends to dominate. It controls not only speech, language, reading, writing, and the skilled operation of the usually dominant right hand, but also our analytical, sequential, and symbolic activities—logic, math, scientific reasoning, and many other forms of thought. The right hemisphere specializes in spatial abilities; for instance, it does better than the left in recognizing faces. It is also the more musical side of the brain, though, of course, it can neither read nor write musical notes. Potters, dancers, and sculptors, among others, seem to favor the right side of their brains.

The two hemispheres differ anatomically, although at first glance they may appear identical. Generally one region of the temporal lobe, the region that contains Wernicke's area, is larger on the left side of the brain than on the right. This is not just the result of acquiring language, for the difference has been found in the brains of stillborn babies and even fetuses; it seems to be an inborn trait. However, the right hemisphere of males sometimes becomes enriched during fetal development as a result of variations in the levels of the male hormone, testosterone; in such cases, some language centers may develop in the right hemisphere, and the child may then be left-handed. About 15 percent of all left-handers—both male and female—have their language centers in the right hemispheres, and another 15 percent show evidence of having language centers both in the left and right hemispheres.

As long as our two half-brains remain connected and undamaged, we have access to the special abilities of both. But sometimes accidents or strokes destroy one hemisphere, and the victim becomes mute. Much depends on the age at which such injuries occur. If a girl's left hemisphere is damaged by a head injury or tumor before the age of five, she can learn to speak again—perhaps after a year of silence—as her right half-brain takes over the job. Some flexibility of this sort is retained up to the age of ten. Later than that, however, dominance for speech and

Two minds in one brain? Yes, say specialists who study the brain—whether through children's art or the composite photographs (above) used to test patients who have undergone split-brain surgery. The operation, very rarely performed, severs the nerves that link the brain's two halves. Such studies reveal that each hemisphere has its own strengths and weaknesses.

In the test pictured above, a split-brain patient briefly viewed a composite image of two faces. When asked what she saw, she replied, "the child"—the half-face that her left, verbal, hemisphere perceived. When asked to point to the photo, her finger indicated the woman's image—seen by her right, nonverbal, hemisphere.

other left-hemisphere functions is fixed. Right-handed adults who have suffered a left-hemisphere stroke can regain some speech only if they have enough undamaged tissue remaining in key areas near the injury.

Sometimes the two half-brains need to be separated for medical reasons—as, for example, when doctors operated on a World War II veteran who had been hit in the head by bomb fragments and had developed such severe epilepsy that nothing seemed to help him. During his frequent seizures he would fall down unconscious, foaming at the mouth. He often hurt himself as he fell. For more than five years his doctors tried every available drug treatment, without success. As a last resort, in order to stop the seizure activity from spreading across the two hemispheres of his brain, they cut through his corpus callosum. The seizures stopped as if by magic. There was a rocky period of recovery, and then the man announced that he felt better than ever. But in some ways he had become two different persons. Sometimes his two hands seemed to fight for control; once he threatened his wife with his left hand while his right hand tried to come to her rescue. When asked to "raise your hand," he could respond only with the right side of his body, which was controlled by his verbal left hemisphere; evidently his right hemisphere did not understand the command.

After studying a number of split-brain patients, Roger Sperry, a Nobel prizewinning psychobiologist, concluded that "each hemisphere . . . has its own perceptions, private sensations, thoughts, and ideas . . . its own private chain of memories and learning experiences that are inaccessible to recall by the other hemisphere . . . a separate 'mind of its own.' " Our sense of identity, then—the feeling that we are just one person—comes, in part, from having an intact corpus callosum.

The Mysteries of Learning

As much as we have learned about some of the mechanisms involved in language, we still know almost nothing about the fundamental questions of why and how human beings learn language in the first place. How do we form the abstract concepts that underlie certain words? How do babies learn to talk? What accounts for the prodigious speed with which young children acquire new words and—more significantly—discover the rules of grammar? Childish mistakes such as "I digged a hole" or "Look at the sheeps" reveal brilliant, creative minds at work; they show that children as young as two or three apply complex grammatical rules—such as adding *ed* to verbs in the past tense or *s* to plural nouns—rules which they must have figured out for themselves. How does the electrochemical

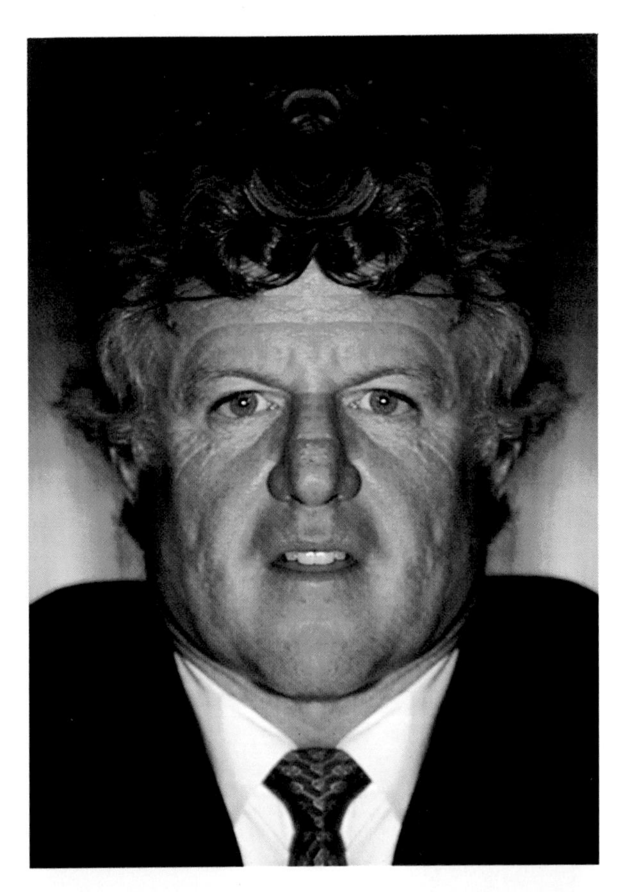

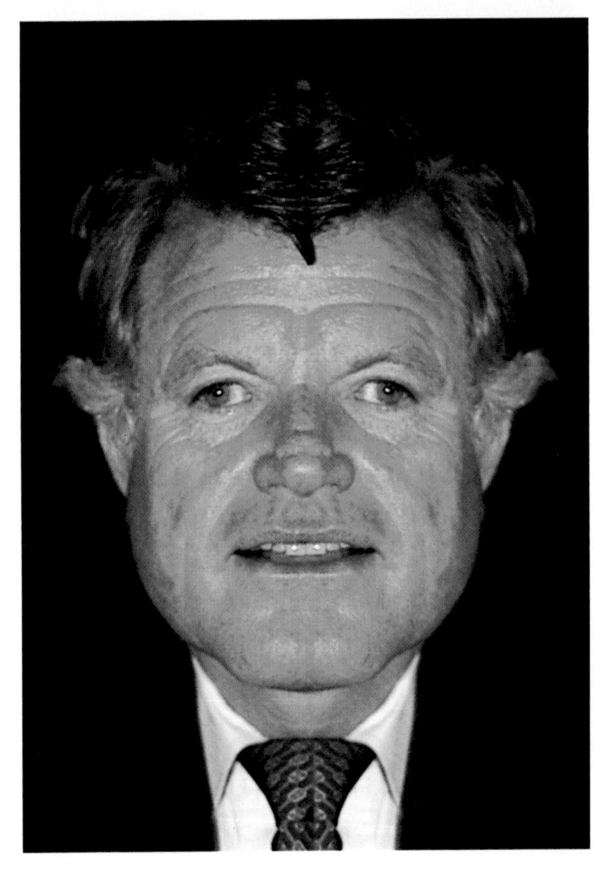

The two hemispheres of a normal brain work together to provide an integrated view of life. But each half specializes: The left usually processes language; the right, visual, spatial, and emotional information. This pattern may vary depending on several factors, including gender.

Asked to draw a "sad" picture (opposite, upper), children—especially girls—tend to place their subjects on the left-hand side of the picture, indicating the dominant role of the brain's right hemisphere in interpreting and expressing sadness. Children of either sex drawing a "happy" picture (opposite, lower) usually center their subject in the middle of the page.

The way we see a face, too, reflects the brain's dual nature. The composite images of Sen. Edward M. Kennedy (both at right) have been pieced together from right-hand (upper) and left-hand (lower) sides of his face. Both composites reveal distinctively different features. Which is more recognizable? For most viewers, the top picture. Why? Because the brain's right hemisphere, which perceives the right-hand side of the senator's image, is usually more adept at recognizing faces.

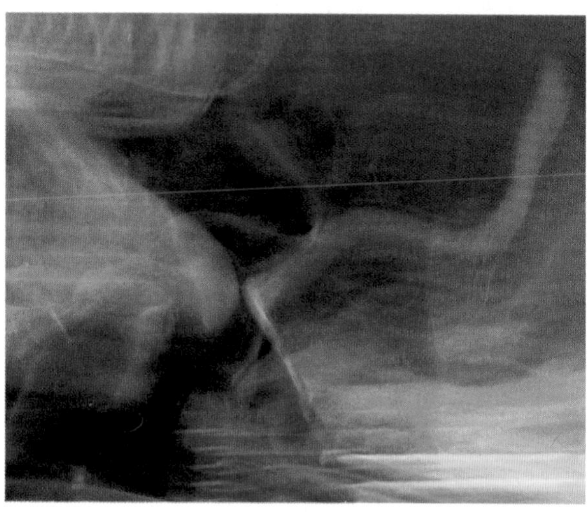

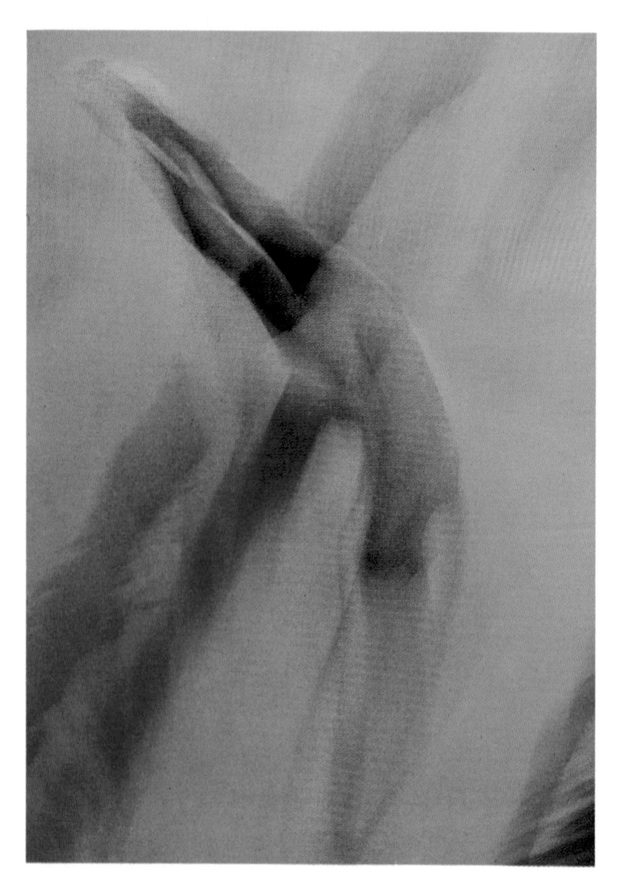

Counseling a police officer in Maryland, a psychologist (left, upper) can use a variety of techniques to soothe a troubled mind.

Scientists know little about the brain—even less about the mind, the sum total of an individual's conscious and unconscious mental activity. To dream, hope, think, will, and reason—all are part of the mind. Psychology, a relatively new and multifaceted discipline, studies the mind and behavior. For some patients with psychological disorders, talk becomes a key to unlocking problems of the mind.

Other patients respond favorably to less traditional methods. The ambiguous, dreamlike photographs (left, center and lower), key to an innovative technique called phototherapy, seek to stimulate an emotional response so that a therapeutic conversation can take place.

Free-form techniques, such as the lifting exercise at the Esalen Institute in Big Sur, California (opposite), attempt to promote awareness and development of the self, rather than tackle any specific psychological problem.

activity we see in the brain's neurons get translated into such knowledge? How do we learn anything—and how do we remember it?

Surprisingly, some forms of learning are possible even without a brain. Individual neurons of the lowly marine snail, which has neither brain nor spine, can learn to ignore a weak stimulus (if it is repeated often enough) or to strengthen all their responses after exposure to a threatening stimulus. They can also be conditioned in the Pavlovian sense. For example, individual neurons can learn to slow down their activity whenever they sense a light.

Such learning alters the structure and chemistry of the neuron's outer membrane. Much of the simple learning done by human beings may follow the same pattern. That is, our brain cells may be structurally altered whenever we learn to ignore the hum of an air conditioner, for instance, without our ever being aware that we have learned it. At the level of individual cells, biochemical mechanisms are pretty much alike in all animals.

Different Kinds of Memory

More complex learning requires a critical mass of interconnected neurons. What distinguishes the human brain is its dense network of linked circuits which involve billions of neurons in different brain regions, giving us an almost unlimited capacity to learn, to assess information and to remember it. Furthermore, human memory is not all of a piece. There is, first, immediate memory, as of passing scenery; this lasts only a second or two and then—if nothing marks the event as important—vanishes without a trace. Short-term memory lasts several minutes; we use it for temporarily important things, such as telephone numbers. Then there is long-term memory, which stores information for hours or for life.

Selected items from short-term memory are transferred to long-term memory through a process we don't yet understand. But a terrible surgical error has shown that two parts of the limbic system—the amygdala and the hippocampus—are essential to the process.

In 1953 a surgeon removed these structures from both the left and right hemispheres of a young man known as H. M. during a brain operation designed to cure him of epilepsy. Since then, H. M. has been unable to make any new memories. Everything vanishes from his mind so rapidly that his own life since the surgery is a total blank to him. He does not know where he lives, remembering only the house of his childhood. He does not recognize people whom he sees every day—not even people he saw just a few minutes earlier. He rereads copies of old magazines, which seem forever new to him. Yet his IQ remains high and he can carry on an

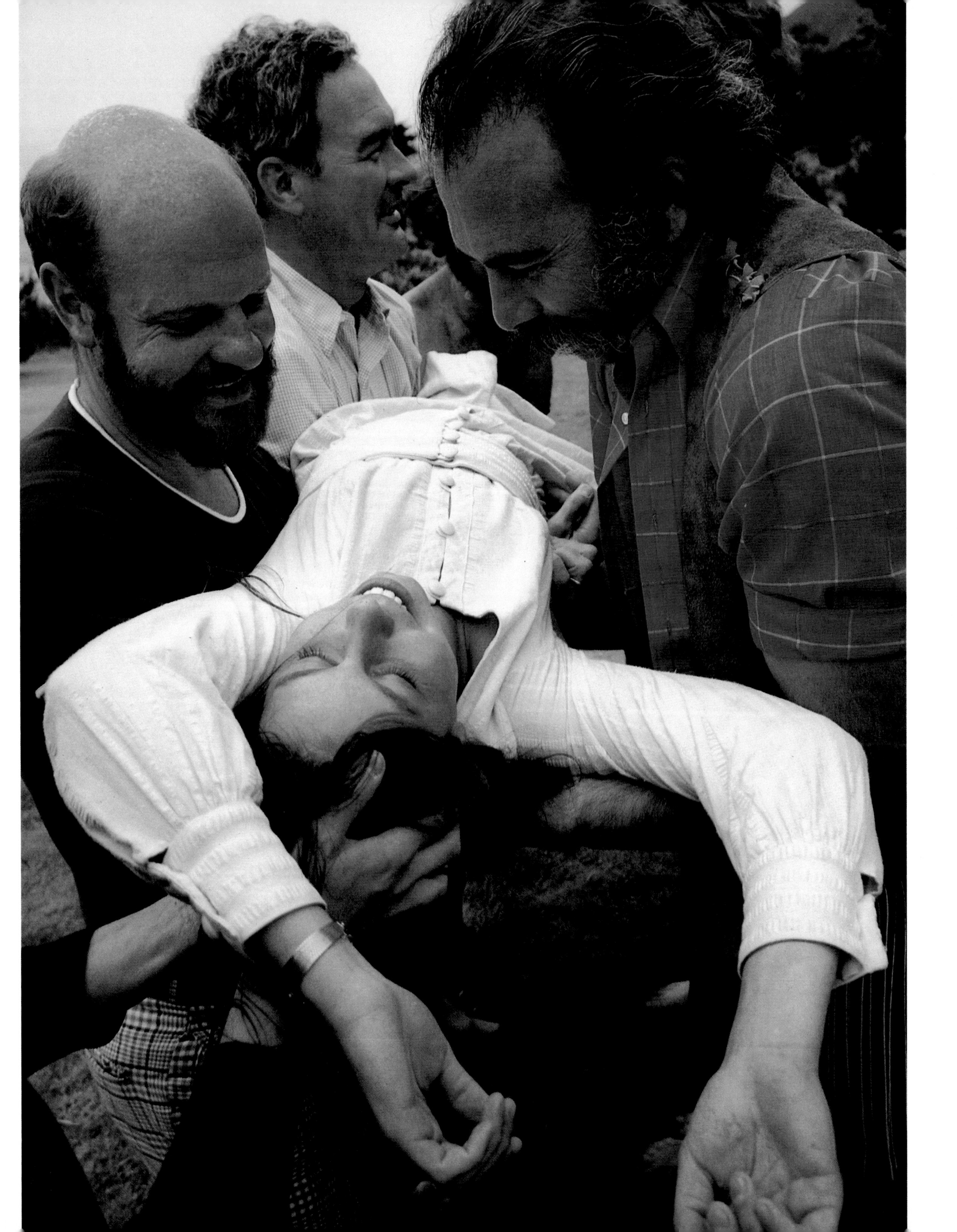

MUSTARD POWDER
FENUGREEK SEED
frankin-sense
Golden seal powdered
GARLIC
CHAPPARRAL

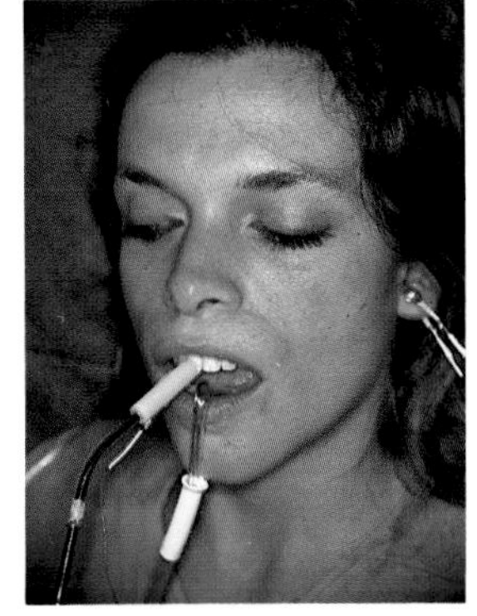

Folk remedies—whether an invigorating curative tea blended in Virginia or the soothing waters of an herbal bath in China—long have been prized for their restorative properties, real or imagined.

But what about sugar pills and other placebos administered purely for psychological relief? These, too, can provide actual pain relief—even after the extraction of wisdom teeth (above). Recent studies indicate that a placebo can indeed reduce pain—if the inert substance tricks the brain into activating the body's natural painkillers. Such studies help scientists better understand the links between mind and body.

intelligent conversation as long as it doesn't deal with anything that happened since his surgery.

The psychologists who tested H. M. were surprised to find that despite his overwhelming handicap he could still learn new skills, such as solving puzzles. He improved with every trial, though he kept insisting that he had never seen the puzzle before. Evidently we have two separate memory networks, one for facts, or declarative memory, and another one for skills and habits, or procedural memory. This second memory network may underlie the simple, unconscious learning that extends in an unbroken line from lower animals to humans.

For years two schools of psychology have done battle—the behaviorists versus the cognitive psychologists—on such questions as how people learn. Do we learn specific responses to stimuli through reinforcement and conditioning (the behaviorist view), or do we acquire information and knowledge in an intentional manner, even without any external rewards (the cognitive view)? It now appears that we are capable of both types of learning, in different circumstances. Skill memory involves lower brain regions such as the basal ganglia and the cerebellum, which generally do not communicate with the conscious mind. Factual memory requires the active cooperation of the limbic system. The two forms of memory may be controlled by different biochemical mechanisms and may have arisen separately in the course of evolution.

In dealing with new factual information, we try to put it in some sort of context, so we construct frameworks of knowledge—highly individual matrices into which we fit the new material so it can be stored efficiently and retrieved easily. Most of this processing appears to be done in the cerebral cortex, where certain permanent memories are stored. At least three-fourths of the cortex on both sides of the brain is taken up with complex activities that are sometimes lumped together under the name "associations" and seem to underlie our intellectual achievements. Whenever we have new ideas, form judgments, make plans, create tools, tell stories, or compose music, we use neurons in association areas of the cortex—particularly those in the prefrontal cortex, an area of the frontal lobes just behind the forehead.

Far from being diffuse and undifferentiated, as once believed, the prefrontal cortex has an exquisitely precise organization. Like other parts of the cortex, it consists of thousands of columns, each one made of about a hundred vertically interrelated neurons and connected to other, similar columns, forming information-processing modules. These modules receive contacts from neat arrays of axons: Those from neurons in the cortex of the opposite hemisphere alternate with those from neurons of the same hemisphere, probably to integrate information, memories, strategies, and plans. The prefrontal cortex also has close connections with three deeper brain areas: the thalamus, the caudate nucleus in the basal ganglia, and the brain stem. It is hard to

The power of faith to cure the sick and heal the lame baffles medical science. In Bahia, Brazil (opposite), witnesses crowd around a faith healer as he ministers to an afflicted woman. At Lourdes, France (right), crutches abandoned near the famous grotto testify to dozens of seemingly miraculous cures.

Why faith healing sometimes works is unknown. Some researchers feel the answer lies with the heightened emotional state of the participant and the physiological changes that accompany it. As one specialist puts it, many doctors "admit the reality of cures of functional illnesses . . . but they cannot see how faith alone can cure an organic disease."

believe that not so long ago, in the 1930s and 1940s, this delicate web of nerve fibers used to be slashed almost blindly during lobotomies, the crude brain operations that pacified thousands of schizophrenic and other mental patients before the discovery of effective drug treatments.

What we choose to store in our long-term memories is closely tied to our emotions: We seek pleasure and shun pain, so if something is rewarding we learn to repeat it, while if it is dangerous or painful we learn to avoid it. Neutral events tend to be forgotten. We learn efficiently only when we are aroused in some way—which may mean that the neurotransmitter norepinephrine is involved in memory.

The recognition that something is dangerous—or else very new and exciting—does more than arouse the mechanism of memory. It also triggers a whole cascade of events in the body through the brain's control of the nervous system and its release of hormones and other chemicals. The brain signals the adrenal gland to release epinephrine (also called adrenaline), making the heart beat faster, raising blood pressure, and directing more blood to the brain and muscles. The entire body is poised for either fight or flight. We are all familiar with some of the results—the racing heart, churning stomach, dry mouth, and trembling hands of fear.

When such hyperarousal is prolonged, it can make us sick. Recurrent fight-or-flight responses to chronic stress may lead to heart disease, for instance, as high levels of epinephrine provoke the release of fats which clog the arteries, impede the flow of blood, and eventually damage the heart. This is just one of the many links between our states of mind and our health. A few conditions such as ulcers and asthma have long been known as psychosomatic illnesses which can develop in response to emotional upsets. And there are some painfully obvious, immediate physical reactions to mental stress such as sweating or vomiting. But now it appears that every aspect of our health may be influenced by our mental state.

The brain can have a direct effect on the immune system, the powerful network that protects us against infections and cancer. Certain white blood cells which play a key role in our immune reactions appear to have specific receptors for brain neurotransmitters. This helps to explain why mental depression can produce ills that range from colds or cold sores to life-threatening diseases: As our immune responses are reduced by brain chemicals, viruses which would normally be destroyed may thrive, while cancer cells which would normally be kept in check may proliferate.

Physical effects such as these can be triggered by mere ideas—for instance, by the arrival of bad news. Our cerebral cortex makes it possible for us not only to plan ahead but also to worry about the future. As a result, we can be stressed just as much by anxious thoughts as by the sight of a real tiger charging at us. Even the mental stress of taking a final exam was sufficient to reduce the

Religion—whether consecrated with a High Mass at St. Stephen's Cathedral in Vienna, Austria, or celebrated as a fervent devotional act in the Andean foothills of Peru—can lift and exalt the human spirit. Through the ages religion has sought answers to life's many riddles. So, too, has science. While science has revealed the electrical and chemical mechanisms by which the brain functions, it has barely begun to tell us how an individual rises to intellectual, spiritual, and moral fulfillment.

immune responses of medical students in a study at Ohio State University. The level of activity of their natural killer cells (white blood cells which instantly recognize and attack foreign cells without having been previously exposed to them) fell significantly on the exam day. In another study, college students who seemed particularly stressed or lonely had a third the amount of natural killer cell activity found in other members of their class.

It's how we evaluate our circumstances that really counts. Generally, challenges that we clearly can control do us no harm—we even seem to thrive on such success—but those that make us feel helpless may be truly dangerous to our health. In some primitive societies, witch doctors punish people who have broken tribal taboos by pointing a bone in their direction. This is supposed to cast a mortal spell. The culprits then believe they are doomed. They fall to the ground, moaning and writhing, refuse contact with others, and die within a few days. Similarly, a number of Americans die each year after taking poison in such small doses that it could not in itself cause death, or after inflicting small, nonlethal wounds on themselves, probably because they are convinced of their doom.

By contrast, the expectation of pain relief can sometimes produce it. If you are told that you will get a shot of morphine to reduce the pain of a tooth extraction, but in fact get only a placebo, you have one chance in three of feeling less pain after the tooth is pulled. There is no magic involved here: The placebo effect may result, in part, from a very real surge of endorphins in the brain. The release of endorphins may be a response to previous conditioning—to the repeated experience that when one is told a drug will stop pain, it does. Whatever the cause, placebos not only make patients feel better but, in some cases, also appear to promote healing.

Another purely mental procedure, hypnosis, can provide powerful pain relief. It can also protect tissue against damage—for example, from burns. It seems to work through a different pathway in the brain than endorphins, possibly by integrating signals at the level of the cerebral cortex, while the endorphin-containing neurons are concentrated in the limbic system.

In view of the many connections between our expectations, our brain chemicals, and our tissues, one might expect that happy thoughts would lead to good health. Just as stress can reduce immunity, can positive thinking increase our resistance to disease? The evidence so far is largely anecdotal. Believers in meditation and other forms of stress management point to some of their results: lower blood pressure, or reduced activity of the sympathetic nervous system. Some researchers claim that cancer patients who have a confident and combative attitude suffer fewer recurrences of their disease than people who feel helpless or hopeless. But it is difficult to tell the cause from the result. Do people feel more confident because their bodies are stronger, or vice versa? The facts are not yet in.

Nevertheless it seems clear that the brain's multitude of interlocking systems is powerful and complex enough to account for any number of physical results. Certain neurotransmitters—for example, norepinephrine—affect both body and brain, influencing such diverse functions as sleep, memory, aggressiveness, stress reactions, pleasure, or rage. The neuropeptides that flow in intricate loops from brain to body, and back to the brain, work to integrate the whole organism.

As in a canvas by Seurat, many points of activity conspire to produce a single effect. But the brain's canvas keeps evolving. Groups of brain cells increase or decrease their activity in constantly shifting combinations and recombinations. The diverse neurotransmitters which pulse through our billions of nerve cells produce such a wealth of themes and variations that we can encompass our past and our future, explore the heavens, and reflect the whole world in our brains.

MAYA PINES

Boy meets ladybug with all the wonder and curiosity of a paleontologist focusing on Earth's fossil record. These qualities of the mind, combined with the insatiable drive to probe beyond the known, distinguish the human brain from all others. Scientist Paul MacLean puts it this way: "An interest in the brain requires no justification other than a curiosity to know why we are here, what we are doing here, and where we are going."

Seeing the Invisible

When Lennart Nilsson was 11 years old, he began taking pictures with a 25-cent camera. In all the years since, he has been peering through lenses of one kind or another, first as a newspaper photographer in his native Sweden, then as a contract photographer for *Life* magazine.

For the past 20 years Nilsson (seated at his scanning electron microscope, right) has specialized in medical and biological photography. He collaborates with Dr. Jan Lindberg (standing, right), a clinical and forensic pathologist who practices at the Karolinska Institute in Stockholm.

The Karolinska Institute, which annually awards the Nobel Prize in Medicine, presented Nilsson with an honorary doctorate in 1976, the first time the award was ever given to a photographer.

Despite such honors, Nilsson still considers himself "just a journalist, not a science photographer. The important thing is to tell a story so that ordinary people can understand it."

Nilsson and Lindberg began working together in the 1960s, preparing stories on heart disease and fetal development for *Life*. They have since produced books and articles, and collaborated on films that have won acclaim in Sweden and America. One film, *The Miracle of Life*, won an Emmy in 1984.

"Lennart is the artist—he sees the views in his head," says Lindberg, who handles the scientific side of the collaboration. Lindberg and Nilsson keep in touch with scientists at the Karolinska Institute and with specialists from abroad, working with them to prepare specimens for microscopic examination.

For the photographs in this book, Nilsson used an array of techniques, as well as equipment ranging from the miniature to the gargantuan: A lens the shape of a penlight captured the inside view of an artery wall (page 146), while Nilsson's scanning electron microscope, which made the virus photograph (page 183), fills most of a small room.

Subjects, such as the fetus on page 37, photographed at actual size or low magnification (up to 50 times) were recorded with standard portrait cameras. For pictures in the medium range (50 to 1,500 times), Nilsson used a specially built light microscope connected to a camera. Pictures taken by this method appear on pages 21 and 234. None of these required great depth of field; that is, the emphasis is on seeing the structure at close range, on a single plane with no background visible.

The three photographs on page 377 were taken with a scanning electron microscope, so named because it uses electrons to produce visual images for the camera. Here one sees, magnified 100 times, 900 times, and 2,000 times, the architecture of human hair, sheathed in its armorlike plates of keratin.

The scanning electron microscope is used for a wide range of magnifications (10 to 100,000 times). Because this microscope uses electrons, rather than light waves, to

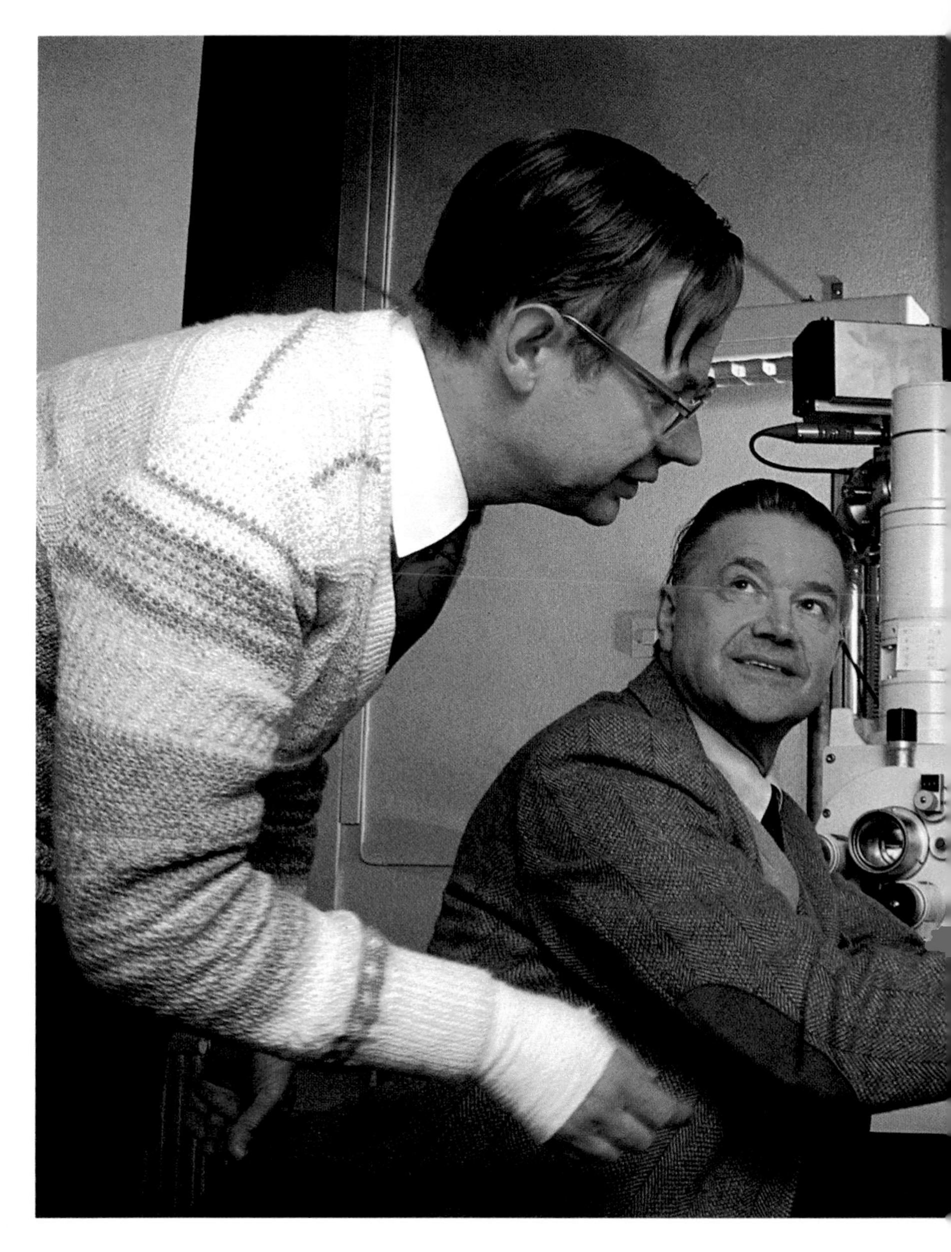

produce images, all such photographs are predominantly gray or blue, depending on the kind of film used. In this book most of the Nilsson photographs were tinted with color after they had already been developed. This was done as an aid to readers, to help distinguish the various parts of cells and tissues.

Subjects for Nilsson's photographs must undergo extensive preparation. His samples are almost always from autopsies and biopsies. Such cells and tissues are dead in one sense, in that they are no longer attached to humans, but alive in another, in that they can live for days or weeks in the controlled conditions of a laboratory.

For instance, in the battle scene involving macrophages and bacteria (pages 170-171), Nilsson and Lindberg took cells and put them in a special nutrient solution to keep them alive. Then they treated the specimens at intervals,

removing one sample, then another, from the culture, and coating each with a fixative solution called glutaraldehyde. Like a snapshot, this process "freezes" each specimen at an instant in time. By preserving several samples over several hours, Nilsson and Lindberg established a sequence for the story, just as a moviemaker would string together scenes to form a plot.

The preserved sample was then dehydrated and coated with an ultrathin layer of gold, a coating that improves the tissue's conductivity in the microscope. The three-step treatment—preservation, dehydration, coating—

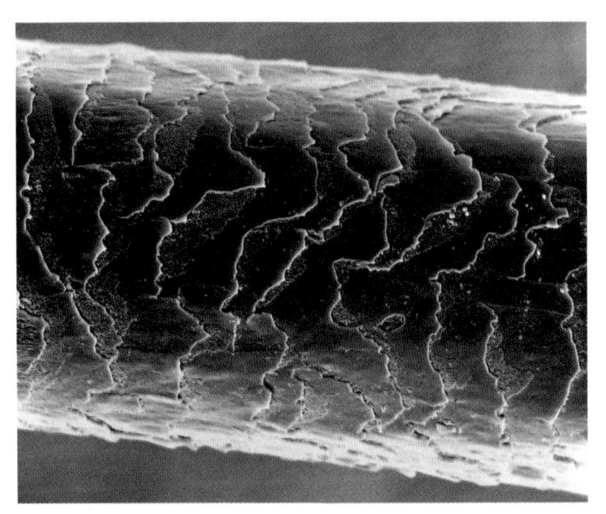

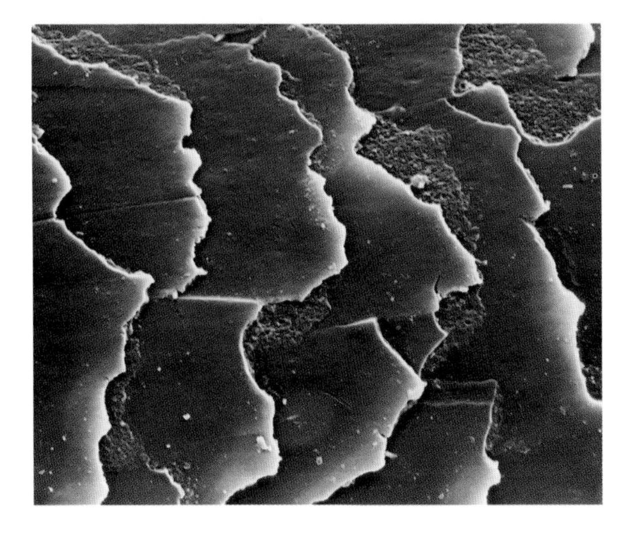

makes samples more durable, so that they can be photographed repeatedly and even transported without damage.

Only after these elaborate preliminaries was the subject ready for a portrait in Nilsson's electron microscope. Mounted on a brass pedestal, the tissue was placed inside a vacuum chamber at the base of the microscope, and then bombarded with electrons shot from a filament at the top of the microscope.

The electrons struck the metallic surface coating the tissue and bounced off, just as light particles rebound from a mirror to convey an image to one's eye. Within the microscope, the bouncing electrons were drawn to a sensor. From the sensor, electrons were converted into light particles, then back into an electronic signal. In a final step, the signal was amplified and sent to a viewing screen at the control panel of Nilsson's microscope.

There Nilsson sits for hours, days, and months, trying one specimen after another until the elements of a story emerge.

"You have to wait and wait and wait until the conditions are right," says Nilsson. "Sometimes I wait one day, sometimes it is one year or several years."

That wait can be trying, but Nilsson has great staying power, according to his friend Lindberg: "Lennart will never give up. He will respect no hours until he gets the story he wants." And that, says Lindberg, is how Nilsson makes "the invisible into the visible."

Illustrations Credits

Abbreviations for terms appearing below:
(t)-top; (c)-center; (b)-bottom; (r)-right; (l)-left; NGP-National Geographic Photographer; NGS-National Geographic Staff.

Cover stamping, Michael A. Hampshire. Page 3, Michel Tcherevkoff. 8, Michel Tcherevkoff.

Beginning the Journey
12, Michel Tcherevkoff. 14-15, Lennart Nilsson, *A Child Is Born;* Dell Publishing Co., New York. 16, Lennart Nilsson. 17-18, Robert J. Demarest. 19, Lennart Nilsson. 20, Robert J. Demarest. 20-27(t), Lennart Nilsson. 27(b), Lennart Nilsson, *A Child Is Born,* Dell Publishing Co., New York. 28-37, Lennart Nilsson. 38, Mickey Pfleger. 39, Enrico Ferorelli, DOT. 40-41, Lennart Nilsson. 42(t), Carl W. Rohrig. 42-47, Kirk Moldoff. 49, Richard Weiss, Peter Arnold, Inc. 50 (lt,lc), Csaba L. Martonyi, University of Michigan. 50 (lb,rb,r), Jan Lindberg. 51, Carl W. Rohrig. 52-53, Dan McCoy.

The Cycle of Energy
54, Michel Tcherevkoff. 56-57, Nathan Benn. 58, James L. Stanfield, NGP. 59, Volkmar Wentzel. 60, © 1984 Jay Maisel. 60-61, Brian Brake, Photo Researchers. 62-63, Bruce Dale, NGP. 63, William S. Weems. 64-65, James L. Stanfield, NGP. 65(t), Yva Momatiuk and John EastCott. 65(b), Andy Levin, Black Star. 66(t), Bruce Dale, NGP. 66(b), © 1984 Jay Maisel. 67, William Rivelli, Image Bank. 68, Sam Abell. 69, Kirk Moldoff. 70-71, Lennart Nilsson. 72-73, Lennart Nilsson, *Behold Man,* Little, Brown and Co. 74(l), Jan Lindberg. 76-77, Lennart Nilsson, *Behold Man,* Little, Brown and Co. 80(t), Jan Lindberg. 83(t), Jean A. Miller. 83(b), Ergun Cagatay, *Life* magazine © 1982 Time Inc. 84-85, Jean A. Miller. 86-87, © 1985 Jay Maisel. 87(t), John McGrail, Wheeler Pictures. 87(b), Michael Melford, Wheeler Pictures. 88-89, Philip A. Harrington, Fran Heyl Assoc. 90-95, Nathan Benn. 96, Cetus Corp.: Photo, Charles O'Rear. 97, George D. Lepp, Bio-Tec Images.

The Powerful River
98, Michel Tcherevkoff. 100-101, Nathan Benn. 102, Jodi Cobb, NGP. 102-103, Dan McCoy. 104-105, Lennart Nilsson, *Behold Man,* Little, Brown and Co. 106-107, Carl W. Rohrig. 107(r), Beckman Laser Institute 109(t), Scott T. Barrows. 109(b), Steven Shames, Visions. 110-111, Lennart Nilsson, *Behold Man,* Little, Brown and Co. 112-113, Nathan Benn. 113, Jeffrey Aaronson, ASPEN. 114-115, Dan McCoy. 116-117, William Strode, Black Star. 118, Kirk Moldoff. 118-119, Lennart Nilsson, *Behold Man,* Little, Brown and Co. 119, Dan McCoy. 120, Arthur J. Olson, Scripps Clinic and Research Foundation. 120-123, Lennart Nilsson. 124, Danielle Pellegrini, Photo Researchers. 125, Anthony Bannister. 126, © 1980 Jay Maisel. 127(t), John Terence Turner, Genetics Systems Corp. 127(b), Lowell Georgia. 128-129, Photo courtesy of NASA. 130, Lennart Nilsson. 130-131, Walter J. Iooss, Jr., © 1984 Fuji Photo Film USA. 132, Scott T. Barrows. 133(l), Art Siegel, University of Pennsylvania. 136, Nathan Benn. 136-137, Dan McCoy. 139(b), Lennart Nilsson, *Behold Man,* Little, Brown and Co. 140, Nathan Benn. 147(t), Dan McCoy. 147(b), Nathan Benn. 148-149, James L. Stanfield, NGP. 150-151, Lennart Nilsson. 152-155, Nathan Benn.

The Healer Within
156, Michel Tcherevkoff. 158(b), Lennart Nilsson, *Behold Man,* Little, Brown and Co. 159, David Louis Olson. 160-161, Lennart Nilsson. 162-163, Carl W. Rohrig. 164-165, Walter J. Iooss, Jr., © 1984 Fuji Photo Film USA. 165, Carl W. Rohrig. 166-167, Lennart Nilsson, *Behold Man,* Little, Brown and Co. 167(l), Lennart Nilsson. 167(r), Kirk Moldoff. 168, Roy A. Underhill. 168-169, Carl W. Rohrig. 172, Kirk Moldoff. 173, Phoebe Dunn. 174, Carl W. Rohrig. 175(l), Lennart Nilsson. 175(r), Arthur J. Olson, Scripps Clinic and Research Foundation. 176-177, Carl W. Rohrig. 178, John Moss, Black Star. 178-179, Eugene J. Gerberg. 180-181, Jeffrey Aaronson, ASPEN. 181, R. J. Johns, Bruce Coleman, Inc. 182, Carl W. Rohrig. 185, Nathan Benn. 186, Victor Lorian, Bronx Lebanon Hospital Center. 187, Nathan Benn. 188, Dan McCoy, Rainbow. 188-189, Kenneth Garrett, Woodfin Camp, Inc. 192-193, Nathan Benn. 194, Manfred von Ardenne. 194-195, Peter Menzel, Stock Boston. 196, Kirk Moldoff. 197, Linda Bartlett, Phototake. 198-200, Dan McCoy. 200(b), David Scharf 1980. 201, Dan McCoy. 202, Lennart Nilsson. 202-205, Nathan Benn. 205, Annie Griffiths. 206, Dan McCoy. 207, Fredrik D. Bodin. 208(t), George F. Mobley, NGP. 208(b), Dan McCoy. 209, Ed Lallo, People Weekly © 1984 Time Inc. 210-211, Martha Cooper. 212-215, Nathan Benn. 216, O. Louis Mazzatenta, NGS.

Stages of Life
220, Michel Tcherevkoff. 222-223, Craig Aurness. 224, Enrico Ferorelli, DOT. 225(t), Richard Howard. 225(b), © Lee Lockwood 1985. 226, Mark Antman, Image Works. 227, Ethan Hoffman, Archive Pictures. 228, Anthony Bannister. 228-229, © 1985 Jay Maisel. 230(t), Edward Lettau, FPG. 230(b), Joe McNally, Wheeler Pictures. 230-231, Bruce Dale, NGP. 232, Mark Perlstein, Black Star. 233, Richard Kalvar, Magnum. 234, Lennart Nilsson. 234-235, Nathan Benn. 236, Bill Hess. 236-237, Mel Wright, Art Resource. 238, Lennart Nilsson. 239, Kirk Moldoff. 240(b), Lennart Nilsson. 240-241, Carl W. Rohrig. 242(lt), Jean A. Miller. 242(lb), Lennart Nilsson. 242(r), Gruppo Editoriale, Fabbri-Bompiani. 243-244, Jan Lindberg. 245-246, Nathan Benn. 246-247, Jan Lindberg. 248, Lennart Nilsson. 249, Nathan Benn. 250-251, Dan McCoy. 252-253, David Hiser, Photographers Aspen. 253, John Launois, Black Star. 254-255, Nathan Benn. 256, Steve Raymer, NGP. 256-257, Nathan Benn. 257, Ethan Hoffman, Archive Pictures. 258-259, Sidney Tabak.

In Touch with the World
260, Michel Tcherevkoff. 262-263, Nathan Benn. 264, Kirk Moldoff. 265-267, Lennart Nilsson. 268, Glenn R. Steiner. 268-269, Phil Schermeister. 270, Walter Wick. Mask by Willa Shallit. 271, Museo Civico, Cremona, Italy. 272, Jim Brandenburg. 272-273, Linda Bartlett. 274, Charles O'Rear. 275, Alexander Tsiaras, Photo Researchers. 276-277, Scott T. Barrows. 278(l), Lennart Nilsson, *Behold Man,* Little, Brown and Co. 278-279, Lennart Nilsson. 279(r), Lennart Nilsson, *Behold Man,* Little, Brown and Co. 280(t), Lennart Nilsson. 280(b)-281, Lennart Nilsson, *Behold Man,* Little, Brown and Co. 282, Galen Rowell, High and Wild Photography. 282-283, Greg Heisler. 284-285, Nathan Benn. 286, Patrick Dyson, University of Florida. 287, Dan McCoy, Rainbow. 288-289(t), Lennart Nilsson, *Behold Man,* Little, Brown and Co. 289(b), Lennart Nilsson. 290, Photo courtesy of NASA. 291, Lynn Johnson, Black Star. 292, Scott T. Barrows. 293, Lennart Nilsson, *Behold Man,* Little, Brown and Co. 294-295, Nathan Benn. 295, Louie Psihoyos. 296, Nathan Benn. 296-301, Carl W. Rohrig. 302-303, Nathan Benn. 304-305, John McGrail, Wheeler Pictures. 305, Scott T. Barrows. 306(t), Lennart Nilsson, *Behold Man,* Little, Brown and Co. 306-309, Jean A. Miller. 310, Nathan Benn. 311, Jean A. Miller. 312-313, Douglas Kirkland, Sygma. 314-315, Bruce Curtis, Peter Arnold, Inc. 315(t), Bruce Hazelton, Focus West. 315(b), Walter J. Iooss, Jr., © 1984 Fuji Photo Film USA. 316, Jan Lindberg. 317(t)-319(t), Lennart Nilsson, *Behold Man,* Little, Brown and Co. 319(b), Lennart Nilsson. 320, Jan Lindberg. 321(t), Created by David Zelter, Cranston Csuri Productions. Photo by Dan McCoy. 321(lb), Jan Lindberg. 321(rc,rb), Dan McCoy. 322, Martin Rogers, Woodfin Camp, Inc. 322-323, Dan McCoy.

Landscapes of the Mind
324, Michel Tcherevkoff. 326(t), Lennart Nilsson. 326(b), Lennart Nilsson, *Behold Man,* Little, Brown and Co. 327, Robert J. Demarest. 328, Lennart Nilsson, *Behold Man,* Little, Brown and Co. 328-330, Lennart Nilsson. 330-332(t), Lennart Nilsson, *Behold Man,* Little, Brown and Co. 332(b), Robert J. Demarest. 332-333, Lennart Nilsson. 334-335, Richard Howard. 336-337, Dan McCoy. 338-339, Dan McCoy, Rainbow. 340, © 1984 Jay Maisel. 340-341, Dan McCoy. 342-343, Michel Siffre. 344, George F. Mobley, NGP. 345, Nathan Benn. 346, Phil Mercurio, Michele Frost, Quantitative Morphology Lab, University of California, San Diego. 347, Galen Rowell, High and Wild Photography. 348, Dan McCoy, Rainbow. 348-349, Dan McCoy. 350, Robert J. Demarest. 351(rt), Dan McCoy. 351(rb), © Howard Sochurek 1984. 352(t), Dan McCoy. 352(b), Thomas Nebbia. 352-353, Nathan Benn. 354(lt,lb), Lennart Nilsson, *Behold Man,* Little, Brown and Co. 354-355, Jodi Cobb, NGP. 355, © Jesse Salb, NeuroScience, Inc. 356, Ronald H. Cohn. 357, Edward Lettau, FPG. 358, Hiroji Kubota, Magnum. 359(t), Suzanne Szasz. 359(b), Robert Burroughs, University of California, San Diego. 360, Thomas J. Abercrombie, NGS. 360-361, Ford Motor Co., Creative Photographer. 362-363, Dan McCoy, Rainbow. 364(l), Roger W. Sperry, California Institue of Technology. 364(rt,rb), Wendy Heller, University of Chicago. 365, Geoffrey Croft, Retna Ltd. 366(t), © Kay Chernush. 366(c,b), Joel L. Walker. 367, Paul Fusco, Magnum. 368, Ira Block. 369(l), Jeffrey Aaronson, ASPEN. 369(r), Richard H. Gracely, National Institutes of Health. 370, © 1984 Jay Maisel. 371, Bruno Barbey, Magnum. 372, William Albert Allard. 372-373, John Launois, Black Star. 374, Mickey Pfleger. 374-375, Kenneth Garrett, Woodfin Camp, Inc. 376-377, Sten Thorold. 377, Lennart Nilsson.

Acknowledgments and Bibliography

We wish to thank the many doctors and other consultants who generously contributed their time and knowledge.

We are grateful for help from Dr. Anthony Casolaro, Dr. Jerry D. Gardner, Dr. Joe B. Harford, Dr. Ralph F. Naunton, and Dr. William C. Roberts, all at the National Institutes of Health. Our thanks go as well to Jules Asher and the staff of the National Institute of Mental Health, and to Jan S. Ehrman and the staff of the National Institute on Aging.

We are also indebted to faculty at a number of universities: Dr. Raphael L. Poritsky at Case Western Reserve University School of Medicine; Dr. Nina Scribanu at The Georgetown University Medical Center; Dr. Kurt E. Johnson, Dr. Marilyn J. Koering, and Dr. Carol L. Ludwig at The George Washington University Medical Center. Also Dr. Yvonne Maldonado, Dr. David A. Newsome, and Dr. Solomon H. Snyder at The Johns Hopkins University School of Medicine; and Dr. Keith L. Moore at the University of Toronto.

We are grateful to Dr. Gordon B. Avery of Children's Hospital National Medical Center, Dr. Barbara Bierer and Dr. Steven J. Burakoff of the Dana-Farber Cancer Institute, Dr. William S. Cain of the John B. Pierce Foundation, and also Dr. Gunter Bahr, Norman Cousins, Dr. James S. Kelley, Dr. Howard H. Patt, and Dr. Joel M. Taubin.

We would also like to thank the Nova Pharmaceutical Corporation and The Rockefeller University for their assistance.

We consulted numerous books and periodicals in the preparation of this book. Listed below is only a sampling of some of the sources we found helpful. Most of these sources are nontechnical, intended for the lay reader.

Books of a general nature include James Bevan, *Anatomy and Physiology;* Helena Curtis, *Biology;* Russell Myles DeCoursey and J. Larry Renfro, *The Human Organism;* Christian de Duve, *A Guided Tour of the Living Cell;* Alvin Silverstein, *Human Anatomy and Physiology;* Sam Singer and Henry R. Hilgard, *The Biology of People;* Eldra Pearl Solomon and P. William Davis, *Human Anatomy and Physiology;* and Alexander P. Spence and Elliott B. Mason, *Human Anatomy and Physiology.*

Sources on the origin of life include Robert Jastrow, *Red Giants and White Dwarfs;* Lynn Margulis, *Early Life;* Norman Myers (editor), *Gaia: An Atlas of Planet Management;* and Lewis Thomas, *The Lives of a Cell.*

Further information on human reproduction and birth can be found in Kurt W. Fischer and Arlyne Lazerson, *Human Development;* Karen Jensen and the editors of U. S. News Books, *Reproduction;* and Lennart Nilsson, Mirjam Furuhjelm, Axel Ingelman-Sundberg, and Claes Wirsén, *A Child Is Born.*

Food and nutrition information is given by Catherine F. Adams, *Nutritive Value of American Foods;* Jane E. Brody, *Jane Brody's Nutrition Book;* L. E. Lloyd, B. E. McDonald, and E. W. Crampton, *Fundamentals of Nutrition;* and the National Research Council, *Recommended Dietary Allowances.*

Sources on the heart include Michael DeBakey and Antonio Gotto, *The Living Heart;* Jake Page and the editors of U. S. News Books, *Blood;* and Brendan Phibbs, *The Human Heart.*

The immune system is the subject of Ronald J. Glasser, *The Body Is the Hero;* Roger Lewin, *In Defense of the Body;* Steven B. Mizel and Peter Jaret, *In Self-Defense;* and Ian R. Tizard, *Immunology.*

More information on aging and the stages of life is found in Robert N. Butler, *Why Survive? Being Old in America;* James F. Fries and Lawrence M. Crapo, *Vitality and Aging;* and David Sinclair, *Human Growth after Birth.*

Sources on the senses include R. L. Gregory, *Eye and Brain;* Peter Nathan, *The Nervous System;* and Robert Rivlin and Karen Gravelle, *Deciphering the Senses.*

For more on the brain see Floyd E. Bloom, Arlyne Lazerson, and Laura Hofstadter, *Brain, Mind, and Behavior;* Richard Restak, *The Brain;* and Scientific American, Inc., *The Brain.*

The following periodicals frequently feature information on new medical and biological research: *American Health, Discover, Mosaic, New Scientist, Science, Science Digest, Science 86, Science News,* and *Scientific American.* In addition, numerous pamphlets on medical and biological subjects are available from the National Institutes of Health in Bethesda, Maryland.

Authors, Photographers, and Illustrators

A graduate of Yale University, JENNIFER GORHAM ACKERMAN is a researcher at the National Geographic Society and a free-lance writer. Her work has appeared in National Geographic books, the *Yale Alumni Magazine,* and other publications.

IAN ANDERSON, is an Australian-born journalist based in California. He contributes commentaries and stories to the Australian Broadcasting Corp. His science articles appear regularly in *New Scientist* and *Science Digest.*

SCOTT THORN BARROWS, a professor at the University of Illinois Medical Center in Chicago, has illustrated more than 20 books with his medical artwork. He has won awards from the Association of Medical Illustrators.

NATHAN BENN is a contract photographer for the National Geographic Society, for which he has covered topics ranging from biblical archaeology to medicinal herbs. His photographs appear in a new book, *God of the Country: A Voyage on the Mississippi.*

ROBERT J. DEMAREST is director of audiovisual services at the College of Physicians and Surgeons of Columbia University. He is an award-winning member of the Association of Medical Illustrators.

At age 16 DAN McCOY began his photojournalism career with the *Johnson City Press Chronicle* in Tennessee. His work has appeared in *Life, Newsweek, Geo,* and other international publications.

After several years as an operating-room nurse, JEAN MILLER returned to school and earned a degree in medical art. She has since worked for numerous clients, including pharmaceutical companies and medical journals. She lives in Ontario.

KIRK I. MOLDOFF of Piermont, New York, is a medical illustrator whose work has appeared in a dozen different publications and journals, including *Sports Illustrated* and *Discover.*

MAYA PINES is the author of several books and many articles about human behavior and science. She was a staff writer for *The New York Times* and *Life.* A contributing editor at *Psychology Today,* she lives in Washington, D. C.

CARL-W. RÖHRIG is an artist living in West Germany. He has illustrated several books and articles, including pieces for *Stern Magazin* and *Geo.*

SUSAN SCHIEFELBEIN writes on medicine and the environment for magazines and television. A former senior editor at *Saturday Review,* she won the National Magazine Award for her story on children and cancer. She lives in New York.

MARGARET SEDEEN is a National Geographic editor and writer. Her many interests include natural history and science, and she has been the chief editor of National Geographic books, such as *Our Universe* and *Great Rivers of the World.*

MICHEL TCHEREVKOFF is a native Parisien now living in New York. His photography has appeared in most major magazines and has won more than 75 advertising awards.

LEWIS THOMAS has won widespread honor for his books and essays on science, including a National Book Award for his best-seller, *The Lives of a Cell.* Thomas is president emeritus of the Memorial Sloan-Kettering Cancer Center in New York City and a university professor at the State University of New York at Stony Brook.

Index

Illustrations appear in **boldface,** picture captions in *italic,* and text references in lightface.

Type composition by National Geographic's Photographic Services. Color separations by Chanticleer Co., Inc., New York, N.Y.; Dai Nippon Printing Company Ltd, Tokyo, Japan; The Lanman Companies, Washington, D.C.; Litho Studios Ltd, Dublin, Ireland. Printed and bound by Kingsport Press, Kingsport, Tenn. Paper by Mead Paper Co., New York, N.Y.

Library of Congress CIP Data:

The Incredible machine.

Bibliography: p.
Includes index.
1. Human physiology—Popular works. 2. Body, Human. I. National Geographic Society (U.S.)
QP38.I53 1986 612 85-29731
ISBN 0-87044-619-3 (alk. paper)
ISBN 0-87044-620-7 (lib. bdg.:alk. paper)
ISBN 0-87044-621-5 (deluxe:alk. paper)